S0-CKO-493

Seedlings of dicotyledons
Structure, development, types.
Descriptions of 150 woody Malesian taxa

Errata

page 15 line 18 from top:... Types A and D, read Types B and D.

page 17 line 4 from bottom; structure, read structures

page 43 line 10 from top: *Sinapsis alba, read Sinapis alba*

page 121 line 4 from top: *Maniltoa brownioides,* read *Maniltoa browneoides*

E. F. de Vogel
Seedlings of dicotyledons. Structure, development, types.
Descriptions of 150 woody Malesian taxa.
Pudoc, Wageningen, the Netherlands, 1980.

Seedlings of dicotyledons

Structure, development, types
Descriptions of 150 woody Malesian taxa

E. F. de Vogel

Centre for Agricultural Publishing and Documentation
Wageningen, 1980

The author works at the Rijksherbarium, Leiden, the Netherlands

This book is published with financial support from
the Indonesian Council of Science (LIPI)
the Netherlands Universities Foundation for International
Co-operation (NUFFIC)
the foundation 'Fonds Landbouw Export Bureau 1916/1918' and
the foundation 'Greshoff's Rumphius Fonds'.

ISBN 90-220-0696-4

© Centre for Agricultural Publishing and Documentation, Wageningen,
the Netherlands, 1980

Printed in the Netherlands

Contents

Plate 1. *Koordersiodendron pinnatum* Merr. (Anacardiaceae), De Vogel 719a.

Plate 2. *Bombax valetonii* Hochr. (Bombacaceae), Dransfield 2360.

Introduction and acknowledgements

Juvenile stages of plants, particularly of trees, are often so strikingly different from the adult stages, that even with good field knowledge of plants it is hard to correlate the seedling with an adult shoot of the same species. Juvenile leaves may be lobed instead of entire or the reverse, simple instead of compound, and the first ones may be opposite whereas the subsequent ones are spirally arranged. Knowledge of seedlings, therefore, requires special studies of all their life stages, based on cultivated material of well identified seed samples. Considerable work has been done to describe and figure the seedlings of herbs, particularly weeds, for obvious economic reasons. Weed control naturally tends to focus on the youngest stages of the weeds. It is evident too, that in forestry the knowledge of seedlings of wild species is of vital importance, especially in the tropics, because of the highly mixed composition of the rain forest. Regeneration depends on seedling growth, which silviculture seeks to influence.

Forced by the need to recognise seedlings in the field, forest botanists have produced some excellent books and articles devoted to tropical tree seedlings: Troup in 1921 (India), Duke in 1965 and 1969 (Puerto Rico), De La Mensbruge in 1966 (Ivory Coast), and Burger in 1972 (Indonesia, mainly Java). These works have also a considerable scientific value, because they make available data about a stage in the life-cycle of the plant of which little is known, which offers many characters useful for taxonomic classification and for morphological and evolutionary considerations.

The coverage of seedling development and seedling morphology by these works is still very limited, even if we consider that a certain type of seedling and the morphology of the young plant is in general more or less the same for all species in a particular (sub)genus. This is understandable because of the difficulties involved in collecting ripe seeds from large trees together with herbarium vouchers which must document the identity. Complete coverage of the subject is, however, desirable. To realise this, two ways are open. First, to collect, grow, describe, and figure seedlings, especially in taxa of trees and lianas where they are unknown, e.g. in the tropics. Second, to compile an index to all data on seedlings scattered in publications. Such an index is in preparation at the Rijksherbarium, Leiden.

The aim of this book is to present general information on seedlings, and specific data on a number of Malesian tree seedlings, the latter partly new to science, for the benefit of foresters and botanists alike. It is estimated that now for ±50% of the Indonesian tree genera the seedling of one or more species is known. However, completion of a seedling bibliography is better postponed a while until more knowledge has been accumulated. Construction of keys for identification of seedlings of genera of Malesian woody plants makes sense only if virtually all sub(genera) have been covered.

The general part of this book contains an account of the seedlings, their parts, and their diversity.

This is followed by a grouping of seedlings according to their characters in a number of seedling types. These types are classified in a morphological scheme, and it is explained how the different seedling types can be thought to be derived from each other. It is reasoned, on morphological grounds, that thin assimilating cotyledons which have been named in this work paracotyledons must be homologous with the first pair of opposite leaves in other seedling types. Some evidence for this hypothesis is found in genera in which more than one seedling type has been encountered.

In other genera, however, indications are present that foodstoring cotyledons may possibly change directly into thin assimilating cotyledons, or the reverse. This matter is by no means settled yet, and requires more investigation.

Chapters are present on the significance of seedlings for taxonomy and ecology, the latter again having silvicultural, that is economic, implications. The general part is concluded with an account of the method which was used in collecting seeds, and a description of a nursery – no shocking novelties, but to my knowledge, hitherto not published in a readily accessible, and freely obtainable form. I think a detailed description of the methods used will be of great help to those who attempt growing seedlings in large quantities.

The special part consists of descriptions and illustrations of 150 seedlings, preceded by some notes about how these descriptions must be interpreted and used.

All these points demonstrate, I hope, that this study of seedlings serves many purposes. This made it a suitable project for cooperation between the botanical institutions of Leiden, The Netherlands, and Bogor, Indonesia, in a development cooperation context.

A 3-year project was set up, sponsored by The Netherlands Universities Foundation for International Co-operation (NUFFIC). To execute this project (coded as RUL 4 under the programme), the present author worked in Indonesia from October 1971 till December 1974. Work included several expeditions to obtain seed samples, and also general herbarium collections: the latter were shared between the two cooperating herbaria. The seeds were cultivated in a special nursery in the Bogor Botanic Gardens and living plants in the sapling stage were offered for planting in this garden. Several other institutions profited as well, including the Forest Research Institute in Bogor. Counterpart staff were trained in collecting techniques and nursery work.

The seedling project RUL 4 has had much benefit from a variety of help, received in all stages of preparation, execution, and follow-up. I can by no means mention all persons and institutions that contributed, but I wish here to mention at least those with which contacts have been most intense.

The project was accepted with enthusiasm by the then director of the Lembaga Biologi Nasional (L.B.N.), Prof. Dr. O. Soemarwoto. The L.B.N. served as counterpart institution, and shared the responsibility for the project with the Rijksherbarium, Leiden. Quite a number of administrative matters were necessary to run the project smoothly. Contact with the Indonesian authorities was handled in a very convenient way by the staff of L.B.N., which smoothened work considerably. This included the provision of working permits, the definition of the administrative status of the Dutch botanist, prolongation of the agreement of the project, and also the provision of letters for permission and travelling for each collecting trip. Thanks are especially due to the subsequent directors of the L.B.N., first Dr. Soeratno Partoatmodjo, later Dr. Setijati Sastrapradja, and also to the assistant director Dr. Soenartono Ardisoemarto.

The project was carried out in the Bogor Botanic Gardens (Kebun Raya) where a plot of land was put at the disposal of the project. Further cooperation was in the form of personnel: technicians who joined the project as counterpart staff, and labour for daily maintenance. Also tools, and materials for maintenance and handling of the plants was supplied. The director of the Gardens, Dr. Didin Sastrapradja, and his staff, are here remembered for all the useful facilities provided. The counterpart staff, Messrs. Aguswara, Abdul Hadi, and Opid Sardiwinata are thanked for their cooperation, and for all efforts they put in the work. In the Herbarium Bogoriense all collected plant material, which was over 3500 numbers, of which most in eight sets, were dried, processed and named. I am much indebted to the Keeper of the Herbarium, Dr. Mien A. Rifai, for his interest in the project, and for the provision of working space and all facilities. The pleasant cooperation with the Herbarium staff was much appreciated. Many of the technical staff of the institution helped in processing the material; I wish to mention here in particular Mr. Nedi, for his unique and universal aptitude in plant identification, from memory and by tenacious comparison.

The excellent drawings are from the hand of Mr. Moehammad Toha, one of the two artists, now retired, who previously proved their skill in making the watercolours for the book by C. G. G. J. van Steenis, The Mountain Flora of Java (1972a). Owing to his perfection in illustrating, and also his working knowledge of plants, Mr. Toha was able to work very largely on his own and with very little consultation. In three years Mr. Toha produced more than 1500 line-drawings and some watercolours, of a quality which may speak for itself.

Most enjoyed were the advice from, and the discussions with Dr. J. Dransfield. His interest in the work during all stages is greatly appreciated. From his own expeditions he invariably brought back seed samples for the project; he and the author of this book made several collecting trips together.

Indonesian Forestry Service (Dinas Kehutanan), and a number of local foresters, are thanked here for their help with the expeditions. The director of the Forest Research Institute at that time, Ir. Soediarto, is remembered for his interest.

The project was backed from The Netherlands in a very pleasant way. From the part of the Rijksherbarium, the responsible institution on the Dutch side, the project was followed closely by letter. The two initiators, Dr. M. Jacobs and Prof. Dr. C. G. G. J. van Steenis, supplied many valuable suggestions that are included in the work. After retirement of Prof. Van Steenis, the pleasant cooperation was continued under his successor Prof. Dr. C. Kalkman, who took over the responsibility for the project on the Dutch side. Many technical, administrative, and scientific aspects of the work were discussed with Dr. M. Jacobs, who visited the nursery during his stay in Bogor. He also advised on the final presentation of the data, and on the whole he was very concerned with the progress of the work. I am much obliged for all his help.

On the Dutch side, the burden of the many administrative matters related to the project was borne by the Bureau of Foreign Affairs of the University of Leiden. Especially Mr. J. F. Jongepier of the Bureau is here mentioned for his efficient work, which contributed much to assure a smooth progress of the project. He, too, visited the nursery in Bogor.

The project owes much to the positive advice to NUFFIC, supplied by the late Dr. A. Thorenaar, The Hague, who during his career in the Forestry Service in Indonesia had developed a personal interest in seedlings, and by Dr. I. A. de Hulster, at the time professor of Tropical Forestry in Wageningen, The Netherlands. Professor De Hulster also succeeded in raising financial support from the Landbouw Export Bureau (L.E.B.) fund for publication of this book. Dr. D. Burger Hzn., at Wassenaar, The Netherlands, who early in his career in the Forestry Service in Indonesia in the 1920's prepared the first book on Malesian seedlings, published in 1972, put his extensive working knowledge on tropical seedlings and their cultivation at the disposal of the author. This was very much appreciated, as without his help the project would not have started so fortunately.

Not in the last place the appearance of this book is due to the NUFFIC, the organisation which supplied all the financial means to carry out the research on the seedlings and for making expeditions outside Java. This organisation also supplied a grant for covering part of the costs for publication of this book. Many others contributed to the project in one way or another, either by suggestions, or with living seeds from their collecting trips. In particular I wish to thank Mr. R. Geesink, Prof. Dr. A. J. G. H. Kostermans, Prof. Dr. R. van der Veen, and Dr. J. F. Veldkamp. Dr. J. H. Wieffering put much time in the translation of an article by Grushvitskyi from the Russian language, for which I am much obliged. Mr. H. Blansjaar did a fine job assembling the plates from the individual drawings. The first draft of the descriptions was converted from handwriting into typescript by Mr. Umar Ali. The final manuscript was typed by Miss M. van Zoelen and Miss E. E. van Nieuwkoop. All three are thanked for their patient collaboration. Finally, the pleasant cooperation with the publishers must here be mentioned.

Without the financial support of the NUFFIC, The Hague, the Landbouw Export Bureau (L.E.B.) fonds, Wageningen, Greshoff's Rumphius fonds, Amsterdam, and a considerable support from the Indonesian Council of Science (LIPI), Jakarta, which makes distribution of the book in Indonesia possible, this work would never have been published.

Seedling glossary

This list is based mainly on seedling literature; some of the definitions have been slightly modified. For general terms see Jackson (1928) and Burger (1972).

Acotylar Without (functional) cotyledons, also used when much reduced scale-like cotyledons are present.

After-ripening The process in which the not yet fully developed embryo in the seemingly ripe seed matures after the fruit is detached from the parent plant.

Anisocotyly, anisocotylar The condition of a seedling in which the (para)-cotyledons are unequal in size or shape, or in which they have a difference in function. Equals: heterocotyly, heterocotylar.

Assimilating 'cotyledons' Thin, green, leaf-like 'cotyledons' of which the function is to provide assimilates for the life-processes in the seedling, on which the juvenile plant is entirely dependent in the period in which further leaves have not yet developed. Equals: photosynthetic 'cotyledons', paracotyledons, see the latter and also under cotyledons.

Blastogeny The study of the mode of germination; also the mode of germination itself.

Cataphyll Here: reduced, or scale-like leaves which are present in certain seedlings (e.g. Horsfieldia type/subtype, 7a) on the lowest stem nodes and sometimes elsewhere on the seedling stem. Equals: scale, scale-leaf, Niederblatt (German lang.).

Caulicle Primary stem of the embryo as found in Angiosperms and Gymnosperms, consisting of either the hypocotyl, or the hypocotyl and several stem internodes.

Collet The basal 'node' of the hypocotyl, the place where root and hypocotyl merge into each other, especially used when the demarcation is externally distinguishable, or a special structure is present.

Cotyledon The primary leaf in the embryo or seedling in Angiosperms and Gymnosperms. Application of this term to leaf-like, assimilating 'cotyledons' may be doubtful. Equals: seed leaf, seed lobe, see also paracotyledons.

Cryptocotylar The condition in which the cotyledons remain enveloped in the persistent fruit wall and/or testa (and if it is present also in the endosperm), together with which they are shed.

Delayed germination The condition that germination of the seed under favourable conditions does not take place until a physiological trigger is activated.

Diaspore The dispersal unit of a plant, which may range from (part of) the entire plant to the seedling itself minus the cotyledons (Rhizophora type, 9), in Dicots mostly consisting of the seed or the fruit.

Dicotyly The condition of having two cotyledons.

Dormancy A period of inactivity of the seed after maturing, before germination commences.

Embryo The juvenile sporophyte in the seed which will develop into the seedling. It consists of the (para)cotyledons, the radicle, and the plumule.

Emergent Used for cotyledons which throw off the envelopments and become exposed.

Envelopments Here: the covering fruit wall and/or testa, and the endosperm if present, around the (para)cotyledons, after germination.

Eophyll(s) The first fully developed foliar leaf or leaves in a seedling above the (para)cotyledons.

Epicotyl Here: the first internode of the stem above the hypocotyl. In literature also used for the entire embryonic axis, consisting of several internodes, above the cotyledonary node. Equals: first internode.

Epigeal, epigean, epigeous Means: Above soil level, for seedlings referring to the position of the (para)cotyledons after germination.

First internode Equals epicotyl.

Foodstoring cotyledon Cotyledon with a low surface/contents ratio, where in the cotyledon tissue food is stored which supports the development of the shoot with leaves during seedling development. In a juvenile stage it has a haustorial function and absorbs the endosperm, which may be absent in the mature seed or almost so, and is then functionless. Foodstoring cotyledons may be emergent or not. Equals: Speicherkotyledon (German lang.).

Germination The process of sprouting of a diaspore, including the emergence of the juvenile plant from the coverings of the diaspore. In plant sociology in The Netherlands sometimes used in a limited sense: starting with the intake of water up to the emergence of the root tip from the envelopments.

Gamocotylar The condition in which the cotyledons are connate with (part of) their adjacent sides.

Haustorial cotyledon Cotyledon in the form of an undifferentiated, colourless suctorial organ which absorbs the nutrients from the here always present endosperm in the diaspore and passes it to the growing regions of the juvenile plant, in an embryonic stage as well as during development of the seedling. Haustorial cotyledons may be emergent or not. Equals: haustorium, suctorial cotyledon.

Heterocotyly The condition of a seedling in which the (para)cotyledons are different in shape, size or function, or a combination of these. Equals: anisocotyly.

Heterophylly The condition of a plant in which two or more different types of leaves are present, each on their own portion of a stem or on different shoots. To be distinguished from anisophylly, the condition where opposite or alternating leaves are unequal.

Hypocotyl The portion of the stem in the embryo or seedling situated between the collet and the cotyledonary node.

Hypogeal, hypogean, hypogeous Means: below soil level. Here used to describe the position of the cotyledons after germination, which are either situated below soil level or on top of the soil, and in addition almost always covered by fruit wall and/or testa.

Metaphyll Mature leaf, as opposed to the juvenile leaf form.

Mesocotyl The internode which develops between the two unequal paracotyledons in certain Gesneriaceae after germination, and which brings the larger one at a higher level than the smaller.

Monocotyly The condition of a seedling where only one (para)cotyledon is present (Monocotyledons).

Paracotyledons Exposed, thin, green, leaf-like, assimilating 'cotyledons', possibly homologous with two opposite first leaves in other seedlings, which through abortion of the true cotyledons have obtained their position and the functions of food provision for the basic life processes in the seedling. Equals: assimilating 'cotyledon', photosynthetic 'cotyledon'. See also cotyledon.

Phanerocotylar The condition of a seedling in which the (para)cotyledons become entirely exposed, free from fruit wall and testa, for a specific period after germination.

Photosynthetic 'cotyledon' Equals: assimilating 'cotyledon', paracotyledon.

Plumular bud The undeveloped terminal bud above the cotyledonary node.

Plumule The somewhat differentiated terminal bud in several embryo types above the cotyledonary node, in which one or more internodes and leaves or scales can be discerned in a primordial stage.

Polycotyly The condition of a seedling with more than two cotyledons.

Polyembryony The condition when more than one embryo develops in the seed.

Pseudocotyledons Two opposite first foliar leaves on top of the first internode, decussate with the place of insertion of the exposed or sometimes enclosed foodstoring or haustorial cotyledons. Mostly pseudocotyledons are more or less different in shape from the subsequent leaves, which in addition are in most cases spirally arranged. Their function is to support further seedling development after the cotyledons are shed, by means of the production of assimilates.

Pseudomonocotyly The condition in Dicot seedlings which have only one (para)-cotyledon.

Radicle The primary root primordium in the embryo.

Resting stage A temporary halt in the development of the epigeal parts of a seedling, which is induced by internal conditions.

Ruminate The condition of cotyledons or endosperm, which due to invaginations show a more or less irregular structure on the surface and in cross-section.

Scale, scale-leaf Here: reduced leaves which are present in certain seedlings (e.g. Horsfieldia type/subtype, 7a), especially on the lowest stem nodes but in some cases also elsewhere on the seedling stem. Equals: cataphyll, Niederblatt (German lang.).

Schizocotyly The condition in seedlings in which the cotyledons are longitudinally split.

Secondary dormancy The condition of inactivity which is sometimes imposed on seeds when they are subjected to adverse conditions.

Secund The condition that organs of one kind on one axis are directed to one side. Here especially referring to cotyledons which have their place of insertion opposite, but together are turned lateral of the stem.

Seedcoat The covering wall of the seed derived from the integuments. Equals: testa.

Seed leaf Equals: cotyledon, seed lobe, see also paracotyledon.

Seedling The juvenile plant, grown from a seed.

Seed lobe Equals: cotyledon, seed leaf, see also paracotyledon.

Suctorial cotyledon Equals: haustorial cotyledon, haustorium.

Supracotyledonary The position of a seedling part which has the place of insertion above the cotyledons.

Taproot The thick and sturdy primary root of a plant which is directly developed from the radicle.

Testa Equals: seedcoat.

Unilateral outgrowth Appendix which is developed to one side. Here especially used for a hook-like projection at the collet.

Viviparous The condition when offspring is produced in the form of a more or less complete, differentiated plantlet. Here used for fruits of which the seed germinates already on the parent plant, and the seedling extends from the fruit while the latter is still attached.

The seedling

Historical review

Quite early in botanical history, attention was paid to seedlings. According to Bernhardi (1832), Cesalpino (1583, publication not seen) gave a description of seedling morphology. He distinguished several parts: the cotyledons and the corculum, the latter consisting of the plumule (the ascending part) and the rostellum (the descending part). Malpighi published seedling descriptions in the works Opera Omnia (1687) and Opera Posthuma (1697). Rumphius gave illustrations of several seedlings in his magnificent work Herbarium Amboinense (1750).

Other early authors on seedlings and embryos are Gaertner (1788–1807), Link (1807), L. C. M. Richard (1808), Mirbel (1809), A. Richard (1819), A. P. de Candolle (1825), and Agardh (1829). At that time, already an extensive knowledge existed about the differences between seedlings. Especially the interpretation of the different parts of embryos and seedlings was the centre of interest. An excellent account of this subject was given by Bernhardi (1832), which covers the different opinions and interpretations. A (literal) translation reads as follows:

> "Linné distinguishes with Caesalpin in the new plant the cotyledons and the heartlet (corculum), the latter according to him consisting of a scaly ascending part, the plumula and a simple descending part, the rostellum. Gaertner names with Adanson the heartlet Embryo, and distinguishes as parts of it the rootlet (radicula), the shaft (scapus) and the plumule.
>
> The two Richards distinguish four main parts in the embryo of the plant: 1) the root body or the rootlet (corps radiculaire), which part is at the lowest end, and which on germination either gives the root its origin or differentiates into the root; 2) the cotyledonary body (corps cotyledonaire); 3) the budlet (gemmule) or the organ which has formerly been called plumule . . . and which is to be interpreted as the first primordium for all ascending parts; 4) the stalklet (tigelle, cauliculus), a not always discernible part which is situated between the cotyledons and the rootlet, and which fuses with these parts.
>
> Mirbel subdivides the embryo in the cotyledonary body and the germ body (blastème), the latter consisting of the rootlet (radicule) and the plumule, consisting seemingly of two germs, which are separated by a central part, the neck (collet) . . . . in the first the caulicle (tigelle) and the gemmule can be occasionally distinguished. . . . In the seedling (plantule) which through germination is developed from the embryo, Mirbel accepts, apart from cotyledons, leaves and ramifications of the root, two main parts, the ascending and the descending stalk (caudex ascendant et descendant), but in the sense that the former cannot entirely be considered as differentiated plumule, and the latter

not merely as differentiated rootlet, for the collet belongs sometimes to the first, sometimes to the other, whether it develops in the direction of the plumule or the radicle.

According to De Candolle, the principal organs of the plant embryo are the rootlet (radicule), the caulicule, usually called plumule, and the cotyledons. The caulicule, according to him, consists of the caulicle proper (tigelle), which extends from the neck up to the cotyledons, and the gemmule, which is situated above the cotyledons. The neck is to him the plane which separates the caulicle from the radicle; it can therefore not be looked for at the place of attachment of the cotyledons.

Link together with Mirbel distinguishes four principal parts of the embryo: 1) the root body (rhizoma) as the lowermost part, from which the root sprouts; 2) the shaftlet (scapellus) or the joining part between root body and cotyledons; 3) the cotyledons or the leaves of the future plantlet (folia seminalia), and 4) the plumule or the bud primordium. The rhizoma is the rostellum of Linné or the radicula of Gärtner; it cannot be considered as the future root, because often pith substance, a character of the stem, is found. On the far end occasionally a small wart is present from which the real root originates. The shaftlet (scapellus s. caulicus, scapus Gärtner, tigelle Richard) is not sharply separated from the rhizoma, and seems not to deserve a special distinction, at least it can only been discerned after germination.

Agardh considers the embryo, the plantlet, the germ in the broad sense (embryo, plantula) to be composed of two essential parts, the heart leaves (cotyledons) and the rostellum (rostellum, radicula). According to him the petioles of the heart leaves are connate and form a single small stem, the heart stem (cauliculus, tigelle), and between the heart leaves develops the bud of the new, proper stem, the stem bud (plumula, gemmula). Rostellum and heart stem (rostellum et cauliculus) form together the germ sensu stricto (corculum)."

Here ends the translation from Bernhardi (1832), who gave his own interpretation of the different parts of the embryo. Three main parts are distinguished. The rostellum or stalklet (cauliculus) consists of the basal part of the embryo from the root tip to the cotyledons. Often no distinction can be made between the part of this axis which is of root nature, and the part with stem nature; moreover, often the largest portion of this axis develops in either a root or a stem. The cotyledons or seed leaves are, in the embryo, only rarely provided with a petiole, but on germination these often enlarge considerably; sometimes they are connate. The cotyledonary petioles sometimes differentiate in a sheath and a petiole proper. Together these parts are named cotyledon body or cotyledon mass. In the stalklet an ascending portion (caudiculus ascendens) and a descending part (caudiculus descendens) can be distinguished, separated by the neck (collum). Bernhardi interpreted the caudiculus ascendens as formed by the fusion of the cotyledonary petioles. Between the cotyledons, on top of the caudiculus ascendens, a bud (gemma) is present, the primordium of a second shoot, which is often already differentiated in the plumule, or develops in a later period.

The modern ideas about the nature of the seedling parts follow closely the conception of A. P. de Candolle (1825), who discussed at length the nature of rootlet, hypocotyl, collet, and cotyledons. He came to the conclusion that the collet is the junction between the descending root part and the ascending stem part (hypocotyl) of the seedling axis. The hypocotyl turns green in light, contains a medullary channel, and often bears hairs, analogous to those on the stem. His main point was that the cotyledons, being the first leaves of the plant, must consequently be borne on the stem. The hypocotyl may be short, but that does not alter its stem nature, and the fact that the collet is sometimes morphologically not discernible he judged, in this respect, to be irrelevant. The nature of the cotyledons was considered of great importance by him. These may have a leaf-like form, or they are swollen, and such differences were found to be linked with different modes of germination. He was probably the first who observed that swollen cotyledons do not possess stomata (which, indeed, is often the case). Their function was judged to be purely foodstoring, whereas the green, leaf-like lobes provide food for the seedling like the normal leaves. Intermediates between the two types are not numerous, but, because of their existence, he concluded that these differences were not of great significance.

In the second part of the nineteenth century, the morphological variation of seedlings was further investigated. Numerous short publications appeared, covering the entire field of the temperate phanerogams. Most publications dealt with the description of one or a few seedlings. Especially German authors were active, the most productive being Irmish, and later A. Winkler. Amongst the others Buchenau, Caspary, and Warming must be mentioned. Winkler published, besides morphological descriptions, also a number of more comparative studies in which he tried to explain the existing variation.

The first thorough attempt at a seedling classification appeared in 1885, when Klebs published his magnificent article 'Beiträge zur Morphologie und Biologie der Keimung'. This was and still is a very important work, especially for European seedlings. Klebs synthesised the then existing knowledge on seedlings. It is subdivided into two parts: an analysis and classification of the main germination forms of seed plants, and a second part on important facets of the biology of germination. Unfortunately Klebs dealt mainly with European plants, of which the large majority is herbaceous, with epigeal seedlings. Consequently he missed the great diversity in hypogeal seedlings, which may be the reason why he attached so much value to only minor variations.

His main subdivision (see Fig. 1) was in a first group comprising most of the Dicotyledons and the Gymnosperms (cotyledons two to numerous), a second group of deviating Dicotyledons of which one or two cotyledons are rudimentary (not further subdivided) and a third group covering the Monocotyledons (cotyledon one). The first group contains seedlings with the cotyledons above the ground (= epigeal seedlings), and those with the cotyledons subterranean (= hypogeal seedlings), the latter not further subdivided. Subsequent subdivision of the epigeal seedlings into five different types was based mainly on the general germination pattern. The shape and development of the roots and the hypocotyl, the occurrence of outgrowths on the collet, the amount and development of the endosperm, and the method by which the cotyledons become exposed, were considered important for classification. Klebs himself stated that the types are all linked by transitions, and that the given classification was purely arbitrary, and only erected to bring order in a confusing mass of

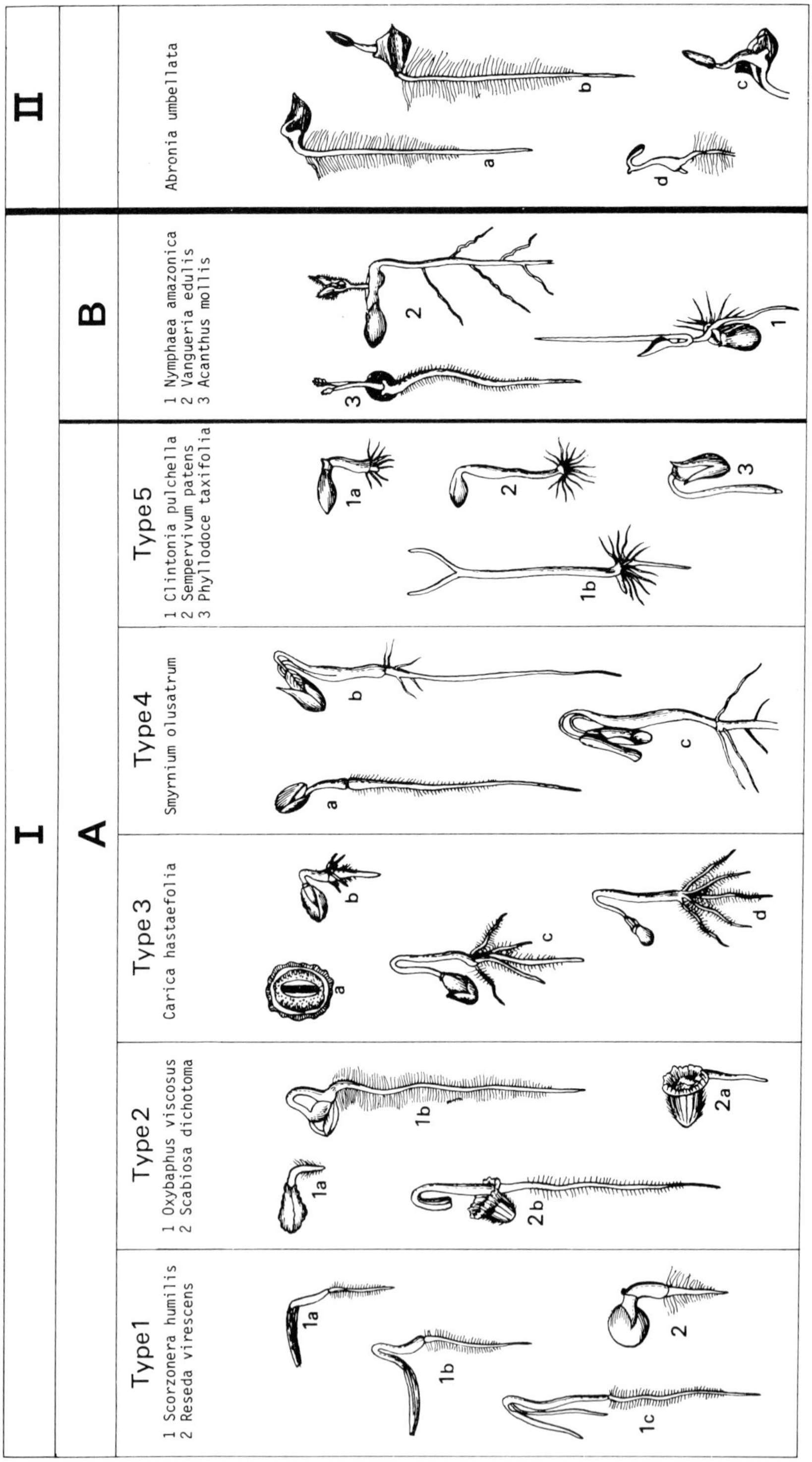

I | II

A | B

Type 1
1 Scorzonera humilis
2 Reseda virescens
1a 1b 1c 2

Type 2
1 Oxybaphus viscosus
2 Scabiosa dichotoma
1a 1b 2a 2b

Type 3
Carica hastaefolia
a b c d

Type 4
Smyrnium olusatrum
a b c

Type 5
1 Clintonia pulchella
2 Sempervivum patens
3 Phyllodoce taxifolia
1a 1b 2 3

1 Nymphaea amazonica
2 Vangueria edulis
3 Acanthus mollis
1 2 3

Abronia umbellata
a b c d

details. Under the types large numbers of examples have been cited. The Monocot group was subdivided into seven types, not to be discussed here because it falls outside the scope of this work.

Unfortunately the excellent work by Klebs has always remained in obscurity, and is hardly known by present-day botanists. On the contrary, the publication of Lubbock 'On Seedlings' (1892) is well known, and up till now has been regarded as a classic by all subsequent workers on seedlings. The two-volume book contains a wealth of seedling descriptions and illustrations, all arranged systematically. Emphasis lies on temperate seedlings, but also quite a number of tropical plants were included. The tropical plants studied by Lubbock are, I presume, those species which were available to him in greenhouses. His strongest point was the presentation of a survey of seedlings, giving hundreds of descriptions of a large diversity of Dicot genera, which enormously extended the existing knowledge of seedling morphology. His general part, and especially the discussion on the relation of the cotyledon shape with the shape of the seed, however, seems in some aspects somewhat futile.

Around 1870, the first anatomical studies of the embryo and the seedling appeared, and regularly more of such pure descriptive works followed in later years. From the beginning of the 20th Century, more emphasis was put on using anatomical characters for phylogenetic considerations. The leading author in this field of study was Ethel Sargeant, who gave the first impetus. Later, she was followed by Gatin, Arber (Agnes), Compton, and Boyd (Lucy), who in their specialised fields contributed much to our understanding of seedlings.

Throughout the 20th Century, occasional workers have devoted much of their time to the study of the morphology of seedlings. In addition, a lot of incidental short publications on seedlings appeared. Again, emphasis was on temperate seedlings, but, now and then, tropical seedlings were included in these works, the latter sometimes even being the subject of voluminous works. Only the most important studies can be mentioned in this connection.

Fritsch (1904) gave a detailed account on seedlings in Gesneriaceae, some of

---

Fig. 1. Classification of Dicot seedlings by Klebs (1885), illustrated by some of his own drawings.

I. Seedplants with two or numerous cotyledons.

A. Cotyledons above the soil.

Type 1. Main root elongating vividly from the first emergence from the seed onwards; the hypocotyl brings the cotyledons out of the seed above the soil; collet not or only relatively little thickened.

Type 2. Germination takes place like in type 1. The collet is distinguished by an exceedingly strong, often unilateral outgrowth.

Type 3. Germination as in type 1, but characterised by the strong, independent development of the endosperm.

Type 4. Main root elongating moderately or strongly; hypocotyl weakly developed. The petioles of the cotyledons pull the latter out of the seed.

Type 5. Main root little or not elongating during germination; at the collet a whorl of long root hairs; further as type 1.

B. Cotyledons subterranean.

II. Dicotyledonous seedplants, of which one or two cotyledons are rudimentary.

III. Seedplants with one cotyledon (Monocotyledons) (7 types, here not further subdivided).

which are peculiar for their anisocotyly. A comparative study of Hypericaceae and Guttiferae was undertaken by Brandza (1908). He described and illustrated the seedlings of a number of species in several genera of both families. For Guttiferae he confirmed the distinction in three types of embryos that had been given earlier by Planchon and Triana (1860–1862), illustrating three seedling types, one of which is typical for the tribe Clusieae, a second one for the tribes Monorobeae and Garcinieae, and a third one for the Calophylleae. His main conclusion was that Hypericaceae and Guttiferae have in its seedlings discriminating characters.

Guillaumin (1910) studied the structure and development of Burseraceae, and paid much attention to the seedlings. He drew taxonomic and phylogenetic conclusions partly based on seedling characters. Several keys to the genera are given in his paper: one on the anatomy, a second on a combination of fruit, seed and seedling characters, a third on the flowers, and a general key based on a combination of the foregoing.

Hickel (1914) published a rather voluminous work on fruits, seeds, and seedlings of the Angiosperm trees and shrubs indigenous or cultivated in France.

In the period 1917–1925, Burkill (one publication in cooperation with Foxworthy) published ten articles on seedlings of several genera of Dipterocarpaceae.

The magnificent work 'The silviculture of Indian trees' by Troup was published in England in 1921. In this book, Troup deals extensively with the cultivation of economically important tree species. He not only gave information on the adult plant, but also described the mode of germination and the seedling in a concise, but very clear way. The three volumes are beautifully illustrated with coloured plates of almost all seedlings described, in many stages of development. In 1977, an announcement was received that the book would be reprinted; the first few families have now appeared.

In the period 1936–1947, Vassilczenko published several articles on seedlings in connection with systematics and phylogeny. Unfortunately, these undoubtedly very interesting articles are in Russian, and the summaries in either German or English only contain the bare essentials. The 1936 article dealt with Angiosperm germination in general. 'Hypogeal' germination was recorded especially for plants of old relict types. For a number of cases Vassilczenko claimed to be able to demonstrate a recapitulation of form and other characters of ancestral types. In other seedlings this recapitulation could not yet be confirmed. Subsequent articles by Vassilczenko dealt with the morphology of germination in certain plant families. In his work on germination in Leguminosae (1937) he tried to supply new evidence on the correlation between hypogeal germination and the morphology of the first leaves. Vassilczenko postulated that the morphology of the first leaves represents the ancestral type, and according to the morphology of the various seedlings different strategies are assumed. Some seedlings have simple leaves as the ancestral form, and compound ones as derived. In others the development is just in the opposite direction, with compound leaves as the ancestral form. In seedlings where all leaves were simple no phylogenetic conclusions could be drawn. In a third article (1941) on Chenopodiaceae the most important observation is that the shape of the cotyledons in this family is constant at either genus or section level. The first two leaves are always opposite.

Plate 3. *Durio dulcis* Becc. (Bombacaceae), De Vogel 759.

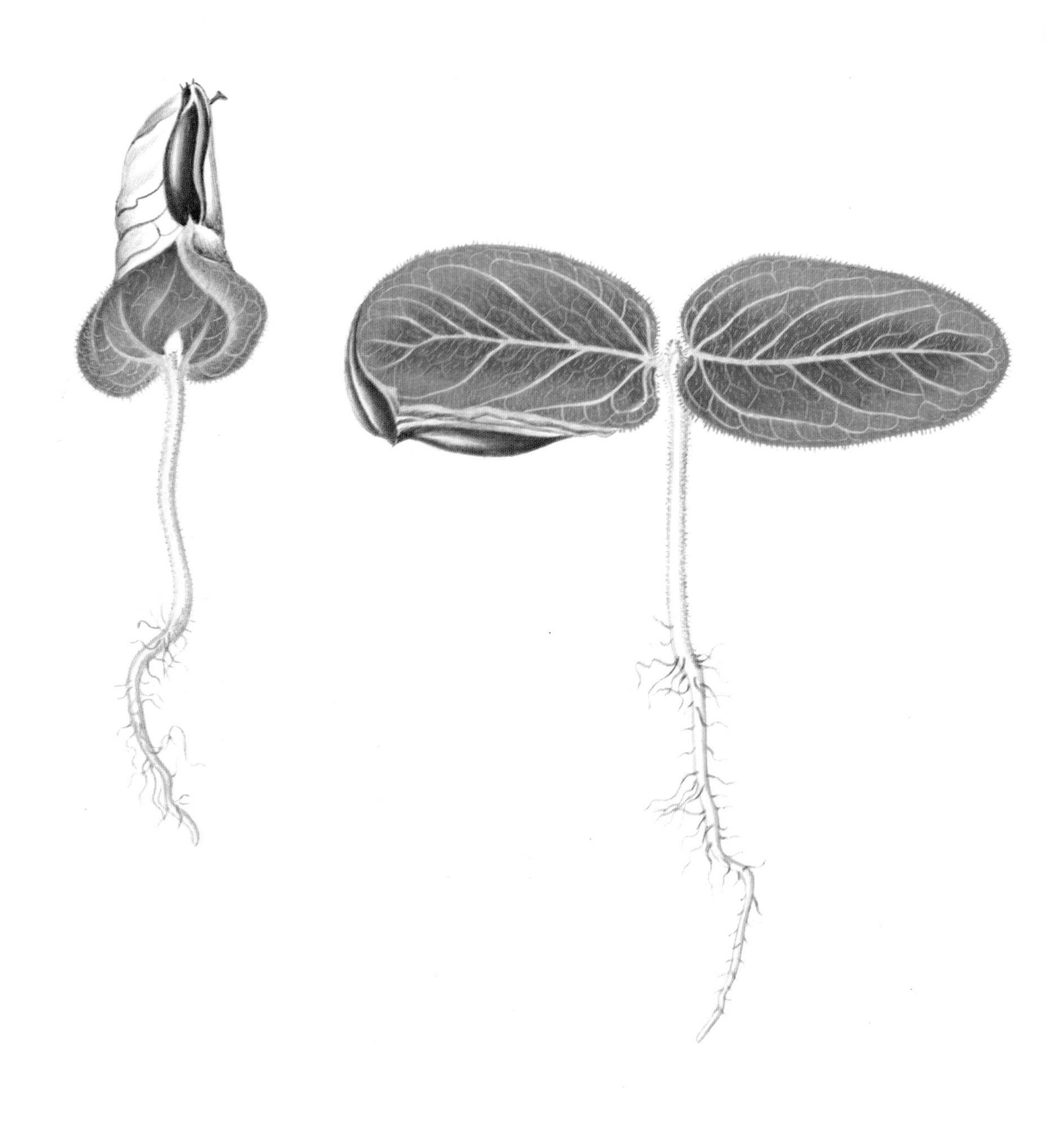

Plate 4. *Neesia altissima* (Bl.) Bl. (Bombacaceae), Dransfield 2531.

Meijer Drees (1941) published a work on seedlings of *Acacia*, which he had studied in Bogor, Indonesia.

A most unusual and important systematic study was undertaken by Léonard. In 1957, his work 'Genera des Cynometreae et des Amherstieae Africaines' appeared, and aroused a lot of interest. In addition to the commonly used systematic characters, and features of pollen and wood, he consistently made use of characters of germination and seedlings, to which he attached great importance. His seedling material was rather diverse and did not always fit the commonly used definitions of epigeal and hypogeal germination. To overcome this, he redefined the terms (Fig. 2):

A. Epigeal germination: cotyledons spreading, after having torn the testa; hypocotyl well developed to almost absent; axis of the shoot-root central in relation to the cotyledons; cotyledons frequently turning green.
  a. cotyledons spreading above the soil ....Type A.
  b. cotyledons spreading at soil level ....Type C.

B. Hypogeal germination: cotyledons remaining jointed and resting in or on the soil, there where the seed has been deposited; hypocotyl not or very little developed; axis of the shoot-root lateral in relation to the cotyledons; cotyledons not turning green ....Types A and D.

(N.B. the difference between B and D is not fundamental and only due to whether the seed is accidentally buried or not before germination.)

In his plate 2, p. 40–41 (Fig. 3) the type with epigeal cotyledons spreading above the soil (epigeal seedlings of former authors) was further subdivided, based on

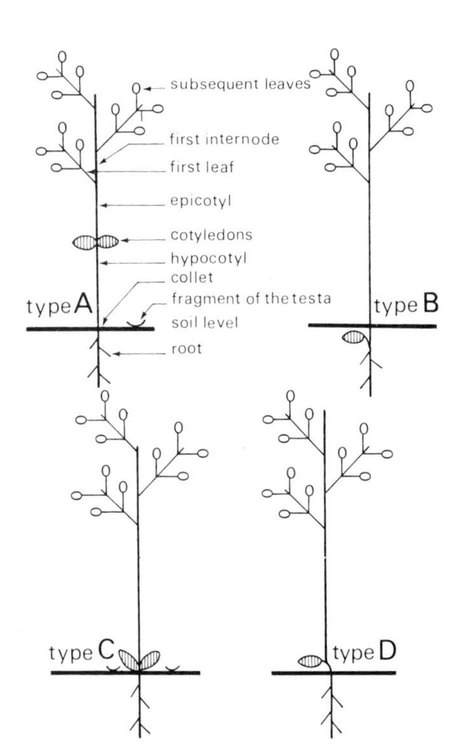

Fig. 2. Léonard's interpretation of the classification in epigeal and hypogeal seedlings.
Type A. Epigeal germination with the cotyledons spreading above soil level.
Type B. Hypogeal germination with the cotyledons remaining covered in the soil.
Type C. Epigeal germination with the cotyledons spreading at soil level.
Type D. Hypogeal germination with the cotyledons remaining on the soil (after Léonard 1957).

Germination epigeal					Germination hypogeal	
Cotyledons spreading above soil level				Cotyledons spreading at soil level	Fruit indehiscent resting on the soil	Seed resting on the soil
First two leaves opposite		First two leaves alternating				
Collet without outgrowth	Collet with outgrowth	Collet without outgrowth	Collet with outgrowth			
Léonard type I	Léonard type II	Léonard type III	Léonard type IV	Léonard type V	Léonard type VI	Léonard type VII
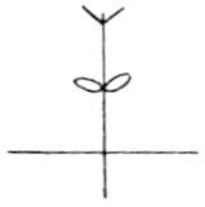	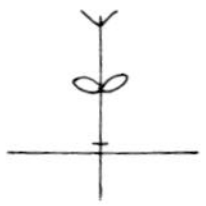	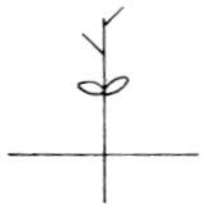		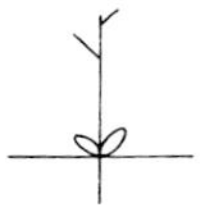	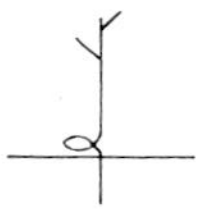	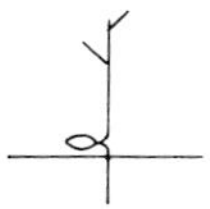
<u>Cynometreae</u>	<u>Cynometreae</u>	<u>Cynometreae</u>	<u>Cynometreae</u>	<u>Cynometreae</u>	<u>Cynometreae</u>	<u>Cynometreae</u>
Afzelia Copaifera (Afr.) Gilletiodendron Schotia (C.Afr.) Talbotiella Tessmannia	Schotia (S.Afr.) Sindoropsis Stemonocoleus	Baikiaea Copaifera (America) Daniellia Detarium Guibourtia (*) Hymenostegia Scorodophloeus Sindora	Guibourtia Hylodendron Trachylobium	Colophospermum	Gossweilerodendron Oxystigma	Crudia
<u>Amherstieae</u>	<u>Amherstieae</u>			<u>Amherstieae</u>		<u>Amherstieae</u>
Aphanocalyx Brachystegia Cryptosepalum Didelotia Gilbertiodendron Julbernardia Monopetalanthus Pellegriniodendron Tamarindus	Paramacrolobium			Berlinia		Anthonotha Isoberlinia

(*) Subgenus Pseudocopaiva

Fig. 3. Classification of seedlings in African Cynometreae and Amherstieae by Léonard (1957).

whether the first two leaves are opposite or alternate. Each of these subgroups in its turn was subdivided on the character whether the collet is in the possession of an outgrowth or not.

Léonard defined the main conclusion of his work as follows: 'The seedlings of all the species of one "good" genus have the same type of structure, or, in other words, only one type of seedling is predominant in each "good" genus.' This was based on the correlation between morphological and blastogenic characters, which he established for the African Cynometreae and Amherstieae that he had studied, in several cases after changing the systematic position of some of the species. This conclusion led him to propose four working hypotheses, of which his own translation in the English language reads:

"1. The establishment of synonymy between genera according to morphological data should be provable by the similarity of their seedlings.
2. Morphologically related genera, which have the same seedlings, may not be generically distinct.
3. The partition of a heterogeneous genus into several genera according to their morphological characters, should be provable by the existence of a particular seedling type for each of them.
4. The existence of several seedling types within one genus may be an indication of a generic heterogeneity that must be checked by other morphological data."

Léonard's classification was very much influenced by the working hypotheses mentioned, and he re-defined several genera accordingly. The validity of these hypotheses is treated at length on pp. 118–122; they have no general application.

Grushvitskyi (1963) published an interesting, but little known article in Russian: 'Subterranean germination and the function of the cotyledon'. In species germinating in a hypogeal manner two different types of cotyledons are encountered. One group has cotyledons with a pure foodstoring function, the other cotyledons enveloped by endosperm with a haustorial function. Up to that time this variation had not been emphasised. Grushvitskyi compiled a list in which all Dicotyledon genera (and species) with a subterranean germination were included, those known from literature as well as from own observation. The conclusion is that subterranean germination in combination with cotyledons with a haustorial function is characteristic for representatives of archaic families of Dicotyledons. Such cotyledons are also found in Cycadales and *Ginkgo*. Subterranean germination in combination with foodstoring cotyledons is found in representatives of 50 families of Dicotyledons, and is evidently more derived. Subterranean and epigeal germination occur so often in one genus, and even sometimes within one species, that the conclusion was drawn that the change-over from the one type to the other probably has repeatedly taken place, on different levels of evolution.

Weberling and Leenhouts (1966) used seedlings for clarification of morphological problems, and as a supplementation to their systematic research in the genus *Canarium* (Burseraceae). In this way they were able to elucidate the nature of the so-called pseudostipules, stipule-like structure at the base of the compound leaf, which must be regarded as representing a (rudimentary) pair of leaflets. Several seedlings of *Canarium* were described and illustrated. The difference was stressed between leaf-like and foodstoring cotyledons, both present in the genus, and each specific for one of the sections studied. Foodstoring cotyledons they defined with the German term 'Speicherkotyledonen' (= storing cotyledons), the opposite first leaves they named 'Pseudokotyledonen'. For the distribution of the seedling characters over the sections of *Canarium* see Table 1.

Duke (1965), in a nicely illustrated article, gave keys for a large number of Puerto Rican tree seedlings. He rejected the terms hypogeal germination and epigeal germination on etymological grounds. Many seeds germinate with the enclosed cotyledons in, or on, the litter above the soil, and he considered it incorrect to describe those as hypogeal. Therefore, he replaced the terms by cryptocotylar and phanerocotylar, terms which found their way into other seedling literature. Duke's new terms are sometimes considered synonyms with the terms they are meant to replace, but as is already stated by Ng (1975), four different combinations are possible using these terms, as their meaning is not identical. In addition, intermediate conditions may occur, and seedlings are known to which neither of these terms is applicable.

A voluminous work on Ivory Coast (Africa) seedlings appeared from the hand of De La Mensbruge (1966). He described and illustrated representatives of 229 genera of woody Dicots, of which quite a number occurs also in Malesia. The descriptions he gave are, unfortunately, not complete, and produced in a tabulated form, which does not permit a clear understanding of the seedling as an entity.

Csapody (1968) published a key to the Dicot seedlings of Central Europe, covering the large majority of the Dicot flora of that region, comprising almost 1500 species, all illustrated. Her first artificial subdivision is in Normal Germination (= epigeal germination) and Abnormal Germination (including hypogeal germination and seedlings without cotyledons or with connate cotyledon parts). That epigeal germination is considered as the normal situation, is evidently due to the

Table 1. Summary of distribution of seedling characters in *Canarium* (Weberling & Leenhouts 1966).

	Sect. *Pimela* (4 sp. studied)	Sect. *Canarium* (2 sp. studied)	Sect. *Africanarium* (1 sp. studied)
Cotyledons:	leaf-like	thick fleshy	leaf-like
	long persistent	dropped after development of two opposite first leaves	long persistent
First leaves:	all spirally arranged	first two opposite, all others spirally arranged	first two opposite, all others spirally arranged

The representative of sect. *Africanarium*, *C. schweinfurthii*, combines characters of both sect. *Pimela* and sect. *Canarium*. In fact the seedling is very peculiar, because, following two long-persistent leaf-like paracotyledons, the first two leaves are opposite and are considered to be pseudocotyledons. This latter feature is characteristic for seedlings of the Sloanea type (2) with foodstoring cotyledons or haustorial cotyledons, and is almost never found in combination with leaf-like paracotyledons. Leenhouts (1966) raised sect. *Africanarium* to the level of subgenus, based on seedling characters as well as on characters of the adult plant.

extremely high percentage of that germination type in the temperate regions. The key is based on artificial characters, and sometimes much-resembling species of one genus are placed wide apart. The work is very practical in the field, and is much used in plant sociology in Europe.

A survey of seedling characters and their occurrence was given by Duke (1969). In addition he gave an illustrated preliminary systematic survey of seedling characters. This deals with the methods of germination of a large number of genera and families. It is the first and only attempt for such a survey, and although very incomplete, as such it is very useful.

Seedling descriptions of 188 species of mainly woody Dicots, representing 138 genera and 51 families from the Malesian region were published in 1972 by Burger. The descriptions are very detailed, and the splendid, mostly full-page plates of line-drawings show several stages of the seedlings. The study on which this work was based had been carried out in the years 1921–1924 in Bogor, Indonesia. At the time no funds for publication were found and 47 years elapsed before the author saw his work in print.

Also in 1972 a practical key on common Malesian sawah weed seedlings was published in cyclostyled form by Pancho and Bardenas.

Very recently the study of seedling physiology has been given a new impetus. Lovell and Moore published two interesting articles: 'A comparative study of cotyledons as assimilatory organs' (1970) and 'A comparative study of the role of the cotyledon in seedling development' (1971), studies on the differences of the physiological strategies in the mode of development of seedlings. The physiology of the cotyledon in Angiosperm (Dicot) seedlings, and its role in seedling development, has also been studied by Marshall and Kozlowski. Their findings have been published in four articles in the years 1973–1977.

Mr. Ng in Kepong, Malaysia, is working on a flora of seedlings covering all Malaysian economic trees. So far, representatives of 15 families have been treated (1975, 1976). Ng uses his own terminology for the seedling parts. A developed hypocotyl lifts the cotyledonary node above the ground, an undeveloped or poorly developed hypocotyl does not. Cotyledons are emergent if they escape from the envelopments, or non-emergent; sometimes they are forced apart with the seedcoat more or less adhering to the outer surface of each cotyledon. Cotyledons may be fleshy or leafy, the intermediate condition is semi-fleshy. In the 1976 paper Ng distinguished four germination types: epigeal germination, hypogeal germination, semi-hypogeal germination, and 'Durian' germination. A regular series of publications from the hand of Mr. Ng are anticipated.

At the time the manuscript of this book was finished and almost in press, an article appeared, 'Seedling morphology of some African Sapotaceae and its taxonomical significance' (Bokdam 1977), in which seedlings of 46 species from 25 genera of that family were dealt with, and in addition a more general account on seedlings was given. Bokdam gave a short discussion on seedling characters of the various seedling parts. He distinguished two stages of development in the seedling: the cotyledon stage, in which the cotyledons are present but foliar leaves have not yet developed, and the eophyll stage, when a shoot has been produced with the first leaves developed. The cotyledon stage was recorded to last from several weeks to some years. Three cotyledon types were recorded for Sapotaceae: two with foliaceous cotyledons, these being either papyraceous with conspicuous venation, or coriaceous in which the venation system is inconspicuous; the third type covers fleshy cotyledons with indistinct nervation. Two basic types of seedlings were distinguished in Sapotaceae:

A. Seedlings with foliaceous cotyledons, developing from seeds with abundant endosperm.

B. Seedlings with fleshy cotyledons, developing from seeds with scanty or no endosperm.

A further subdivision of these types, based on whether the cotyledons escape from the testa or not, and the development of the nervation, resulted in the distinction of four seedling types in Sapotaceae: A1 Omphalocarpum type; A2 Argania type; B1 Tieghemella type; B2 Butyrospermum type. In addition to these types, three seedling types were distinguished in Magnoliales, viz. Pycnanthus type, Monodora type, and Cananga type, and one in Asterales, the Asterales type. For the characterisation of the types see Fig. 4, which represents Bokdam's table 1, in which the features are represented in a tabulated form, accompanied by a schematic illustration. The seedling types were given the name of one of the genera in which the type occurs, similarly as done in the present work, of which Bokdam had foreknowledge. By doing so, these names actually have now priority over the ones proposed here if they cover the same seedling variation. But most of his seedling types do not correspond with the ones distinguished here, or are considered as a subtype (see also pp. 53–54). Moreover, Bokdam did not distinguish development types but used a mixture of germination features and morphological characters, which makes his types not comparable with the presently proposed ones. Therefore, none of his names has been used for the presently distinguished seedling types.

The last major general work to be mentioned is a kind of seedling flora on Northwest European seedlings (Muller 1978). Short diagnoses of the seedlings are given, a key is present, and nearly all species treated (over 1200) are illustrated.

Mrs. G. Maury (Paris) is finishing a manuscript in which she describes almost one hundred species of SE. Asian Dipterocarpaceae seedlings which she has studied in Malaya. Basing her work on features like cotyledon shape, form of the embryo, etc. she tries to find clues in which way evolution has taken place in that family, and her ideas go in the direction of a new subdivision of the Dipterocarpaceae.

At Bandung (W. Java), Mrs. Sri Hajati Widodo is working on the ecology of seedlings in natural vegetations.

During the course of time, changes have occurred in the aims with which the various authors undertook their seedling studies. The earliest articles simply described the unknown young stages of the plants, such articles are published still. They are very valuable as they increase our knowledge of the morphological variation in seedlings and the differences in germination and seedling development. This was soon followed by attempts to interpret the morphological variation, and to understand the nature of the seedling parts. At first, only the external morphology was taken into consideration; in the second half of the nineteenth century also the anatomy of the seedling was studied. By studying the morphological variation, attempts were made to find discriminating systematic characters in seed-

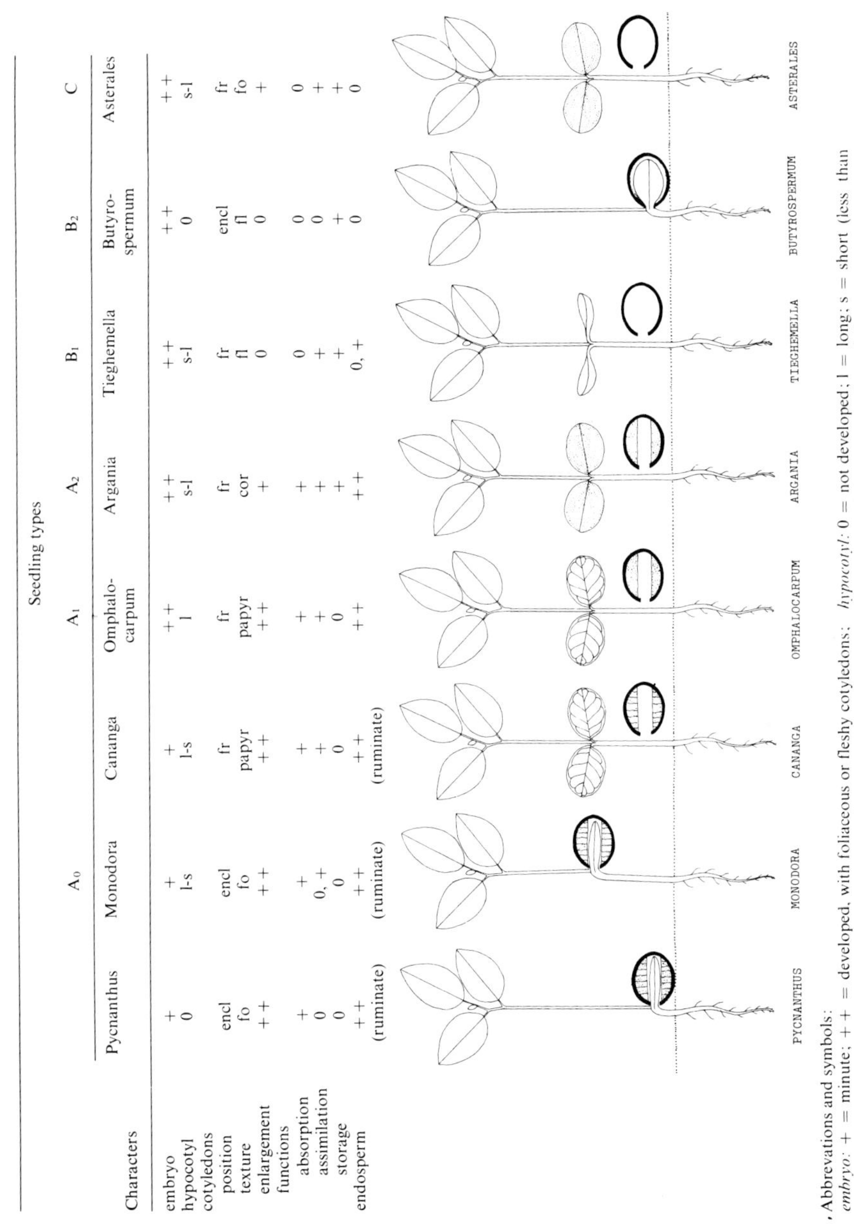

Characters	Seedling types							
		A_0		A_1	A_2	B_1	B_2	C
	Pycnanthus	Monodora	Cananga	Omphalo-carpum	Argania	Tieghemella	Butyro-spermum	Asterales
embryo	+	+	+	++	++	++	++	++
hypocotyl	0	l-s	l-s	l	s-l	s-l	0	s-l
cotyledons								
position	encl	encl	fr	fr	fr	fr	encl	fr
texture	fo	fo	papyr	papyr	cor	fl	fl	fo
enlargement	++	++	++	++	+	0	0	+
functions								
absorption	+	+	+	+	+	0	0	0
assimilation	0	0, +	+	+	+	+	0	+
storage	0	0	0	0	+	+	+	+
endosperm	++ (ruminate)	++ (ruminate)	++ (ruminate)	++	++	0, +	0	0

Abbrevations and symbols:
embryo: + = minute; ++ = developed, with foliaceous or fleshy cotyledons; *hypocotyl:* 0 = not developed; l = long; s = short (less than 4 cm long); *cotyledons position:* encl = enclosed in the testa; fr = liberated, unfolded; *cotyledons texture:* fo = foliaceous; papyr = papyraceous; cor = coriaceous; fl = fleshy; *cotyledons enlargement:* ++ = strongly enlarged; + = enlarged; 0 = not enlarged; *endosperm:* ++ = abundant endosperm; + = scanty endosperm; 0 = no endosperm.

Fig. 4. Classification of Dicot seedlings by Bokdam (1977).

lings, and occasionally seedling characters were used in the delimination of genera.

In pure as well as in applied botany the recognition of seedlings has always been a desideratum. This must, however, be based on knowledge of the complete seedling variation in an area. For several regions this knowledge is now available, e.g. for Europe, and floras with keys for the recognition of seedlings are now being published. In other areas, like in the tropics, such floras are also much needed for silviculture, and for regeneration studies. Knowledge there is, however, very incomplete, and the study of seedlings is still in the exploring stage. In the course of time occasional attempts have been made to make use of seedling characters for phylogenetic considerations. External morphological, as well as anatomical, characters have been occasionally used in genera, tribes or families. A recent development is the study of the physiology of the seedling and its parts, especially of the cotyledons, and their function in the development of the juvenile plant.

Definition of the seedling

The seedling is the very juvenile stage of the plant after germination. Jackson (1928) defined the seedlings as: (1) a plant produced from seed, in distinction to a plant propagated artificially, (2) a young plant so produced. Burger (1972), in his extensive seedling glossary, did not define the term. There is of course a general understanding of what a seedling is, but a sharp definition does not exist. In literature, for instance, the term seedling is used for woody plants from the beginning of germination up to a stage where it is 25 to 50 cm high. When larger, it is called a young plant, until it is over about one metre high and then it is called sapling.

The definition, if possible, should be based on a clearly delimited phase of life in the juvenile plant. A distinct change occurs in seedlings of juvenile plants with foodstoring or with haustorial cotyledons and endosperm when the cotyledons are shed. In the early stages these organs play an important role in the food supply, and they support the development of a shoot with leaves. As soon as they are shed, the plantlet must rely upon food production by organs which were not yet present in the embryo and have been formed after germination. As soon as the cotyledons are shed, juvenile plants of this category are beyond the stage of seedling. Morphologically this stage is easy to recognise, because when the cotyledons are shed their former place of attachment can always be found by means of the scars.

In other seedlings such a distinct change does not occur. In the Barringtonia type (13) and in the Garcinia type (14) the food is stored in the hypocotyl. In the former, the hypocotyl remains part of the seedling axis, and it is not clear when the food there is exhausted and subsequently the plantlet relies entirely on assimilates produced by the leaves. In the latter, the hypocotyl together with the primary root, are situated laterally to the axis formed by the shoot and the secondary root. Here the hypocotyl is often persistent for an extensive period, possibly long after the food is exhausted. After the development of the secondary root, the hypocotyl and primary root eventually die off, and no longer form part of the seedling axis.

Seedlings of the Macaranga type (1) and the Cyclamen type (5) have a physiology which is again different. Leaf-like assimilating 'cotyledons' (paracotyledons), which become extracted from the seed, are present. The juvenile plant contains a food reserve only sufficient for the initial development and elongation of the seedling, after which the juvenile plant must rely, for further development, on the assimilates which are produced by the paracotyledons. Such paracotyledons have a similar assimilatory function as the foliar leaves, and are usually very long persistent. They are still present when a large number of leaves have developed and in some cases almost to the stage in which the plant is flowering. Arguments can be found that paracotyledons are directly derived from the first foliar leaves through abortion of the 'true' cotyledons (see pp. 99–104). Seedlings with paracotyledons are, therefore, devoid of true cotyledons. Endosperm may be present or absent. If, for the Macaranga type and the Cyclamen type, the same argumentation for the definition of the seedling is used as for those with true cotyledons, then such seedlings would be considered to be beyond the stage of seedling as soon as the testa and endosperm are shed and the paracotyledons are unfolded. However, it is hardly admissible to describe a juvenile plant with exposed paracotyledons as being beyond the seedling stage. Therefore, the definition of the seedling as given in the beginning of this chapter must remain rather obscure.

A Dicot seedling consists of a root, a hypocotyl, (para)cotyledons, and a plumular bud. Any of these parts may be well developed, be present in a more or less reduced form, or may even be entirely aborted. Stem and leaves are formed through secondary development, although in some seedling types a first internode or internodes and two or more first leaves may already be present in the seed in the form of a plumule. Directly after germination this plumule enlarges (e.g. Sloanea type, 2; Horsfieldia type/subtype, 7a; Heliciopsis type/subtype, 6a), after which growth of the epigeal parts halts temporarily, for a certain, often specific period. In seedlings of the Macaranga type (1), between the two green paracotyledons an undifferentiated bud is present which starts its development only after a prolonged resting stage.

Seedlings have mostly been classified into two main groups, based on the position of the (para)cotyledons and whether or not the testa and fruit wall are shed. These groups represent germination types, and, according to Ng (1978, in a colloquium on juvenile forms of Spermatophytes in Toulouse) should not be confused with morphological types and with development types. The (para)cotyledons can be raised above the soil and become exposed (epigeal seedlings), or they remain in the soil, lateral of the stem, covered by the fruit wall and/or the testa (hypogeal seedlings). Duke (1965) proposed a subdivision into two other groups: cryptocotylar and phanerocotylar, which also represent germination types. The former have the cotyledons enclosed in the fruit wall and/or testa; in the latter the (para)cotyledons become entirely exposed. Both classifications are not entirely satisfactory because they give only limited information on the seedling, as they do not reflect other criteria which seem equally important. More detailed classifications of seedlings have been given by Klebs (1885) who distinguished morphological types, and by Léonard (1957) who gave developmental types, but these have not been followed in other publications I have seen.

A new classification, based on seedling development, and on functional criteria of the organs, in which all parts of the seedling are taken into account, seems possible and desirable, and is here proposed.

Very important appears the position and function of the (para)cotyledons. The fundamental distinction here recognised is between seedlings with either foodstoring or haustorial cotyledons, those with thin, leaf-like, green assimilating paracotyledons, and a third group consisting of seedlings that have no seedleaves at all.

Paradoxically, the very large group with paracotyledons is homogeneous, whereas the very small group without cotyledons, and the larger groups with foodstoring and haustorial cotyledons, can be subdivided into several seedling types. Some of the types distinguished are rather common, others are extremely rare. Some of the types in the proposed subdivision can be classified as epigeal seedlings, while others are hypogeal. In this work these terms are used in a purely descriptive way, and not as a typification of the seedling. In seedlings with epigeal (para)cotyledons, phanerocotylar as well as cryptocotylar seedlings can be distinguished if only the emergence or non-emergence of these organs is taken into account. Altogether, 16 different types of seedling development (seedling types) are distinguished here.

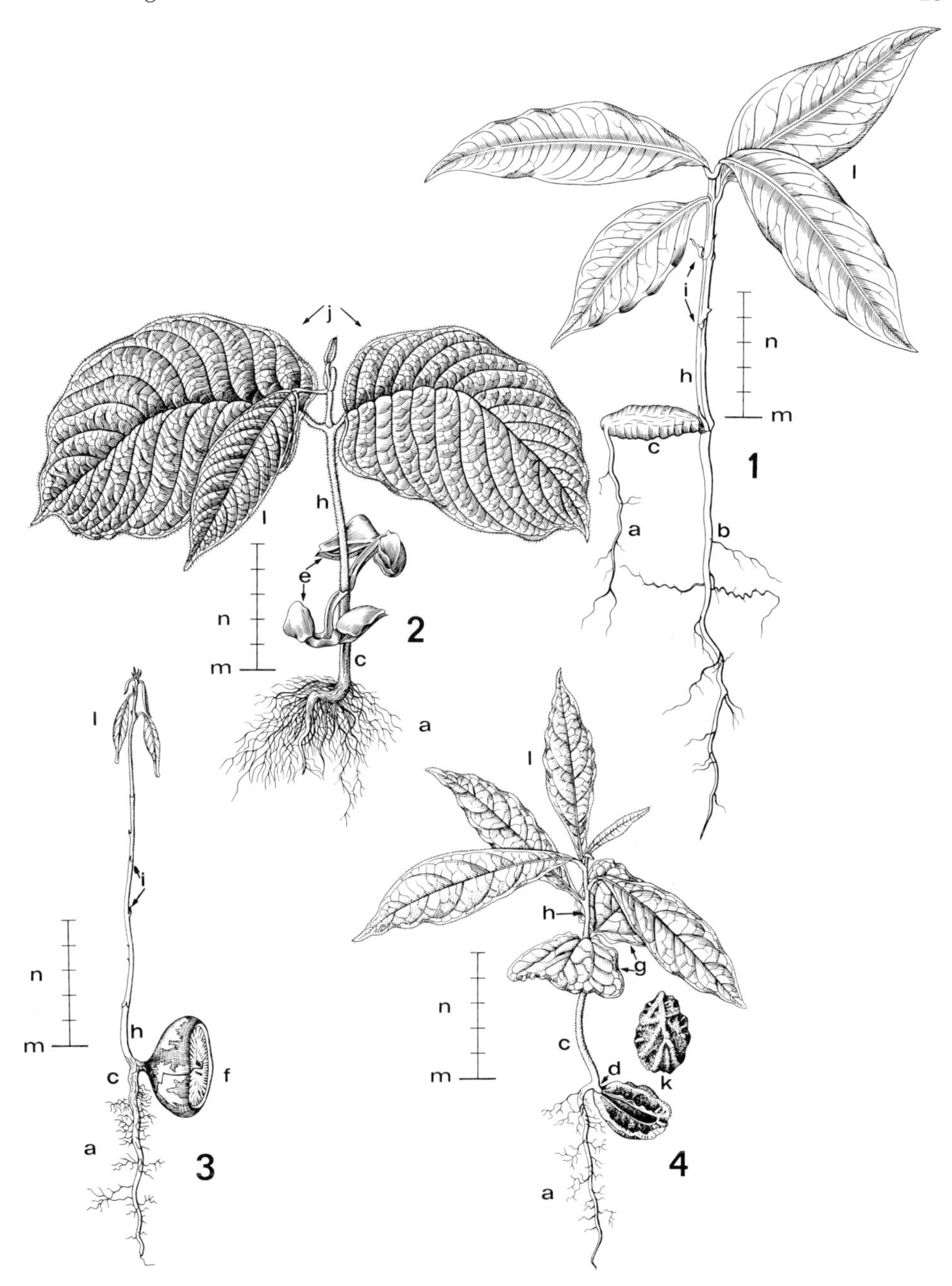

Fig. 5. Nomenclature of the seedling parts. 1. *Garcinia sp.* (Guttiferae, Bogor Bot. Gard. VI. C. 244); 2. *Shorea compressa* (Dipterocarpaceae, De V. 2392); 3. *Lithocarpus korthalsii* (Fagaceae, Bogor Bot. Gard. VIII. B. 39); 4. *Terminalia foetidissima* (Combretaceae, Bogor Bot. Gard. VII. F. 32). a. primary root system; b. secondary root; c. hypocotyl; d. unilateral outgrowth of the hypocotyl; e. free foodstoring cotyledons; f. cotyledons enclosed in the fruit wall; g. paracotyledons; h. first internode; i. cataphylls; j. pseudocotyledons; k. fruit wall; l. leaves; m. soil level indication; n. scale in cm.

Structure, function, and variation of the seedling parts

The present chapter is based on personal observation as well as on the study of seedling literature. Most of the described variation in seedlings has been actually observed by the author in living specimens. It seemed useful also to summarise the very scattered seedling literature, to supplement the observations on the variability, and to present an analysis of seedling morphology which is as complete as possible. Such an analysis has never been given. Where necessary, the author's own interpretations and points of view are given on the literature. It is assumed that the present analysis covers almost all variations of the seedling parts, not only for Malesia, the area under special consideration from which most examples were taken, but worldwide.

In the following paragraphs of the present chapter the seedling parts are successively discussed from a morphological point of view. The next chapter is devoted to a synthesis of these data, and describes the 16 different types of seedling development (seedling types) that can be distinguished (not to be confused with germination types which describe only the method of sprouting, or morphological types which also take the shape of the seedling parts into account). Then follows a chapter in which the seedling types are arranged according to their similarities and dissimilarities. An analysis leads to a number of morphological conclusions. References to the seedling types are given by the indication of their type names and numbers (see pp. 56–76, e.g. Horsfieldia type/Pseuduvaria subtype, 7b). References to species descriptions and illustrations in the special part of this work are by means of indication of the figure number (e.g. *Alangium javanicum*, Fig. 29).

The root

In the embryo the root is represented by the radicle. The root is the descending part of the seedling axis. It is separated from the ascending part of this axis by the collet. In species germinating in an epigeal manner the root, on germination, behaves differently from those germinating in a hypogeal manner.

In the embryo of species germinating in an epigeal manner the radicle is the meristematic tip of the stem-like projection below the cotyledons. On germination the radicle pierces the fruit wall and/or the testa through elongation of the hypocotyl, and is pushed free from these envelopments. The radicle also starts growing when exposed, and accelerates after the young root has touched the soil. Entry in the soil is facilitated by a constant circumnutating movement of the root tip (Darwin 1880). Establishment of the young plant is through extensive growth of the root. A firm foothold is gained by elongation of the primary root, the development of lateral roots or accessory roots, the formation of root-hairs, or the slimy dissolving of epidermal cells (Klebs 1885).

In the embryo of species germinating in a hypogeal manner the radicle consists, in most cases, of almost the entire part of the axis below the cotyledons. A short to very short hypocotyl is present, consisting of the transitional zone between the radicle and the cotyledonary node. On germination, initial growth of the axis below the cotyledons consists mainly of elongation of the young root. This pierces the envelopments, or emerges from an opening. Establishment of the young plant is facilitated by the development of a sturdy taproot. Root-hairs and dissolving of epidermal cells are most often not present in hypogeal tree seedlings; branching or the development of accessory or lateral roots takes place later in the development.

Positive geotropism causes the young root or the entire axis hypocotyl-root to curve to the soil. The rootlet quickly enters the soil, thus reducing the chance that the young meristematic tissue of the tip dries up. Species with small seeds are generally speaking adapted to germination in open soil. If these seeds come in a situation where they are scattered amongst dead leaves, the relatively small root must seek its way to the mineral soil but is exposed to a heavy desiccation risk. Often it will not reach the soil, and the seedling ultimately dies. Big-seeded seedlings are much better adapted to this situation as they have a more sturdy root system.

Some big-seeded genera such as *Shorea*, *Hopea*, *Dipterocarpus*, etc. (Dipterocarpaceae), *Parishia* and *Swintonia* (Anacardiaceae), and other forest genera show a germination pattern which can even cause the death of the seedling under adverse circumstances. The fruit is provided with large wing-like projections which represent the persistent sepals, that keep the fruit tip pointing more or less oblique to above. On germination, the hypocotyl and root emerge from this tip, and have to turn a relatively large semi-circle to reach the soil. In this process they are often hampered by the wing-like projections. For a prolonged period the root is then free in the air, and a dry spell will result in the loss of the majority of a germinating seedcrop. Apparently a germination like this can only have developed under constant everwet conditions.

Early development of the radicle in tree seedlings is into an unbranched taproot, lateral roots normally develop later. The robustness of the taproot can vary according to the species. To some extent this is also influenced by ecological factors, e.g. texture and fertility of the soil. In addition to its two main functions, water-supply and foothold, each root serves, to some extent, as a foodstoring organ. Rarely it is specialised in this direction and then it is usually a spindle-shaped organ. In Fagaceae this has been recorded for several species of *Quercus* (Hickel 1914, Troup 1921), and *Castanopsis* (Burger 1972). In the present investigation a similar condition was found for *Stelechocarpus burahol* (Annonaceae).

Normally, the primary root develops into the main root of the plant. Reduction or even abortion of the primary root has been recorded, but is rare. In a number of herbaceous temperate seedlings, e.g. *Clintonia pulchella* (Campanulaceae) the primary root starts its elongation not in the early stages of germination, but much later. In others, it is replaced almost entirely by accessory roots, e.g. *Batrachium heterophyllum* (Ranunculaceae, Klebs 1885, his type 5). For tree seedlings this is less common, but in *Bhesa robusta* (Celastraceae, Fig. 57) all roots from the collet are alike and none is obviously the primary root.

Strong reduction or abortion of the main root takes place in Guttiferae, the tribes

Monorobeae and Garcinieae. The seed contains an undifferentiated solid body, the hypocotyl, from one pole of which sprouts the primary root. Mostly, its growth is restricted to the early stages of seedling development, but sometimes even this initial growth of the primary root is suppressed. Soon after appearance of the shoot from the other pole of the hypocotyl, a sturdy secondary root develops from the junction of this shoot and the hypocotyl. This grows quickly into a much sturdier root than the primary root, and takes over its function as main axis. Thus, the hypocotyl and primary root are situated lateral to the new main axis, and may be long persistent as the upper 'lateral' root, before they finally disappear.

In *Hodgsonia macrocarpa* (Cucurbitaceae, Fig. 65) the primary root never emerges from the seed. On the nodes of the emergent thin, trailing stem, small additional roots develop which anchor in the soil. When the stem reaches support and starts climbing they attach the stem to the substratum. In several other climbing genera also additional roots from the nodes are recorded in seedlings (e.g. Cremers 1974).

Lateral roots in tree seedlings normally appear when the primary root has developed into a fair-sized taproot. Their number may be small or large, they may appear rather soon or relatively late, and they have often a specific arrangement. Four longitudinal rows seem not uncommon. In *Aesculus* (Hippocastanaceae) eight distinct rows are present. Due to soil conditions this arrangement is, however, often more or less distorted. A whorl of small lateral roots at the collet is a rather common feature, in epigeal as well as in hypogeal seedlings. The rootlets may attach themselves to the fruit wall or testa and thus aid the seedling in the extraction from the envelopments during germination. Lateral roots too may be branched to various degrees.

Root hairs are common in epigeal seedlings (Klebs 1885), macroscopic ones being present on the young parts close to the tips of main and lateral roots. They often disappear rather quickly with senescence, while more towards the growing tip new ones appear. Thus, a rather limited zone of closely set root hairs is present near the end of the root. It is peculiar that in primary forest seedlings macroscopic root hairs are rarely encountered, especially not in hypogeal seedlings. It is possible that such seedlings develop a mycorrhiza in a later stage (according to Kozlowski, 1971, root hairs are absent on roots bearing mycorrhizae), but in rather young seedlings so far no sign of this has been observed. Troup (1921) recorded root hairs for seedlings of *Adina cordifolia* and *Stephegyne parvifolia* (both Rubiaceae). In the present investigation they were only encountered in the tree seedlings of *Bombax valetonii* (Bombacaceae), two species of *Parartocarpus* (Moraceae), *Melanochyla fulvinervis* (Anacardiaceae), and *Trewia nudiflora* (Euphorbiaceae). All of these, except *Parartocarpus* and *Melanochyla*, have assimilating epigeal cotyledons and are more or less light demanding. Seedlings of *Parartocarpus*, however, were found growing under leaf cover in undisturbed evergreen forests. Microscopic investigations on the occurrence of root hairs in tree seedlings are few, this field requires much more study.

Root nodules occur in a restricted number of families. Where present, they only develop when the roots of the plant are infected by bacteria, in a rather young seedling stage or when the plant is older. Root nodules often serve in fixing nitrogen. They are common in the Leguminosae – Papilionaceae where they result from invasion of the root by *Rhizobium* bacteria, and have been recorded (also in the seedling) in many genera. They also occur in the temperate tree genus *Alnus* (Betulaceae), which has thoroughly been studied on this characteristic. Nodules have also been illustrated in an older *Podocarpus imbricatus* seedling (Podocarpaceae, Burger 1972), but he apparently overlooked this feature, as these tubercles are not mentioned in the seedling description. Kozlowski (1971) recorded root nodules in Gymnosperm genera like *Agathis*, *Araucaria*, *Libocedrus*, *Phyllocladus*, *Dacrydium*, and *Sciadopitys*. Allen and Allen (1965) confirmed their presence in 8 families of Dicots for a large number of species, especially in the genera *Alnus*, *Casuarina*, *Coriaria*, *Elaeagnus*, *Hippophaë*, *Shepherdia*, *Comptonia*, *Ceanothus* and *Purshia* (after Kozlowski, publication not seen).

In hemi-parasites like *Ximenia* (Olacaceae), in Loranthaceae (Klebs 1885), and also in *Santalum* (Santalaceae, Burger 1972) haustoria were described in early stages of the seedling (see also Kuijt 1969). In parasites like Orobanchaceae (Caspary 1854) the primary root of the seedling elongates somewhat during germination, but it does not differentiate unless a foot of a host plant is encountered. Then a disc is formed which attaches the parasite to the host.

Root characters will not often be used for identification of the seedling, but sometimes they have diagnostic value. In *Diospyros* (Ebenaceae) all evidence confirms that the roots are blackish. In various genera and species root tips may have a colour different from, and contrasting to, that of the remainder of the root. Often the tips are white or cream-coloured while the roots are brownish. Burger (1972) recorded for three species of *Artocarpus* (Moraceae) that the root of the seedling turns orange to orange-red, while the tips are yellow or white. Occasionally the smell of the root may also have diagnostic value.

The hypocotyl

The hypocotyl is the basal portion of the ascending axis, and consists of one 'internode' only. It is situated between the (para)cotyledonary node and the collet. In species germinating in an epigeal manner the hypocotyl is already pronounced in the embryonic stage, and, on germination, will develop into an elongated hypocotyl. In the embryonic stage it is called caulicle. In species germinating in a hypogeal manner the hypocotyl is much reduced, in the embryo as well as in the seedling.

The hypocotyl contains a transitional zone where the exarch stele of the root changes into the endarch stele of the ascending axis. This transitional zone may be either long or short, and it even may occupy the entire hypocotyl. Where it is restricted to a small zone it is usually found in the upper part of the hypocotyl. Especially in hypogeal seedlings where the hypocotyl is short, the transition zone is sometimes extended to the lower portion of the first internode (Fahn 1967).

The hypocotyl is equivalent with an internode, and consequently not branching. None the less, Roeper (1824) recorded that three species of *Euphorbia* (Euphorbiaceae) possess branching hypocotyls. These branches develop laterally from the hypocotyl at non-specific places. Bernhardi (1832), in addition, recorded branches sprouting especially from the collet for species of *Linaria* (Scrophulariaceae), but these are also formed at irregular places along the hypocotyl (Csapody 1968).

Branching of the hypocotyl, at non-specific places, may possibly suggest root nature of the branching part. On the other hand, it is possible that the hypocotyl in certain seedlings is a composite organ. If the assumption is true, that Macaranga type seedlings (1) are derived from Sloanea type seedlings (2), as discussed on pp. 99–104, then the hypocotyl of the former is equivalent with the hypocotyl and the first internode of the latter. That may explain the occasional branching of the hypocotyl. The consequence of this assumption is, that hypocotyls in the different seedling types may not be homologous. Evidence to support this view has, however, not yet been brought forward by seedling anatomists. So far no definite nodes have been recorded in hypocotyls, although a transition zone has been demonstrated.

The collet is the junction between the root and the stem. Sometimes it is a very pronounced structure, in other cases almost indistinguishable. The collet is often regarded as a node, although it never bears a leaf; definitely it is not homologous with a stem node which has an entirely different structure. It may occur as a swollen portion at the base of the hypocotyl, in the form of a ring, or, as an unilateral outgrowth. Klebs (1885) based his germination type 2 on the presence of such an outgrowth. In temperate epigeal herbaceous seedlings the presence of a thickened collet is rather common. In those seedlings it serves as an aid in the extraction of the paracotyledons from the fruit or seed.

In seedlings of tropical trees the presence of a thickened collet is rare. It is found in most species of *Terminalia* (Combretaceae). During germination it hooks on the inside of the lower fruit valve. Thus, the curving hypocotyl obtains a foothold, and pushes up the upper fruit valve like a jack. By its subsequent elongation the convolute paracotyledons are withdrawn upside-down from the fruit valves (Fig. 6). A similar outgrowth is found in *Morinda* (Rubiaceae). During the emergence of the root the outgrowth is firmly clasped between the tips of the two persistent seed valves. The hypocotyl emerges laterally from the testa, and carries the paracotyledons free above the soil. Due to the persistent empty testa, on the base of the hypocotyl, the seedling much resembles the Horsfieldia type/subtype (7a). Compton (1912) recorded a thickened collet for *Mimosa pudica*, *Leucaena glauca* (both Leguminosae/Mimosaceae), and *Petalostylis labichaeoides* (Leguminosae/Caesalpiniaceae).

The (para)cotyledonary node terminates the hypocotyl. This node may be more or less swollen, and is sometimes provided with small appendages which may be of stipular nature and have no apparent function. In most cases it has no special function during germination, but in species with peltate foodstoring cotyledons the cotyledonary node may thicken and pushes the cotyledons apart during germination (e.g. Endertia type, 11).

The hypocotyl of seedlings germinating in an epigeal manner is usually already distinctly developed in the embryo. During germination it elongates, pushing the radicle out of the seed. In many cases the hypocotyl curves into a loop above the ground, with both the root, and the hypocotyl tip with the (para)cotyledons in the seed, remaining in the soil. The hypocotyl often stays arrested in this position for a considerable period. After the resting period, but in other cases soon after emergence, it will stretch, and thrust the (para)cotyledons (resp. either free and

Plate 5. a. *Santiria tomentosa* Bl. (Burseraceae), De Vogel 747.
b. *Rourea minor* (Gaertn.) Leenh. (Connaraceae), Dransfield 2523.

Plate 6. *Bhesa robusta* (Roxb.) Hou (Celastraceae), Dransfield 2392.

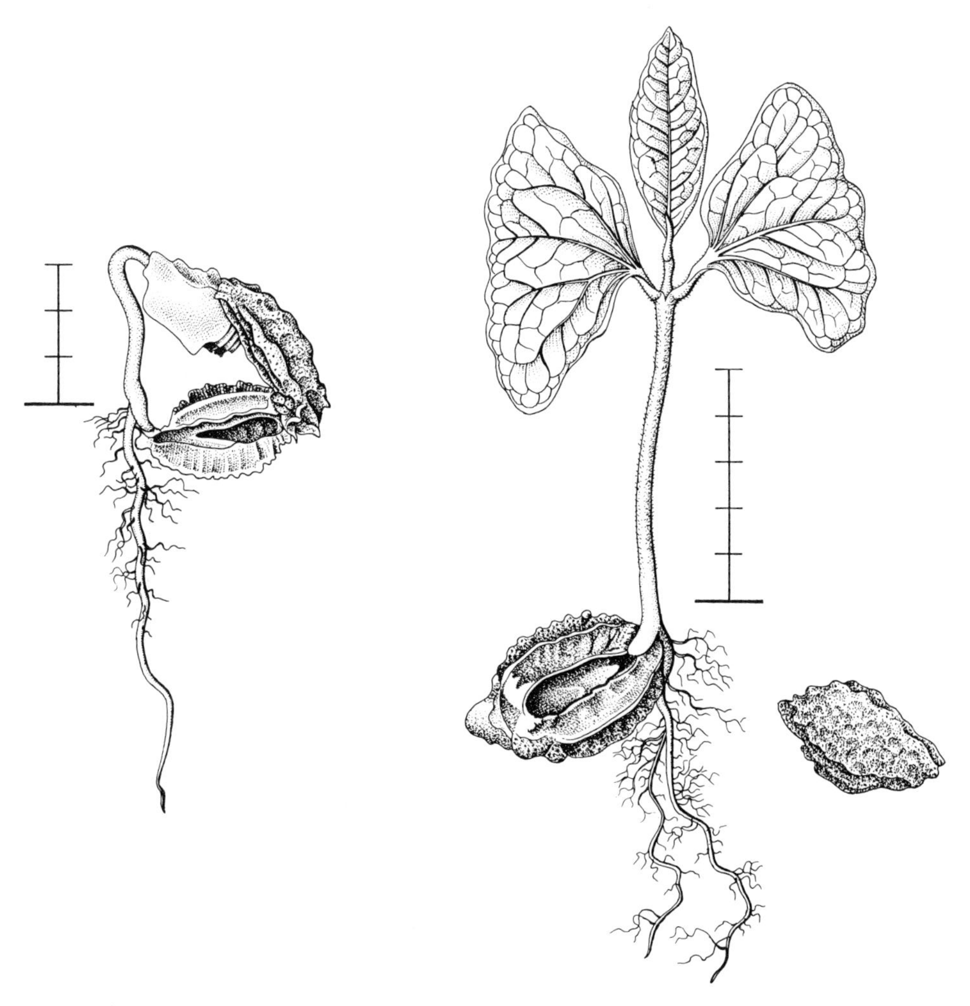

Fig. 6. *Terminalia catappa* (Combretaceae), De V. 1379. Method by which the seedling emerges from the fruit.

exposed or still enclosed in the envelopments) high above soil level. A hypocotyl with such characteristics is present in several seedling types (e.g. Sloanea type, 2; Macaranga type, 1). Its main function is to bring the cotyledons above the soil, to give them an exposed position. In a number of cases the hypocotyl has a pronounced foodstoring function. Evidently in such cases the fact that the cotyledons are carried above the soil is of minor importance, because these remain in the envelopments and never become exposed. The food in the cotyledons or the endosperm is conveyed to the hypocotyl. There it is stored, and mobilised again as soon as the shoot starts elongating, after the cotyledons with the envelopments are shed. This is found in seedlings of the Blumeodendron type (8), the Coscinium type (10), and the Rhizophora type (9). In the Ternstroemia type (4) the cotyledons are much reduced or absent, a certain amount of food is stored in the hypocotyl. In most seedlings of these types the hypocotyl is a swollen, fusiform body. In the Rhizophora type, transfer of the food is already taking place while the fruit is still on the tree. In many cases the hypocotyl is the dispersal unit; it is detached from the fruit which remains attached on the parent plant.

In the embryo the hypocotyl of seedlings germinating in a hypogeal manner is scarcely developed, and during germination it will hardly elongate. In the seedling it cannot be easily distinguished from the root, and there is never a recognisable collet. Elongation is slightly more pronounced in some cases, but, nevertheless, the cotyledons remain ± at soil level. A specialised situation is found in the hypogeal seedlings of the Barringtonia type (13), and the Garcinia type (14) where cotyledons are not present, or are very much reduced. The seed or fruit is entirely filled with the swollen hypocotyl. On germination this hypocotyl does not enlarge, it remains within the envelopments, and the root and the shoot emerge from opposite poles of the seed. In the Barringtonia type the hypocotyl is persistent, as it remains part of the seedling axis. In the Garcinia type the hypocotyl (and the primary root) are long persistent but do not enlarge, and ultimately disappear.

Hypocotyl characters are sometimes useful in generic delimitation or in distinction at species level. Measurements (length and diameter and general outline may be rather constant), indumentum, colour, and specific ornamentation, such as ridges, may be diagnostic. The features of the hypocotyl may be the same as in the rest of the stem, or they may be different in those parts.

The seed leaves

Seed leaves, seed lobes or cotyledons, in the broad classical meaning, are the lowermost leaves of the juvenile plant. In the embryonic stage they are already present in a rather definite shape. In literature these three terms have always been used as synonyms, but Muller (1978) applied the term seed leaves exclusively to those seed lobes which become exposed. The term cotyledons he reserved for seed lobes which never leave the envelopments; seedlings with such seed lobes were described as devoid of seed leaves.

Cotyledons, in the classical meaning, occur in three different manifestations. They may be either foodstoring, haustorial, or photosynthetic at the moment of germination; each type may have a haustorial function during the (early) development of the embryo. In the present work (pp. 99–104), indications are found that leaf-like photosynthetic 'cotyledons' are possibly of different origin than the other types. At least in a number of examples photosynthetic cotyledons must be regarded as homologous with the lowermost stem leaves, and not directly with haustorial or foodstoring cotyledons, which in those cases evidently have aborted. Therefore, it is better to avoid the term cotyledon for this sort of primary seedling leaves, and to apply another term: paracotyledons.

In the Angiospermopsida the subclasses Monocotyledoneae and Dicotyledoneae are, amongst others, defined by the number of the cotyledons: Monocots one, Dicots two. If the present distinction between seed lobe types is made, then Dicotyledons cannot be described any more as possessing two cotyledons; the large majority does not have cotyledons but paracotyledons. Even within one genus these two different types of seed lobes may occur.

Real cotyledons, which are either foodstoring or haustorial, are present in a large number of families (see list on pp. 78–92). Their number is almost always two. However, in *Terminalia megalocarpa* (Combretaceae) the number is always three (Coode, pers. comm.), and in *Gustavia angusta* (Lecythidaceae) the number is two, incidentally three, while also partially divided cotyledons occur (personal observation, all seeds from the same parent). Abortion of cotyledons is present in quite a number of seedling types. In the Barringtonia type (13), the Garcinia type (14), the Ternstroemia type (4), and the Orobanche type (16) that is quite obvious, because either the cotyledons are entirely absent, or they are represented by much reduced scales. In the embryos of those types no well developed leaf-like organs are present. In the Macaranga type (1), and in the Cyclamen type (5), however, in the embryonic stage leaf-like organs are present in a more or less primordial stage which have always been regarded as homologous with real cotyledons. But, at least in some genera of those seedling types, the real cotyledons evidently have been aborted, and it seems acceptable to assume abortion of the cotyledons in all seedlings of these two types. That also accounts for the variation in the number of the paracotyledons, which is relatively common in these types. In several genera the number of the paracotyledons is variable, but then each species has a specific number. For instance in *Capparis* two (Duke 1965, 1969) and three paracotyledons (pers. obs.) have been recorded. And in the species of *Terminalia* with paracotyledons the number is specific and may vary between two and four (Coode, pers. comm.). Variation between specimens of the same species may also occur. In *Degeneria vitiensis* (Degeneriaceae, Swamy 1949) 87% of the specimens studied possessed three paracotyledons, the remaining 13% four. In many species normally with two paracotyledons, incidental deviations occur in which their number is more than two.

Differentiation of (para)cotyledons takes place soon after the first cell divisions in the zygote. At the tip of the very juvenile embryo some terminal cells are formed, which through further divisions will produce the (para)cotyledons and the plumule or the terminal bud (Maheshwari 1950). It is not known to me whether differences exist in the ontogeny of cotyledons and paracotyledons. In the almost mature embryo the (para)cotyledons attain their definite form and shape, although in Macaranga type (1) seedlings the paracotyledons usually enlarge considerably after germination.

As a rule the fully developed cotyledons and paracotyledons have a different shape compared with the foliar leaves of the seedling. Real cotyledons, however, deviate much more in this respect than paracotyledons.

The position of the (para)cotyledons is on top of the hypocotyl, inserted on the (para)cotyledonary node. The normal situation in Dicots is that the place of attachment of the two cotyledons is opposite. In seedlings germinating in an epigeal manner, the (para)cotyledonary blades are situated opposite on either side of the stem (e.g. Macaranga type, 1; Sloanea type, 2). In species germinating in a hypogeal manner, the blades are often borne lateral of the stem (secund) although their place of attachment is opposite, like in the Horsfieldia type/subtype (7a), or they are opposite on either side of the stem, as in the Endertia type/Chisocheton subtype (11b). Rarely (para)cotyledons are inserted on the seedling axis at different levels; in that case the (para)cotyledonary node is oblique (*Sauropus*, Euphorbiaceae, Fig. 87), or a mesocotyl is produced after germination (Gesneriaceae, Fritsch 1904). Where more than two (para)cotyledons are present these are arranged in a whorl.

The duration of the persistence of (para)cotyledons on the seedling depends on the seedling type, and on the function these organs have in food production. Paracotyledons provide assimilates, and they are normally very long persistent (Macaranga type, 1). The period of activity of foodstoring cotyledons is dependent on the amount of food stored in the tissues, and the speed this food is used by the growing plantlet. In seedlings of the Sloanea type (2), the cotyledons are soon shed after the plumule has developed into the first internode with two opposite leaves. In e.g. the Heliciopsis type/subtype (6a) they are shed when a complete shoot has been formed with often a specific, usually relatively large number of internodes and several developed leaves. In occasional examples foodstoring cotyledons may last much longer than one year (*Eusideroxylon*, Lauraceae, Fig. 102). In seedlings of the Blumeodendron type (8), the Coscinium type (10), and the Rhizophora type (9) the cotyledons are shed as soon as the food is transferred from the cotyledons to the hypocotyl. In the Endertia type (11) the number of developed leaves supported by the cotyledons may vary, the cotyledons are usually long persistent.

Haustorial cotyledons are persistent as long as the endosperm still provides food; they usually last relatively short, and they are shed together with the envelopments. In some cases representing the Sloanea type (e.g. *Bhesa*, Celastraceae, Fig. 57) the endosperm and testa are shed and the haustorial cotyledons become exposed for a short period, after which they are dropped. In the Sterculia stipulata type (3) only the testa is shed while the endosperm remains adhering to the lower surface of the exposed cotyledonary blades.

The position of the (para)cotyledons in the seed is usually in one plane, the blades lying straight with the upper sides facing and in close contact. Deviating positions occur in several seedling types, but especially in seedlings of which the (para)cotyledons become exposed. Duke (1969) gave a survey of the different types of folding. Conduplicate-induplicate paracotyledons are recorded for *Avicennia* (Verbenaceae) and others, plicate ones in *Cordia* (Boraginaceae), while in *Terminalia* (Combretaceae) and other genera they are convolute. More intricate is the position in *Petiveria* (Phytolaccaceae) where one paracotyledon is reclinate and in addition involute, and the other convolute around the first. *Cariniana* (Lecythidaceae) paracotyledons are complexly circinnate. In addition to these examples the embryos of Burseraceae (especially *Canarium*) must be mentioned, where the (para)cotyledon lobes are with the upper sides facing, but in addition irregularly interwoven. The always enclosed, relatively thin cotyledons in *Dipterocarpus* (Dipterocarpaceae) have the same position but they remain much folded in the seed and fruit.

On germination the (para)cotyledons are extracted from the envelopments or not. On germination free foodstoring cotyledons more or less spread unless they are peltate; usually they turn more or less oblique to the stem. Unfolding paracotyledons have various methods of shedding the endosperm and/or testa, and the blades turn more or less in a horizontal plane. In *Canarium hirsutum* (Burseraceae, Fig. 7 and 51) the folded palmate paracotyledons first stretch vertically while the upper sides are still facing each other, then, the central leaflets spread, followed, a few days later, by the lateral leaflets.

Stipules

Stipules at the base of real cotyledons are very rare, in the present work they are recorded for *Koompassia* (Leguminosae/Caesalpiniaceae). Paracotyledonary stipules are known from a very limited number of families in which the presence of stipules is a character of the adult plant. In a general note on (para)cotyledonary stipules Duke (1969) gives, as example, that these occur in the three (sub)families of Leguminosae and in Rubiaceae. As can be concluded from the present, and other studies, the presence of paracotyledonary stipules is probably the rule in Rubiaceae. They are in general interpetiolarly connate as are those of the foliar leaves. However, Burger (1972) recorded their absence in *Neonauclea obtusa*. The statement of Duke about presence of (para)cotyledonary stipules in Leguminosae needs some clarification, because he only specifically mentioned *Prosopis juliflora* (Leguminosae/Mimosaceae). From the present study, and from Burger (1972), it can be concluded that they are only common in the Leguminosae/Mimosaceae group, relatively rare in the other groups, and even sometimes within a genus they

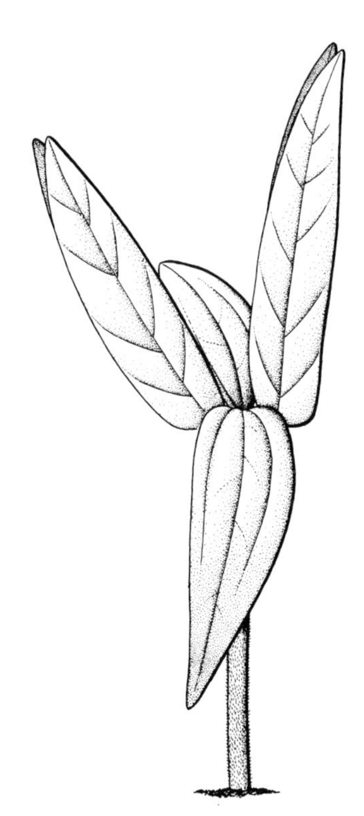

Fig. 7. *Canarium hirsutum* (Burseraceae), De V. 1397. The paracotyledons are withdrawn from the fruit wall with the upper sides together. First the midlobes turn in a horizontal plane, a few days later followed by the lateral ones.

may be present or not (Table 2). Other students, including Lubbock (1892) who covered 72 genera of Leguminosae, unfortunately did not pay attention to this feature. As far as is known to me paracotyledonary stipules have further only been recorded for *Euonymus* (Celastraceae, Burger 1972), and *Cappâris* (Capparidaceae).

Table 2. Records on (para)cotyledonary stipules in Leguminosae (after Burger 1972 and personal observations)

	(para)cotyledonary stipules
Caesalpiniaceae	
Afzelia	–
Caesalpinia	–
Cassia	in 2 species out of 4
Cynometra	–
Endertia	–
Koompassia	+
Peltophorum	–
Phanera	–
Piliostigma	+
Saraca	–
Tamarindus	–
Mimosaceae	
Acacia	in 1 species out of 6
Adenanthera	stipules in the form of a tuft of hairs
Albizia	in 5 species out of 6. Stipules wart-shaped, hair-like, in the form of a tuft of hairs, or scale-like
Dichrostachys	stipules hair-shaped
Leucaena	stipules hair-shaped
Parenterolobium	–
Parkia	stipules hair-shaped
Pithecellobium	–
Samanea	+
Papilionaceae	
Butea	–
Clitoria	–
Crotalaria	–
Dalbergia	–
Desmodium	–
Erythrina	–
Indigofera	+
Ormosia	–
Pterocarpus	+
Sophora	–
Tephrosia	–

The petiole

In (para)cotyledons often a petiole can be distinguished, although sessile (para)-cotyledons are not uncommon. Sheath-like petioles have not been recorded in Dicots. Usually the base of the petiole is somewhat thickened, slightly clasping the node. Almost always they are free structures, terete, or semi-orbicular in cross-section, and then, often channelled above. Very often a similar indument as on the stem is found on the petioles, which in several cases extends on part of the lower portion of the (para)cotyledon blade, especially on the nerves.

In seedlings with paracotyledons the petioles may have a wide variation in length, although in most they are relatively short, or even the blade is sessile. Lubbock (1892) suggested that in epigeal seedlings sessile or shortly petiolate paracotyledons are correlated with long hypocotyls, and that long petioles belong with short hypocotyls, but he cited also examples which do not show this combination. As far as can be concluded from literature and own observation, the combination long (para)-cotyledonary petioles and a short hypocotyl is rare.

In seedlings with exposed foodstoring or haustorial cotyledons the petioles are mostly short, or the blade is sessile. In seedlings with covered cotyledons, whether these are carried above the soil or not, the petioles are well pronounced but not exceedingly long, and have a function during germination in pushing radicle and plumule out of the seed or fruit. In the seedling they are often more or less curved, due to the lateral (secund) position of the cotyledons, while their place of attachment is opposite (e.g. *Horsfieldia*, Myristicaceae, Fig. 136). In hypogeal species with peltate cotyledons clasping the stem (e.g. *Dysoxylum*, Meliaceae, Fig. 127) the petioles hardly elongate; they are pushed apart by thickening of the cotyledonary node.

Fused (para)cotyledonary petioles are rare, and occur especially in a limited number of genera of temperate herbaceous plants (e.g. Bernhardi 1832, Csapody 1968). The fusion may be complete, or the tip of the petioles may be free over a short distance. The developing shoot either breaks through the tube, often at a non-specific place, or pushes its way up and appears between the cotyledons, in the latter case much resembling a proper epigeal seedling, with the fused petioles resembling the hypocotyl. They have been recorded in e.g. *Aconitum*, *Anemone*, *Clematis*, and *Delphinium* (all Ranunculaceae), *Bunium*, *Chaerophyllum*, *Ferulago*, and *Prangos* (all Umbelliferae), *Polygonum* (Polygonaceae), *Dodecatheon* (Primulaceae), and *Serratula* (Compositae); in these genera fused as well as non-fused petioles occur. In *Eranthis* (Ranunculaceae) and *Smyrnium* (Umbelliferae) seedlings of the three species described all have connate petioles.

Marah (Cucurbitaceae), a perennial from subtropical seasonally arid areas in N. America, possesses foodstoring cotyledons on up to 20 cm long fused petioles that push the radicle and plumule deep into the soil on germination. This has been interpreted as an adaptation to fire (Schlising 1969).

Fused (para)cotyledonary petioles are extremely rare in woody plants. Exell and Stace (1972) recorded these for African *Combretum* (Combretaceae), in three species out of 100 (*C. collinum*, *C. fragrans*, and *C. zeyheri*), all three from savanna regions liable to fire. They speculated that this feature may be advantageous where burning or heavy grazing occurs. The feature has not been recorded for any rainforest species in the family. Other possibilities in the genus are: (1) the hypocotyl long and the paracotyledonary petioles short, (2) the cotyledons remain subterranean, (3) the hypocotyl short and subterranean and the paracotyledons borne on long petioles.

The tree *Quercus semecarpifolia* (Fagaceae), which occurs at high altitude, has seedlings germinating in a hypogeal manner, with fused petioles 2.5–10 cm long (Troup 1921). Troup interpreted the petiolar tube as a protecting device to bring the radicle through a thick growth of grass and litter into the soil.

The blade

The (para)cotyledonary blades, the most important part of the cotyledons in respect to the food provision to the juvenile plant, can be classified in three categories, according to their form and function in the mature embryo and in the germinated seedling: foodstoring cotyledons, haustorial cotyledons, and assimilating paracotyledons. In the first, reserve food is stored in the cotyledon tissues; in the second, it is stored in the endosperm from which it is drawn by the cotyledon; and in the third, very little or no reserve food is stored after the blade has expanded, but assimilates are synthesised through photosynthesis.

Almost always the (para)cotyledons of a seedling are equal in function, shape, and size. However, in *Streblus* and in some *Artocarpus* species (Moraceae, Endertia type/Streblus subtype, 11c) one of the foodstoring cotyledons is small, and on germination it lifts from the larger like a lid (Troup 1921). In *Sauropus* (Euphorbiaceae, Fig. 87) the foodstoring cotyledons are unequal in size and placed on different levels along the stem. Lubbock (1892) recorded (slightly) unequal paracotyledons for *Brassica* and *Raphanus* (Cruciferae), *Pachira* (Bombacaceae), and *Cereus* (Cactaceae). Fritsch (1904) described very unequal paracotyledons for the genera *Streptocarpus*, *Monophyllaea*, *Roettlera*, *Klugia*, and in the seedlings in some of the species of *Ramondia*, *Saintpaulia*, and *Drymonia* (all Gesneriaceae). On germination the paracotyledons are more or less equal in size. One remains arrested, or changes only little, and is often soon dropped, the other enlarges, sometimes considerably so. In several species of *Streptocarpus* and *Monophyllaea* the much enlarged paracotyledon remains the only photosynthetic organ of the plant, in the others additional foliar leaves develop. In some of these seedlings an internode (mesocotyl) develops between the unequal paracotyledons after germination, which brings the larger one in a position above the smaller. In some species of *Peperomia* (Piperaceae) one cotyledon remains in the seed and maintains a haustorial function, whereas the other is withdrawn and becomes exposed (A. W. Hill 1906).

Foodstoring cotyledons

Foodstoring cotyledons are usually massive and succulent, and in most cases with a hard, fleshy consistency. They contain a certain amount of food, which in the course of senescence is withdrawn, the blade decreasing in volume until it shrivels and is shed. Foodstoring cotyledons may be green, but often they have a different, mostly dark colour. In general the surface/contents ratio is relatively low, which means that in exposed ones photosynthesis, if present, can only be of minor importance (e.g. Lovell and Moore 1970). Lovel and Moore recorded little or no increase of cotyledon area in foodstoring cotyledons of the temperate *Lupinus albus*, *L. angustifolius*, *Phaseolus multiflorus, Phaseolus vulgaris,* and *Pisum sativum*, all Leguminosae/ Papilionaceae. In the present investigation, a similar situation has been observed for all species studied with foodstoring cotyledons (covered as well as free). The volume of the cotyledon blade increases slightly or not on germination, it slowly decreases during senescence.

In the immature embryo foodstoring cotyledons often have a haustorial function, absorbing the endosperm until this has entirely disappeared. In several cases, e.g. in Lauraceae, small remains of endosperm are still present in the mature seed (Kostermans pers. comm.); evidently these have no function in the food provision for the germinating seedling.

Foodstoring cotyledons may remain in the enveloping fruit wall and/or testa, or are withdrawn and become exposed. Enclosed cotyledons are normally not fused, but partial to entire fusion has been recorded in *Eugenia* (Myrtaceae, Henderson 1949), in *Quercus* and *Lithocarpus* (Fagaceae, Camus 1942), *Eusideroxylon* (Lauraceae, Fig. 102) and *Gluta* (Anacardiaceae, Hou 1978). Exposed cotyledons occur in seedlings of the Sloanea type (2a, b); in most cases they are opposite, borne on a long epigeal hypocotyl, situated above the soil level; more rarely they are situated at soil level. On germination, such cotyledons remain within the envelopments, often for a considerable period. They are either in the soil, with the hypocotyl forming a loop above the ground, or they are thrusted above the soil on a stretched hypocotyl. After a resting period the envelopments are thrown off and the cotyledons become exposed.

The blade of most exposed foodstoring cotyledons is elliptic or ovate in outline, and semi-elliptic in cross-section. In the seed the upper surfaces are flat and face each other. Deviating shapes are e.g. the palmate cotyledons of many Burseraceae (Fig. 51–52) and the two-winged cotyledons in many Dipterocarpaceae (Fig. 72). In *Lupinus sulphureus* (Leguminosae/Papilionaceae, Lubbock 1892) the cotyledonary blades are fused at the base, forming a peltate structure on top of the hypocotyl.

In the Endertia type (11a, 11b) and the Cynometra ramiflora type (12) the envelopments are persistent on the usually peltate foodstoring cotyledons, but split along the margins of the latter. By growth of the cotyledonary node, the cotyledons are pushed aside, thus, more or less, exposing the free upper surfaces. In the Endertia type/subtype (11a) they are borne above soil level on an epigeal hypocotyl. In the Endertia type/Chisocheton subtype (11b) and the Cynometra ramiflora type (12) they are borne at soil level. In the Endertia type/Streblus subtype (11c) the smaller cotyledon lifts from the larger like a lid, thus exposing the upper surfaces, or one cotyledon is withdrawn from the seed while the other one remains covered.

Haustorial cotyledons

Haustorial cotyledons are relatively thin, and their consistency chartaceous to rather soft-fleshy. They serve in withdrawing the food present in the endosperm, which is mobilised through hydrolysis. Their colour is usually opaque, although in exposed ones also light colours occur like pinkish, creamy-white, and whitish-green.

The haustorial function already commences in the embryo, and continues until they are shed. In a few cases the envelopments are shed and the cotyledons become exposed for a short period. Rather often the dropped endosperm still seems to contain a considerable amount of food.

It is peculiar that seedlings with enclosed haustorial and foodstoring cotyledons have the same external morphology. They differ only in the function and the morphology of the hidden cotyledons while the morphological structure of the exposed parts is the same. Therefore, in several types foodstoring as well as haustorial cotyledons with endosperm have been classified together. The Horsfieldia type/subtype (7a), the Heliciopsis type/subtype (6a), and the Endertia type/Streblus subtype (11c) may have either foodstoring cotyledons or haustorial ones in endosperm. In seedlings with enclosed cotyledons borne above the soil such as the Horsfieldia type/Pseuduvaria subtype (7b), and the Blumeodendron type (8), foodstoring as well as haustorial cotyledons in endosperm are encountered. In the Sloanea type (2a, b) both types of cotyledons become ultimately exposed. In the Heliciopsis type/Koordersiodendron subtype (6b) and the Coscinium type (10) so far only one of the two cotyledon types has been observed, but both are expected to occur. The explanation may be that the two kinds of cotyledons are fundamentally not very different. Haustorial cotyledons absorb food and pass it directly on to other seedling parts. Foodstoring cotyledons absorb food from the endosperm in the embryonic stage and store it in the cotyledon tissue, and transfer to the other seedling parts is postponed until germination.

Haustorial cotyledons remain, in most cases, enclosed in the enveloping fruit wall and/or testa and in the endosperm, lateral of the stem. Often they start as rather small blades which grow into the endosperm, thus cleaving this into two parts. The enclosed type is most often borne at soil level (Heliciopsis type/subtype, 6a; Horsfieldia type/subtype, 7a). The blades are in general flat, with the upper sides facing each other and the lower surfaces in close contact with the endosperm (see Fig. 8), which may be ruminate or not. Sometimes the margins of the blades are fringed, and extend into the endosperm. On germination the cotyledonary petioles elongate and bring the plumule and radicle out of the envelopments, in a way similar to the examples with foodstoring cotyledons.

Enclosed haustorial cotyledons borne above soil level on an elongated hypocotyl (Horsfieldia type/Pseuduvaria subtype, 7b) are rarely encountered. The cotyledons are borne lateral of the stem. In the Blumeodendron type (8) and the Coscinium type (10) the enclosed haustorial cotyledons in endosperm are borne on top of the epigeal hypocotyl, blocking shoot development until they are shed. In the Endertia type/Streblus subtype (11c) one cotyledon remains in the seed at soil level and has a haustorial function, while the other one is withdrawn and becomes exposed (e.g. *Peperomia peruviana*, Piperaceae, Fig. 12, see p. 71; A. W. Hill 1906). In the Sterculia stipulata type (3) the cotyledons are for some period retained in the seed, either above the soil on a stretched hypocotyl, or in the soil with the hypocotyl curved in a loop. Subsequently the testa is shed to expose a compound organ formed from the cotyledons with firmly adherent endosperm attached to the lower surface. These organs much resemble foodstoring cotyledons as the cotyledon with the adhering endosperm form a massive succulent structure, but the colour is whitish or pale

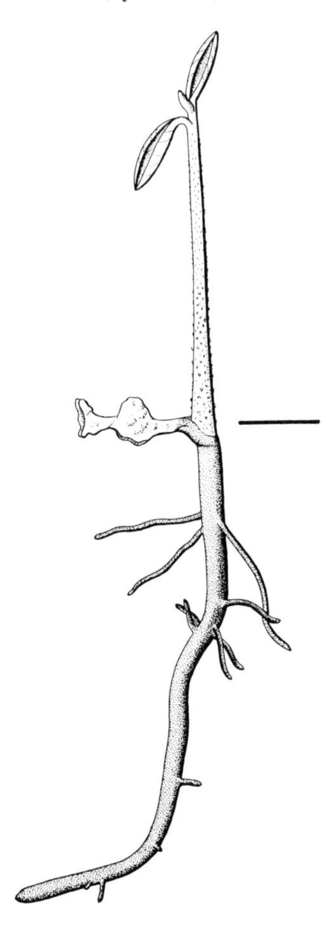

Fig. 8. *Polyalthia sp.* (Annonaceae), De V. 1603. The testa and endosperm are removed to illustrate the haustorial cotyledons, which under natural circumstances never emerge.

cream-coloured with a brown lower surface, which is never found in foodstoring cotyledons (see Fig. 9). These organs become increasingly revolute in form through shrinking of the endosperm due to re-absorption by the cotyledons, and ultimately they are shed.

Complete exposure of haustorial cotyledons is found in the Sloanea type/subtype (2a). At first the enclosed cotyledons are carried above the soil on a stretched hypocotyl, or they remain in the soil with the hypocotyl curved in a loop which stretches later. Then the testa and the endosperm are both shed, exposing the cotyledons. In the exposed position they last only a short period, usually not more than a few days, during which they start shrivelling and become revolute, after which they are dropped. Extraction of these haustorial cotyledons, therefore, seems functionless (e.g. *Bhesa robusta,* Celastraceae, Fig. 57).

Paracotyledons

Paracotyledons (foliaceous assimilating 'cotyledons') are relatively thin, but usually 2–7 times thicker than the subsequent foliar leaves (Marshall and Kozlowski

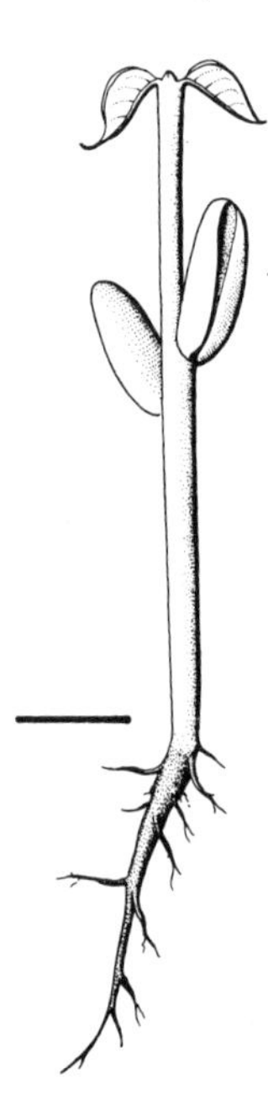

Fig. 9. *Sterculia cordata* (Sterculiaceae), De V. 1569. The endosperm remains persistent on the abaxial (lower) surface of the cotyledons after the testa is shed. The margins of these composed bodies turn revolute.

1976, 1977). The texture is membranous, herbaceous or somewhat coriaceous, the colour always green. Seedlings with paracotyledons may grow from seeds with or without endosperm, the dispersal unit may be a seed or a fruit. In seeds containing endosperm the paracotyledons at first have an absorbing function, although when they become exposed the endosperm often does not appear to be entirely exhausted (e.g. many Rubiaceae, Ebenaceae, etc.). Marshall and Kozlowski (1976) demonstrated that the total amounts of lipids, protein, and nonstructural carbohydrate contents in embryonic cotyledons (= paracotyledons) of exalbuminous tree species (*Acer*, Aceraceae; *Robinia*, Leguminosae/Papilionaceae) were considerably higher than in albuminous species (*Ailanthus*, Simaroubaceae; *Fraxinus*, Oleaceae). In the seedlings of these species they found that, as reserves in the paracotyledons decreased (within a period of ± 6 days), mesophyll cells became vacuolate and chlorophyll was synthesised.

Whether endosperm is present or absent in the seed, the amount of food in the paracotyledons is just sufficient to support expansion of these organs when they become exposed, after which the juvenile plant is dependent on photosynthesis. In seeds without endosperm the paracotyledons rapidly become exposed. According to Marshall and Kozlowski (1976) chlorophyll synthesis in such species (*Acer*, *Robinia*) begins earlier than in those from endosperm containing seeds (day 2–4 in *Acer*, *Robinia* 4–6, *Fraxinus* 6–8, *Ailanthus* 6–10). Also the emergence of the cotyledons begins earlier. In seeds containing endosperm the paracotyledons are first retained within the envelopments, either on top of the stretched hypocotyl, or in the soil with the hypocotyl curved in a loop. Ultimately the paracotyledons always become exposed and are borne above soil level (Macaranga type, 1), sometimes with the remains of the endosperm adhering for some time to one of the blades (e.g. *Neesia*,

Bombacaceae). Rarely the hypocotyl is short and the paracotyledons are borne on long petioles. The blades, which function as those of foliar leaves and much resemble them in gross morphology, spread ± horizontally, and during the first few weeks they are the only food-producing organs on the seedling. Then internodes with foliar leaves are produced. The paracotyledons turn yellow and are dropped when a more or less specific, usually rather large, number have been formed.

Considerable enlargement of the surface area of the paracotyledons takes place after extraction from the seed. Lovell and Moore (1970) recorded that enlargement in *Cucurbita pepo* (Cucurbitaceae), *Helianthus annuus* (Compositae), *Ricinus communis* (Euphorbiaceae), *Sinapsis alba* (Cruciferae), and *Trifolium pratense* (Leguminosae/Papilionaceae), all temperate species, was due to mere cell stretching, whereas in *Cucumis sativa* (Cucurbitaceae) increase in paracotyledon area was caused by a combination of cell stretching and a considerable increase in the number of cells through cell division. Pronounced intercalary growth at the base of both paracotyledons has been recorded for the genera *Clarkia, Eucharidium*, and *Oenothera* (Onagraceae, Lubbock 1892). In these genera the shape of the enlarged blade may become entirely different from that of the newly emerged paracotyledon.

Unequal development of the paracotyledons takes place in the genera *Streptocarpus*, *Monophyllaea*, *Roettlera*, *Klugia*, and in several species of *Ramondia*, *Saintpaulia*, and *Drymonia* (all Gesneriaceae, Fritsch 1904, see p. 38). Pseudomonocotyly in Dicots is present in the Cyclamen type (5), and has so far only been recorded for temperate herbaceous plants. On germination one epigeal leaf-like organ is produced from the small, somewhat swollen subterranean hypocotyl, and serves as the only assimilating organ for a prolonged period (e.g. *Cyclamen*, Primulaceae; several *Corydalis* species, Fumariaceae). This leaf-like organ, which, in most examples, is rather similar to the later developed leaves and fits with their developmental sequence, has often been regarded as the one 'cotyledon' in the examples mentioned. However, similar seedlings are also known which differ only in having two very small real cotyledons besides the one developed leaf (e.g. *Dentaria*, Cruciferae; *Sanguinaria*, Papaveraceae; *Isopyrum*, Ranunculaceae, and some species of *Symphytum*, Boraginaceae and *Anemone*, Ranunculaceae, Csapody 1968). The minute cotyledons remain subterranean and become free, sometimes they are retardedly freed from the seedcoat. They have hardly any function in seedling development. Comparing these seedlings with those of *Cyclamen* and *Corydalis*, one is forced to conclude that in this group of seedlings a process of reduction of the cotyledons is present. In a number of species, the true cotyledons are still present, although small, and almost functionless. In others, the process has led to complete reduction of the cotyledons, and the one paracotyledon is the only leaf-like organ present on the seedling just after germination. In several of the cited genera (e.g. *Anemone*), besides seedlings with pseudomonocotyly, also species occur in which the seedlings have two paracotyledons.

Indumentum

Indumentum is almost always absent on the blade of true cotyledons, whether these are exposed or hidden in the envelopments. The cotyledonary petioles, however, may be hairy, usually slightly so, and then indumentum is also present on the seedling axis. In *Dacryodes rugosa* (Burseraceae), the indumentum on the cotyledon blade is similar to that on the hypocotyl. Paracotyledons are more often provided with indumentum, which, in several cases, may cover the entire blade. In others, only the upper or the lower surface, but in general it is restricted to the paracotyledonary petiole, with slight traces on the base of the blade, especially on the lower surface and the nerves. Mostly, however, the paracotyledons are entirely glabrous, especially in tropical primary forest seedlings. In *Neesia altissima* (Bombacaceae) the blade is provided with hairs similar to those on the subsequent leaves, but different from the indumentum on the hypocotyl and the stem.

Emergentia

Emergentia are rare in (para)cotyledons, they have never been recorded in tropical primary forest seedlings. In *Saxifraga*, sections *Euaizoonia*, *Porphyrion*, and *Xanthizoon*, an epithem hydathode similar to that on the foliar leaves is present on the tip of the paracotyledons; in the other sections not (Engler and Irmscher 1919).

(Para)cotyledons are very important for seedling identification. The place of attachment, the coverings and the method by which they emerge, their nature, shape, dimensions, position, consistency, ornamentation, indumentum and colouration may be specific, and give clues for the identity.

(Para)cotyledonary buds

(Para)cotyledonary buds are situated in the axils of these organs. Species germinating in a hypogeal manner sometimes have these buds well developed and pronounced, but more often, in true cotyledons as well as in paracotyledons, they are represented only by an axillary meristem. The number of the buds is usually one per axil. Sometimes more are present, and then they may have a specific arrangement.

In normal seedling development the buds and meristems remain dormant, and never develop into a shoot. Their function is to repair damage when the main axis of the seedling is lost in a young stage. In seedlings with a food reserve this form of regeneration may be effective (e.g. when foodstoring cotyledons are present or with haustorial cotyledons in endosperm), as such reserves support the elongation of (one or more) accessory shoots, one of which will ultimately take over the function of the main shoot. In seedlings with foodstoring cotyledons this feature sometimes has an essential survival value in the normal development of the seedling. In e.g. *Quercus semecarpifolia* (Fagaceae, Troup 1921) the shoot often dies during severe winters, sometimes several years in succession. Each spring season it is then replaced by a new shoot from either a cotyledonary bud or from one of the lower stem buds. Also in tropical primary forest species dying back of the main shoot occurs, followed by replacement through enlargement of cotyledonary buds (e.g. *Spatholobus*, Fig. 118 and *Whitfordiodendron*, Fig. 119, both Leguminosae/Papilionaceae). Seedlings with paracotyledons are much more sensitive to loss of the stem, after which they usually die. In the genera *Lotus, Scorpiurus*, and *Securigera* (Leguminosae/Papilionaceae) a terminal bud is absent. A number of buds are present in the axil of each (?para)-cotyledon, in an alternating arrangement left and right of its midrib, and decreasing acropetally in size. Four of these, two from each (?para)cotyledon axil, develop into a shoot (Dormer 1945).

The stem

The stem is the ascending axis of the plant. It consists of the hypocotyl (pp. 29–32), nodes, and internodes, serves to bring the leaves in an exposed position, and has a conducting function for water, minerals, and assimilates.

In the embryo, the hypocotyl may or may not be distinctly developed. In the first case (embryos producing epigeal seedlings) either the hypocotyl is the only developed stem portion, the subsequent stem portion being represented by a terminal bud between the (para)cotyledons, or a plumule is present with, in almost all cases, one internode primordium and two opposite leaf primordia. If the hypocotyl is hardly developed in the embryo, a plumule is always present, consisting of one or a few undifferentiated internodes with scales or undeveloped leaves.

During and after germination the stem may develop gradually, or in flushes when, more or less, a specific number of internodes are present. After being formed, each internode elongates to attain its full length, or, when growth is in flushes, several internodes are formed which may elongate simultaneously. Resting stages in the development of the stem take place in several seedling types, in others development is gradual. According to the type of seedling, the first internode may be similar to the subsequent ones, or much longer.

The orientation of the axis in seedlings of trees, shrubs, and herbs is always erect. Such an orientation is also present in the lower portion of the stem in almost all seedlings of climbers and lianas. In these plants a marked difference exists in the length between the early, stiff and erect internodes, and the later, flexible ones that are much longer. Cremers (1973, 1974) compared seedlings of 20 liana species with the architectural models described by Hallé and Oldeman (1970). Of these, four species match the 'Mangeot-model', two species the 'Roux-model', whereas three other species match the 'Leeuwenberg-', 'Massart-', and 'Cook-model'. In addition to his own observations, Cremers gave a survey of the architecture in lianas known from literature; eight more architectural models can be distinguished, the model of Corner, Tomlinson, Chamberlain, Schoute, Petit, Nozeran, Champagnat, and Troll, making the total number thirteen out of the 24 described for trees by Hallé and Oldeman. Eleven of the species Cremers studied do not match described tree architectural models, and are classified into the following architectural groups:

- the juvenile form is orthotropic (erect); the liana axis is a monopodium; sexual organs are borne lateral.
- the juvenile form is orthotropic (erect); the liana axis is composed of a series of chained sympodia.

A third model regarded unique for lianas was based on representatives only known to him from the literature, each with a very individual architecture. This was defined as:

- the juvenile form is plagiotropic (having the direction of growth oblique or horizontal), hooked on to the supporting tree by the adventitious roots.

These three architectural types were not given names, because Cremers judged it premature to define architectural models specific for lianas. The group with the axis flexible in all internodes and trailing from the beginning, is even amongst climbers relatively rare. In the present work this peculiar architecture of the seedling was observed in *Hodgsonia* (Cucurbitaceae, Fig. 65) and *Whitfordiodendron* (Leguminosae/Papilionaceae, Fig. 119). Cremers wrongly attributed an entirely flexible stem to *Hedera helix* (Araliaceae); according to Lubbock (1892) and Csapody (1968) the seedling stem in that species has an erect poise. In the present work architectural types could not be distinguished because during the investigations most seedlings did not reach a stage in which these types could be recognised.

The number of erect, stiff lower internodes in most seedlings of climbers may be ± constant for a certain species, but definitive data have not yet been collected. The flexible, climbing portion of the stem may be produced in different ways. The main axis may, at a specific stage, produce elongated flexible internodes (e.g. *Connarus*, Connaraceae). In others, e.g. *Landolphia* (Apocynaceae, Cremers 1973), the growth of the main axis terminates, and flexible lateral branches are produced. Monopodial growth as well as sympodial growth have been recorded in lianas.

Development of the seedling axis is variable in the different seedling types; especially differences in the resting stages are important in this context.

In seedlings of which the (para)cotyledons become exposed, two different modes of development of the stem can be distinguished. In seedlings of the Macaranga type (1) stem development commences with elongation of the hypocotyl (which may soon turn erect, or remains long curved in a loop before it turns erect, or rarely remains short), after which growth temporarily halts. The paracotyledons are then still covered by the envelopments. When growth is resumed the paracotyledons become exposed, and a second rest takes place when these are fully enlarged and spread. In that stage the terminal bud between the paracotyledons is still at rest. After this second resting stage in the stem development (depending on the species lasting some days up to several weeks) growth is resumed by the sprouting of the terminal bud. Internodes may be produced one by one, or several in a flush, with a short rest after development of each flush of internodes and leaves. The first internode and two opposite first leaves, if present, are never produced in the same growth period in which the paracotyledons emerge and expand.

In seedlings of the Sloanea type (2) and the Sterculia stipulata type (3) the hypocotyl may elongate during germination, or rarely it remains relatively short. A first temporary rest occurs with the cotyledons (foodstoring or haustorial ones) still covered in the envelopments, like in the Macaranga type (1). In the subsequent development the envelopments are shed and the cotyledons become exposed, but, unlike the Macaranga type, here the plumule, which was somewhat differentiated in the seed develops at once. It grows into a more or less long first internode which may even become as long as the hypocotyl and which bears in almost all cases two opposite leaves on top. A second resting stage only commences when these leaves are fully enlarged. This rest may be relatively short, up to several weeks. Subsequent development of the stem (and leaves) may again be gradual, or may take place in flushes; the then developed internodes are usually much smaller than the first

Plate 7. *Terminalia copelandii* Elm. (Combretaceae), De Vogel 1325.

Plate 8. *Dryobalanops* cf. *lanceolata* Burck (Dipterocarpaceae), Bogor Bot. Gard., without number.

internode. A (long) first internode with two opposite first leaves before the second rest is almost a rule in these two seedling types with exposed true cotyledons; it is much less common in other seedling types, and is almost never found in the Macaranga type.

In seedlings with cotyledons which remain enclosed, the hypocotyl may be a well developed structure, or, more commonly, it remains small. In the latter case (species germinating in a hypogeal manner) the hypocotyl remains at or below soil level. The stem is represented by the plumule, which is the primordium of the ascending axis consisting of the undifferentiated stem internodes and the corresponding undeveloped leaves. More rarely the plumule contains only an initial first internode with two undeveloped first leaves on top. On germination, the plumule is pushed free from the blocking fruit wall and/or testa by elongation of the cotyledonary petioles (Heliciopsis type/subtype, 6a; Horsfieldia type/subtype, 7a). In the Endertia type/Chisocheton subtype (11b), the Endertia type/Streblus subtype (11c), and the Cynometra ramiflora type (12) the stem primordium becomes exposed and free to develop because the cotyledons are pushed aside by swelling of the cotyledonary node, the persistent fruit wall and/or the testa being torn along the margins of the cotyledons.

In all above mentioned seedling types the plumule starts elongation in the early stages of germination. All internodes present take part in this first elongation process, but the lowermost internodes usually remain relatively small when fully elongated. These are provided with scale-leaves or partly aborted leaves, whereas the higher ones of the same flush are larger, with fully developed leaves. Stem development then comes to a temporary rest. When the plumule consists only of an undifferentiated first internode and two undeveloped leaves, the temporary rest takes place when these organs have fully elongated. Subsequent development may take place in flushes, each flush producing several internodes with – at least in the higher ones of each flush – fully developed leaves. Elongation may also be gradual, with one internodium and one leaf developing at a time. Within one genus, plumules consisting of one internode with two leaf primordia, as well as with more internode and leaf primordia, may be encountered (e.g. in *Dysoxylum*, Fig. 127, 128, Meliaceae).

In the Heliciopsis type/Koordersiodendron subtype (6b), the Horsfieldia type/ Pseuduvaria subtype (7b), the Endertia type/subtype (11a), the Blumeodendron type (8), the Coscinium type (10), and the Rhizophora type (9) a well developed epigeal hypocotyl is present after germination. In the first two types stem development is by immediate withdrawal of the plumule from the envelopments while the enclosed cotyledons remain functional. Further growth is similar to that of the Heliciopsis type/subtype (6a) and the Horsfieldia type/subtype (7a). The development of the shoot in the Endertia type/subtype (11a) is the same as in the Endertia type/Chisocheton subtype (11b) except for the elongation of the epigeal hypocotyl.

In the latter three types a distinct rest in the development of the stem takes place after elongation of the hypocotyl. The cotyledons remain enclosed in the envelopments, blocking the tip of the hypocotyl; in the meantime the food from these organs is transferred to the hypocotyl. When exhausted, the cotyledons are shed, or they are pushed off by the shoot which then resumes growth. Stem development may be through the gradual production of several short internodes with all leaves spirally arranged (e.g. *Durio*, Bombacaceae, several species, Fig. 49), or by production of a long internode with two opposite first leaves, and then subsequent development occurs after a temporary rest (*Blumeodendron*, Euphorbiaceae, Fig. 81).

Stem development in the Barringtonia type (13), the Garcinia type (14), and the Hodgsonia type (15) is gradual or in flushes, while in the Cyclamen type (5) either the formation of a stem is postponed for a long period or a stem is not formed (rosette forming plants).

Absence of a terminal bud in seedlings of *Lotus*, *Scorpiurus*, and *Securigera* (all Leguminosae/Papilionaceae) was recorded by Dormer (1945). Several buds are present in the (?para)cotyledonary axil, from the largest of which four equivalent shoots develop, from each (?para)cotyledon axil two, which are produced by axillary buds. Two of these buds are considerably larger than the two others; they produce shoots which may be erect with long internodes, or form a rosette and have short internodes.

From the foregoing it is clear that characters of the stem may be useful for seedling identification. Clues to the identity of a seedling may be found in the developmental stages, especially important are the occurrence of resting phases, and the morphology of the parts. Additional supporting characters for specific recognition may be present in appendages (tendrils, thorns, ridges), indumentum (e.g. specific hair types, such as hammer-shaped hairs in Malpighiaceae etc., stellate or dendroid hairs) and colour. Such characters are sometimes characteristic for whole families, and can be observed on the seedling parts as well as on the twigs of the adult plant.

The leaves

In the embryo fully developed foliar leaves are never present. In the Macaranga type (1) mostly an undifferentiated bud is present instead of a plumule. Development of this bud commences only after the paracotyledons have emerged, and a temporary halt in the development has taken place after these organs are fully developed. Then the shoot elongates, and the leaf primordia produce well-differentiated leaves. In several seedling types, e.g. the Sloanea type (2), the Horsfieldia type (7), and the Heliciopsis type (6), a plumule is present in which primordial leaves or scales can be discerned. These primordia develop into full-grown leaves or scales directly after appearance of the shoot. A temporary rest precedes further development.

The leaves in a seedling do not always have the same morphology. Depending on the seedling type, they may be aborted in the lowest internodes, they may there be present in the form of cataphylls, or as fully developed leaves. Cataphylls are the scale-like, reduced leaves present in many seedlings which germinate in a hypogeal manner, e.g. often in the Horsfieldia type/subtype (7a). They may be present at the base of the seedling stem, a remnant from the juvenile stage in which they protected the plumular bud. Often they are also found at the base of each flush of internodes and leaves, where they served the same function of protection. Cataphylls may be caducous or persistent. They may be equivalent with the blade of a leaf and are then more or less triangular and sessile (e.g. *Artocarpus elasticus*, Moraceae, Fig. 133), of petiolar nature, and then the leaf blade is absent (a.o. *Knema*, Myristicaceae), or even be of stipular origin, and then either or not differentiated (e.g. *Endertia*, Leguminosae/Caesalpiniaceae, Fig. 110a).

Foliar leaves are always different from the preceding cotyledons or paracotyledons, in shape as well as in texture, although the resemblance of leaves with paracotyledons may still be rather striking. In most cases there is an acropetal sequence of smaller leaves to larger ones. Two first leaves with an opposite position are always slightly different in shape from the later leaves which are in a spiral phyllotaxy (amongst others, Sloanea type, 2). Cataphylls may gradually pass into fully differentiated leaves, or the change may be abrupt.

The phyllotaxy in seedlings is usually specific. In most seedlings all leaves are arranged in a spiral, in others they may be decussate or alternating. With a spiral phyllotaxy the number of leaves per turn of the spiral may be the same in all parts of the stem, e.g. in many seedlings of the Macaranga type (1), or in the lower part of the stem the scales or leaves are placed closer together and more are present per turn, as in many seedlings of the Horsfieldia type/subtype (7a). Almost specific for the Sloanea type (2), but also found in other seedling types, is the arrangement where the first two leaves are opposite while the subsequent leaves are arranged in a spiral.

Incisions in the leaves may be present from the first seedling leaves onwards, or the first few leaves are entire and higher ones more or less deeply incised (e.g. *Artocarpus elasticus*, Moraceae, Fig. 133). Examples are known where the first juvenile leaves are entire, the higher ones incised, while the leaves of the adult plant are entire again (*Heliciopsis*, Proteaceae, Fig. 143).

Compound leaves in the seedling may be the result of a gradual sequence from simple leaves to more complex ones, or the first seedling leaves may already be compound. When the adult plant has leaves with a large number of leaflets, or is complexly compound, these are mostly the result of a gradual adding of (pairs of) leaflets, or a gradual increase in complexity. Exceptions to this sequence occur e.g. in *Dysoxylum densiflorum* (Meliaceae, Burger 1972) where trifoliolate first leaves are succeeded by simple ones, and still higher leaves are trifoliolate again. A similar sequence was recorded for *Ulex* (Leguminosae/Papilionaceae, Duke 1969). In *Acacia oraria* (Leguminosae/Mimosaceae, Burger 1972) the first leaf is simply pinnate, the second one double compound with two pinnae, the third leaf shows a much reduced blade and a webbed petiole, and higher leaves are in the form of an undivided phyllode.

Imparipinnate leaves (odd-pinnate; compound leaves with an odd terminal leaflet) in seedlings are often preceded by a sequence of simple leaves which gradually increase in size (e.g. many *Dysoxylum* species, Meliaceae, Fig. 127; *Canarium*, Burseraceae, Fig. 51, 52). When a more or less specific size is exceeded, an incision appears in the blade which partly or entirely splits off one (or two) basal leaflets. This leaflet may be fully differentiated into a petiolule and a blade, or be still partly connate to the rhachis. When one leaflet is separated the terminal leaflet is asymmetric at the base. Addition of leaflets in higher leaves is usually gradual, at the expense of the size of the terminal leaflet which becomes relatively smaller. Differentiation of the leaflets into compound structures, by which the compound leaf becomes double compound, proceeds in a similar way. In each higher leaf one or more leaflets may become partly or entirely compound (e.g. *Leea*, Leeaceae; *Melia*, Meliaceae, Burger 1972). When the first leaves are already compound the sequence to leaves with more leaflets may be gradual or abrupt. In *Harpullia arborea* (Sapindaceae, Fig. 164) the cataphylls at the base of the stem pass directly into compound leaves.

Paripinnate leaves (even-pinnate; compound leaves with a terminal pair of leaflets) in seedlings are in general not preceded by fully developed simple leaves; mostly all leaves are compound. In higher leaves gradually one or more pairs of leaflets may be added (e.g. *Cassia*, *Peltophorum*, Leguminosae/Caesalpiniaceae, Burger 1972). Higher leaves may become more complexly compound by addition of pinnae. In Leguminosae/Mimosaceae the main rhachis of the leaf bears either all simple leaflets, or pinnae, never a mixture of the two (e.g. *Albizia*, *Dichrostachys*, etc., Leguminosae/Mimosaceae, Burger 1972). In some cases, however, in the lowest parts of the seedling stem some cataphylls are present, which pass abruptly into paripinnate leaves. In *Endertia* and *Cynometra* (both Leguminosae/Caesalpiniaceae, Fig. 108, 110) and others the cataphylls are of stipular nature, the leaves being aborted in the lowest internodes.

In seedlings of several Sapindaceae genera, the lowest leaves may be imparipinnate whereas the higher ones are paripinnate. In *Lepisanthes* (Fig. 166), *Erioglossum*, *Pometia*, and *Schleichera* (Burger 1972) the lowest, opposite leaves bear a terminal leaflet and all subsequent leaves do not. In *Nephelium* sometimes one of the first opposite leaves is imparipinnate while the other is paripinnate like the subsequent leaves. *Cubilia*, *Dimocarpus*, and some species of *Harpullia* (e.g. *H. arborea*, Fig. 164) have all (developed) leaves paripinnate.

Palmate leaves in the seedling are, as far as is known to me, never preceded by simple leaves or scales. All leaves are well developed; the lowermost ones may have three leaflets (e.g. *Bombax*, Bombacaceae, Burger 1972), to 5–8 (*Sterculia foetida*, Sterculiaceae, Burger 1972). In higher leaves the number of leaflets may increase, at the expense of the lateral leaflets. Sometimes the most lateral leaflets are not entirely free, but in the basal portion connate to the next one (*Sterculia foetida*).

Stipules may already be present in the (para)cotyledons (e.g. most Rubiaceae genera, some genera of Leguminosae, see pp. 35–36). Commonly, however, they appear with the first foliar leaves (many genera of Leguminosae, Sterculiaceae, etc.). When these are opposite the stipules may be interpetiolarly connate (e.g. *Spatholobus*, Leguminosae/Papilionaceae, Fig. 118), or they are not fused at all (e.g. *Caesalpinia bonduc*, Leguminosae/Caesalpiniaceae, Fig. 107). In *Parenterolobium* (Leguminosae/Mimosaceae) the cotyledons and the first pair of opposite leaves do not have stipules, but the higher leaves do.

Pseudostipules, a feature peculiar for *Canarium* and other genera of Burseraceae, occur at the base of the petiole, but rarely so in the lower seedling leaves. Where two opposite leaves are present, pseudostipules are always absent (a.o. *Canarium littorale*, Fig. 52). These structures are in some cases already present in the lower simple leaves (*C. hirsutum*, Fig. 51), in others not before a large number of simple and compound leaves have been produced (*C. schweinfurthii*, beyond the 15th leaf,

Weberling and Leenhouts 1966). Obviously, the appearance of pseudostipules is not connected with a certain stage in the development of the seedling. Weberling and Leenhouts (l.c.) interpreted these structures as possibly representing a rudimentary pair of leaflets, which has descended to the base of the petiole.

Characters of the seedling leaves are very important for seedling identification. The arrangement is often specific: in seedlings of the Sloanea type (2) and also in other types the first two leaves are opposite while the subsequent ones are all spirally arranged. When the adult leaves are opposite, all seedling leaves have this arrangement too (a.o. *Eugenia*, Myrtaceae). The shape of the leaves in the very juvenile stage of the seedling is often different from that of the adult shoot, but the older seedling leaves may be more or less similar to the adult leaves. In certain seedlings (e.g. Horsfieldia type/subtype, 7a), a specific change exists from the lowest scale-leaves to the fully developed seedling leaves. With growth in flushes often the lowest leaves of each flush are scale-like while the higher ones are fully developed. Two opposite first leaves have mostly a more or less different shape compared with the subsequent leaves.

Imparipinnate leaves in the older seedling and in the adult shoot are mostly preceded by unifoliolate leaves in the very juvenile seedling. With paripinnate leaves mostly all seedling leaves are paripinnate.

Stipules, indumentum, glands, and pellucid dots, when present, are usually found in the lowest seedling leaves as well as in the older parts; often they are very specific, and present characters which can facilitate seedling identification. Domatia, however, are often not present in the very first leaves, but they may occur in subsequent leaves.

The colour of juvenile seedling leaves is sometimes very characteristic. They may be whitish or cream-coloured, but more often they have a specific red colour (e.g. *Shorea sp*., Dipterocarpaceae, Fig. 72; *Barringtonia racemosa*, Lecythidaceae, Fig. 13, see p. 73); this may also be the case in all newly developed leaves of the plant, but there this feature is less conspicuous due to the presence of a large number of older, green leaves. Many other details may have diagnostic value, and can be used for identification at family, genus, or species level, like the presence or absence of exudate, the smell of crushed leaves, the texture, etc.

Seedling classification

A classification of seedlings should cover the entire variation in morphology and structure. Such a classification can be based on germination features, on morphological characteristics, on details of the development of the juvenile plant, as well as on a combination of these.

The generally accepted classification in two groups, which has been long in use, is based on characteristics of germination and the position of the (para)cotyledons:

1. Seedlings with epigeal germination. The (para)cotyledons are carried above the soil, borne on a long hypocotyl, and become exposed to light.
2. Seedlings with hypogeal germination. The cotyledons are borne at soil level and normally remain together in the fruit wall and/or testa.

Additional characteristics are that, in epigeal seedlings, the axis is situated centrally between the (para)cotyledons, which often turn green and have a photosynthetic function; in hypogeal seedlings the cotyledons are borne laterally, do not turn green, and function for storage of food or as haustoria.

Not all seedlings are covered by these two terms. That led Léonard (1957) to revise the definitions. He considered the position of the seedling axis after germination more important than whether the cotyledons are lifted from the soil or not. By epigeal germination he understood:

Cotyledons spreading after having torn the testa; hypocotyl well developed to almost absent; seedling axis central in relation to the cotyledons; cotyledons frequently turning green. Hypogeal germination was circumscribed as:

Cotyledons remaining together, in or on the soil where the seed has been deposited; hypocotyl absent or almost so; seedling axis lateral in relation to the cotyledons; cotyledons do not turn green. Although under these latter definitions a larger variation of seedlings is covered, still not all seedlings fit. Examples are those with enclosed cotyledons borne above the soil (Léonard mentioned two examples in a note), all acotylar seedlings, and seedlings in which the cotyledons spread but the testa remains adherent to the abaxial (lower) surface.

The same objection holds for the classification proposed by Duke (1965), who only took into consideration the characteristic whether the (para)cotyledons become exposed or not. Phanerocotylar seedlings are characterised by the (para)cotyledons escaping the testa after germination, whereas in cryptocotylar seedlings the cotyledons remain in the testa. The term subcryptocotylar germination was used in his descriptions in the case of exposed cotyledons at soil level. Acotylar seedlings, however, do not fit in. The terms proposed by Duke are not equivalent with the terms epigeal and hypogeal which they are meant to replace. A seedling of *Blumeodendron* (Euphorbiaceae, Fig. 81), for instance, has the cotyledons borne above the soil on an elongated hypocotyl, but they never leave the testa. Consequently it is epigeal in the classical meaning (in broad sense), and cryptocotylar in the classification by Duke. Nonetheless the terms are generally regarded as synonymous, and used in that way.

Klebs (1885, see also pp. 11–13) attempted a classification based not only on the cotyledons but on the entire structural variation of all seedling parts. Many more combinations are possible, and he recognised seven types in Dicotyledons alone. Due to the kind of material he studied, mainly seedlings of European herbaceous plants, the largest variation was recorded in seedlings with the exposed cotyledons above the soil. In the classification in the present work these all belong in either the Macaranga type (1) or the Sloanea type (2). Seedlings with subterranean cotyledons were hardly covered by Klebs, and consequently he missed the large variation in that group. The Dicots of which one or two cotyledons are rudimentary, are, in the present work, divided over several types, but Klebs put them all together in one type. The main underlying reason for the differences with the presently proposed classification is that it is based on the development of the seedling and its parts, whereas the classification of Klebs was based on morphological details. Moreover, the nature of the seed leaves, which can be leaf-like paracotyledons, foodstoring cotyledons, or haustoria, is here regarded as important, whereas Klebs mainly studied material of the first category. Variation in that group is small, which accounts for the fact that his types could not be clearly circumscribed, which he explicitly reports. Similar or comparable variation as that on which Klebs based his types with exposed cotyledons has also been found within several seedling types of the present classification. Klebs did not name his seedling types, but numbered them, which may be the cause that his classification has not gained acceptance.

Also Léonard (1957, see also pp. 15–17) distinguished seedling types on combinations of characters, although he kept to the subdivision in epigeal and hypogeal, which he re-defined. Based on the study of seedlings of 37 genera in the tribes Cynometreae and Amherstieae (Leguminosae/Caesalpiniaceae) he recognised seven seedling types. The characters used for delimitation are: the persistence of the fruit wall and/or testa versus the cotyledons eventually exposed. Epigeal seedlings were subdivided further on the spiral or opposite position of the first leaves, and the presence of an outgrowth on the collet. These characters were also judged important in the present classification, except for the last criterion, which seems less useful because of its rarity in Dicots, and because a pronounced collet is linked with intermediates to not-pronounced ones. Léonard's failure to distinguish his seedling types by name may have contributed to their neglect in other seedling studies.

Bokdam (1977) distinguished eight types of seedlings, based on a mixture of morphological and functional features, presence or absence of endosperm and whether or not this is ruminate, and the size of the embryo (see also pp. 19–21). His Omphalocarpum type, Cananga type, Argania type, and Asterales type have, in the present work, been classified under the Macaranga type (1), based on the foliaceous nature of the paracotyledons, which have a pure photosynthetic function when exposed. Bokdam's distinction of the first two types is based on the difference of the endosperm, which is ruminate in the Cananga type, in the Omphalocarpum type not. Endosperm is, however, not part of the seedling, and it has an entirely different origin. Therefore, the morphological characteristics of this tissue should not be used for the distinction of seedling types. Omphalocarpum type and Argania type differ in the nature of the paracotyledons, which are papyraceous in the first, and coriaceous in the second. Both types of paracotyledon have a photosynthetic function when they are exposed and fully developed, and the food-reserve which was present in the tissues is exhausted during the development. In my opinion such differences in texture are insufficient for the distinction of seedling types. In the Omphalocarpum type and the Asterales type a slight difference is present in the texture of the paracotyledons and in the Asterales type endosperm is absent in the mature seed. The differences between these types are, in my opinion, not of a sufficient level to distinguish seedling types.

Bokdam's Pycnanthus type and Butyrospermum type have in the present work been classified together under the Horsfieldia type/subtype (7a). The size of the embryo, whether large or minute, is of no importance for seedling classification, and the transition from haustorial cotyledons to foodstoring ones has been judged in this work as relatively simple. The Monodora type is equivalent with the Horsfieldia type/Pseuduvaria subtype (7b) for seedlings in which the cotyledons remain in the envelopments, but if the cotyledons eventually escape the seedling is here classified as Sloanea type/subtype (2a). Finally, the Tieghemella type is the same as the Sloanea type/subtype (2a) in the present work, with the exception that under this latter type also seedlings with emergent haustorial cotyledons are classified. Tieghemella has been chosen by Bokdam as the typical representative of this type. This deviates from the large majority of seedlings of this type by the alternate to subopposite position of the first leaves, whereas most seedlings of this type have two opposite first leaves.

In the present work, the solution for seedling classification is sought in a distinction of seedling types based on combinations of characters which express the mode of development of the seedling. The shape and the functions of the composing parts of seedlings can vary to a limited degree. Not all combinations are possible, and not all characters are of equal importance. The characters on which the proposed classification is based are:

Fruit wall Partly (endocarp) or entirely persistent around the cotyledons or not. Interpretation cannot fail, although rarely these tissues may disappear early through insect attack.

Testa Persistent around the cotyledons or not. Interpretation cannot fail, although in some seedlings the testa is rather loose around the cotyledons after germination, and may occasionally be torn and the cotyledons may be visible.

Endosperm Persistent or not in the mature seed and during germination. In the juvenile seed it is present, but may be absorbed in the course of development. In combination with haustorial cotyledons it is persistent as long as the cotyledons or it may be dropped while the cotyledons become exposed for a short period; with paracotyledons it may be present or not but is much shorter persistent than these organs; with foodstoring cotyledons it is absent or reduced and functionless.

Hypocotyl Length and position in relation to the soil level, function (foodstorage). The hypocotyl determines the place where the cotyledons are attached: clearly above the soil, or at or below soil level. The length of a hypocotyl may be variable within a genus. It has been recorded that the difference between a hypogeal hypocotyl and an epigeal one within a genus may be due to only one gen (*Phaseolus*, Lamprecht 1948a, b), but this is not necessarily the case in all genera where both types are present (Ruggles Gates 1951).

(Para)cotyledons Type (haustorial or foodstoring cotyledons vs. paracotyledons); presence or absence, reductions, emergent or not, position in relation to the soil and the seedling axis. The type of cotyledon and its behaviour are most significant for seedling classification; rarely difficulties are met when determining its nature.

Seedling axis Differentiation of the terminal bud after emergence; development and resting stages.

Leaves Presence, reductions, and especially the arrangement of the first two leaves (opposite or alternate) in relation to the subsequent leaves.

See for the existing variations in morphology and function of the seedling parts the chapter 'Structure, function, and variation of the seedling parts' (pp. 26–51).

The characters used are not of equal significance. The position of the (para)cotyledons in relation to the seedling axis, or the persistence of the envelopments are more fundamental than e.g. the length of the hypocotyl. Some characters may be significant for only one seedling type, and are not present in other seedlings.

In several of the recognised types, a few of the representatives may show features which are characteristic for other seedling types. In seedlings of the Horsfieldia type/subtype, for instance, the covering seedcoat around the cotyledons in some cases is rather loose, and may easily become broken, thus exposing part of the cotyledons. Such a seedling may then be mistaken for the Sloanea type/Palaquium subtype (e.g. *Harpullia arborea*, Fig. 164). A seedling of the Macaranga type with two opposite first leaves may be mistaken for the Sloanea type/subtype, but in the former the paracotyledons remain persistent very long, and the enlargement of the plumule takes place only after a prolonged rest with only the paracotyledons exposed. In most seedling types, however, such intermediates do not occur.

The existence of intermediates may somewhat obscure the differences between the types. However, the large majority of seedlings can be classified without doubt, and this was taken as a sufficient basis for the classification here proposed.

Foodstoring and haustorial cotyledons differ rather much in shape, consistency, and function, in the mature seed and on the developed seedling. They are, however, homologous structures, and differentiation takes place during the ontogeny in the juvenile seed. With both types of cotyledon endosperm is formed in the juvenile seed. Foodstoring cotyledons absorb that tissue during their development, until in the mature seed it is entirely absorbed, or only some small remains are left which have no further food providing function. The absorbed nutrients are stored in the tissues of the cotyledon until after germination, and are then mobilised again to support further growth of the seedling. Haustorial cotyledons also absorb nutrients from the endosperm during development of the immature embryo, but do not store these in the cotyledon tissue. The endosperm does not become exhausted, and on germination a considerable amount of endosperm tissue is still present in the seed.

During and after emergence of the seedling from the seed, the haustorial cotyledons continue to draw nutrients from the endosperm until this is exhausted.

The difference between foodstoring cotyledons and haustorial cotyledons is consequently of less importance than the difference of both with paracotyledons. They are homologous, and the differences in morphology are caused by a dissimilarity of the intake and subsequent transport of nutrients. The gross morphology of many seedlings with foodstoring or haustorial cotyledons is exactly the same, the only difference being the nature of the cotyledons and the presence or absence of endosperm in the mature seed. Therefore, seedlings with foodstoring cotyledons and with haustorial cotyledons, in the present work, have been classified together in several seedling types. These are the Horsfieldia type/subtype (7a), the Horsfieldia type/Pseuduvaria subtype (7b), the Heliciopsis type/subtype (6a), the Blumeodendron type (8), the Sloanea type (2), and the Endertia type/Streblus subtype (11c). In the Heliciopsis type/Koordersiodendron subtype (6b) so far only foodstoring cotyledons have been recorded, in the Coscinium type (10) only haustorial cotyledons with endosperm, but it is expected that the other cotyledon type occurs there also.

The seedling types have been based on the mode of development of the juvenile plant which is expressed by the morphological characteristics. That means that some of the types can be recognised already in an early stage, just after germination, others only in a more advanced stage. In seedling types in which the (para)cotyledons eventually become exposed and are borne on an epigeal hypocotyl, a development stage is often present in which these organs are borne above the soil but are still covered. The seedling then resembles the Blumeodendron type (8) or the Coscinium type (10). Further development results in the emergence of the (para)cotyledons (e.g. Macaranga type, 1; Sloanea type, 2). Therefore, it is advisable to observe seedlings during their entire development, or, in the field, to study seedling specimens of different age and development, from the beginning of germination to the stage when the (para)cotyledons are shed.

The seedling types described hereafter are based on data derived from the present investigation as well as on seedling descriptions from literature. No claim is made here that all the variation in Dicot seedlings is covered. Only a limited number of genera and species has been observed personally in a living state. The seedling literature covers only a relatively small number of species and not all publications could be studied. Nevertheless the author assumes that most of the existing variation in the development of seedlings has been covered.

Key to the types of Dicot seedlings (development types)

1. (Para)cotyledons both eventually emergent and exposed due to shedding of fruit wall and testa (and the endosperm, except in the Sterculia stipulata type), or both absent or much reduced and then eventually the entire seedling without coverings.
   2. (Para)cotyledons well developed.
      3. Paracotyledons thin and leaf-like (small ones rarely somewhat succulent or subulate, especially in Cactaceae and halophytes), green with a pure assimilat-

ing function after germination, nearly always long-persistent, mostly on a long hypocotyl, or on long petioles. All foliar leaves arranged in a spiral (exceptionally the first two are opposite whereas the subsequent ones are in a spiral) or decussate, produced after a temporary rest with only the paracotyledons developed. *1. Macaranga type,* p. 59

3. Cotyledons, either massive and fleshy, with an obvious foodstoring function, and colour often not green, or relatively thin, rather soft-fleshy or chartaceous pale haustoria. The plumule almost always develops without rest into a first internode with one pair (rarely more pairs) of opposite leaves. After a temporary rest the subsequent spirally arranged leaves are produced (rarely all leaves are in decussate pairs), and usually the cotyledons are then shed.

4. Cotyledons either foodstoring, or haustorial with endosperm, becoming entirely free and exposed due to shedding of all envelopments. *2. Sloanea type* (with 2 subtypes), p. 61

5. Hypocotyl strongly elongating, epigeal; the cotyledons borne above soil level. *2a. Sloanea type and subtype*, p. 61

5. Hypoctyl not or hardly elongating, ± subterranean; the cotyledons borne at or below soil level. *2b. Sloanea type/Palaquium subtype*, p. 61

4. Cotyledons haustorial, when exposed the endosperm persistent on the lower (abaxial) side of the cotyledon (resembling a foodstoring cotyledon). *3. Sterculia stipulata type*, p. 62

2. Cotyledons entirely absent or much reduced.

6. (Para)cotyledons absent or more or less scale-like, borne above soil level by the epigeal hypocotyl. Soon after germination a shoot with (scales and) leaves is produced. *4. Ternstroemia type*, p. 63

6. Cotyledons absent, or much reduced and then (sometimes retardedly) freed from the seedcoat, ± subterranean on a small, often swollen hypogeal hypocotyl or root part. One leaf-like organ developing after germination, long serving as a single assimilating paracotyledon; the shoot (if developing) produced in a much later stage. *5. Cyclamen type*, p. 63

1. Cotyledons never entirely free from fruit wall and/or testa, both (rarely one emergent and free) entirely or at least on the lower (abaxial) side lastingly covered, or both absent or much reduced and then the fruit wall and/or testa persistent around the foodstoring hypocotyl.

7. Fruit wall (or at least the endocarp) as well as the testa persistent around the cotyledons; in acotylar seedlings long-persistent around the hypocotyl.

8. Seedling viviparous, the hypocotyl elongating into a long fusiform organ which already in the young fruit breaks through the fruit wall while still on the parent plant. *9. Rhizophora type*, p. 67

8. Seedling as a rule developing after the fruit has dropped from the parent plant, the hypocotyl before not protruding from the fruit.

9. Cotyledons absent or much reduced; the fruit contains one seed with a massive hypocotyl, with at the apex some minute scales. On germination the root emerges from the apical pole of the fruit, the shoot pierces the fruit wall at the basal pole. *13. Barringtonia type*, p. 72

9. Cotyledons present, well developed.
  10. Cotyledons pushed apart by thickening of the cotyledonary node. Fruit wall and testa splitting along the margins of the cotyledons, the halves remaining adnate to the blade. The root and the short hypocotyl emerge from below, the shoot above through the opening between the opposite cotyledons, which remain at soil level on either side of the seedling axis.
  12. Cynometra ramiflora type, p. 70
  10. Cotyledons remaining together within the fruit wall and testa, never borne opposite on either side of the stem after germination.
    11. Cotyledons secund. The plumule is pushed free from the blocking envelopments by elongation of the cotyledonary petioles, and develops into a shoot before the cotyledons are shed.
    6. Heliciopsis type (with 2 subtypes), p. 64
    12. Hypocotyl not or hardly elongating, ± subterranean; the cotyledons are borne at or below soil level.
    6a. Heliciopsis type and subtype, p. 64
    12. Hypocotyl strongly elongating, epigeal; the cotyledons are borne either above soil level, or the cotyledonary petioles are attached above soil level, but due to their length the enveloped cotyledons rest on the soil.
    6b. Heliciopsis type/Koordersiodendron subtype, p. 65
    11. Cotyledons persistent on top of the hypocotyl, covering the plumule and blocking its development into a shoot until they are shed.
    10. Coscinium type, p. 67

7. Seed free from the fruit wall, but the testa persistent around the cotyledons, or in acotylar seedlings the testa long-persistent around the hypocotyl.
  13. Cotyledons absent or much reduced.
    14. Seed containing a massive hypocotyl; primary root and shoot emerging from opposite poles. Taproot more or less retarded in growth (or entirely abortive) and soon overtopped by an accessory root from the apical node of the hypocotyl which develops in the main root of the seedling.
    14. Garcinia type, p. 72
    14. Seed minute, producing a small filament which develops into a swollen haustorium when in contact with a root of a host plant (for small seeds with a swollen hypocotyl and reduced cotyledons which are retardedly freed from the envelopments see Cyclamen type). *16. Orobanche type*, p. 74
  13. Cotyledons present, well developed.
    15. Cotyledons pushed apart by thickening of the cotyledonary node, becoming opposite. The root and the short hypocotyl emerge from below, the shoot above from between the opposite cotyledons.
    11. Endertia type (with 3 subtypes), p. 69
    16. Cotyledons unequal in size (foodstoring ones) with the halves of the testa persistent on the cotyledon blade, or rather equal (haustorial ones) but one withdrawn from the seed becoming free, the other remaining covered. The seedling thus presents two unequal cotyledon bodies at soil level. *11c. Endertia type/Streblus subtype*, p. 70

16. Cotyledons always foodstoring, ± equal in size, the testa halves persistent on the blade.
  17. Hypocotyl not or only slightly enlarging, ± subterranean; the cotyledons borne at soil level.
  11b. Endertia type/Chisocheton subtype, p. 70
  17. Hypocotyl distinctly enlarging and epigeal; the more or less spreading cotyledons borne above the soil.
  11a. Endertia type and subtype, p. 69
15. Cotyledons remaining together within the testa, never borne opposite on either side of the stem after germination.
  18. Hypocotyl and taproot never emerging from the seed; shoot creeping, rooting on the nodes. *15. Hodgsonia type*, p. 74
  18. Taproot, or both taproot and hypocotyl well developed, emerging from the seed on germination; shoot almost always erect, at least in the lower internodes.
    19. Cotyledons persistent on top of the hypocotyl, covering the plumule and blocking its development into a shoot until they are shed.
    8. Blumeodendron type, p. 66
    19. Cotyledons secund, the plumule is pushed free from the blocking testa by elongation of the cotyledonary petioles, and develops into a shoot before the cotyledons are shed.
    7. Horsfieldia type (with 2 subtypes), p. 65
      20. Hypocotyl not or hardly elongating, ± subterranean; the cotyledons are borne at or below soil level.
      7a. Horsfieldia type and subtype, p. 65
      20. Hypocotyl strongly elongating, epigeal; the cotyledons are borne above soil level. *7b. Horsfieldia type/Pseuduvaria subtype*, p. 65

Description of the seedling types

1. Macaranga type

During germination the seedling becomes eventually free from all envelopments. Germination commences with the emergence of the radicle, which elongates and is pushed out of the envelopments by the stretching hypocotyl. During and after establishment of the seedling the hypocotyl usually continues enlargement. In many cases it is first curved in a loop above the soil, with its tip and base ± subterranean, but ultimately it turns erect and lifts the paracotyledons (with or without coverings) above soil level. Sometimes it remains short, and then usually the paracotyledons are lifted above soil level by elongation of the petioles which may be free or (partly) connate into a tube.

In seeds with endosperm the paracotyledons are for a prolonged period enclosed within the coverings. In many cases they remain so while they are carried above the soil on top of the elongating hypocotyl. After a more or less prolonged rest in this position they spread, and shed all envelopments, thus becoming entirely free. In others, the envelopments remain in the soil during elongation of the hypocotyl, and the paracotyledons are thus withdrawn. In seedlings from endosperm containing seeds the paracotyledons have first a haustorial function, during which they withdraw nutrients from the endosperm. On exposure they acquire the assimilatory function typical for these organs. In seeds without endosperm the paracotyledons are, in most cases, more or less folded in the seed. After emergence of the root and hypocotyl, and after a short temporary rest, they unfold, and throw off the testa in a rather early stage; they do not have at first a haustorial function.

Paracotyledons are thin, green, and leaf-like. In the course of germination the blades expand and become exposed. In Macaranga type seedlings no additional food source is present on the seedling after the paracotyledons have become exposed, and consequently the juvenile plant is almost entirely dependent on the assimilates they produce. Paracotyledons are relatively long persistent on the plantlet. While foliar leaves are produced and become active they remain functional, and are shed when a specific (often relatively large) number of leaves have developed. After the first resting stage with the paracotyledons enclosed in the envelopments, a second temporary rest takes place with these organs fully developed and exposed. The terminal bud remains dormant until growth is resumed, after which it develops into a shoot with leaves. The first few leaves are, in the majority of cases, well developed and, in most examples, all spirally arranged (in some cases all leaves are decussate); only exceptionally the first two are opposite while the subsequent ones are arranged in a spiral (as typical for seedlings of the Sloanea type).

The Macaranga type is the most common seedling type. In herbaceous Dicots only very few seedlings are not of the Macaranga type. Also in woody Dicots the type is very common; it is present in more than 50% of Malesian woody genera of which the seedling has been studied. A large number of families have Macaranga type seedlings in all or most genera, e.g. Compositae, Cruciferae, Labiatae, Ranunculaceae, Rubiaceae, and Umbelliferae. In the present work many examples of the Macaranga type are described and illustrated, too many to cite them all. See *Alangium javanicum*, Fig. 29, *Canarium hirsutum*, Fig. 51, *Macaranga hispida*, Fig. 85, *Engelhardia spicata*, Fig. 98, and *Coffea liberica*, Fig. 149 for an impression of the variation in morphology.

Notes: The principal character of the Macaranga type is the long-persistent leaf-like paracotyledons, which become free and spread in the air and have a photosynthetic function. Characteristic are the two temporary rests, the first with the cotyledons enclosed by the envelopments and with the root and hypocotyl exserted, the second with the entire seedling free of envelopments and only the paracotyledons developed. In some species the hypocotyl remains short, and the paracotyledons are borne on long petioles. These may be either free over their entire length, or rarely they are connate into a tube, which then resembles a real hypocotyl. In the Macaranga type I have refrained from basing subtypes on the place of attachment of the paracotyledons. Even if they are attached at soil level the blades are borne above soil level on long petioles, and many intermediates between a minute subterranean hypocotyl and a distinct epigeal one are present, which obscures the distinction.

2. Sloanea type

During germination the seedling becomes eventually free from all envelopments. Germination commences with the emergence of the radicle which is pushed out of the envelopments by the elongating hypocotyl, as in the Macaranga type. In the Sloanea type the hypocotyl remains rarely small and subterranean, in most cases it is epigeal and enlargement usually still continues after establishment of the seedling. Often it is first curved into a loop with the tip and base ± subterranean, but ultimately it turns erect and lifts the cotyledons (with or without coverings) above soil level. A first (often short) rest in the development of the seedling takes place while the cotyledons are still enclosed.

In seeds without endosperm the cotyledons are, generally speaking, relatively thick, and have a foodstoring function. After the first rest they become exposed and are relatively short-persistent (much shorter than paracotyledons). During senescence they shrivel relatively slowly until they are shed, which is usually soon after a second temporary rest (with two leaves developed in addition to the cotyledons). In seeds containing endosperm the cotyledons are haustorial, rather soft, pale, and usually not green, and are thinner than foodstoring cotyledons but thicker than paracotyledons. They remain, for the largest part of their existence, in a covered position, absorbing the food from the endosperm, sometimes when still in the soil, more often on top of the erect hypocotyl. After this first rest in the development they become exposed; in this position they rather quickly become revolute and are shed, often before the second temporary rest (in which two leaves are developed in addition to the cotyledons).

The plumule, a more or less developed bud in which an internode and two juvenile leaves can be distinguished, is, unlike in the Macaranga type, already present in the seed. During germination it develops after the first rest without a pause and forms a more or less long first internode, with two fully developed leaves (rarely these leaves are more or less reduced), or more than one pair is present. Exceptionally the first leaves are not opposite. In, for example, *Hymenaea* and *Guibourtia* (both Leguminosae/Caesalpiniaceae) the two seemingly opposite, entire leaves must be interpreted as one compound leaf of which the rhachis is reduced (see e.g. Léonard 1957). After the development of the plumule into a leaf-bearing shoot, the second rest takes place, differing from that in the Macaranga type where it occurs after only the paracotyledons have developed.

Further development is by growth of the terminal bud, which produces internodes and leaves. The cotyledons are then soon shed (foodstoring cotyledons) or they have been already shed (haustorial cotyledons). In this newly formed stem the leaves are, in almost all cases, arranged in a spiral, unlike the first leaves which are opposite (rarely all leaves are decussate); in addition, the shape of these leaves is slightly different from that of the first leaves.

Two subtypes can be recognised. The Sloanea subtype (2a) is characterised by an elongated, epigeal hypocotyl. The cotyledons are borne above soil level. In the Palaquium subtype (2b) the opposite to more or less secund free cotyledons are borne at, or below soil level, the hypocotyl being short and subterranean.

The Sloanea subtype (2a) is present in a large number of genera, but is much less common than the Macaranga type, to which it is second in frequency. In herbaceous Dicots it is rarely encountered, but in woody Dicots it is quite common, especially in the three (sub)families of Leguminosae, in Anarcardiaceae, and Dipterocarpaceae. In the present work many examples of the Sloanea subtype are described and illustrated, too many to cite them all. See *Canarium littorale*, Fig. 52, *Bhesa robusta*, Fig. 57, *Shorea sp*., Fig. 72, *Sloanea javanica*, Fig. 79, *Sauropus rhamnoides*, Fig. 87, *Limonia acidissima*, Fig. 160, and *Parartocarpus bracteatus*, Fig. 135, for an impression of the variation in morphology.

The Palaquium subtype (2b) is rare, and has for Malesian genera only been recorded for Burseraceae (*Dacryodes*), Leguminosae/Caesalpiniaceae (*Bauhinia*, and *Lysiphyllum*), Leguminosae/Mimosaceae (*Inga*), Leguminosae/Papilionaceae (*Erythrina*, *Ormosia*), Oleaceae (*Linociera*), Sapotaceae (*Madhuca*, *Palaquium*), and Sterculiaceae (*Sterculia* possibly). Often more than one seedling type is present in these genera. The examples in this book are: *Ormosia sp*., Fig. 112 and *Palaquium philippinense*, Fig. 168.

Notes: The principal characteristic of the Sloanea type is the early development, before the second rest, of the first internode and the two opposite first leaves. A peculiar allocation of tasks is present between the cotyledons and the first leaves. The former support the development of the shoot with two leaves, and they are soon shed after the seedling resumes growth. Even if they are green, their photosynthetic function is of minor importance. The first leaves provide assimilates during and after the second rest; they remain persistent when new leaves develop and the cotyledons are shed.

3. Sterculia stipulata type

During germination the seedling becomes eventually free from all envelopments except the persisting endosperm. Germination commences like in the Sloanea type, the hypocotyl first curves and finally carries the enclosed cotyledons and endosperm above the soil. After a first resting stage the testa is shed, and the pale haustorial cotyledons spread. The endosperm, which is split in two halves, remains persistent on the lower surface of the cotyledon blades. These compound structures have a rather fleshy appearance, and resemble foodstoring cotyledons. Rather soon after emergence these bodies become revolute, and are shed, with the endosperm still adhering. Development and morphology further much resembles the Sloanea type, with two opposite first leaves. A second resting stage takes place after these first leaves have developed. Subsequent leaves are all spirally arranged.

The Sterculia stipulata type is very rare, and has so far only been recorded in Sterculiaceae. In *Pterygota* (Mensbruge 1966) besides this type also the Macaranga type has been recorded. In *Sterculia* the seedlings represent the Macaranga type, the Horsfieldia type/subtype, possibly the Sloanea type, and the Sterculia stipulata type. The example in this book is *Sterculia stipulata*, Fig. 173; see also Fig. 9, p. 42 *Sterculia cordata*.

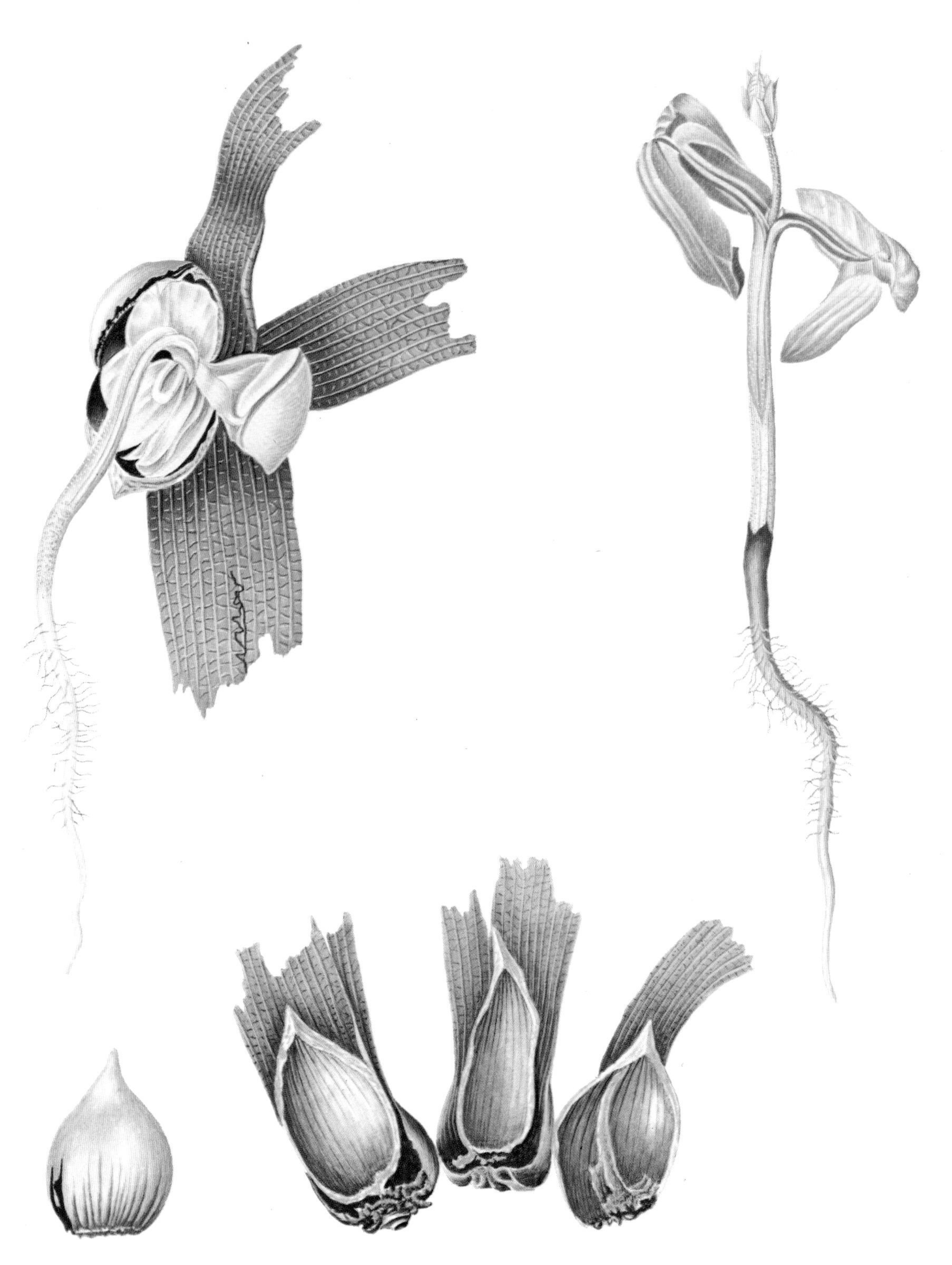

Plate 9. *Shorea sp*. (Dipterocarpaceae), Bogor Bot. Gard. VII.C.1. See also plate 10.

Plate 10. *Shorea sp*. (Dipterocarpaceae), Bogor Bot. Gard. VII.C.1. See also plate 9.

Notes: The Sterculia stipulata type resembles the Sloanea type, but is different in the persistence of the endosperm on the lower (abaxial) surface of the cotyledon blade.

4. Ternstroemia type

During germination the seedling becomes free from all envelopments. The fusiform foodstoring hypocotyl swells or uncoils, forcing itself out of the fruit wall and/or testa, and the seedling is then entirely free from all envelopments. The cotyledons are very small and scale-like, or entirely wanting. The food-storage in the hypocotyl supports the development of the shoot, which takes place soon after germination. Resting stages do not occur in the development of the seedling. The lowest internodes are short, higher up the stem they become larger. The lowest leaves may be scale-like or well developed, all are spirally arranged or the lowest two are opposite and all higher ones spirally arranged.

The Ternstroemia type has so far only been recorded in *Planchonia*, Lecythidaceae (Burger 1972, but there the first two leaves were interpreted as cotyledons), and in *Ternstroemia*, Theaceae. The example in this book is *Ternstroemia elongata*, Fig. 175.

5. Cyclamen type

During germination the seedling becomes eventually free from all envelopments. A roundish to ellipsoid subterranean hypocotyl or swollen root part emerges from the seed, or develops later. The cotyledons may be present or not, if present they are at most small and remain subterranean, and sometimes they are only retardedly freed from the envelopments. The food in the hypocotyl or root part supports the development of a single leaf. In seedlings where two cotyledons are present, the foliar nature of this leaf is obvious. In the ones without cotyledons the single leaf has often been interpreted as a single cotyledon. This is, however, not much different from the subsequent leaves, and fits their sequence, similarly like in those seedlings in which the cotyledons are small but present. One can conclude that in the former the true cotyledons have entirely disappeared, while in the latter such a reduction process is not yet completed (see also p. 43). The single leaf serves for a long time as the only assimilating paracotyledon, in some cases even during the entire first season. Further leaves are often arranged in a rosette, or the first leaf is replaced by a second one, and more leaves are produced only later in the development.

The Cyclamen type has been recorded, so far, only for herbaceous Dicots from temperate regions. Several examples occur in genera in which also Macaranga type (1) seedlings have been recorded. The type is encountered in e.g.: *Corydalis* and *Sanguinaria* (both Papaveraceae), *Anemone* (partly, also Macaranga type), and *Isopyrum* (both Ranunculaceae), *Cyclamen* (Primulaceae), *Dentaria* (= *Cardamine* p.p.) (Cruciferae), and *Symphytum* (partly, also Macaranga type) (Boraginaceae). For examples see Fig. 10; no further examples are described in this book.

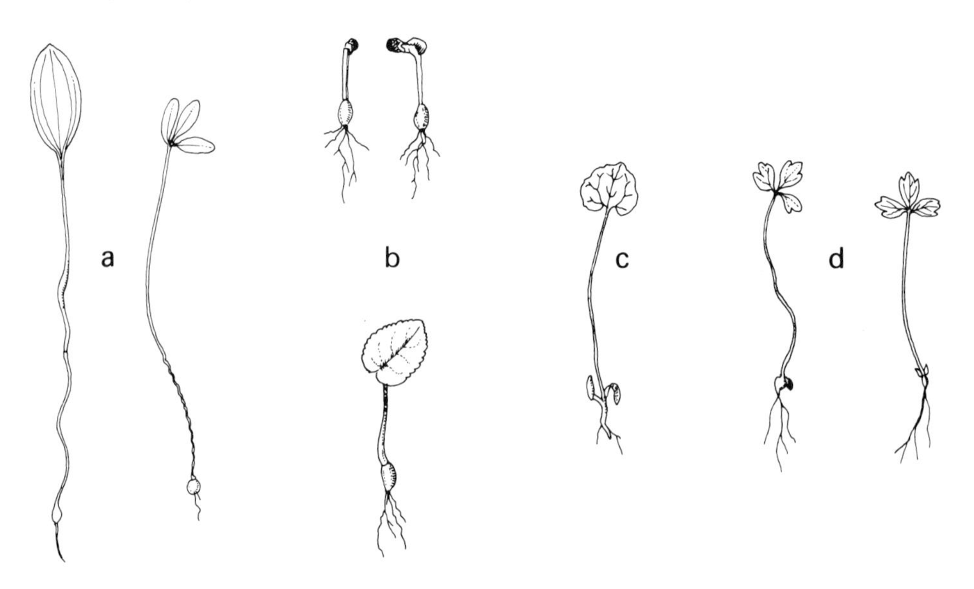

Fig. 10. Cyclamen type (5). – a. *Corydalis cava* (Papaveraceae); b. *Cyclamen persicum* (Primulaceae); c. *Sanguinaria canadensis* (Papaveraceae); d. *Anemone nemorosa* (Ranunculaceae) (all after Csapody 1968).

6. Heliciopsis type

During germination the seedling does not become free from the envelopments; both the fruit wall (at least the endocarp) and the testa remain persistent around the cotyledons, and are shed together with these organs. On germination, the taproot is the first part to emerge; it is pushed out of the fruit by elongation of the cotyledonary petioles. These elongate until they are, for a large part, extending from the envelopments, in this way bringing radicle, hypocotyl and plumule in an exposed position. The taproot develops into a sturdy root system. The hypocotyl remains subterranean and is small to almost wanting (only rarely it becomes well developed and epigeal).

The cotyledons are well developed; either thick foodstoring organs, or haustorial and embedded in the endosperm (for argumentation why foodstoring as well as haustorial cotyledons have been taken together in one type, see pp. 55–56). They remain together, clasped in the persistent envelopments with their upper sides facing each other; the entire body is borne lateral of the stem (secund). Already in the mature fruit the plumule is somewhat developed; as soon as it is pushed free it starts elongating into a shoot. The shoot often bears in its lower part some reduced leaves or scales which pass gradually into the developed leaves; also all leaves may be fully developed. Rarely the first two leaves are opposite (e.g. *Semecarpus curtisii*, Anacardiaceae, coll. Herb. Leiden).

Two subtypes can be recognised. In the Heliciopsis subtype (6a) the cotyledons are attached and borne at or below soil level, the hypocotyl being short and subterranean. The enclosed cotyledons are situated in or on the soil, dependent on where the fruit was deposited. The Koordersiodendron subtype (6b) is characterised by an elongated, epigeal hypocotyl. The cotyledonary petioles are consequently attached high above soil level. Dependent on the length and rigidness of the cotyledonary petioles, the enclosed cotyledons may either be borne high above the ground (*Koordersiodendron*, Anacardiaceae), or they remain resting on the soil (e.g. *Dipterocarpus*, Dipterocarpaceae; *Semecarpus*, Anacardiaceae). So far in this subtype only foodstoring cotyledons have been encountered; haustorial cotyledons with endosperm are, however, expected to occur.

The Heliciopsis subtype (6a) is rather common in tropical woody Dicots, and has been recorded for 45 out of 453 Malesian woody genera. Many examples are present in Anacardiaceae, Fagaceae, Lauraceae, Leguminosae/Papilionaceae, Rosaceae, and others. A few examples amongst the many in this book are: *Connarus odoratus*, Fig. 62, *Eusideroxylon zwageri*, Fig. 102, *Spatholobus platypterus*, Fig. 118, *Heliciopsis velutina*, Fig. 143, *Parinari elmeri*, Fig. 146, etc.

The Koordersiodendron subtype (6b) is known only from *Koordersiodendron* and *Swintonia* (Anacardiaceae), and *Dipterocarpus* and *Vatica* p.p. (Dipterocarpaceae). Examples in this book are *Koordersiodendron pinnatum*, Fig. 31, and *Dipterocarpus palembanicus*, Fig. 67.

7. *Horsfieldia type*

During germination the seedling does not become free from all envelopments; the testa remains persistent around the cotyledons and is shed together with these organs. Germination and morphology are similar as in the Heliciopsis type. On germination the taproot, hypocotyl, and plumule are pushed free from the testa by elongation of the cotyledonary petioles. Here, too, a small hypocotyl is most common, rarely a long epigeal hypocotyl is present. Foodstoring cotyledons, as well as haustorial ones in endosperm, have been recorded (for argumentation why foodstoring as well as haustorial cotyledons have been taken together in one type, see pp. 55–56). They remain together covered in the persistent testa, lateral to the stem (secund). The shoot develops while the cotyledons are still present. All leaves may be well developed or the lowest ones may be scale-like. In most cases they are all arranged in a spiral, sometimes the first two are opposite and the subsequent ones are spirally arranged.

Two subtypes can be recognised. In the Horsfieldia subtype (7a) the cotyledons are attached and borne at soil level, the hypocotyl being short and subterranean. The enclosed cotyledons are situated in or on the soil, dependent on where the seed was deposited. The Pseuduvaria subtype (7b) has an elongated, epigeal hypocotyl. Unlike in the Heliciopsis type/Koordersiodendron subtype (6b), foodstoring as well as haustorial cotyledons in endosperm have been recorded. The cotyledonary petioles observed are all short and rigid; these are attached above soil level and the cotyledons are situated above the soil.

The Horsfieldia subtype (7a) is rather common in tropical woody Dicots, and has been recorded for 51 out of 453 Malesian woody genera. Many examples are present in Annonaceae, Leguminosae/Papilionaceae, Meliaceae, Myristicaceae, Rutaceae, Sapindaceae, and others. A few examples amongst the many in this book are: *Caesalpinia bonduc*, Fig. 107, *Artocarpus elasticus*, Fig. 133, *Horsfieldia wallichii*, Fig. 136, *Harpullia arborea*, Fig. 164, and *Lepisanthes alata*, Fig. 166.

The Pseuduvaria subtype (7b) is very rare. It has been recorded for *Pseuduvaria* (Annonaceae), *Durio* (Bombacaceae), *Pangium* (Flacourtiaceae), *Phytocrene* (Icacinaceae), and is possibly also found in *Ardisia* (Myrsinaceae). Examples in this book are *Pseuduvaria reticulata*, Fig. 41 and *Durio excelsus*, Fig. 48.

Notes: The difference with the Heliciopsis type (6) is that there the fruit wall as well as the testa remain persistent; further the two types are largely similar. Persistence of the fruit wall in the latter is not solely due to one-seededness of the fruit, although almost always in the Heliciopsis type only one seed is present in the fruit (rarely polyembryony may occur, especially in cultivated fruit trees, like *Mangifera*, Anacardiaceae). In Myristicaceae (*Horsfieldia*, *Knema*, *Myristica*), however, one-seeded fruits produce Horsfieldia type seedlings. More often, however, Horsfieldia type seedlings result from fruits with more than one seed which become free from the fruit wall. Although the two types are morphologically closely related, the characteristic presence or absence of the fruit wall is a sufficient difference to keep them apart. Both are rather common in woody Dicots, but only in relatively few families the two types occur together.

8. Blumeodendron type

During germination the seedling does not become free from all envelopments; the testa remains persistent around the cotyledons on top of the hypocotyl, and is shed together with these organs. Germination commences with the enlargement of the hypocotyl (and the cotyledonary petioles). This brings the radicle in a free position, after which it develops into a sturdy root system. The usually sturdy, fusiform hypocotyl is mostly curved in a loop above the ground before it turns erect and carries the enclosed cotyledons above the soil; sometimes it remains subterranean. The cotyledons (either foodstoring or haustorial in endosperm) are well developed and petiolate; they remain clasped together in the persistent testa with the flat upper sides facing each other, covering the tip of the hypocotyl, and blocking the elongation of the shoot. During a prolonged resting stage in this position nutrients are transferred from the cotyledons or the endosperm towards the hypocotyl. After the cotyledons are shed, this food supports the development of the shoot and leaves. The leaves may be all spirally arranged, the lower ones of which may be scale-like, or they are all decussate, and then the shoot shows a gradual development. When the first two leaves are opposite on top of an elongated first internode, a second rest takes place when these are fully developed; subsequent leaves are then all spirally arranged.

The Blumeodendron type has so far only been recorded in Annonaceae (*Mezzettia*, *Mezzettiopsis*, *Xylopia*, and others), Bombacaceae (most species of *Durio*), Ebenaceae (*Diospyros*), Euphorbiaceae (*Blumeodendron*), Flacourtiaceae (*Pangium*), Myrsinaceae (*Ardisia* possibly), Olacaceae (*Strombosia*), Santalaceae

(*Santalum*), and Sterculiaceae (*Leptonychia* possibly). Examples in this book are *Mezzettia leptopoda*, Fig. 38, *Durio oxleyanus*, Fig. 49, and *Blumeodendron tokbrai*, Fig. 81.

9. Rhizophora type

During germination the seedling does not become free from the envelopments before the cotyledons are shed. It is viviparous; while still on the parent plant germination commences with elongation of the strongly developing, fusiform, food-storing hypocotyl, which pierces the testa and fruit wall of the still young fruit. For a long period it remains in this position, slowly enlarging and accumulating food. When the fruit is mature, the hypocotyl is dropped because the cotyledons are detached, or the fruit is dropped together with the hypocotyl.

The cotyledons serve to pass the food from the parent plant to the growing seedling before the hypocotyl is detached. They are rather reduced. In *Ceriops* and *Rhizophora* they are entirely fused into one body, of which the petiolar part forms a tube enclosing the plumule. In *Bruguiera* the petioles are fused into a similar tubular body, in which, at the tip, the cotyledons are represented by 3–5 free lobes. The fused petioles do not extend from the fruit. At maturity of the seedling, the petioles are detached from the hypocotyl which then falls off, leaving the empty fruit with the cotyledons on the tree (*Ceriops*, *Rhizophora*), or the hypocotyl is dropped together with the fruit. In the latter case the cotyledons are then soon shed, and the cotyledonary body enclosed in testa and fruit wall is pushed off by the plumule (*Bruguiera*). The plumule develops only when the cotyledons are not longer present; it produces a sturdy shoot. The first internode may be much longer than the subsequent ones or not. All leaves are decussate; sometimes the lowest pairs may be reduced or scale-like. Only rarely the fruit body is quite persistent and then this is pierced by the growing point and retained clasping the stem on top of the hypocotyl (*Bruguiera parviflora*, Burger 1972, *Aegiceras*, Chapman 1976).

The Rhizophora type is only known from the mangrove genera of the Rhizophoraceae, viz. *Bruguiera*, *Ceriops*, *Kandelia*, and *Rhizophora*, and from *Aegiceras*, Myrsinaceae. Several examples can be found in Burger (1972), see also Fig. 11; no further examples are described in this book.

Notes: The principal character of the Rhizophora type is the vivipary. The type resembles the Coscinium type (10) which is not viviparous. The hypocotyl may be 'planted' in the soft mud when it falls down, but more often it will fall into the water, wash ashore, and is anchored by quick development of the root system.

10. Coscinium type

During germination the seedling does not become free from the envelopments; the testa and the fruit wall remain persistent around the cotyledons on top of the hypocotyl, and are dropped together with these organs. Germination commences with the enlargement of the hypocotyl. Thus, the radicle is pushed out of the fruit,

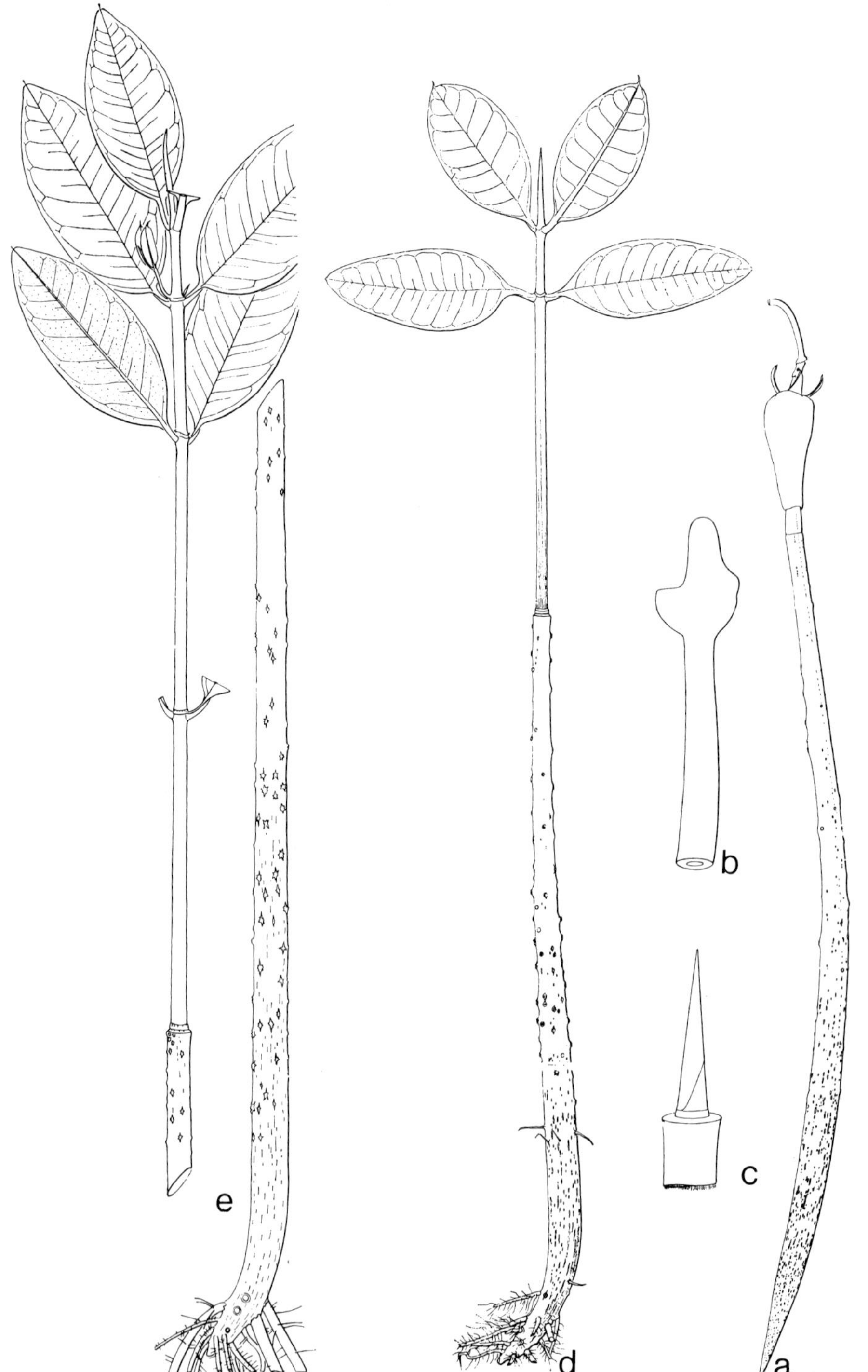

Fig. 11. Rhizophora type (9). *Rhizophora mucronata* (Rhizophoraceae). – a. mature fruit; b. cotyledon body; c. terminal bud shortly after detachment; d and e. seedling (after Burger 1972).

after which it is able to establish. The hypocotyl enlarges into a sturdy organ, which is often first curved into a loop above the ground before it elongates.

In this type only haustorial cotyledons in endosperm have been observed, but it is expected that also foodstoring cotyledons may occur. The cotyledons remain in the fruit wall and testa covering the tip of the hypocotyl, thus blocking the enlargement of the shoot. During that resting period nutrients are transferred from the endosperm towards the hypocotyl, where the food is stored. After the cotyledons are shed, this food supports the development of the shoot and leaves. In the only example known all leaves are spirally arranged, and the shoot develops gradually.

The Coscinium type has so far only been recorded in Menispermaceae, *Coscinium*. The example in this book is *C. fenestratum*, Fig. 130.

Notes: The Conscinium type resembles the Rhizophora type (9) but the seedling is not viviparous, and the Blumeodendron type (8), but there only the testa is persistent.

11. Endertia type

During germination the seedling does not become free from all envelopments; the testa remains persistent on the abaxial (lower) surface of the (foodstoring) cotyledons, or one (haustorial) cotyledon is withdrawn from the seed, whereas the other remains covered. In the first case, on germination the cotyledonary node between the small radicle and plumule enlarges, and, in combination with the pressure of the swelling cotyledons or endosperm, the testa splits. Through the opening the root emerges from below and develops into a rather sturdy taproot. The hypocotyl may remain small and subterranean, or develops into an epigeal organ.

The cotyledons are well developed, thick foodstoring organs (in all three subtypes), or haustoria embedded in endosperm (in part of the Streblus subtype). Those of the foodstoring category become free from each other on germination, thus, more or less, exposing the free upper surface. The testa, which is split along the margins of the cotyledons, remains adhering to the lower surface of the cotyledon blade. The cotyledons are equal or rarely unequal in size; in the first case they are just pushed apart by thickening of the cotyledonary node, in the latter case the smaller one opens like a lid from the larger one which remains on or in the soil. With haustorial cotyledons one is withdrawn from the seed, the second one remains covered. Due to the method of opening of the seed, the separated cotyledons are opposite, with the plumule centrally between them. The shoot appears from above between the cotyledons, opposite to the root which projects from below; its development may be gradual or in flushes. The leaves may be arranged all in a spiral or are all decussate, the basal ones being scale-like or well developed, or the first two leaves are opposite on a long first internode, and the subsequent spirally arranged leaves are produced after a temporary rest in the development.

Three subtypes can be recognised. The Endertia subtype (11a) has the foodstoring cotyledons above the ground on an epigeal hypocotyl; they spread somewhat, and the flat, free upper surfaces are more or less exposed. In the Chisocheton subtype (11b) the hypocotyl is small and subterranean, the foodstoring cotyledons are at soil level. Often they are peltate, have a semi-ellipsoid shape, with flat facing sides. They do not spread, at most they are pushed somewhat apart by swelling of the cotyledonary node. The Streblus subtype (11c) has the cotyledons (foodstoring or haustorial in endosperm) borne at soil level. The foodstoring ones are unequal in size and shape, the smaller one opens like a lid from the larger one and spreads widely, thus exposing the upper sides of both cotyledons. With haustorial cotyledons one is withdrawn from the seed and becomes exposed, while the other remains within the seed and endosperm as a haustorium. The exposed cotyledon may be different from the covered one or not, but the enclosed one together with the coverings is much larger.

The Endertia subtype (11a) is known from Celastraceae (*Lophopetalum*), Connaraceae (*Connarus*), Leguminosae/Caesalpiniaceae (*Endertia*, *Saraca*), Leguminosae/Mimosaceae (*Pithecellobium*), Leguminosae/Papilionaceae (*Ormosia*), and Meliaceae (*Sandoricum*). The examples in this book are, amongst others, *Lophopetalum javanicum*, Fig. 58, *Endertia spectabilis*, Fig. 110a, b, and *Sandoricum koetjape*, Fig. 129.

The Chisocheton subtype (11b) has been recorded from Leguminosae /Papilionaceae (*Whitfordiodendron*), Meliaceae (*Aglaia, Chisocheton, Dysoxylum* and *Trichilia*), Myrtaceae (*Eugenia*), Polygalaceae (*Xanthophyllum*), and Sapotaceae (*Lucuma*). Examples in this book are a.o. *Whitfordiodendron myrianthum,* Fig. 119, *Chisocheton pentandrus*, Fig. 126, and *Dysoxylum sp*., Fig. 127 and 128.

The Streblus subtype (11c) is known from Moraceae (*Artocarpus* and *Streblus*) and Piperaceae (*Peperomia*). See Fig. 12; no further examples are described in this book.

Note: in the Streblus subtype (11c) the seedlings with haustorial cotyledons have only tentatively been included. The method of emergence of the free (? haustorial) cotyledon is quite different from that of the foodstoring cotyledon in the other examples. In addition it is not clear to me whether the free cotyledon must be interpreted as a real haustorium or as a foliar leaf.

12. Cynometra ramiflora type

During germination the seedling does not become free from the envelopments; the fruit wall (at least the endocarp) and the testa are both persistent on the abaxial (lower) surface of the (foodstoring) cotyledons. On germination the cotyledonary node situated between the small radicle and plumule, enlarges, and, in combination with the pressure of the swelling cotyledons, both testa and fruit wall split rather regularly along the margins of the cotyledon blades, to which the halves remain adhering. Through the opening the root emerges from below, and develops into a sturdy root system. The short hypocotyl remains ± subterranean.

The massive foodstoring cotyledons are ± equal, symmetric, well developed, more or less peltate, with rather reduced petioles. Due to the method of opening of the

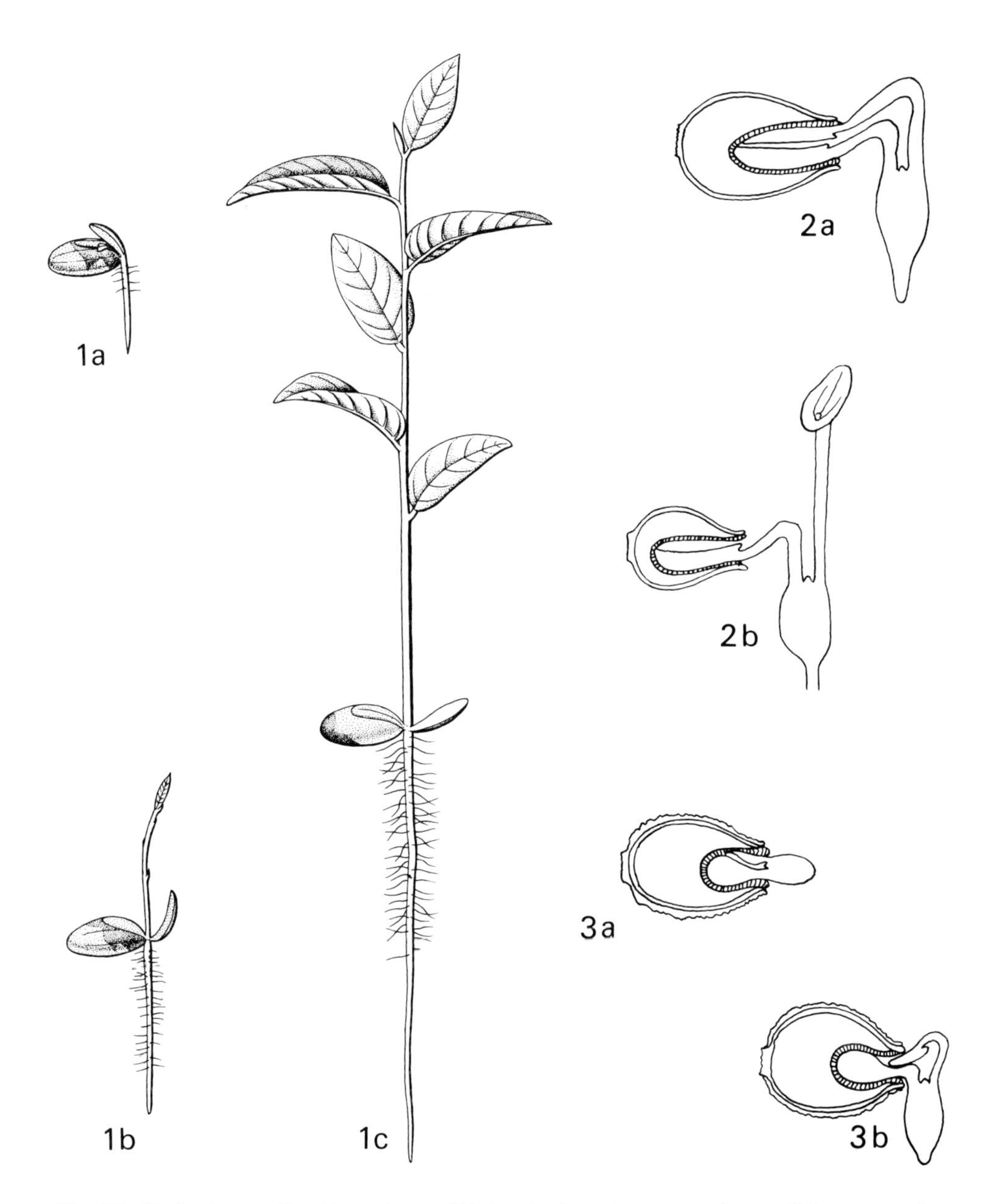

Fig. 12. Endertia type/Streblus subtype (11c). – 1a, b, c. *Artocarpus integer* (Moraceae, after Troup 1921); 2a, b. *Peperomia peruviana*; 3a, b. *Peperomia parvifolia* (both Piperaceae, after Hill 1906).

seed, the separated cotyledons are opposite, with the plumule situated centrally between them. Through the opening thus formed, the shoot appears from above, opposite to the root which projects from below; its development is gradual. The leaves are all spirally arranged; at the base some scale-leaves are present which gradually pass into the developed leaves.

The Cynometra ramiflora type has, so far, only been recorded from Lauraceae (*Cinnamomum* and *Litsea*) and Leguminosae/Caesalpiniaceae (*Cynometra*). Examples in this book are *Litsea noronhae*, Fig. 104 and *Cynometra ramiflora*, Fig. 108.

Notes: The Cynometra ramiflora type resembles the Endertia type/Chisocheton subtype (11b), but there only the testa is persistent. The difference is not solely due to one-seededness of the fruit. In e.g. *Eugenia* (Myrtaceae) the fruit contains only one seed; there the seedlings may represent the Horsfieldia type/subtype (7a), the Heliciopsis type/subtype (6a), the Endertia type/Chisocheton subtype (11b), and the Sloanea type/subtype (2a).

13. Barringtonia type

During germination the seedling does not become free from the envelopments; the fruit wall (at least the endocarp) as well as the testa remain persistent around the hypocotyl. Germination commences with the emergence of the taproot from the apex of the fruit; it establishes and develops into a sturdy root system. The hypocotyl is a massive, foodstoring body completely filling the fruit wall and testa which remain persistent till long after germination.

The cotyledons are rudimentary or absent. Sometimes two minute opposite scales are present on top of the hypocotyl, possibly representing the cotyledons. These scales may be followed by a number of spirally arranged scales; in other cases the top of the hypocotyl bears only a spiral of scales. The shoot emerges opposite to the root, from the base of the fruit. In its lowest portion it bears several scales, arranged in a spiral, which gradually pass into the developed leaves.

The Barringtonia type is known from *Barringtonia* (Lecythidaceae) only. See Fig. 13; no further examples are described in this book.

14. Garcinia type

During germination the seedling does not become free from all envelopments; the testa is persistent around the hypocotyl (rarely the testa disintegrates early and the hypocotyl becomes exposed). Germination commences mostly with the emergence of the taproot from one pole of the seed. This root is sooner or later restricted in its development; sometimes the primary root does not develop at all. The hypocotyl is a swollen, foodstoring body completely filling the seed. It remains mostly covered by the adnate testa till long after germination.

The cotyledons are either absent or they are rudimentary and then represented

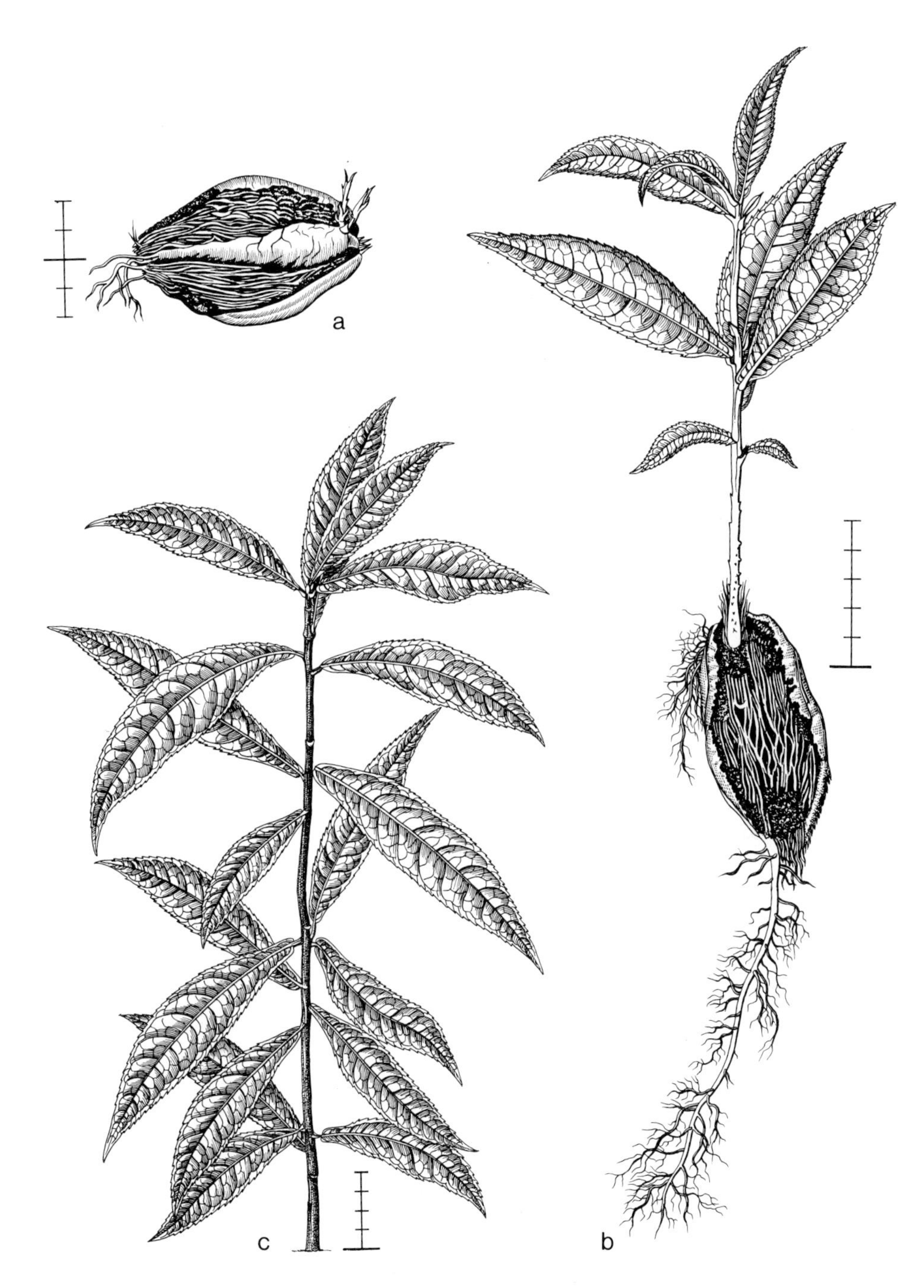

Fig. 13. Barringtonia type (13). *Barringtonia racemosa* (Lecythidaceae), Bogor Bot. Gard. IX. D. 36, collected and planted i-1973. – a. date?; b. 28-iii-1973; c. 30-viii-1974.

by two small fleshy warts at the apex of the hypocotyl; they do not develop. The shoot appears from the seed pole opposite to the primary root (when the primary root is abortive from the same pole as the main secundary root). Soon after the emergence of the shoot an accessory root develops from the junction between this shoot and the hypocotyl. This secondary root soon overtops the primary root, and takes over its function of main descending axis of the young plant. The primary root together with the hypocotyl are situated lateral to the stem (secund), and may remain long persistent in this position, but ultimately they disintegrate. The shoot develops gradually. The leaves are decussate; the lowest pairs are often much reduced and scale-like, and pass gradually in the developed leaves.

The Garcinia type is known only from Guttiferae, and is according to Brandza (1908) typical for the tribes Monorobeae and Garcinieae (e.g. *Garcinia*, possibly also *Ochrocarpus*). For examples see Fig. 14; no further examples are given in this book.

15. Hodgsonia type

During germination the seedling does not become free from all envelopments; the testa remains persistent around the foodstoring cotyledons, the hypocotyl as well as the primary root. All three never emerge. On germination only the shoot emerges from an opening at a pole of the seed. This shoot, which develops gradually, is flexible in all internodes, lies on the ground, and when support is encountered it climbs up. On the nodes accessory roots are formed. The leaves are all arranged in a spiral; the lowest ones are much reduced and scale-like and pass gradually into the developed leaves.

The Hodgsonia type is known from *Hodgsonia* (Cucurbitaceae) only. The example in this book is *Hodgsonia macrocarpa*, Fig. 65.

16. Orobanche type

During germination the (minute) seedling does not become free from all envelopments at the end of germination; the testa remains persistent around the tip of the developing small axis. The type is only found in (hemi)parasitic plants. On germination a (more or less slender) unbranched axis develops. Differentiation between root and hypocotyl is not present, and a normal root is not formed. When this axis comes into contact with the root of a host plant it pierces the epidermis of that organ, after which it fuses with the tissues. Subsequently the tissues of the host plant invade the seedling axis, which becomes much swollen. The apical tip of the axis remains long undifferentiated and clasped by the testa and endosperm, of which the latter gradually becomes absorbed. Finally these envelopments are shed, and the growing tip becomes exposed. Only then a few first scale leaves are formed (of which the lower ones are in two opposite pairs and the higher ones spirally arranged in *Orobanche*), followed by developed leaves.

The Orobanche type has so far only been recorded for Orobanchaceae

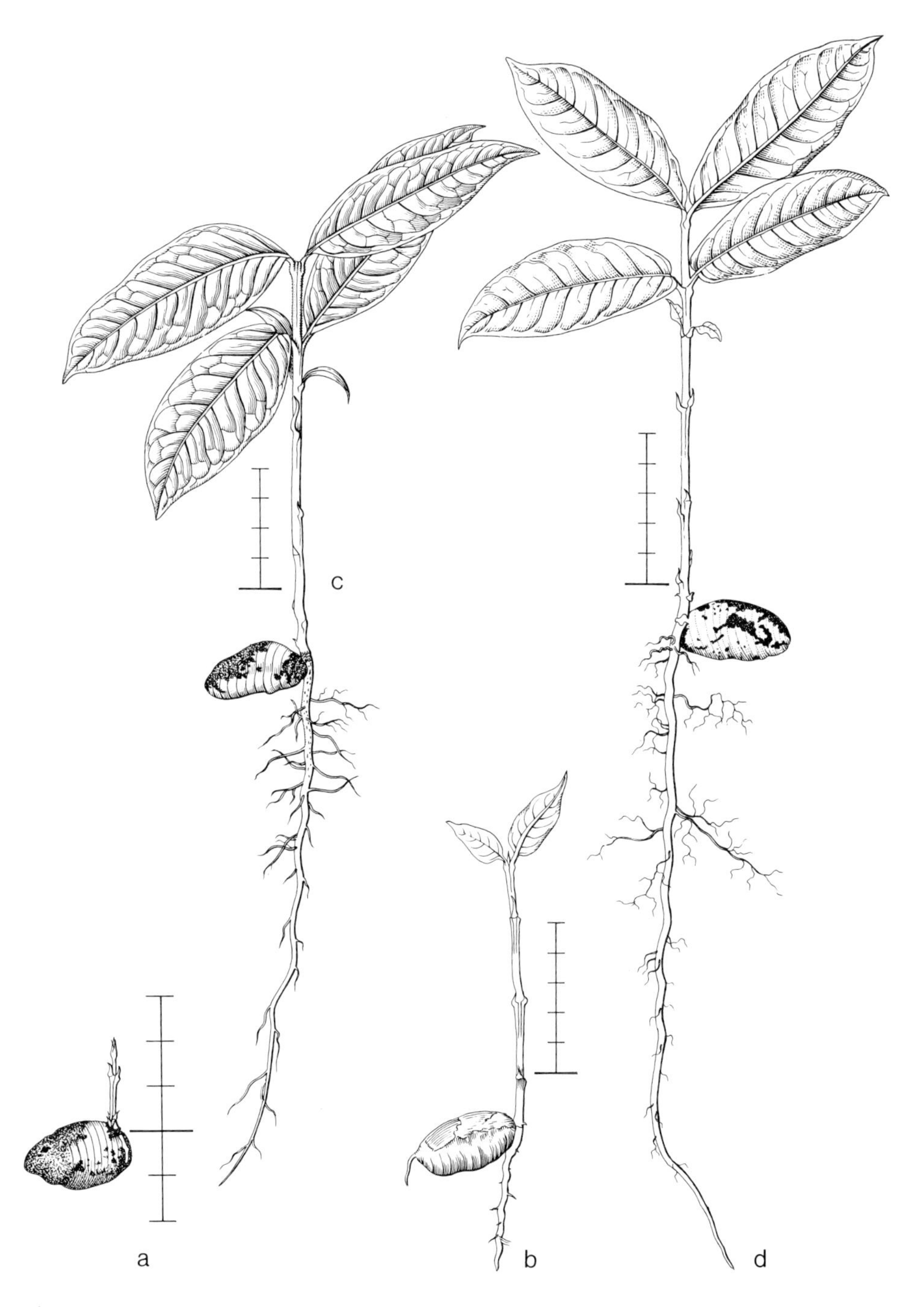

Fig. 14. Garcinia type (14). *Garcinia dulcis* (Guttiferae), De V. 1403, collected 25-vi-1972, planted 30-vi-1972. – a, b, c, and d date?

(*Orobanche*, *Phelipaea*). Possibly *Cuscuta* (Convolvulaceae), *Moneses* and *Orthilia* (Pyrolaceae), and *Cynomorium* and *Balanophora* (Balanophoraceae) can also be classified in this type, but more details about the seedlings are needed. See Fig. 15; no further examples are given in this book.

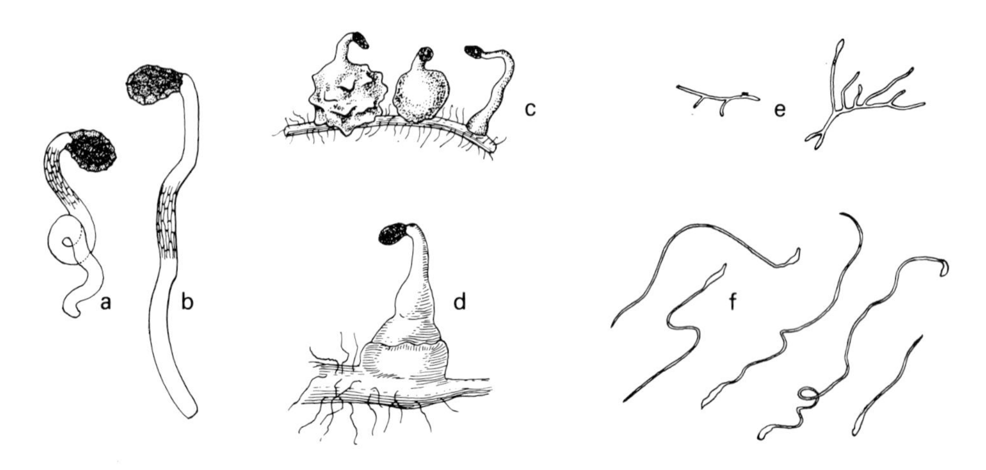

Fig. 15. Orobanche type (16). – a, b, and c. *Phelipaea ramosa* in different stages of development; d. *Orobanche minor* (both Orobanchaceae, after Caspary 1854); e. *Moneses uniflora* (Pirolaceae); f. *Cuscuta campestris* (Convolvulaceae) (both after Csapody 1968).

In Table 3 the types which here have been recognised are compared with the classifications proposed by other authors.

Table 3. Comparison of the proposed seedling types with former classifications of Dicot seedlings.

Bokdam's classification	Duke's classification	Léonard's classification	Klebs' classification	Classical classification	Proposed classification
			Seedlings with one or both cotyledons rudimentary		Ternstroemia type (4)
				EPIGEAL SEEDLINGS — in strict sense (Para) cotyledons free	Cyclamen type (5)
Cananga type Omphalocarpum type Argania type Asterales type	Phanerocotylar	Type I Type II Type III Type IV Type V	Type 1 Type 2 Type 3 Type 4 Type 5		Macaranga type (1)
Tieghemella type					Sloanea type/subtype (2a)
	? Phanerocotylar			in broad sense (Para) cotyledons partly or entirely covered	Sterculia stipulata type (3)
	CRYPTOCOTYLAR				Blumeodendron type (8)
			Seedlings with one or both cotyledons rudimentary		Rhizophora type (9)
					Coscinium type (10)
Monodora type					Horsfieldia type/ Pseuduvaria subtype (7b)
					Heliciopsis type/ Koordersiodendron subtype (6b)
					Endertia type/subtype (11a)
	Subcryptocotylar			Epigeal sensu Léonard	Sloanea type/Palaquium subtype (2b)
	? Subcryptocotylar				Endertia type /Streblus subtype (11c)
	CRYPTOCOTYLAR	Type V	Cotyledons subterranean	HYPOGEAL SEEDLINGS	Endertia type/Chisocheton subtype (11b)
					Cynometra ramiflora type (12)
		Type VI, VII			Heliciopsis type/subtype (6a)
Butyrospermum type Pycnanthus type					Horsfieldia type/subtype (7a)
					Hodgsonia type (15)
			Seedlings with one or both cotyledons rudimentary		Barringtonia type (13)
					Garcinia type (14)
					Orobanche type (16)

Malesian woody Dicot genera, with indication of seedling type and seedling literature

The numbers correspond with those of the seedling types described on pp. 59–76. The references deal not necessarily with Malesian representatives. 'De Vogel obs.' means that the author has observed seedlings, but these have not been described in this work.

1, 6a	ACERACEAE
1, 6a	*Acer* L. – Amann; Csapody; Hickel; Lubbock; Schopmeyer; Troup
1	ALANGIACEAE
1	*Alangium* Lmk – Ng 1975; De Vogel
1, 2a, 6a, 6b	ANACARDIACEAE
2a	*Buchanania* Spreng. – Burger; Lubbock; Troup
2a	*Dracontomelon* Bl. – Burger; De Vogel obs.
6a	*Gluta* L. – Burger; Troup; De Vogel
6b	*Koordersiodendron* Engl. – Meijer; De Vogel
?2a	*Lannea* Rich. – Lubbock; Taylor
6a	*Mangifera* L. – Duke 1965, 1969; Lubbock; Pierre; Troup; De Vogel
6a	*Melanochyla* Hook. f. – De Vogel
	Melanorrhoea Wall. = *Gluta*
	Odina Roxb. = *Lannea*
2a	*Parishia* Hook. f. – De Vogel
2a, ?6a	*Pistacia* L. – Hickel
1, 2a, ?6a	*Rhus* L. – Csapody; Hickel; Lubbock; Schopmeyer
6a	*Semecarpus* L. f. – (Herb. Leiden)
2a	*Spondias* L. – Duke 1965, 1969; Mensbruge; Troup; De Vogel
6b	*Swintonia* Griff. – De Vogel
1	*Toxicodendron* Mill. – Gillis
1	ANCISTROCLADACEAE
1	*Ancistrocladus* Wall. – Keng.
1, 2a, 7a, 7b, 8	ANNONACEAE
7a	*Artabotrys* R. Br. – Cremers 1973
1	*Cananga* Hook. f. & Th. – Burger; De Vogel obs.
1	*Cyathocalyx* Champ. ex Hook. f. & Th. – De Vogel
8	*Mezzettia* Becc. – De Vogel
8	*Mezzettiopsis* Ridl. – De Vogel
7a, 8	*Polyalthia* Bl. – Grushvitskyi; De Vogel

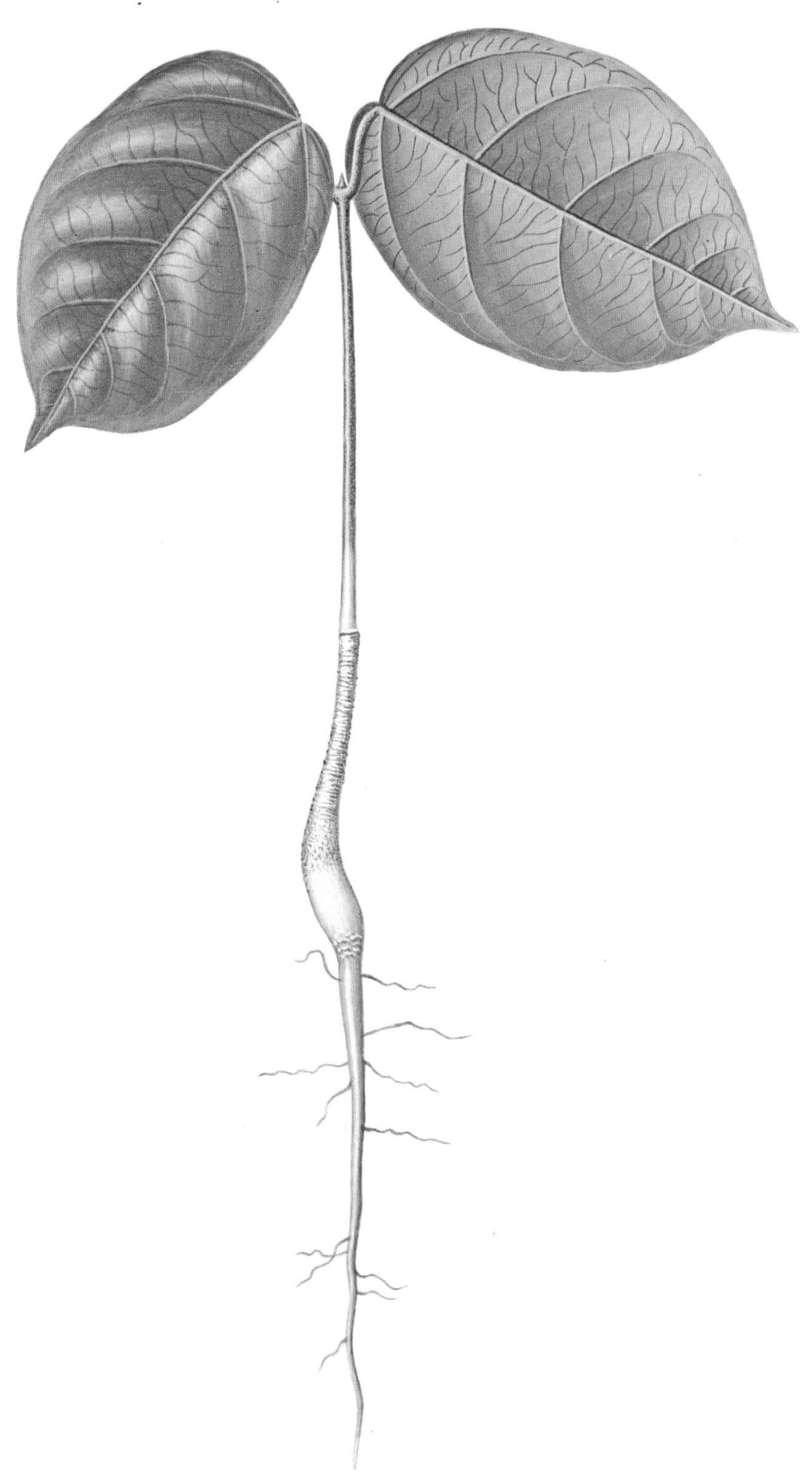

Plate 11. *Blumeodendron tokbrai* (Bl.) Kurz (Euphorbiaceae), De Vogel 703.

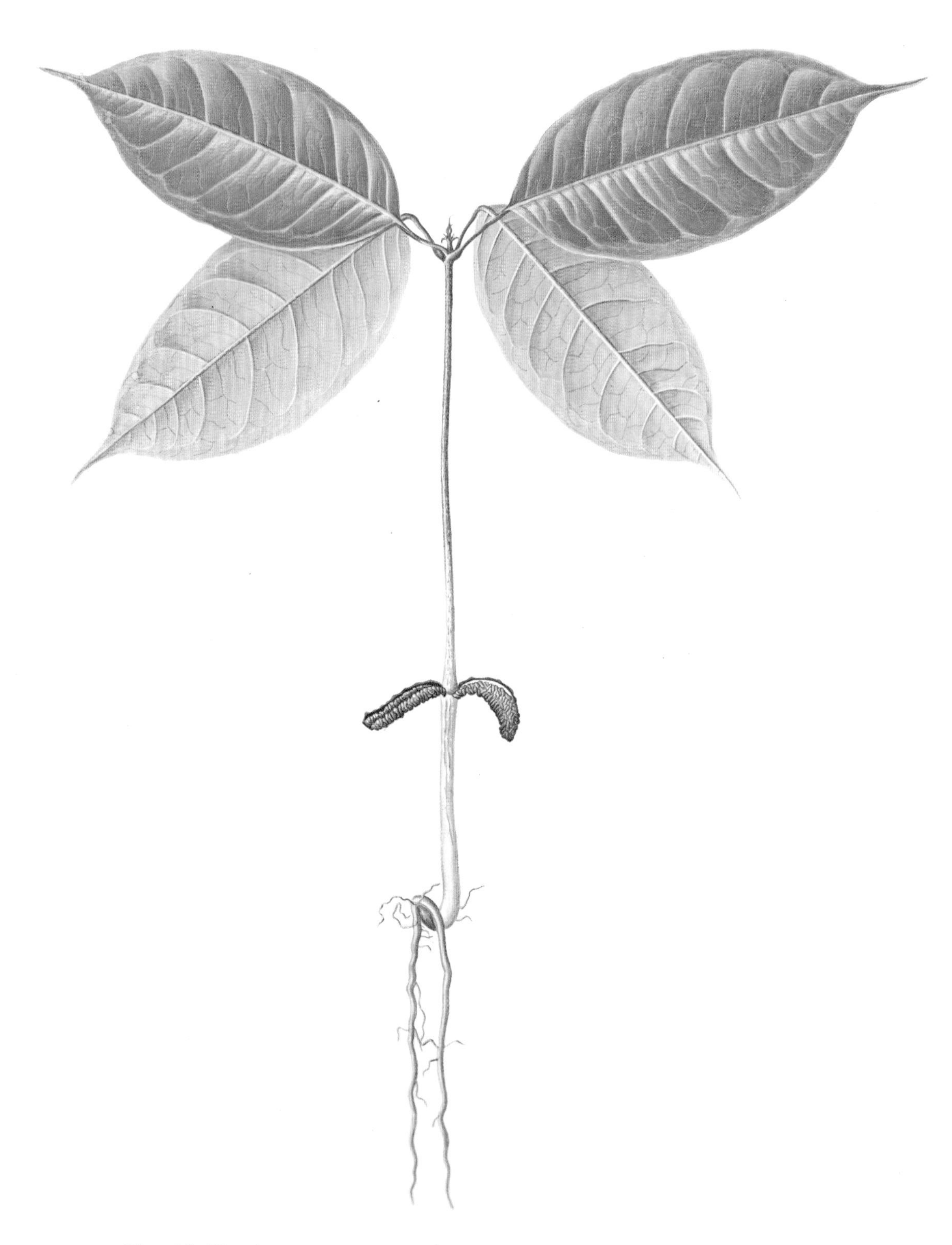

Plate 12. *Elateriospermum tapos* Bl. (Euphorbiaceae), De Vogel 813.

7b	*Pseuduvaria* Miq. – De Vogel
1	*Saccopetalum* Benn. – Burger
7a	*Stelechocarpus* Hook. f. & Th. – De Vogel
1	*Trivalvaria* Miq. – De Vogel
2a, 8	*Xylopia* L. em. St. Hil. – Mensbruge; Taylor
1, 2b, 6a	APOCYNACEAE
1	*Alstonia* R. Br. – Burger; Mensbruge; Taylor
1	*Anodendron* A. DC. – Lubbock
6a	*Cerbera* L. – De Vogel
	Echites Bl. non L. = *Anodendron*
1	*Ervatamia* Stapf – Lubbock
1	*Holarrhena* R. Br. – Mensbruge; Troup
1	*Hunteria* Roxb. – Mensbruge
6a	*Kopsia* Bl. – De Vogel
1	*Ochrosia* Juss. – De Vogel
1	*Parameria* Benth. – Lubbock
	Tabernaemontana pro mal. = *Ervatamia*
1	*Voacanga* Thouars – Mensbruge
2b	*Willughbeia* Roxb. – De Vogel
1	*Wrightia* R. Br. – Troup
1	AQUIFOLIACEAE
1	*Ilex* L. – Csapody; Duke 1965; Hickel; Lubbock
1	ARALIACEAE
1	*Aralia* L. – Csapody; Hickel; Lubbock
1	*Dendropanax* Decne & Planch. – Duke 1965
1	*Harmsiopanax* Warb. – Lubbock
	Schubertia Bl. = *Harmsiopanax*
1	*Tupidanthus* Hook. f. & Th. – Lubbock
1	ASCLEPIADACEAE
1	*Sarcostemma* R. Br. – Lubbock
	ASTERACEAE = COMPOSITAE
1	BERBERIDACEAE
1	*Berberis L.* – Csapody; Hickel; Lubbock; Schopmeyer
1	*Mahonia* Nutt. – Csapody
1	BIGNONIACEAE
1	*Oroxylum* Vent. – Troup
1	*Radermachera* Z. & M. – Burger
1	*Stereospermum* Cham. – Mensbruge; Troup
1	BIXACEAE
1	*Bixa* L. – Duke 1969
1, 7b, 8	BOMBACACEAE
1	*Bombax* L. – Burger; Mensbruge; Taylor; Troup; De Vogel
7b, 8	*Durio* Adans. – Burger; Meijer; De Vogel
1	*Kostermansia* Soegeng – Soegeng
1	*Neesia* Bl. – De Vogel

1	BORAGINACEAE
1	*Cordia* L. – Duke 1965; Lubbock; Mensbruge; Troup
1	*Tournefortia* L. – Lubbock
1	BUDDLEJACEAE
1	*Buddleja* L. – Csapody
1, 2a, 2b, 6a	BURSERACEAE
1, 2a	*Canarium* L. – Guillaumin; Lubbock; Mensbruge; Meijer; Ng 1975; Rumphius; Taylor; De Vogel; Voorhoeve; Weberling & Leenhouts
2a, 2b	*Dacryodes* Vahl – Duke 1965, 1969; Mensbruge; Ng 1975; De Vogel; Voorhoeve
1	*Garuga* Roxb. – Burger; Guillaumin; Troup
6a	*Haplolobus* H. J. Lam – Leenhouts & Widodo
1	*Protium* Burm. f. – Burger; Duke 1969; Guillaumin
2a, 6a	*Santiria* Bl. – Ng 1975; De Vogel
1	*Scutinanthe* Thw. – Ng 1975
1	*Triomma* Hook. f. – Ng 1975
1	BUXACEAE
1	*Buxus* L. – Baillon; Csapody; Hickel; Lubbock
1, 2a	CAPPARIDACEAE
1	*Capparis* L. – Duke 1965, 1969; Lubbock; De Vogel
2a	*Crateva* L. – Troup
1	CAPRIFOLIACEAE
1	*Lonicera* L. – Csapody; Hickel; Lubbock; Schopmeyer
1	*Sambucus* L. – Csapody; Hickel; Lubbock; Schopmeyer
1	*Viburnum* L. – Csapody; Hickel; Lubbock; Schopmeyer
?1	CARICACEAE
?1	*Carica* L. – Duke 1965
1	CASUARINACEAE
1	*Casuarina* L. – Burger; Duke 1965; Hickel; Lubbock; Ng 1976
1, 2a, 7a, 11a	CELASTRACEAE
2a	*Bhesa* Buch.-Ham. ex Arn. – De Vogel
1	*Celastrus* L. – Csapody; Lubbock; Schopmeyer
1	*Euonymus* L. – Burger; Csapody; Hickel; Lubbock; Schopmeyer
11a	*Lophopetalum* Wight – Meijer; De Vogel
7a	*Salacia* L. – Mensbruge; De Vogel
1	*Siphonodon* Griff. – De Vogel
1	CLETHRACEAE
1	*Clethra* Gaertn. – Lubbock
1	COCHLOSPERMACEAE
1	*Cochlospermum* Kunth – Duke 1969; Lubbock
1, 2a, 6a, ?7a	COMBRETACEAE
1, ?7a	*Combretum* Loefl. – Exell & Stace; Lubbock

1	*Lumnitzera* Willd. – Duke 1965
1, 2a, 6a	*Terminalia* L. – Burger; Duke 1965; Mensbruge; Taylor; Troup; De Vogel obs.; Voorhoeve
1	COMPOSITAE
1	*Pluchea* Cass. – Burger
6a, 7a, ?11a	CONNARACEAE
6a, ?11a	*Connarus* L. – Duke 1969; De Vogel
7a	*Rourea* Aubl. – Duke 1965; De Vogel
1	CORNACEAE
1	*Mastixia* Bl. – De Vogel
15	CUCURBITACEAE
15	*Hodgsonia* Hook. f. & Th. – De Vogel
?7a	DICHAPETALACEAE
?7a	*Dichapetalum* Thouars – Mensbruge
1	DILLENIACEAE
1	*Dillenia* L. – Burger; Ng 1975; Troup; De Vogel
1	*Hibbertia* Andrews – Lubbock
2a, 6a, 6b	DIPTEROCARPACEAE
2a	*Anisoptera* Korth. – Burkill 1917, 1920b
	Balanocarpus Bedd. p.p. = *Neobalanocarpus*
6b	*Dipterocarpus* Gaertn. f. – Burkill 1920b; Gilg; Meijer; Pierre; Troup; De Vogel
2a	*Dryobalanops* Gaertn. f. – Burkill 1920b; Meijer; De Vogel
2a	*Hopea* Roxb. – Burkill 1920b, 1923; Pierre; Troup; De Vogel
	Isoptera Scheff. ex Burck = *Shorea* sect. *Shorea* subsect. *Shorea*
2a	*Neobalanocarpus* Ashton – Burkill 1920a, b, 1923
	Pachynocarpus Hook. f. = *Vatica* sect. *Vatica*
6a	*Pentacme* A. DC. – Pierre
	Retinodendron Korth. = *Vatica* sect. *Vatica*
2a, 6a	*Shorea* Roxb. – Burkill 1917, 1918, 1920a, b, 1922, 1923, 1925; Meijer; Roxburgh; Troup; De Vogel
2a, 6b	*Vatica* L. – Burkill 1920b; Meijer; De Vogel
1, 2a, 8	EBENACEAE
1, 2a, 8	*Diospyros* L. – Csapody; Hickel; Lubbock; Mensbruge; Ng 1976; Schopmeyer; Taylor; Troup; De Vogel; Wright
2a	ELAEAGNACEAE
2a	*Elaeagnus* L. – Csapody; Hickel; Lubbock
1, 2a	ELAEOCARPACEAE
1	*Elaeocarpus* L. – Lubbock; De Vogel
2a	*Sloanea* L. – Duke 1965; De Vogel
1	ERICACEAE

1	*Gaultheria* L. – Schopmeyer
1	*Rhododendron* L. – Csapody; Hickel; Lubbock; Schopmeyer; Troup
1	*Vaccinium* L. – Csapody
1	ERYTHROXYLACEAE
1	*Erythroxylum* P. Br. – Duke 1965
1, 2a, 8	EUPHORBIACEAE
1	*Acalypha* L. – Lubbock
1	*Alchornea* Sw. – Duke 1965
1	*Aleurites* J. R. & G. Forst. – Baillon; Burger; Duke 1965
1	*Andrachne* L. – Csapody
1	*Antidesma* L. – Burger
1	*Aporusa* Bl. – De Vogel
1	*Baliospermum* Bl. – Burger
1	*Bischofia* Bl. – Burger; Troup
8	*Blumeodendron* Kurz – De Vogel
1	*Breynia* J. R. & G. Forst. – Burger
1	*Bridelia* Willd. – Burger; Mensbruge; Troup
1	*Croton* L. – Duke 1965; Mensbruge; De Vogel
1	*Drypetes* Vahl – Duke 1965; Mensbruge; Troup; De Vogel
2a	*Elateriospermum* Bl. – De Vogel
1	*Excoecaria* L. – Troup
1	*Glochidion* J. R. & G. Forst. – Burger
1	*Macaranga* Thouars – Burger; Mensbruge; De Vogel
1	*Mallotus* Lour. – Troup; De Vogel
1	*Melanolepis* Rchb. f. & Zoll. – Burger
1	*Microdesmis* Planch. – Mensbruge
1	*Petalostigma* F. v. M. – Lubbock
1	*Phyllanthus* L. – Burger; Lubbock; Mensbruge; Troup
	Putranjiva Wall. = *Drypetes*
1	*Sapium* P. Br. – Duke 1965; Mensbruge
1	*Securinega* Juss. – Burger
2a	*Sauropus* Bl. – De Vogel
1	*Trewia* L. – Troup; De Vogel
1, 6a	FAGACEAE
6a	*Castanea* L. – Borbas; Csapody
6a	*Castanopsis* Spach – Burger; Schopmeyer
6a	*Lithocarpus* Bl. – Burger; Schopmeyer; De Vogel obs.
1	*Nothofagus* Bl. – Hickel
6a	*Quercus* L. – Borbas; Burger; Csapody; Hickel; Lubbock; Schopmeyer; Troup; De Vogel obs.
1	*Trigonobalanus* Forman – Hou 1971
1, 7b, 8	FLACOURTIACEAE
1	*Casearia* Jacq. – Duke 1965

1	*Flacourtia* l' Hérit. – Burger
1	*Homalium* Jacq. – Burger; Duke 1965
1	*Hydnocarpus* Gaertn. – Troup; De Vogel
7b, 8	*Pangium* Reinw. – Burger; De Vogel obs.
7a	GONYSTYLACEAE
7a	*Gonystylus* T. & B. ex Miq. – De Vogel
1	GOODENIACEAE
1	*Scaevola* L. – De Vogel
6a, 7a, 14	GUTTIFERAE
6a	*Calophyllum* L. – Brandza; Burger; Duke 1965; Lubbock; De Vogel obs.
14	*Garcinia* L. – Brandza; Burger; Duke 1965; Lubbock; Pierre; Taylor; De Vogel obs.
7a	*Mammea* L. – Brandza; Mensbruge; Taylor; De Vogel; Voorhoeve
7a	*Mesua* L. – Brandza
7a	*Ochrocarpus* Thouars – Brandza; Engler; Pierre
	Xanthochymus Roxb. = *Garcinia*
1	HAMAMELIDACEAE
1	*Altingia* Norona – Burger
6a	HERNANDIACEAE
6a	*Hernandia* L. – Duke 1965; De Vogel
1, 6a, 7b	ICACINACEAE
1	*Gomphandra* Wall. ex Lindl. – De Vogel
6a	*Gonocaryum* Miq. – De Vogel
?1	*Iodes* Bl. – Cremers 1974
7b	*Phytocrene* Wall. – Sleumer 1942
?6a	*Pyrenacantha* Wight – Cremers 1974
1	*Stemonurus* Bl. – Sleumer 1971; De Vogel
1	JUGLANDACEAE
1	*Engelhardia* Lesch. ex Bl. – De Vogel
1	LABIATAE
1	*Salvia* L. – Burger; Lubbock
6a, 12	LAURACEAE
6a	*Beilschmiedia* Nees – Mensbruge; De Vogel
6a, 12	*Cinnamomum* Bl. – Burger; De Vogel
6a	*Cryptocarya* R. Br. – De Vogel
6a	*Eusideroxylon* T. & B. – Meijer; De Vogel
6a	*Lindera* Thunb. – Schopmeyer
6a, 12	*Litsea* Lmk – Burger; De Vogel
6a	*Machilus* Nees – Troup
6a	*Persea* Mill. f. – Duke 1965
1, 4, 13	LECYTHIDACEAE
13	*Barringtonia* J. R. & G. Forst. – Burger; Chibber; Payens; De Vogel obs.
1	*Combretodendron* A. Chev. – Mensbruge; Taylor

4	*Planchonia* Bl. – Burger; De Vogel obs.
1	LEEACEAE
1	*Leea* L. – Burger; De Vogel obs.
1, 2a, 2b, 6a, 7a, 11a, 12	LEGUMINOSAE/CAESALPINIACEAE
1	*Acrocarpus* Wight ex Arn. – Troup
2a	*Afzelia* Sm. – Léonard; Mensbruge; Taylor; De Vogel; Voorhoeve
1, 2a, 2b, 7a	*Bauhinia* L. – De Candolle; Compton; Duke 1965; Lubbock; Troup
1, 2a, 7a	*Caesalpinia* L. – Compton; Hickel; Lubbock; De Vogel
1, 2a	*Cassia* L. – Burger; De Candolle; Compton; Duke 1965; Mensbruge; Troup; De Vogel obs.
	Coulteria H. B. K. = *Caesalpinia*
1, ?2a	*Crotalaria* L. – De Candolle; Compton; Lubbock
7a	*Crudia* Schreb. – Léonard; Mensbruge
2a, 12	*Cynometra* L. – Mensbruge; Taylor; De Vogel; Voorhoeve
1	*Delonix* Rafin. – De Candolle; Lubbock
2a	*Dialium* L. – Mensbruge; Taylor
11a	*Endertia* Steen. & De Wit – De Vogel
7a	*Entada* Adans. – De Candolle
2a	*Erythrophleum* Afz. ex R. Br. – Taylor
1	*Gleditsia* L. – De Candolle; Hickel
	Hardwickia Roxb. (p.p., sensu Troup) = *Kingiodendron*
?2a	*Intsia* Thouars – Compton
6a	*Kingiodendron* Harms – Troup
2a	*Koompassia* Maing. – Meijer; De Vogel
2a	*Lysiphyllum* De Wit – Compton
?2a	*Maniltoa* Scheff. – Knaap-van Meeuwen
2a	*Parkinsonia* L. – Duke 1965
1	*Peltophorum* Walp. – Burger
2a	*Phanera* Lour. – De Vogel
1	*Piliostigma* Hochst. – Burger; Troup
	Poinciana L. = *Delonix*
11a	*Saraca* L. – Compton; De Vogel
2a	*Sindora* Miq. – Léonard
2a	*Tamarindus* L. – Burger; De Candolle; Compton; Duke 1965; Léonard; Troup
1, 2a, 2b, 7a, 11a, 11b	LEGUMINOSAE/MIMOSACEAE
11b	*Abarema* Pittier – De Vogel
2a	*Acacia* Mill. – Burger; Compton; Duke 1965; Hickel; Lubbock; Troup
2a	*Adenanthera* L. – Burger; Compton; Duke 1965; Lubbock

2a	*Albizia* Durazz. – Burger; Compton; Csapody; Duke 1965; Hickel; Mensbruge; Schopmeyer; Taylor; Troup; De Vogel obs.
1	*Dichrostachys* W. & A. – Burger; Lubbock
	Enterolobium Mart. p.p. = *Samanea*
2a, 2b	*Inga* Mill. – De Candolle; Duke 1965, 1969
1	*Mimosa* L. – Compton; Csapody; Troup
2a	*Parenterolobium* Kosterm. – De Vogel
2a	*Parkia* R. Br. – Burger; Duke 1965; Mensbruge; Taylor; De Vogel
2a	*Piptadenia* Benth. – Taylor
7a, 11a	*Pithecellobium* Mart. – Burger; Compton; Duke 1965, 1969; De Vogel obs.
1, ?2a	*Prosopis* L. – Duke 1965; Lubbock; Troup
2a	*Samanea* Benth. – Burger; Duke 1965; Mensbruge
1, 2a, 2b, 6a, 7a, 11a, 11b	LEGUMINOSAE/PAPILIONACEAE
2a	*Abrus* L. – Compton
	Afrormosia Harms = *Pericopsis*
2a	*Baphia* Afzel. – Mensbruge
6a	*Butea* Koen. ex Roxb. – Burger; Troup
1, 2a	*Dalbergia* L. f. – Burger; De Candolle; Compton; Lubbock; Troup
1	*Derris* Lour. – Compton
1, 2a, 6a	*Desmodium* Desv. – De Candolle; Compton; Lubbock; Ohashi; De Vogel
2a, 2b, 7a	*Erythrina* L. – Burger; De Candolle; Compton; Duke 1965; Mensbruge; Taylor; De Vogel obs.
7a	*Flemingia* Roxb. – Lubbock
2a	*Indigofera* L. – Burger; De Candolle; Compton; Csapody; Hickel
2a	*Millettia* W. & A. (p.p. = *Whitfordiodendron)* – Geesink pers. comm.; Mensbruge
7a	*Mucuna* Adans. – Sastrapradja c.s. 1972, 1975
2b, 7a, 11a	*Ormosia* Jackson – Duke 1965, 1969; De Vogel
2a	*Pericopsis* Thw. – Mensbruge; Taylor
6a, 7a	*Pongamia* Vent. – Troup; De Vogel obs.
1, 2a, 6a	*Pterocarpus* L. – Burger; Duke 1965, 1969; Lubbock; Mensbruge; Troup; De Vogel obs.
1	*Sesbania* Pers. – Lubbock
2a, 7a	*Sophora* L. – De Candolle; Compton; Csapody; Hickel; Lubbock; De Vogel
6a	*Spatholobus* Hassk. – De Vogel
1	*Tephrosia* Pers. – Burger; De Candolle
11b	*Whitfordiodendron* Merr. – De Vogel
1	LOGANIACEAE
1	*Fagraea* Thunb. – Burger

1	*Strychnos* L. – Cremers 1973; Leeuwenberg; Troup; De Vogel
1	LYTHRACEAE
1	*Lagerstroemia* L. – Burger; Ng 1975; Troup; De Vogel
1	MAGNOLIACEAE
1	*Illicium* L. – Hickel
1	*Magnolia* L. – Csapody; Hickel; Schopmeyer
1	*Manglietia* Bl. – Burger
1	*Michelia* L. – Burger
1	*Talauma* Juss. – De Vogel
1	MALPIGHIACEAE
1	*Aspidopterys* Juss. – De Vogel
1	MALVACEAE
1	*Abutilon* Mill. – Csapody; Lubbock
1	*Hibiscus* L. – Burger; Csapody; Duke 1965, 1969; Hickel; Lubbock
1	*Thespesia* Soland. ex Correa – Burger; Duke 1965; De Vogel obs.
1, 6a	MELASTOMATACEAE
1, 6a	*Memecylon* L. – Mensbruge; De Vogel
1, 2a, 7a, 11a, 11b	MELIACEAE
11b	*Aglaia* Lour. – Burger; De Vogel
2a	*Azadirachta* Juss. – Burger; Troup
	Carapa Aubl. = *Xylocarpus*
1	*Cedrela* L. – Burger; Duke 1969; Troup
	Chickrassia A. Juss. = *Chukrasia*
11b	*Chisocheton* Bl. – De Vogel
1	*Chukrasia* A. Juss. – Troup
2a, 7a, 11b	*Dysoxylum* Bl. – Burger; Pennington & Styles; Pierre; De Vogel
1, 2a	*Melia* L. – Burger; Duke 1965, 1969; Hickel; Lubbock; Troup
11a	*Sandoricum* Cav. – De Vogel
1	*Swietenia* Jacq. – Burger; Duke 1965, 1969; Troup
1	*Toona* Roem. – Burger; Duke 1969; Hickel; Troup
2a, 11b	*Trichilia* P. Br. – Duke 1965, 1969; Mensbruge
7a	*Walsura* Roxb. – Lubbock
7a	*Xylocarpus* Koen. ex Juss. – Burger; Chapman; Duke 1969; Mensbruge; Troup
1, 10	MENISPERMACEAE
1	*Anamirta* Colebr. – De Vogel obs.
1	*Cissampelos* L. – Duke 1965
10	*Coscinium* Colebr. – De Vogel
1	*Tinomiscium* Miers – De Vogel
1	MONIMIACEAE

1	*Kibara* Endl. – De Vogel
	Mollinedia Pr. stirp. pap. = *Wilkiea*
1	*Wilkiea* F. v. M. – Duke 1969
1, 2a, 7a, 11c	MORACEAE
7a	*Antiaris* Lesch. – Taylor
7a, 11c	*Artocarpus* J. R. & G. Forst. – Burger; Duke 1965; Jarrett; Troup; De Vogel
1	*Broussonetia* l' Hérit. ex Vent. – Hickel
1	*Ficus* L. – Csapody; Duke 1965, 1969; Lubbock; Troup
1	*Morus* L. – Csapody; Hickel; Mensbruge; Taylor
2a	*Parartocarpus* Baill. – Jarrett; De Vogel
11c	*Streblus* Lour. – Burger
1	MYRICACEAE
1	*Myrica* L. – Duke 1965; Lubbock; Schopmeyer
7a	MYRISTICACEAE
7a	*Horsfieldia* Willd. – De Vogel
7a	*Knema* Lour. – De Vogel
7a	*Myristica* L. – De Vogel
1, ?7b, ?8, 9	MYRSINACEAE
9	*Aegiceras* Gaertn. – Chapman
1, ?7b, ?8	*Ardisia* Sw. – Burger; Duke 1965, 1969; Lubbock; De Vogel
1	*Embelia* Burm. – Lubbock
1	*Maesa* Forsk. – Lubbock
1	*Rapanea* Aubl. – Duke 1965
1, 2a, 6a, 7a, 11a, 11b	MYRTACEAE
1, 2a,	*Eucalyptus* l' Hérit. – Burger; Csapody; Hickel; Lubbock; Schopmeyer
2a, 6a, 7a, 11a, 11b	*Eugenia* L. – Burger; Duke 1965, 1969; Lubbock; Troup; De Vogel obs.
1	*Melaleuca* L. – Lubbock
1	*Rhodamnia* Jack – Burger
1	*Rhodomyrtus* Reich. – Lubbock
	Syzygium Gaertn. = *Eugenia*
1	*Tristania* R. Br. – Burger; Lubbock; De Vogel
1	NYSSACEAE
1	*Nyssa* L. – Burger; Csapody; Hickel; Schopmeyer
2a, 6a	OCHNACEAE
6a	*Ochna* Schreb. – Lubbock; Mensbruge
2a	*Ouratea* Aubl. – Duke 1969
1, 7a, 8	OLACACEAE
1, 8	*Strombosia* Bl. – Heckel 1901; Kuijt; Mensbruge; Sleumer 1935; Taylor; De Vogel; Voorhoeve
7a	*Ximenia* L. – Duke 1965; Sleumer 1935

1, 2a, 2b, 6a	OLEACEAE
	Chionanthus L. = *Linociera*
1	*Fraxinus* L. – Csapody; Hickel; Lubbock; Schopmeyer
1	*Jasminum* L. – Csapody; Hickel
1	*Ligustrum* L. – Csapody; Hickel; Lubbock; Schopmeyer
1, 2a, 2b, 6a	*Linociera* Sw. – Duke 1965; Mensbruge; Schopmeyer
1	*Olea* L. – Hickel; Lubbock
2a	*Schrebera* Roxb. – Mensbruge
1	OXALIDACEAE
1	*Averrhoa* L. – Van Balgooy pers. comm.
1, 7a	PASSIFLORACEAE
1	*Adenia* Forsk. – Lubbock
	Modecca Lmk = *Adenia*
1, 7a	*Passiflora* L. – Csapody; Duke 1969; Hickel; Lubbock
1	PITTOSPORACEAE
1	*Pittosporum* Banks ex Gaertn. – Burger; Csapody; Lubbock; Ng 1976
2a, 6a, 11b	POLYGALACEAE
6a	*Securidaca* L. – Duke 1965, 1969
2a, 11b	*Xanthophyllum* Roxb. – Ng 1975
1, ?2a, 6a	PROTEACEAE
1	*Banksia* L. f. – Lubbock
1	*Grevillea* R. Br. – Lubbock
6a	*Helicia* Lour. – De Vogel
6a	*Heliciopsis* Sleum. – De Vogel
?2a	*Stenocarpus* R. Br. – Lubbock
1	PUNICACEAE
1	*Punica* L. – Csapody
1, 7a	RANUNCULACEAE
1, 7a	*Clematis* L. – Csapody; Hickel; Lubbock
1, 2a, 6a	RHAMNACEAE
1	*Colubrina* Rich. ex Brongn. – Duke 1965, 1969
1	*Gouania* L. – Cremers 1974
1	*Rhamnus* L. – Csapody; Hickel; Lubbock; Schopmeyer
?2a, 6a	*Ventilago* Gaertn. – (?Cremers 1973); De Vogel
1, 2a	*Ziziphus* Juss. – Csapody; Duke 1965, 1969; Hickel; Troup; De Vogel
1, 9	RHIZOPHORACEAE
9	*Bruguiera* Lmk – Burger; Chapman; Hou 1958; De Vogel obs.
1	*Carallia* Roxb. – Burger
9	*Ceriops* Arn. – Burger
9	*Kandelia* W. & A. – Ulbrich
9	*Rhizophora* L. – Burger; Chapman; Duke 1965, 1969; Gill & Tomlinson; Hou 1958; Mensbruge; Rumphius; De Vogel obs.

1, 2a, 6a	ROSACEAE
2a, 6a	*Parinari* Aubl. – Mensbruge; Taylor; De Vogel; Voorhoeve
1	*Photinia* Lindl. – Schopmeyer
2a, 6a	*Prunus* L. – Csapody; Schopmeyer; Troup; De Vogel
2a	*Pyrus* L. – Csapody; Lubbock; Schopmeyer
1	*Rosa* L. – Csapody; Lubbock; Schopmeyer
1	*Rubus* L. – Hickel
	Sorbus L. = *Pyrus*
1	RUBIACEAE
1	*Adina* Salisb. – Troup
1	*Anthocephalus* A. C. Rich. – Troup
1	*Canthium* Lmk – Lubbock; Mensbruge; Taylor; De Vogel
1	*Cephaelis* Sw. – Duke 1969
1	*Coffea* L. – Csapody; Duke 1965; Lubbock; De Vogel
1	*Diplospora* DC. – De Vogel
1	*Gaertnera* Lmk – De Vogel
1	*Gardenia* Ellis – Hallé; Troup; De Vogel
	Grumilea Gaertn. = *Psychotria*
1	*Guettarda* L. – Duke 1965; De Vogel
1	*Hymenodictyon* Wall. – Troup
1	*Hypobathrum* Bl. – Burger
1	*Ixora* L. – Duke 1965
1	*Lasianthus* Jack – Duke 1965
1	*Mitragyna* Korth. – Mensbruge; Troup
1	*Morinda* L. – Lubbock; Mensbruge; De Vogel
1	*Nauclea* L. – Burger; Mensbruge; Taylor; Troup
1	*Neonauclea* Merr. – Burger
	Oxyanthus Hassk. non DC. = *Randia*
1	*Pavetta* L. – Lubbock; De Vogel
	Plectronia L. = *Canthium*
	Posoqueria Bl. non Aubl. = *Randia*
1	*Psychotria* L. – Duke 1965; Lubbock; Mensbruge; De Vogel
1	*Randia* L. – Duke 1965, 1969; Lubbock; Mensbruge
	Stephegyne Korth. = *Mitragyna*
1, 2a, 7a	RUTACEAE
7a	*Chloroxylon* DC. – Troup
2a, 7a	*Citrus* L. – Csapody; Duke 1965, 1969; Hickel; Lubbock; De Vogel
7a	*Clausena* Burm. f. – De Vogel
?2a	*Evodia* Forst. – Csapody
	Fagara L. = *Zanthoxylum*

2a	*Limonia* L. – Lubbock; De Vogel
1	*Micromelum* Bl. – Burger
7a	*Murraya* L. – Burger; Duke 1969
1	*Skimmia* Thunb. – Csapody
1, 2a	*Zanthoxylum* L. – Csapody; Duke 1965; Hickel; Mensbruge; Schopmeyer; Taylor
1	SABIACEAE
1	*Meliosma* Bl. – Duke 1965, 1969; Lubbock; De Vogel
1	SALICACEAE
1	*Salix* L. – Csapody; Hickel; Lubbock; Troup
1, 8	SANTALACEAE
1	*Exocarpus* Labill. – Kuijt
8	*Santalum* L. – Burger; Lubbock; Pilger; Troup
1	*Thesium* L. – Csapody
1, 2a, 7a	SAPINDACEAE
2a, 7a	*Allophylus* L. – Burger; Duke 1965, 1969; Mensbruge
7a	*Atalaya* Bl. – Lubbock
7a	*Cubilia* Bl. – De Vogel
7a	*Dimocarpus* Lour. – De Vogel
1	*Dodonaea* Mill. – Burger; Duke 1965, 1969; Lubbock
7a	*Erioglossum* Bl. – Burger
2a	*Ganophyllum* Bl. – Burger
7a	*Harpullia* Roxb. – De Vogel
7a	*Lepisanthes* Bl. – De Vogel
7a	*Litchi* Sonn. – (Herb. Leiden)
7a	*Nephelium* L. – Meijer; De Vogel
2a	*Pometia* J. R. & G. Forst. – Burger
2a, 7a	*Sapindus* L. – Duke 1965, 1969; Hickel; Lubbock; Troup
2a	*Schleichera* Willd. – Burger; Troup
1, 2a, 2b, 11b	SAPOTACEAE
	Bassia Koen. ex L. non All. = *Madhuca*
1	*Chrysophyllum* L. – Mensbruge; Taylor
11b	*Lucuma* Molina – Lubbock
2b	*Madhuca* Gmel. – Troup
1	*Manilkara* Adans. – Burger; Duke 1965; Mensbruge
1, 2a	*Mimusops* L. – Lubbock; Mensbruge; Taylor; Troup
2b	*Palaquium* Blanco – Burger; De Vogel
1	*Planchonella* Pierre – De Vogel
6a	SARCOSPERMATACEAE
6a	*Sarcosperma* Hook. f. – Ng 1975
1	SAXIFRAGACEAE
1	*Deutzia* Thunb. – Csapody
1	*Hydrangea* L. – Csapody
1	SCYPHOSTEGIACEAE
1	*Scyphostegia* Stapf – Hou 1972

1, 2a	SIMAROUBACEAE
1	*Ailanthus* Desf. – Csapody; Hickel; Lubbock; Nooteboom; Troup
2a	*Irvingia* Hook. f. – Mensbruge; Meijer; Pierre; Taylor; De Vogel
1	SOLANACEAE
1	*Brunfelsia* L. – Lubbock
1	*Cestrum* L. – Duke 1965, 1969
1	SONNERATIACEAE
1	*Duabanga* Buch.-Ham. – Troup
1	*Sonneratia* L. f. – Chapman; De Vogel
1	STAPHYLEACEAE
1	*Turpinia* Vent. – Duke 1965, 1969
1, ?2a, 3, 6a, 7a, ?8	STERCULIACEAE
	Erythropsis Lindl. = *Firmiana*
1	*Firmiana* Marsigli – Taylor
1	*Guazuma* Mill. – Burger
1	*Helicteres* L. – Burger
6a	*Heritiera* Dry. – Burger; Pierre; Taylor; Troup
1	*Hildegardia* Schott & Endl. – Mensbruge
1	*Kleinhovia* L. – Burger
?8	*Leptonychia* Turcz. – Mensbruge
1	*Pentapetes* L. – Lubbock
1	*Pterospermum* Schreb. – Burger; Troup
1, 3	*Pterygota* Schott & Endl. – Mensbruge; Taylor
1	*Scaphium* Schott & Endl. – De Vogel
1, ?2a, 3, 7a	*Sterculia* L. – Burger; Duke 1965, 1969; Hickel; Lubbock; Mensbruge; Taylor; De Vogel
	Tarrietia Bl. = *Heritiera*
1, 2a	STYRACACEAE
1, 2a	*Styrax* L. – Burger; Hickel; Lubbock; Ng 1976; De Vogel obs.
?2a	SYMPLOCACEAE
?2a	*Symplocos* L. – Lubbock
1, 4, 7a	THEACEAE
	Camellia L. = *Thea*
1	*Gordonia* Ellis – Burger
1	*Laplacea* H. B. K. – Burger
1	*Pyrenaria* Bl. – De Vogel
1	*Schima* Reinw. ex Bl. – Burger
4	*Ternstroemia* Mutis ex L. – De Vogel
7a	*Thea* L. – Lubbock
1, 6a, ?7a	THYMELAEACEAE
1, ?7a	*Daphne* L. – Csapody; Hickel
6a	*Phaleria* Jack – De Vogel
1	TILIACEAE

1	*Berrya* Roxb. – Lubbock; Troup
1	*Grewia* L. – Burger
1	*Microcos* L. – Mensbruge
	Omphacarpus Korth. = *Microcos*
1	*Schoutenia* Korth. – Burger
1	*Triumfetta* L. – Duke 1965
2a	TRIGONIACEAE
2a	*Trigoniastrum* Miq. – Ng 1975
1	TURNERACEAE
1	*Turnera* L. – Lubbock
1, ?2a	ULMACEAE
1	*Aphananthe* Pl. – Hickel
1, ?2a	*Celtis* L. – Csapody; Hickel; Lubbock; Mensbruge; Rumphius; Schopmeyer; Taylor; Troup
1	*Trema* Lour. – Burger; Mensbruge; Taylor
1	*Ulmus* L. – Csapody; Hickel; Lubbock; Schopmeyer
1	URTICACEAE
	Conocephalus Bl. non Hill = *Poikilospermum*
1	*Boehmeria* Jacq. – Csapody
1	*Poikilospermum* Zipp. ex Miq. – Lubbock
1, 2a, 6a, 7a	VERBENACEAE
2a	*Avicennia* L. – Chapman; Duke 1965; Mensbruge
1	*Callicarpa* L. – Lubbock
7a	*Clerodendron* L. em. R. Br. – Burger; Csapody; Hickel; Lubbock
6a	*Gmelina* L. – Burger; Troup
1	*Peronema* Jack – Burger
1	*Premna* L. – Burger
1	*Tectona* L. f. – Burger; Duke 1965, 1969; Troup
6a	*Teysmanniodendron* Koord. – Kostermans; De Vogel
1, 2a	*Vitex* L. – Burger; Csapody; Hickel; Lubbock; Mensbruge; Schopmeyer
1	VIOLACEAE
1	*Rinorea* Aubl. – Mensbruge; Ng 1975
1	VITACEAE
1	*Ampelopsis* Michx. em. Planch. – Hickel
1	*Parthenocissus* Planch. – Csapody; Lubbock; Schopmeyer
1	*Tetrastigma* Planch. – De Vogel
1	*Vitis* L. sensu Planch. – Csapody; Hickel; Lubbock
1	WINTERACEAE
1	*Drimys* J. R. & G. Forst. – Lubbock

Classification of the seedling types

It is postulated that all seedling types of Dicots can be derived from each other through differential changes in the gross morphology and development of the parts. The question is which seedling type must be regarded as the most primitive, and which as derived. Most authors, to date, regard the Macaranga type (1), with thin assimilating paracotyledons, as the ancestral type. Assimilating paracotyledons were supposed to have changed into foodstoring or haustorial cotyledons. This point of view has, however, never been supported by documentation. Authors dealing with the subject had at their disposal mainly seedlings from the temperate regions, where the Macaranga type is the rule and other types are rare; this may have influenced their ideas. Only Grushvitskyi (1963), who studied temperate as well as tropical seedlings, was of the opinion that haustorial cotyledons which remain within the endosperm and testa are the most primitive, and foodstoring cotyledons and leaf-like assimilating 'cotyledons' (paracotyledons) are derived. He based his opinion on correlations between the systematic position and the seedling type. Haustorial cotyledons in endosperm and fruit wall and/or testa were found only in 'archaic' families of Dicots, and were, therefore, considered as the primitive type. Non-emergent foodstoring cotyledons, present in representatives of 50 families of Dicots, are also found in more evolved groups; it was recorded that epigeal as well as hypogeal germination may occur within a genus and sometimes even within one species. Epigeal assimilating cotyledons were recorded for 'archaic' as well as for evolved families; in the latter they are dominant to almost exclusive. Hypogeal and epigeal germination were considered by Grushvitskyi to be unstable characteristics, and it was assumed that the change from one type to the other could have taken place at various stages in the progress of evolution. This idea was based on the fact that hypogeal germination with foodstoring cotyledons is often present in only one or a few non-related genera in a family, sometimes even in one or few species within a genus.

For the present attempt to classify the seedling types it is important to determine the most primitive one amongst them. Two ways are open: 1. to make a survey of seedling types in taxa which are commonly regarded as 'primitive', 2. to arrange the seedling types according to their similarities and deduct the most primitive seedling type by comparison.

The recent Dicot families which are usually considered to be most primitive, all belong to the Polycarpicae (sensu Takhtajan, 1959). The orders from which seedlings are known and which may be important for the question of the ancestral seedling type, are the Magnoliales, the Laurales and the Piperales. Genera in these orders, of which seedlings have been described, and the types which these represent, are given in Table 4.

Table 4. Seedling types in genera of the Magnoliales, the Laurales and the Piperales. Summarised after Burger (1972), Csapody (1968), Duke (1965, 1969), Lubbock (1892), De la Mensbruge (1968), Schopmeyer (1974), Troup (1921), and the present work.

	Macaranga type (1)	Sloanea type (2a)	Heliciopsis type/subtype (6a)	Horsfieldia type/subtype (7a)	Horsfieldia type/ Pseuduvaria subtype (7b)	Blumeodendron type (8)	Endertia type/ Streblus subtype (11c)	Cynometra ramiflora type (12)
Magnoliales								
Magnoliaceae								
Liriodendron	1							
Magnolia	1							
Manglietia	1							
Michelia	1							
Talauma	1							
Degeneriaceae								
Degeneria	1							
Annonaceae								
Annona						8		
Artabotrys				7a				
Asimina						8		
Cananga	1							
Cleistopholis	1							
Cyathocalyx	1							
Enantia						8		
Guatteria	1							
Hexalobus	1							
Mezzettia						8		
Mezzettiopsis						8		
Monodora						8		
Pachypodanthium						8		
Polyalthia				7a		8		
Pseuduvaria					7b			
Saccopetalum	1							
Trivalvaria	1							
Uvariastrum	1							
Xylopia		2a				8		

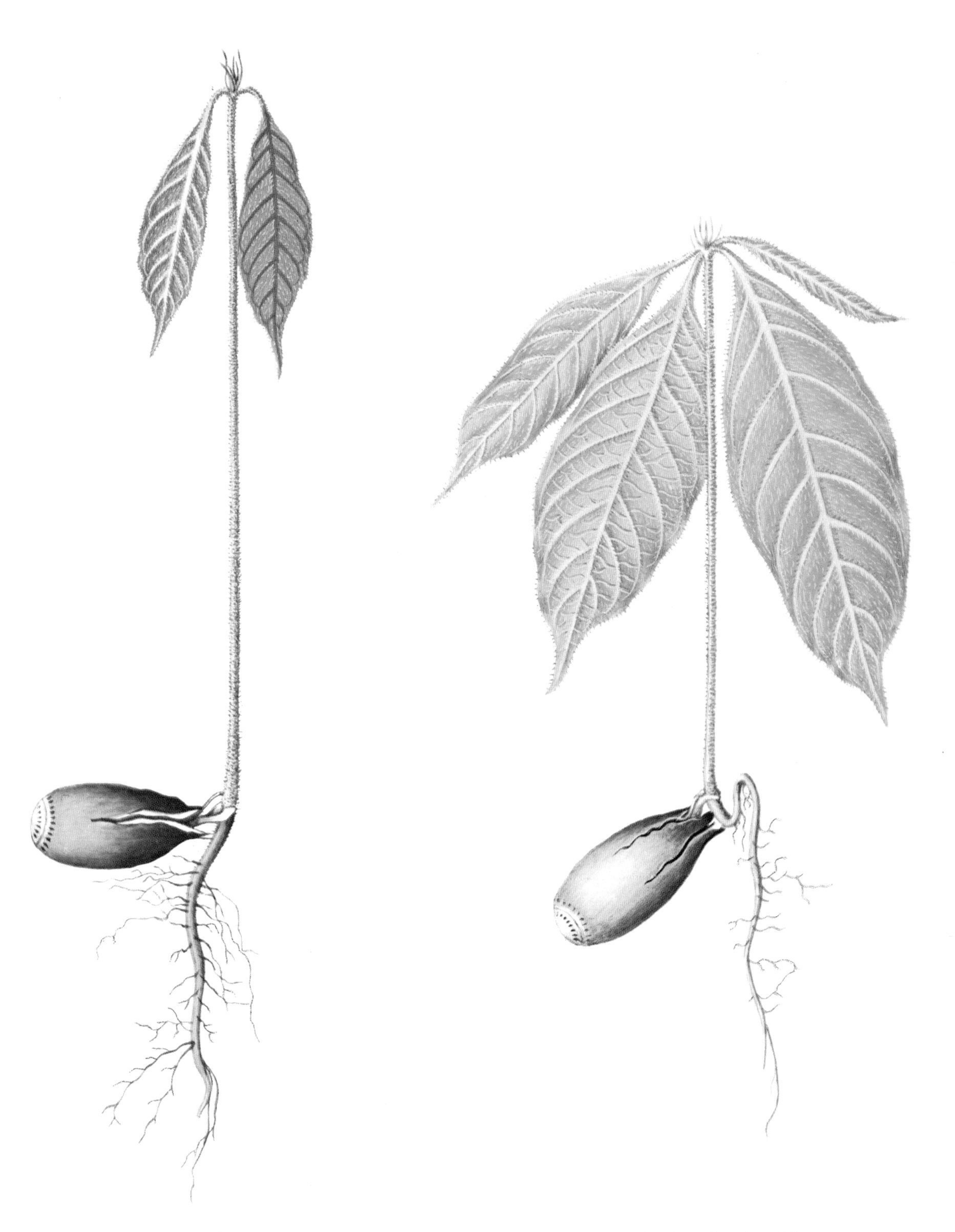

Plate 13. *Quercus turbinata* Roxb. (Fagaceae), Bogor Bot. Gard. VIII.B.18. See also plate 14.

Plate 14. *Quercus turbinata* Roxb. (Fagaceae), Bogor Bot. Gard. VIII.B.18. See also plate 13.

Table 4—*continued*

	Macaranga type (1)	Sloanea type (2a)	Heliciopsis type/subtype (6a)	Horsfieldia type/subtype (7a)	Horsfieldia type/ Pseuduvaria subtype (7b)	Blumeodendron type (8)	Endertia type/ Streblus subtype (11c)	Cynometra ramiflora type (12)
Myristicaceae								
Coelocaryon				7a				
Horsfieldia				7a				
Knema				7a				
Myristica				7a				
Pycnanthus				7a				
Virola					7b			
Laurales								
Monimiaceae								
Kibara	1							
Mollinedia	1							
Siparuna	1							
Wilkiea	1							
Lauraceae								
Beilschmiedia			6a					
Cinnamomum			6a					12
Cryptocarya			6a					
Eusideroxylon			6a					
Laurus								12
Lindera			6a					
Litsea			6a					12
Machilus			6a					
Nectandra			6a					
Ocotea								12
Persea			?6a					12
Phoebe			6a					
Umbellularia			6a					
Hernandiaceae								
Hernandia			6a					
Piperales								
Piperaceae								
Peperomia	1						11c	
Piper	1							

This survey shows that in families commonly regarded as 'primitive' a great diversity is present in the seedlings. Some families are homogeneous whereas, in others, several types are encountered. Assimilating paracotyledons, haustorial cotyledons as well as foodstoring cotyledons are present. The Macaranga type (1) is typical for Magnoliaceae, Degeneriaceae, and Monimiaceae. The Horsfieldia type/subtype (7a) with haustorial cotyledons which, in the opinion of Grushvitskyi, represents the most primitive condition, is specific for Myristicaceae. In Annonaceae both types occur, together with the Blumeodendron type (8) which also agrees with the primitive characters proposed by Grushvitskyi. A clue to the ancestral seedling type can, therefore, not be obtained from the survey in Table 4.

In this respect it may be worthwhile to investigate the occurrence of seedling types of taxa which are classified lower in the hierarchical system of plants, i.e. in the Gymnosperms. It is possible that the Angiosperms are derived from the Pteridospermae, which may also have given origin to the other Gymnosperms. Of the latter, the Cycadopsida are possibly more related with the Pteridosperms than the Coniferopsida. A survey of seedling types in Gymnosperms is given in Table 5.

The table shows that the variation in seedling types in Gymnosperms is much less than that in the primitive extant Dicots. Two types are present: the Macaranga type (1) and the Horsfieldia type/subtype (7a). This statement would implicate that cotyledons and paracotyledons in Gymnosperms are homologous with the equivalents in Dicots. This may, indeed, be the case for the non-emergent cotyledons, of which two are present, sometimes fused into one body, of which the function, number and position is comparable. The number of paracotyledons in Gymnosperms ranges from 2 to 14, whereas in Dicots almost always two are present; their function

Table 5. Seedling types in Gymnosperms. 1. Macaranga type; 7a. Horsfieldia type and subtype. Mainly summarised after Hill & De Fraine, Seedling structure of Gymnosperms I–IV (1908–1910), and De Ferré, Les formes de jeunesse des Abiétacées, Ontogénie-Phylogénie (1952); the system is after Kalkman (1972).

CYCADOPSIDA			
Cycadales[1]			
Cycadaceae			
Cycas	(7a)		
Zamiaceae			
Bowenia	(7a)	*Macrozamia*	(7a)[2]
Ceratozamia	(7a)	*Microcycas*	(7a)
Dioon	(7a)	*Stangeria*	(7a)
Encephalartos	(7a)		
CONIFEROPSIDA			
Ginkgoales			
Ginkgoaceae			
Ginkgo	(7a)[2]		

Coniferales			
Pinaceae			
Abies	(1)	*Pinus*	(1)
Cedrus	(1)	*Pseudolarix*	(1)
Keteleeria	(7a)	*Pseudotsuga*	(1)
Larix	(1)	*Tsuga*	(1)
Picea	(1)		
Araucariaceae			
Araucaria sect. *Eutacta*	(1)		
sect. *Colymbea*	(7a)[2]		
Taxodiaceae			
Athrotaxis	(1)	*Sciadopitys*	(1)
Cryptomeria	(1)	*Sequoia*	(1)
Cunninghamia	(1)	*Taiwania*	(1)
Glyptostrobus	(1)	*Taxodium*	(1)
Metasequoia	(1)	*Wellingtonia*	(1)
Cupressaceae			
Actinostrobus	(1)	*Juniperus*	(1)
Callitris	(1)	*Libocedrus*	(1)
Chamaecyperus	(1)	*Tetraclinis*	(1)
Cupressus	(1)	*Thuja*	(1)
Frenela	(1)	*Widdringtonia*	(1)
Podocarpaceae			
Podocarpus	(1)		
Cephalotaxaceae			
Cephalotaxus	(1)		
Taxales			
Taxaceae			
Taxus	(1)		
Torreya	(7a)		
GNETOPSIDA			
Ephedrales			
Ephedraceae			
Ephedra	(1)		
Gnetales			
Gnetaceae			
Gnetum	Not comparable to a Dicot type		
Welwitschiales			
Welwitschiaceae			
Welwitschia	Not comparable to a Dicot type		

[1] According to Hill and De Fraine (1909b) stomata are generally present on the cotyledons which remain enclosed in the persistent seedcoat and endosperm.
[2] Presence of stomata on the enclosed cotyledons has been specifically mentioned.

and position in both groups is the same. The question remains whether one paracotyledon in Angiosperms is homologous with only one, or with more than one, paracotyledon in Gymnosperms. For both opinions arguments have been given by earlier authors, but the question is not yet solved. Another difference is the nature of the endosperm, which, in both groups, serves as a food source for the embryo and seedling. In Gymnosperms this tissue is haploid (n), and equivalent with the macroprothallium (female gametophyte). In Angiosperms it is a unique tissue (secondary endosperm) which originates from the union of two (or more) polar nuclei with the nucleus of one of the two sperm cells of the pollen (3n or more). Because the endosperm is not a part of the seedling, these differences are of no consequence for the question whether Gymnosperm and Angiosperm cotyledons are homologous.

The Horsfieldia type/subtype is typical for the groups which are placed lowest in the systematic hierarchy, viz. the Cycadopsida and the Ginkgoales, and occurs in some genera or sections in the other groups. In all, the cotyledons are of the haustorial type, and remain within the endosperm. A peculiar feature is that stomata are present on many of these cotyledons, which has been considered to support the idea that hypogeal cotyledons are derived from epigeal and emergent ones.

The interpretation of the seedlings of *Welwitschia* and *Gnetum* poses some problems, because of the unique nature of the foot or feeder, a cylindrical organ produced by the hypocotyl, an equivalent of which does not exist in Angiosperms, Cycadopsida and Coniferopsida. This organ develops much later than the cotyledon primordia, but serves a major function in the food supply of the embryo and the seedling. The latter much resembles the Horsfieldia type/subtype. The leaf-like cotyledons in *Gnetum* are similar to the subsequent leaves in morphology and texture; in some cases they remain reduced and are not functional. The seedlings of *Gnetum* and *Welwitschia* are not readily comparable to a Dicot seedling type.

Fossil Gymnosperm seeds are known, but preserved embryos are rare, and seedlings have never been found. Wieland (a.o. 1911) provided illustrations of *Cycadeoidea* (= *Bennettites*) seeds with embryo, in which a massive hypocotyl and two thick cotyledons can be distinguished which completely fill the seed, endosperm being absent. It is probable that the seedling was hypogeal, most likely of the Horsfieldia type. The cotyledons are sometimes connate at the base. Camberlain (1935) speculated that the seedling of Cycadofilicales (= Pteridosperms) was hypogeal, with the cotyledons remaining within the seed and the endosperm, serving as haustoria (Fig. 16).

It is evident that from the recent Gymnosperms and the scanty information on fossils no clue can be derived to the question which seedling type is ancestral in Dicots. The common occurrence of the Horsfieldia type in the most primitive groups, however, may be a basis for reserve to regard the Macaranga type as the ancestral type, and makes an approach from a different angle worthwhile.

Comparison of the different mode of development in the seedling types recognised may throw new light on the question. The underlying thoughts are that the seedling type which was present in the ancient Dicots can still be found amongst living plants, and that it must be possible to deduce the most primitive type by comparison of seedlings.

Based on morphological similarities and differences, the seedling types can be

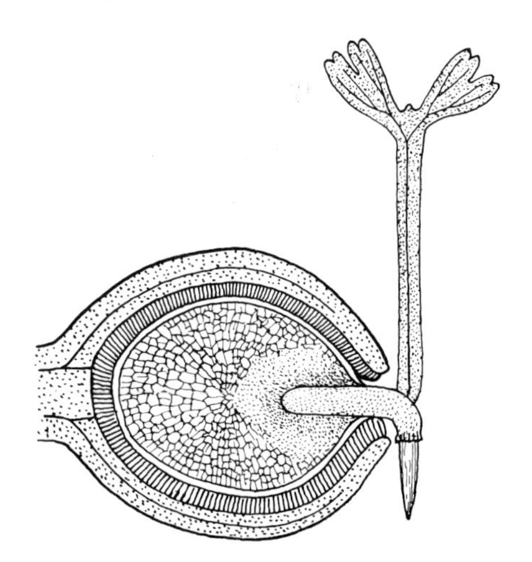

Fig. 16. Hypothetical mode of germination of a Pteridosperm seed (after Chamberlain 1935).

arranged into a model in which their morphological relationships are expressed (Fig. 17). The drawings are schematical and represent the most typical features, they are indicated by the number of the type or subtype. Seedling types which have most features in common are placed next to each other. The Blumeodendron type (8) and the Coscinium type (10) and the subtypes of the Horsfieldia type (7a, 7b) and the Heliciopsis type (6a, 6b) are represented by two illustrations: one with haustorial cotyledons in endosperm, one with foodstoring cotyledons. The scheme is entirely based on morphological considerations and on differences in the development. Resemblance of the types does not necessarily indicate phylogenetic relationships, although this is evidently the case between systematically related representatives of the types. Phylogenetical arguments have not been taken into account in the construction of this model.

Almost all seedling types represented in the model have actually been found, except the Heliciopsis type/Koordersiodendron subtype (6b) with haustorial cotyledons and the Coscinium type (10) with foodstoring cotyledons, both indicated with *. The existence of these can however be postulated.

Morphological series may be read either way. Therefore, arguments must be provided as to why a certain type is chosen as the starting point from which the others are derived. Two opposing views are present. One is that epigeal seedlings are primitive. No distinction was made between exposed paracotyledons and foodstoring cotyledons, but evidently the first category was regarded as primitive because exposed foodstoring cotyledons are relatively rare in Europe where the idea was developed. The opposing idea is that enclosed haustorial cotyledons represent the primitive type; the representatives which Grushvitskyi (1963) mentioned belong to the Horsfieldia type/subtype (7a), the Horsfieldia type/Pseuduvaria subtype (7b), and the Blumeodendron type (8). By suggesting that the Macaranga type is the most primitive, and deducing the series of steps necessary for the change towards other types, it may be possible to evaluate whether this supposition is acceptable or not.

The type which is nearest in all respects to the Macaranga type is the Sloanea type/subtype (2a) (Fig. 18). The cotyledons become exposed and are borne above the soil on an elongated hypocotyl, whereas in all other types the cotyledons are not emergent or no cotyledons are present. If the Macaranga type is considered as ancestral, necessarily the Sloanea type/subtype must be directly derived. The changes involved are, however, rather complex, which can be demonstrated when comparing the actual differences between the two types (Table 6).

An evaluation of the changes involved in the derivation of the Sloanea type/subtype from the Macaranga type brings me to the conclusion that former students may have too lightly assumed that this step is simple. A major reason to reject a derivation

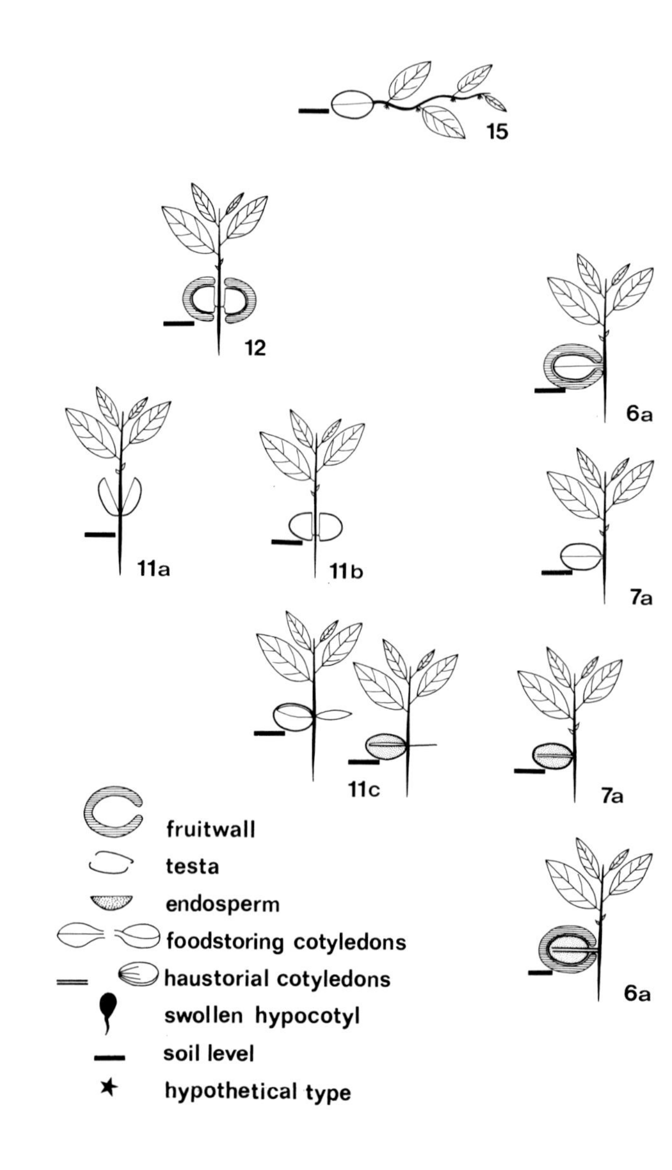

in this direction is that the combined change, of assimilating paracotyledons into haustorial or foodstoring cotyledons, and the shift in the resting stage after emergence of these organs, is too great. In the Sloanea type, before the second rest takes place, photosynthetic leaf surface is produced next to the emerged cotyledons. It must be concluded that this photosynthetic area is indispensable for the development of seedlings of that type. It is, therefore, illogical to propose a loss of the photosynthetic function of the paracotyledons. These must become short-lived, and they support in their changed form only the development of the first two leaves which

Fig. 17. Model in which the recognised seedling types are arranged according to morphological similarities.

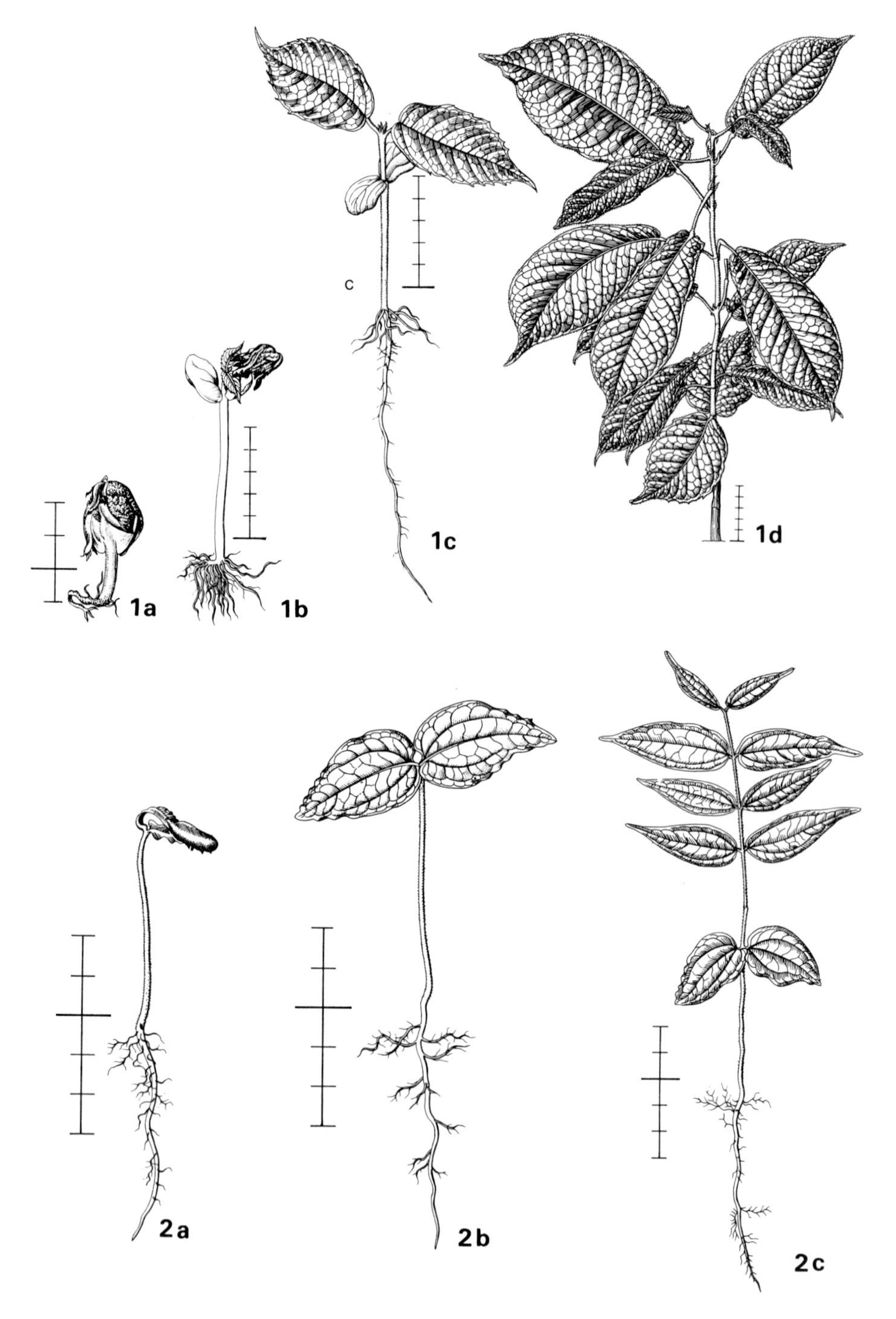

Fig. 18. Comparison of differences in the development between seedlings of the Sloanea type/subtype (1. *Sloanea javanica*, Elaeocarpaceae, De V. 2212 and 2365) and the Macaranga type (2. *Strychnos sp*., Loganiaceae, De V. 2363).

Table 6. Differences between the Macaranga type (1) and the Sloanea type/subtype (2a).

	Macaranga type	Sloanea type/subtype	
		Haustorial cotyledons	Foodstoring cotyledons
Endosperm in the germinating seedling	present in the mature seed, partly absorbed tissue shed by the emerging paracotyledons, or absent	present in the mature seed, partly absorbed tissue shed by the emerging cotyledons	extensive or entire absorption of endosperm in the embryonic stage, absent or functionless at germination
Second temporary rest in the development of the epigeal parts	after emergence and expansion of the paracotyledons	after emergence of the cotyledons and development and expansion of the first internode and two opposite leaves	after emergence of the cotyledons and development and expansion of the first internode and two opposite leaves
(Para)cotyledons	after emergence distinct enlargement (cell division and/or stretching)	no distinct enlargement	no distinct enlargement
	very long persistent after second rest	shed before or shortly after second rest	shed shortly after second rest
	thin to coriaceous	somewhat succulent, chartaceous to soft-fleshy	succulent, hard-fleshy
	no reserve food after emergence and enlargement	some reserve food in the cotyledon tissue	large amount of reserve food in the cotyledon tissue
	chlorophyll present	chlorophyll absent or almost	chlorophyll absent or present
	often haustorial function when enclosed, pure photo synthetic function after emergence	pure haustorial function when enclosed, functionless after emergence	foodstoring function, photosynthetic function after emergence of minor importance or absent
Leaves	all arranged in a spiral (or all decussate)	first two opposite, subsequent ones in a spiral	first two opposite, subsequent ones in a spiral (? or all decussate)

serve exactly the same function as the paracotyledons. In addition, the temporary rest in the development of the epigeal parts must be postponed until these leaves have developed, which are, in addition, opposite, even in plants where the normal phyllotaxis is spiral.

In fact, the change from the Macaranga type to the Sloanea type would involve so many alterations in structure and development that it is rather unlikely that it has taken place. A derivation in the reverse direction seems more plausible. If it is assumed that in the Sloanea type the cotyledons become aborted, a seedling results with two opposite, leaf-like organs, and the plantlet exactly matches the Macaranga type in structure. Assimilating 'cotyledons' of the latter type are, in this view, equivalent with normal stem leaves. The differences in the resting stages of the two types are, under this hypothesis, fully comprehensible. Both types show a resting stage in a comparable phase of development, with two stem leaves fully developed. In the Sloanea type below the leaves the two real cotyledons are present, which are not developed in the Macaranga type. Also, the relative similarity of assimilating 'cotyledons' with stem leaves is thus explained. The former are quite distinct from the latter, but the differences in morphology and function are of a much lower category than with foodstoring or haustorial cotyledons.

Seedlings with two kinds of cotyledons have been classified in the Sloanea type: haustorial cotyledons with endosperm, and foodstoring cotyledons. Through abortion of the first kind the seedling would turn into the Macaranga type where in the seed endosperm is present. With those of the second category no endosperm would be present. The actual situation is that Macaranga type seedlings are produced by endosperm containing seeds as well as endospermless seeds.

The cardinal point in this hypothetical derivation is the abortion of the true cotyledons. That extensive reduction or abortion actually has taken place in seedlings is demonstrated in quite a number of seedling types: Rhizophora type (9), Barringtonia type (13), Garcinia type (14), Ternstroemia type (4), and Cyclamen type (5). Anatomical evidence can possibly provide arguments for or against this hypothesis. Studies should be made in natural entities in which both types occur, so far such information is almost wanting.

An alternative possibility is that a transference of function has taken place from the first foliar leaves to the cotyledons, by which the latter have attained a photosynthetic function, and acquired a leaf-like appearance. Such a transition would make the early development of the first two leaves superfluous because assimilates are then produced by the cotyledons, and development of the plumule could be postponed till after the second rest.

Quite a number of genera are heterogeneous with respect to the seedlings. A large proportion of these are in the possession of both Macaranga type and Sloanea type/subtype; other combinations are less frequent (see Table 7). A discussion of some examples may serve to underline the proposed hypothesis that the Macaranga type may be regarded as derived from the Sloanea type/subtype.

Within a genus it may be assumed that, if different seedling types are present, these have either been derived from a common ancestral type, or from each other, possibly directly, or through intermediates which have disappeared. The occurrence of differ-

Table 7. Genera of Malesian woody plants with more than one seedling type.

Seedling type:	1	2a	3	4	6a	6b	7a	7b	8	9	11a	11b	11c	12
Acer (Acerac.)	1				6a									
Allophilus (Sapindac.)		2a					7a							
Ardisia (Myrsinac.)	1			?4				7b						
Artocarpus (Morac.)							7a						11c	
Bauhinia (Legum./ Caesalp.)	1	2a					7a							
Caesalpinia (Legum./ Caesalp.)	1	2a												
Canarium (Burserac.)	1	2a												
Cassia (Legum./ Caesalp.)	1	2a												
Celtis (Ulmac.)	1	?2a												
Cinnamomum (Laurac.)					6a									12
Citrus (Rutac.)		2a					7a							
Clematis (Ranunculac.)	1						7a							
Connarus (Connarac.)					6a						11a			
Crotalaria (Legum./ Caesalp.)	1	?2a												
Cynometra (Legum./ Caesalp.)		2a												12
Dalbergia (Legum./ Papil.)	1	2a												
Desmodium (Legum./ Papil.)	1	2a			6a									
Diospyros (Ebenac.)	1	2a							8					
Durio (Bombacac.)								7b	8					
Dysoxylum (Meliac.)		2a					7a					11b		
Erythrina (Legum./ Papil.)		2a					7a							
Eucalyptus (Myrtac.)	1	2a												
Eugenia (Myrtac.)		2a			6a		7a				11a	11b	11c	
Fagara (Rutac.)	1	2a												
Linociera (Oleac.)	1	2a			6a									
Litsea (Laurac.)					6a									12
Melia (Meliac.)	1	2a												
Memecylon (Melastomatac.)	1				6a									
Mimusops (Sapotac.)	1	2a												
Ormosia (Legum./ Papil.)		2a					7a							
Pangium (Flacourtiac.)								7b	8					
Parinari (Rosac.)		2a			6a									
Passiflora (Passiflorac.)	1						7a							
Pithecellobium (Legum./ Mimos.)							7a				11a			
Polyalthia (Annonac.)							7a	8						
Pongamia (Legum./ Papil.)					6a		7a							
Prosopis (Legum./ Mimos.)	1	?2a												
Prunus (Rosac.)		2a			6a									
Pterocarpus (Legum./ Papil.)	1	2a			6a									
Pterygota (Sterculiac.)	1		3											
Rhus (Anacardiac.)	1	2a			?6a									
Santiria (Burserac.)		2a			6a									

Table 7—*continued*

Seedling type:	1	2a	3	4	6a	6b	7a	7b	8	9	11a	11b	11c	12
Sapindus (Sapindac.)		2a					7a							
Shorea (Dipterocarpac.)		2a			6a									
Sophora (Legum./ Papil.)		2a					7a							
Sterculia (Sterculiac.)	1	?	3				7a							
Strombosia (Olacac.)	1								8					
Styrax (Styracac.)	1	2a												
Terminalia (Combretac.)	1	2a			6a									
Trichilia (Meliac.)		2a										11b		
Vatica (Dipterocarpac.)		2a			6a									
Vitex (Verbenac.)	1	2a												
Xanthophyllum (Polygalac.)		2a										11b		
Xylopia (Annonac.)		2a							8					
Ziziphus (Rhamnac.)	1	2a												

ent seedling types is discussed for three cases: *Diospyros*, for which extensive documentation exists, *Sterculia*, and *Guibourtia/Colophospermum.*

In the genus *Diospyros* (Ebenaceae) either haustorial cotyledons are present, or thin assimilating paracotyledons, which both serve to draw nutrients from the endosperm in the early period of seedling development. Three seedling types are known in the genus: the Blumeodendron type (8), the Sloanea type/subtype (2a) and the Macaranga type (1), see Fig. 19. The genus is a natural entity, and the assumption that these three types represent a pathway of changes appears an adequate explanation for the existing differences in morphology. All three types have a first resting stage with the (para)cotyledons hidden in the testa and endosperm. The second temporary rest in the Blumeodendron type is usually with two opposite developed leaves, or with several leaves arranged in a spiral; the cotyledons are shed when the shoot with leaves starts elongation. This stage is equivalent with that in the Sloanea type/subtype in which the cotyledons are emerged and the first internode with two developed leaves are present. In the Macaranga type the important difference is that the second rest phase occurs when only the paracotyledons have developed.

Additional data are provided by Wright (1904), who studied embryos and seedlings of 20 species of Ceylonese *Diospyros*, and was already intrigued by the differences in the seedlings of this genus. He noted early detachment of the cotyledons in *Diospyros quesita* (Blumeodendron type) and 15 other species (Blumeodendron type and Sloanea type/subtype) in a stage when still a large amount of endosperm was present. A high death rate of seedlings was recorded as a consequence of a diseased condition of the apex of the epicotyledonary axis, which was due to death and decomposition of the cells of the detached cotyledons and the endosperm. In commenting on this 'suicidal mode of development' he stressed the development of the first internode with two opposite leaves in the embryo. This feature is absent in seedlings of which the 'cotyledons' are long-lived and have an assimilatory function. He described that such epicotyl leaves:

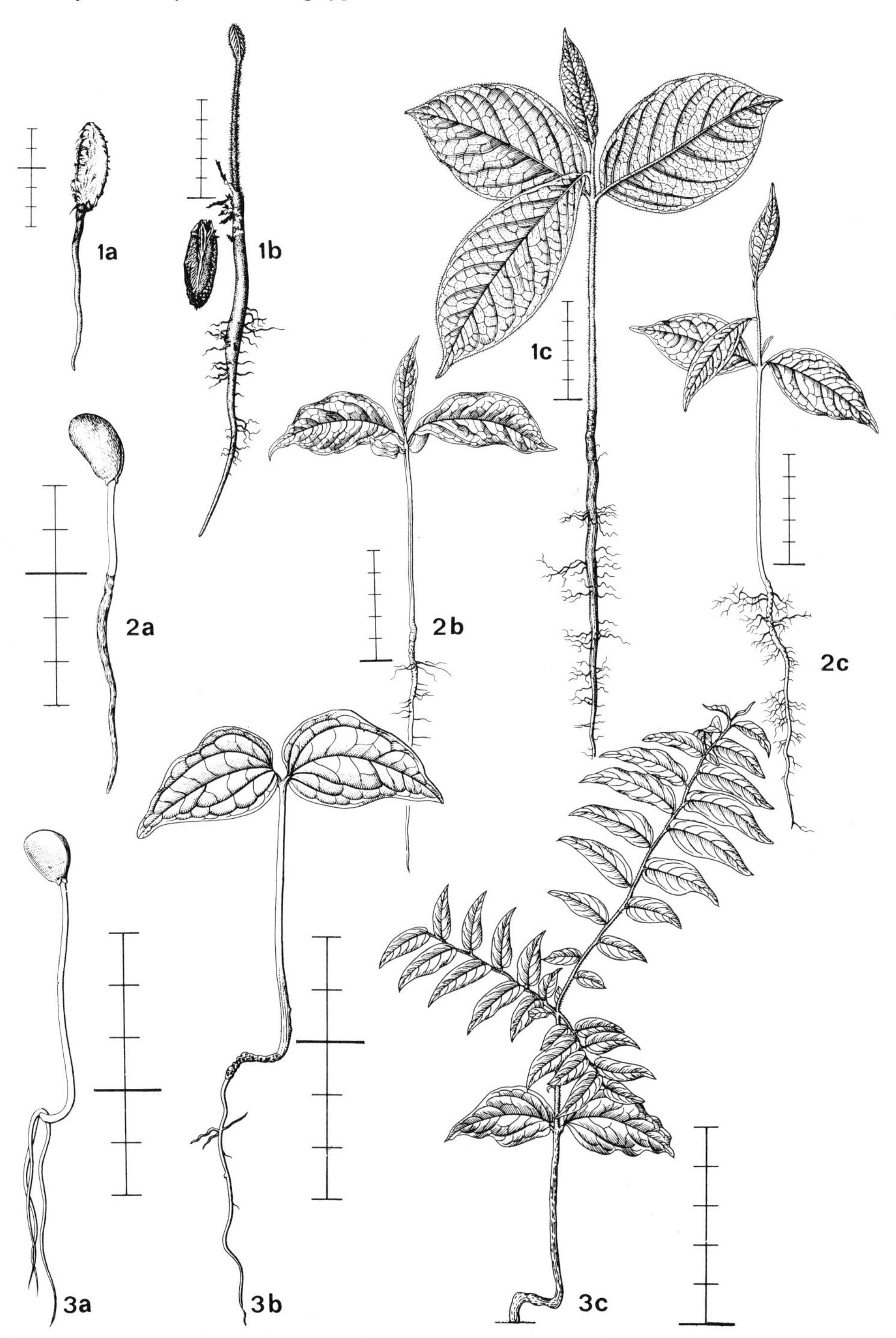

Fig. 19. Comparison of differences in the development between species of *Diospyros* (Ebenaceae). – 1. *Diospyros macrophylla*; 2. *D. curraniopsis*; 3. *D. sp*. (De V. 2556).

"usually form an opposite pair of interlocked leaves, and thus remind one still more forcibly of cotyledons. When the epicotyl leaves form an opposite pair these are usually persistent alone for many months; this suggests most strikingly that their enhanced development has really been an effort, and internal evidence indicated that it is associated with abortion of cotyledons, which appears to have been taking place through many generations. The majority of our [= Ceylonese] species show this detachment of cotyledons and enhanced development of epicotyledonary leaves which take on the work usually assigned to cotyledons. The cotyledons never acquire palissade tissue, but this layer is laid down in the epicotyledonary leaves long before they are exposed to light."

The differences in anatomy between Macaranga type seedlings on the one hand, and Sloanea type/subtype and Blumeodendron type on the other, are worth noticing (Fig. 20). In the latter two, cotyledon traces as well as traces from two epicotyledonary leaves are continued through the hypocotyl into the primary root. There are three traces from the cotyledons in all 16 species with haustorial cotyledons studied, except in *Diospyros ovalifolia* where this number is two. Wright recorded a tendency of the xylem and phloem of the median trace to abort in the vicinity of the collet. The number of traces from the epicotyledonary leaves in the hypocotyl is two, from each leaf one. In Macaranga type seedlings from each of the paracotyledons two traces are continued through the hypocotyl in the primary root, whereas the epicotyledonary traces (one per leaf) die away immediately below the paracotyledonary node. Each of the strands, from the cotyledons, the paracotyledons as well as the epicotyledonary leaves, may separate in the hypocotyl in two, three or more parts. In the collet area of

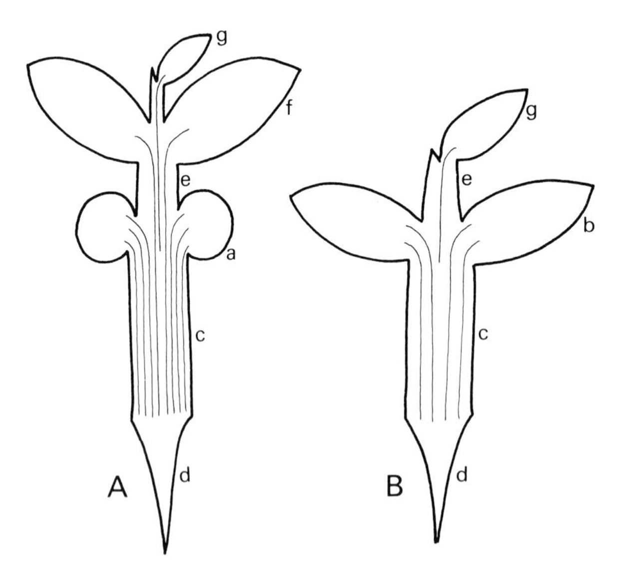

Fig. 20. Differences in vascular arrangement in the hypocotyl between seedlings representing: A. Sloanea type/subtype (2a) and B. Macaranga type (1) in *Diospyros* (Ebenaceae, schematic, after data by Wright 1904). – a. cotyledons; b. paracotyledons; c. hypocotyl; d. root; e. first internode; f. opposite first leaves; g. first of the spirally arranged leaves.

Diospyros embryopteris (Sloanea type/subtype) Wright counted as many as twenty-two groups of xylem resulting from the splitting of the original eight traces.

Interpretation of these facts, regarding the resting phases, anatomy and morphology, is not simple. If, in *Diospyros*, the Macaranga type would be ancestral to the Sloanea type/subtype and Blumeodendron type, quite a number of fundamental changes are necessary. Morphological and anatomical changes are needed for the differentiation of long-persistent assimilating 'cotyledons' into soon caducous, non-photosynthetic haustoria, in which palissade tissue is wanting. A third, median trace must develop from the then haustorial cotyledons which are non-functional outside the seed, and in most cases do not even emerge. In addition early development of the plumule must take place, before the second rest in the development of the stem, in combination with production of vascular strands from the first leaves which are prolonged through the hypocotyl to the primary root.

If the Sloanea type/subtype (or the Blumeodendron type) would be ancestral to the Macaranga type two possibilities for this change could be realised: a transition of the assimilatory function from the leaves to the cotyledons, or abortion of the cotyledons, the function of these organs being taken over by the epicotyl leaves.

In the case of transference of function of the epicotyl leaves to the haustorial cotyledons, morphological and anatomical changes are required for the differentiation of soon caducous haustoria into long-persistent assimilating 'cotyledons'. Palissade tissue must be formed in the embryonic stage, and in addition chlorophyll. The median trace of the cotyledons must abort. The differentiation of the terminal bud in the embryo must be delayed till after the second rest in the development of the shoot, and the first two leaves must become alternate instead of opposite. The vascular strands from the epicotyledonary leaves must abort in the hypocotyl.

If abortion of the haustorial cotyledons has taken place, the epicotyledonary leaves take over the position and food-providing function of the former. The necessary changes involved are complete abortion of the cotyledons together with their vascular strands in the hypocotyl. Position and development of the epicotyl leaves remain the same (if started from seedlings with opposite first leaves) and their photosynthetic function is maintained. The vascular strands of these leaves must separate over their entire length in the hypocotyl. Resting periods in the development of the axis remain the same.

An evaluation of the changes involved renders the derival of the Sloanea type (or the Blumeodendron type) from the Macaranga type as less acceptable, due to the many alterations which are needed. Derivation in the reverse direction, through transference of function, would require also an intricate pattern of changes which must then take place simultaneously. The simplest and least involved differentiation of the seedling is by abortion of the true cotyledons in the Sloanea type/subtype, and the epicotyledonary leaves acting as substitutes for these organs. Early development of leaves in the embryo occurs in *Diospyros* in those species where the cotyledons are soon aborted. Continuation of this process would lead to entire abortion of the cotyledons and thus the first leaves may acquire the position of the true cotyledons. Detailed anatomical research on *Diospyros* seedlings is however necessary to test this hypothesis.

In *Sterculia* (Sterculiac.) the seedlings represent the Horsfieldia type/subtype (7a), the Sterculia stipulata type (3), the Sloanea type/subtype (2a) and the Macaranga type (1). In this case too it is more plausible to accept abortion of the haustorial cotyledons which leads to the Macaranga type, than a change of the assimilating paracotyledons into haustoria, with early production of two opposite leaves as a consequence.

In seedlings of the genera *Guibourtia* and *Colophospermum* (Leguminosae/Caesalpiniaceae, see Fig. 21) a case of probable abortion of foodstoring cotyledons can be demonstrated. Léonard (1957) separated these genera partly on the basis of seedling characters and flower morphology. In seedlings of *Guibourtia* the foodstoring cotyledons become exposed and two large, opposite, ± orbicular simple 'leaves' are produced before the second temporary rest takes place. They are borne on a short first internode, and situated in the same plane as the cotyledons. The two 'leaves', which are in the same plane as the cotyledons, must be interpreted as two leaflets of the first compound leaf, of which the rhachis is (almost) aborted. In some species of this genus opposite to these two leaflets a minute caducous cusp is present. This has been interpreted as a reduced second leaf, which was opposite to the one still present. They serve a similar function as normal opposite leaves, and the seedling can be classified as belonging to the Sloanea type/subtype. The two-foliolate leaves produced later are all spirally arranged along the stem. In the genus *Colophospermum* no foodstoring cotyledons are found. Two large opposite leaf-like paracotyledons are present, borne ± at soil level, closely resembling the two first 'leaves' in *Guibourtia* seedlings. Also the morphology of the subsequent leaves, and their arrangement, is almost exactly the same as in *Guibourtia*. Interpreting the close resemblance between the seedlings of the two related genera as purely accidental is less acceptable in the light of the alternative possibility of abortion of the true cotyledons in the Sloanea type seedling of *Guibourtia*, which would result in the

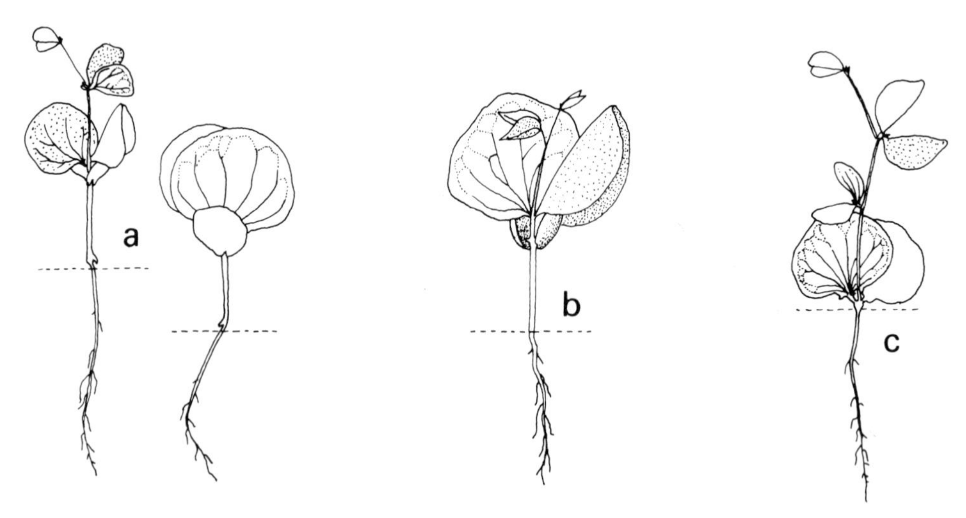

Fig. 21. Seedlings of the genera *Guibourtia* and *Colophospermum*. A. *Guibourtia conjugata*; B. *Guibourtia coleosperma*; C. *Colophospermum mopane* (Leguminosae/Caesalp., after Léonard 1957).

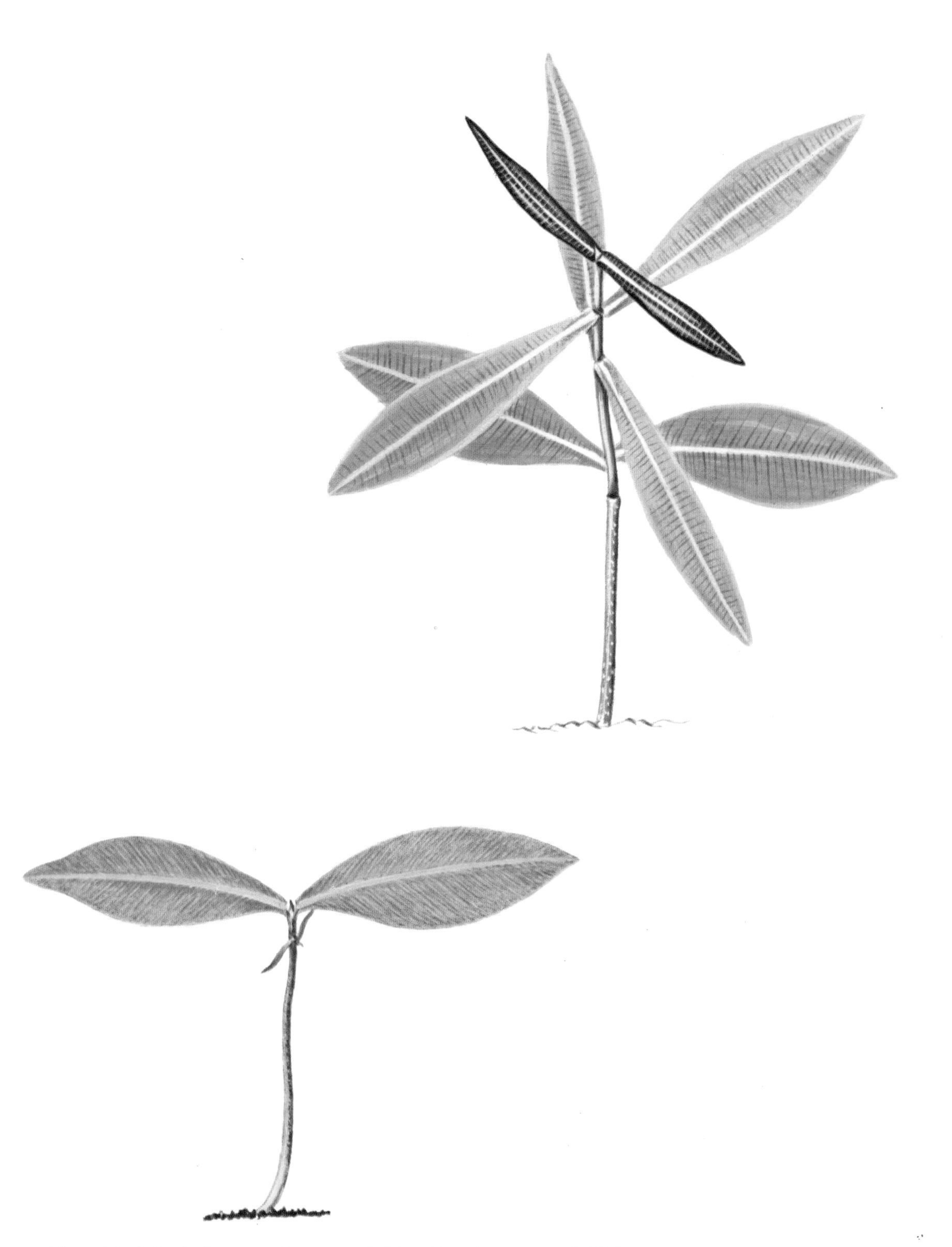

Plate 15. *Calophyllum venulosum* Zoll. (Guttiferae), De Vogel 1071.

Plate 16. *Barringtonia racemosa* (L.) Spreng. (Lecythidaceae), Bogor Bot. Gard. IX.D.36.

Macaranga type seedling of *Colophospermum*. The fact that reduction processes are present in the first leaves further supports the assumption of abortion of the cotyledons.

From the foregoing analysis it can be concluded that many arguments can be given that, at least in a number of cases, the Sloanea type/subtype is ancestral to the Macaranga type. The same argumentation can be applied to other genera in which both types are present (see Table 8), and also, although with slightly more reserve, to seedlings of different genera of one family in which both types occur.

Arguments against the hypothesis outlined above are, however, also present. It cannot be ignored that in some seedlings stomata are present in non-emergent cotyledons. This has been interpreted as a proof that such cotyledons have been emergent in a preceding period (Hill and De Fraine 1913, Grushvitskyi 1963). In Sapotaceae (Bokdam, pers. comm.) and Burseraceae, *Canarium* (Leenhouts. pers. comm.), Macaranga type seedlings are found in species in which the most primitive characters are present, whereas the Sloanea type/subtype occurs in species with more advanced features.

The genus *Canarium* has features which cannot be explained by the present hypothesis. In this genus foodstoring cotyledons as well as leaf-like paracotyledons are both palmate. In the Sloanea type/subtype the first leaves above the foodstoring cotyledons are entire. If the paracotyledons in the Macaranga type would be

Table 8. Quantitative seedling type distribution over different genera of some families (Malesian woody genera only).

Seedling type:	1	2a	3	6a	6b	7a	7b	8	9	11a	11b	11c	12	14
Anacardiac.	2	7		2	2									
Annonac.	4	1				3	2	3						
Apocynac.	9	1		2										
Bombacac.	3						1	1						
Burserac.	5	4		1										
Dipterocarpac.		5		1	2									
Euphorbiac.	24	2						1						
Guttiferae				1		3								1
Laurac.				9									2	
Legum./Caesalp.	9	17		1		3				2			1	
Legum./Mimos.	3	8				2				1				
Legum./Papil.	6	12		5		5				1	1			
Meliac.	5	4				3				1	4			
Morac.	4	1				2						2		
Myrtac.	5	2		1		1				1	1			
Oleac.	5	2		1										
Rhizophorac.	1								4					
Rosac.	3	3		2										
Rutac.	4	4				4								
Sapindac.	1	5				12								
Sapotac.	4	2		1							1			
Sterculiac.	11	?1	2	1		1		?1						
Verbenac.	5	2		2		1								

homologous with the first leaves in the Sloanea type, then it is hard to explain that they show a similar palmate lobation as the foodstoring cotyledons. Transference of function (and morphological characteristics), in which the property of an organ has shifted to another site, is well-known in plants (Corner 1958), but in his examples this shift is to organs placed lower on the axis, not higher. Moreover, *Canarium schweinfurthii* has palmate, leaf-like paracotyledons (a Macaranga type feature) and two opposite first leaves (typical for the Sloanea type). This feature may be used to support an argument that leaf-like paracotyledons can indeed change into foodstoring ones.

It is clearly not possible to draw a unanimous conclusion on the evidence available. To my mind it is, however, evident that in a number of cases the Sloanea type/subtype (2a) must be considered ancestral to the Macaranga type. A model can be constructed based on semophyletic considerations (Merkmalsphylogenie sensu Zimmermann 1930, phylogenetic derivations by means of characters of organs), in which the various seedling types are placed in morphological series. It must be realised that this is different from a phylogeny of existing taxa, which would represent a line of evolution.

The Sloanea type cannot be regarded as the basal type from which *all* other types have originated. It is too specialised. In addition the Sloanea type hardly occurs amongst the seedling types in the most 'primitive' taxa (see Table 4, p. 94). I feel more inclined to endorse the opinion of Grushvitskyi (1963), who considered subterranean germination with haustorial cotyledons in endosperm as characteristic for the first seedplants. Haustorial cotyledons in endosperm have also been suggested as the most primitive in Monocots (Boyd 1932). The type which best meets the requirements of primitiveness is the Horsfieldia type/subtype (7a) with haustorial cotyledons in endosperm. All other seedling types can be derived from this type by such processes as: shift in the foodstoring function from the endosperm to the cotyledons or the hypocotyl; reduction of the life-span of cotyledons up to complete abortion, the function of food-provision in the meantime being taken over by other seedling parts; differentiation in the method of emergence from the seed or fruit; variation in the length of the hypocotyl.

From the Horsfieldia type/subtype (7a, see the description on pp. 65–66) with haustorial cotyledons in endosperm, four pathways of differentiation emerge in which through relatively simple steps in the gross morphology other (sub)types originate (see Fig. 22). Some of these pathways are short, and lead to only one or a few seedling types. Other pathways are much longer, and some of the types represented may be the starting point of a new differentiation of the line into separate branches. Each of the four pathways are fully described, alternative ones are dealt with at the end of each leading group. For the sake of convenience the shortest leads are described first.

A. The cotyledons, situated at soil level, remain haustorial and endosperm is present in the mature seed and during seedling development. One cotyledon is withdrawn from the seed and becomes exposed, the other one remains covered, the seedling axis develops centrally between them (11c).

B. The cotyledons, situated at soil level lateral of the seedling axis, remain haustorial, endosperm is present in the mature seed and during seedling development, and the fruit wall remains persistent around the cotyledons, the endosperm and the testa (6a); the hypocotyl enlarges and carries the secund cotyledons enclosed in endosperm, seedcoat and fruit wall above soil level (6b, hypothetical for seedlings with haustorial cotyledons, but the equivalents are known from other pathways); the hypocotyl enlarges and becomes epigeal, the cotyledons enclosed in endosperm, seedcoat and fruit wall remain covering its tip, either in the soil or borne epigeally, where they remain blocking shoot development until they are shed (10).

C. The secund cotyledons remain haustorial and endosperm is present in the mature seed and during seedling development, the hypocotyl enlarges and carries the cotyledons enclosed in endosperm and testa above soil level (7b). From this type two different pathways emerge:

C1. The hypocotyl enlarges and becomes epigeal, the cotyledons enclosed in endosperm and seedcoat remain covering its tip, either in the soil or borne epigeally, where they remain blocking shoot development until they are shed (8).

C2. The hypocotyl enlarges and becomes epigeal, the cotyledons enclosed in endosperm and testa covering its tip, either in the soil or borne epigeally, in a subsequent growth period testa and endosperm are shed and the opposite haustorial cotyledons become exposed, while in this same growth period the plumule develops centrally between them, producing (almost always) a first internode with two developed leaves, further development is by production of a shoot with leaves (2a) (an alternative development is that the hypocotyl remains short and the free opposite haustorial cotyledons are borne ± at soil level, 2b, or only the testa is shed and the endosperm remains persistent on the abaxial, lower surface of the opposite cotyledons, 3); abortion of the true cotyledons, the hypocotyl enlarges and becomes epigeal, the paracotyledons enclosed in endosperm and testa covering its tip, either in the soil or borne epigeally, in a subsequent growth period testa and endosperm are shed and the opposite paracotyledons become exposed, further development is by production of a shoot with leaves (1).

D. The cotyledons, situated at soil level lateral of the seedling axis, change into foodstoring ones, endosperm is present in the immature seed, but is entirely resorbed or almost so during maturing of the seed, the seedcoat remains persistent around the cotyledons (7a). From this type with foodstoring cotyledons five different pathways emerge:

D1. The cotyledons remain foodstoring, covered by the testa, and are situated at soil level, hypocotyl and primary root do not emerge from the seed, the shoot which is produced creeps over the soil and roots on the nodes (15).

D2. The unequal cotyledons, situated at soil level, remain foodstoring, the smaller one opens like a lid from the larger, and both remain covered on the abaxial side by the persistent seedcoat, the seedling axis is central between them (11c).

D3. The cotyledons, situated at soil level, remain foodstoring. They are pushed aside by swelling of the cotyledonary node, the seedling axis centrally between them, the testa splits along the margins of the cotyledons to which it remains adhering on the abaxial side (11b); the hypocotyl enlarges and carries the opposite cotyledons with the persistent seedcoat which are situated opposite on either side of the seedling axis above soil level (11a).

D4. The cotyledons, situated at the soil level lateral of the seedling axis, remain foodstoring, the fruit wall remains persistent around the cotyledons and the testa (6a). From this type two different pathways emerge:

D4i. The cotyledons, situated at soil level, remain foodstoring, they are pushed aside by swelling of the cotyledonary node with the seedling axis centrally between them, the testa and fruit wall split along the margins of the cotyledons to which they remain adhering on the abaxial side (12).

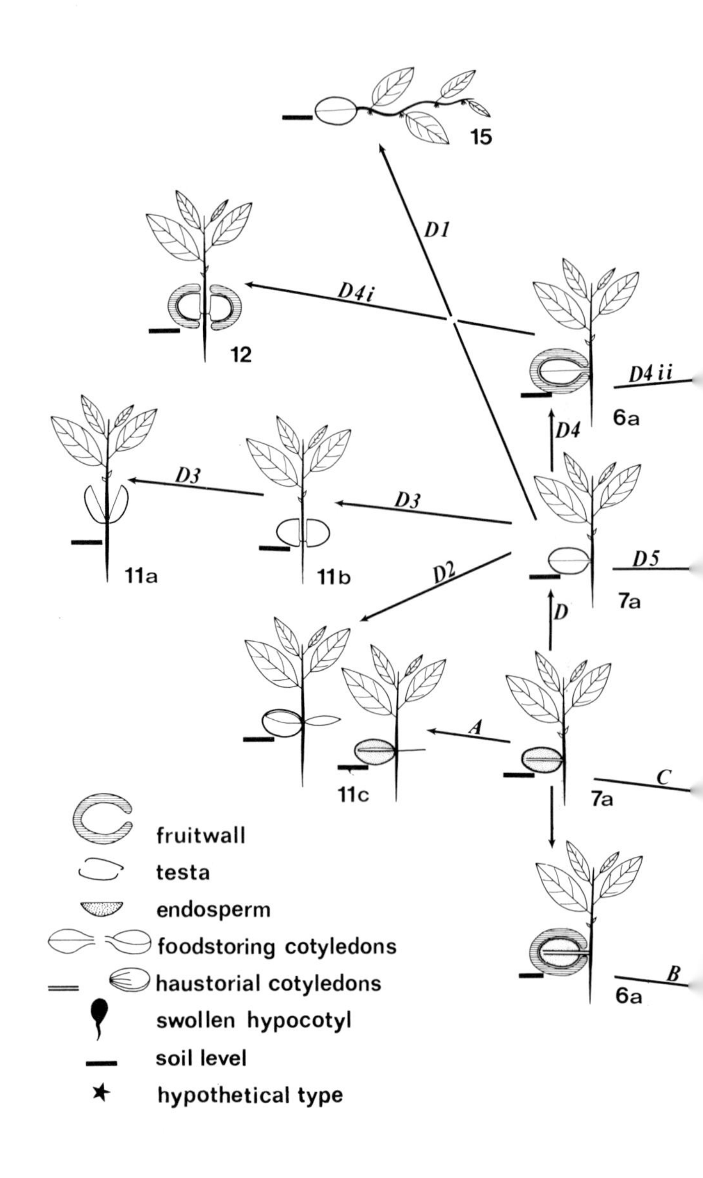

D4ii. The cotyledons, borne lateral of the stem, remain foodstoring, covered by the persistent testa and fruit wall, the hypocotyl enlarges and carries the enclosed cotyledons above soil level (6b); the hypocotyl enlarges and becomes epigeal, the cotyledons enclosed in the persistent testa and fruit wall remain covering its tip, either at soil level or borne epigeally, where they remain blocking shoot development until they are shed (10, hypothetical for seedlings with foodstoring cotyledons, but the equivalents are known from other pathways). From this hypothetical type two different pathways emerge:

D4iia. The fruit becomes viviparous, the cotyledons become reduced and more or

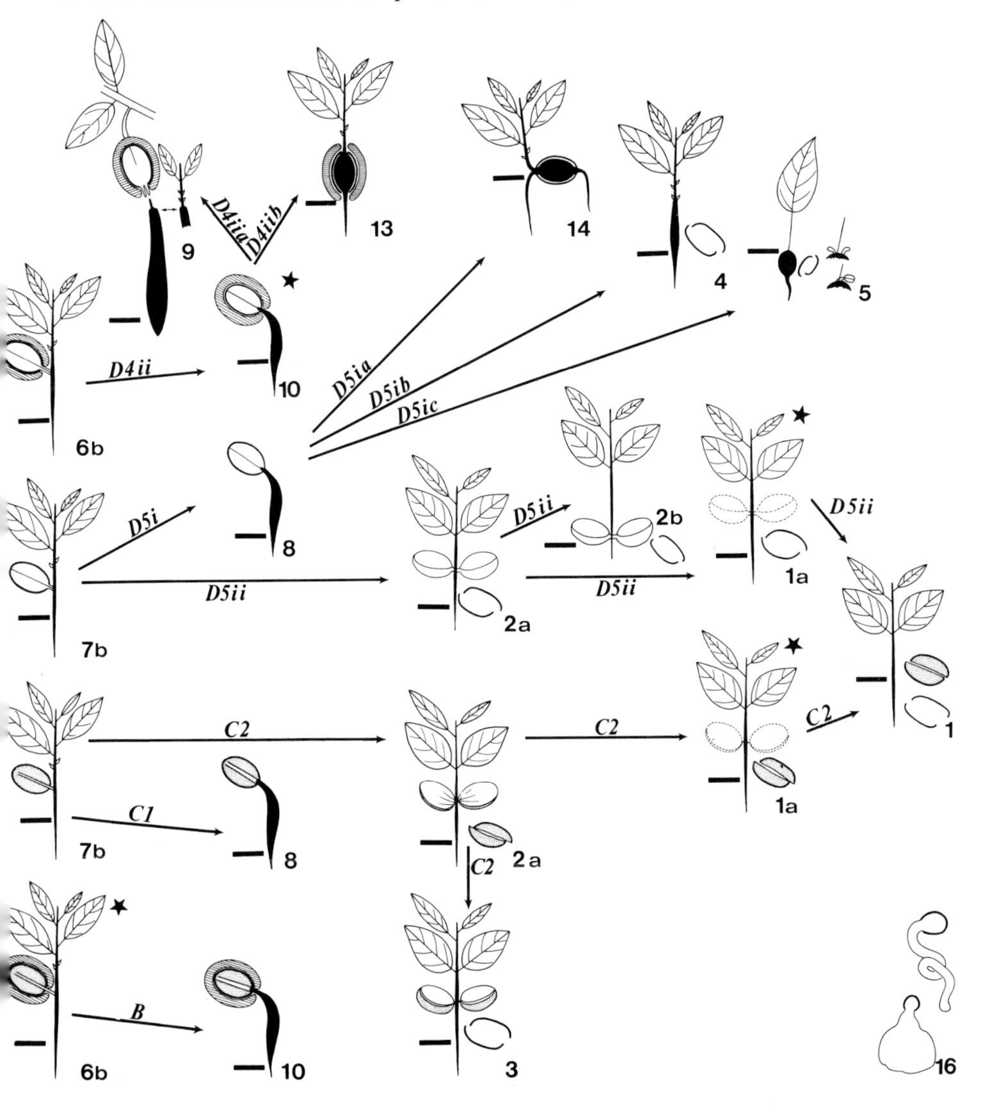

Fig. 22. Semophyletic model showing hypothetical pathways of derivation between the different seedling types and subtypes.

less fused, and during maturing of the fruit they serve to pass food from the parent plant to the hypocotyl which pierces the fruit wall and becomes exposed, and is either dropped from the fruit, or the fruit drops with the hypocotyl and remains covering its tip, blocking shoot development until the cotyledons with the coverings are shed (9).

D4iib. The cotyledons become almost entirely reduced or are aborted, a large amount of food is stored in the massive hypocotyl, the root emerges from one pole of the fruit, the shoot from the opposite pole (13).

D5. The cotyledons remain foodstoring, the hypocotyl enlarges and carries the cotyledons enclosed in the testa above soil level (7b). From this type two different pathways emerge:

D5i. The hypocotyl enlarges and becomes epigeal, the cotyledons enclosed in the testa remain covering its tip, either in the soil or borne epigeal, where they remain blocking shoot development until they are shed (8). From this type three different pathways emerge:

D5ia. The cotyledons become almost entirely reduced or are aborted, a large amount of food is stored in the massive hypocotyl, the primary root emerges from one pole of the seed, the shoot from the opposite pole, an accessory root develops from the top of the hypocotyl and soon overtops the primary root which ceases growth (14).

D5ib. The cotyledons become almost entirely reduced or are aborted, food is stored in the more or less massive hypocotyl, which becomes entirely free from the seedcoat, and produces a shoot with leaves without delay (4).

D5ic. The cotyledons become reduced or are aborted, some food is stored in the relatively small but somewhat swollen hypocotyl or root part. One leaf is produced which serves long as the only assimilating organ (5).

D5ii. The cotyledons remain foodstoring, the hypocotyl enlarges and becomes epigeal, the cotyledons enclosed in the testa covering its tip, either in the soil or borne epigeal, in a subsequent growth period the testa is shed and the foodstoring cotyledons become exposed, while in this same growth period the plumule develops into a first internode with two developed leaves, further development is by production of a shoot with leaves (2a) (an alternative development is that the hypocotyl remains short and the free foodstoring cotyledons are borne at soil level, 2b); the true cotyledons are aborted, the hypocotyl enlarges and carries the paracotyledons enclosed in the testa above soil level, in a subsequent growth period the testa is shed and the paracotyledons become exposed, further development is by production of a shoot with leaves (1).

The place of the Orobanche type in this scheme of derivations is not clear. The reduced nature of this type provides no clue to the morphologically closest type. Possibly this is the Macaranga type; the systematic position of the representatives of the Orobanche type makes this probable.

In the chart, only one pathway is indicated for the change from haustorial cotyledons to foodstoring ones, in the Horsfieldia type/subtype (7a). It is, however, imaginable that this change can also take place in the Horsfieldia type/Pseuduvaria subtype (7b), the Blumeodendron type (8), the Heliciopsis type/subtype (6a), the

Heliciopsis type/Koordersiodendron subtype (6b) and the Coscinium type (10). For the sake of clarity these pathways have not been indicated in the scheme.

The pathways of differentiation in the scheme are based on the assumption that if a feature has changed, it will not change back to its original state. Foodstoring cotyledons, once they have obtained this function, will not change back to haustorial ones, and in such seedlings functional endosperm will not return. If in the course of differentiation the cotyledons have become emergent, they will not become enclosed again in a subsequent type. Variations of the hypocotyl length, however, may occur in all types; in some this part may become almost aborted.

If this view is not accepted, several alternative pathways are possible, and then a number of types can be placed in other lines of derivation. The Endertia type/subtype (11a), for example, may be directly derived from the Sloanea type/subtype (2a) through persistence of the testa on the abaxial (lower) surface. This pathway may be continued to the Endertia type/Chisocheton subtype (11b) through reduction of the hypocotyl. In my opinion, however, the change of an organ is rather definitive, and a return to the original situation is less likely. Therefore alternative possibilities will not be discussed further.

Seedlings and taxonomy

In taxonomic as well as seedling literature occasionally remarks are found concerning the importance of seedlings for taxonomic considerations. Differences in seedling morphology have sometimes been a basis for re-arrangement of classifications. It has been assumed, sometimes tacitly, that in natural genera consequently the seedlings must be similar.

An explicit supporter of this view was Léonard (1957), who adduced many facts for it. In African Cynometreae and Amherstieae (Leguminosae/Caesalpiniaceae) he discovered that differences in blastogeny and morphology of the seedling are correlated with differences in the adult plants. This led him to the conclusion that in those groups 'the seedlings of all the species of one "good" genus have the same structure, or, in other words, only one type of seedling is predominant in each "Good" genus'. On this basis he developed four working hypotheses indicating the value of seedlings for classification in the groups he dealt with. These hypotheses, in the present work already mentioned on p. 17, as quoted from Léonard's summary in English, p. 294, read:

"1. The establishment of synonymy between genera according to morphological data should be provable by the similarity of their seedlings.
2. Morphologically related genera, which have the same seedlings, may not be generically distinct.
3. The partition of a heterogeneous genus into several genera according to their morphological characters, should be provable by the existence of a particular seedling type for each of them.
4. The existence of several seedling types within one genus may be an indication of a generic heterogeneity that must be checked by other morphological data."

The term 'morphological' was used by Léonard in the sense of 'characteristics of the adult plant', as distinct from characteristics of the seedling. The term 'heterogeneous' has been used in the meaning of 'artificial' or 'of polyphyletic origin'.

On the strength of correlations of characters of the adult plants with features of the seedlings, Léonard made extensive alterations in the taxonomic delimitation in the Cynometreae and Amherstieae, including the re-establishment of four 'old' genera, the description of seven new genera, and the reduction of eight genera. In broad lines this classification has been followed by Brenan (1967) for the Caesalpinioideae in the Flora of Tropical East Africa, as far as those groups occur in that area. The fact that Léonard's taxonomic ideas have been adopted by subsequent authors is an indication that, at least for these tribes, the proposed hypotheses on seedlings seem to be applicable.

A relatively large number of genera in woody plants have more than one seedling type (in Malesian woody Dicots 58 out of 453). Amongst these, the genera *Canarium* (Burseraceae, Leenhouts 1956, 1959), *Sterculia* (Sterculiaceae, Tantra 1976) and *Desmodium* (Leguminosae/Papilionaceae, Ohashi 1973) have been subject to recent revisions.

In the genus *Canarium*, subgenera as well as sections have been distinguished. According to Leenhouts and earlier authors, the genus is a natural entity, and the morphological differences between the subgenera and sections are not sufficient to raise these to the rank of genus. At section level the genus is diverse in the seedlings (Table 9, based on Weberling and Leenhouts 1966).

Table 9. Seedling types in the genus *Canarium*.

Subgenus *Canarium*	Sect. *Canarium*:	Sloanea type/subtype (2a)
	Sect. *Pimela* :	Macaranga type (1)
Subgenus *Africanarium*	:	Macaranga type (1)
Subgenus *Canariellum*	:	Seedlings not known

Weberling and Leenhouts rejected the propositions by Léonard that dissimilarity in the seedlings should always be the basis for raising subgenera or sections to the rank of genus, although they confirm that in *Canarium* differences in the seedlings coincide with sections.

Tantra (1976) revised the Malesian species of the genus *Sterculia*. He refrained from subdividing it into sections, because of the assumed reticulate affinities between the species, although he distinguished several groups without a taxonomic status based on natural affinities. He was very definite about the homogeneity of the genus, and there is no difference in opinion amongst recent authors that *Sterculia* is a natural entity. Although aware of Léonard's hypotheses on the importance of seedlings for taxonomic classifications, and the presence of four seedling types in *Sterculia*, viz. Macaranga type (1), possibly Sloanea type/subtype (2a), Sterculia stipulata type (3) and Horsfieldia type/subtype (7a), Tantra saw no grounds to raise the groups to generic level (pers. comm.).

The Asiatic *Desmodium* complex (Leguminosae/Papilionaceae) has been subject to a revision by Ohashi (1973). In *Desmodium* sensu stricto, seven subgenera were distinguished. Although the genus 'is an assemblage of considerably heterogeneous species-groups (it) cannot be divided into smaller genera due to the continuity in main characters'. Ohashi recorded hypogeal germination in the subgenus *Podocarpium*, and epigeal germination in the six other subgenera (Table 10); illustrations of seedlings were only given for sect. *Catenaria*, *Dollinera*, *Podocarpium* and *Sagotia*.

The subgenus *Podocarpium*, which Ohashi considered the most advanced group in *Desmodium* s.s., has never been given the status of genus by earlier monographers, although one of its sections (*Monarthrocarpus*) had been recognised as a genus by Merrill, based on the one-seeded pod. Ohashi pointed to other literature in which deviations are reported from the general rule that 'the mode of germination is fixed throughout a family or at least a genus', viz. Hara (1946, *Phaseolus*,

Table 10. Seedling types in the genus *Desmodium* s.s. (after Ohashi 1973).

Subgenus *Catenaria*	Sloanea type/subtype (2a)	(1 sp. illustrated)
Desmodium	Sloanea type/subtype (2a)	
Dollinera	Sloanea type/subtype (2a)	(2 sp. illustrated)
Hanslia	Sloanea type/subtype (2a)	
Ougeinia	Sloanea type/subtype (2a)	
Podocarpium	Heliciopsis type/subtype (6a)	(3 sp. illustrated)
Sagotia	Sloanea type/subtype (2a)	(2 sp. illustrated)

Leguminosae), Hara and Kurosawa (1963, *Rubia*, Rubiaceae), and Conde and Stone (1970, Juglandaceae), but seems to have been unaware of the publication by Léonard.

The difference of opinion among the various authors concerning the heterogeneity of seedlings within genera deserves some discussion. Léonard attributed much value to this feature, and was inclined to raise the various groups with different seedling types to the rank of genus, if additional morphological characters could be found. The other authors did not object to such a heterogeneity among the seedlings within a genus, and stressed the homogeneity of the group they dealt with. It is, however, noteworthy that in *Canarium* as well as in *Desmodium* at section level the seedlings are uniform as far as is known from literature.

The differences in opinion outlined above seem due to variation in the interpretation of the term heterogeneous. Léonard used it in the sense of 'artificial', or 'of polyphyletic origin'. In that case the groups of a heterogeneous genus have no natural relationships. Diversity of seedlings may then underline the differences between groups within such a genus, and be indicative of species or species groups of which the systematic position must be reconsidered. If, on the contrary, the genus is a natural entity, it may still be possible to distinguish subgeneric groups. In the case when different seedling types are present these are characteristic for e.g. subgenera or sections, and the heterogeneity expressed by the seedlings can not be used as a basis to split the genus up into new genera.

The seedling types distinguished by Léonard are all found in a number of genera, and even in different families. This occurrence is often not correlated with taxonomic relationships. No reason can be presented to presume that, when a heterogeneous genus is split up, 'this should be provable by the existence of a particular seedling type for each of them' (hypothesis 3). Léonard stated explicitly that he was 'still convinced of the pre-eminence of morphological (= adult) characters in systematics', but in fact he used characteristics of seedlings as diagnostic, and features of the adult plant as supporting characters.

The value of seedlings for systematic considerations varies largely with the taxonomic level at which they are used. Seedlings of one species are as a rule very uniform, although they may differ in such characteristics as speed of development, measurements of the parts, etc., as these may be influenced by environmental factors. Exceptions to this rule are known, however. For *Durio zibethinus* (Bombacaceae) Soepadmo and Eow (1976) recorded elongation of the cotyledonary petioles and development of the plumule into a stem with leaves, the covered cotyledons remaining persistent (Horsfieldia type/Pseuduvaria subtype, 7b). All other authors observed that the enclosed cotyledons remain covering the tip of the hypocotyl and shoot development only commencing when the cotyledons are shed (Blumeodendron type, 8). Knaap-van Meeuwen (1970) described a distinct, epigeal hypocotyl in two seedlings of *Maniltoa brownioides* (Leguminosae/Caesalpiniaceae), whereas in a third this organ was hardly developed and subterranean. The plantlets were of different origin, and no voucher specimens were indicated, which leaves their identity uncertain.

In the large majority of genera only one seedling type is present; for woody Malesian genera this is ± 87% (see list on pp. 78–92). In fact this percentage may be lower, because for most genera information on seedlings is limited to one or a few species only. Usually the features of congeneric seedlings are so similar that they can be recognised as belonging together, but seedlings of other genera may be very similar too. In genera where more than one seedling type is present (for woody Malesian genera ± 13%) the differences in morphological structure are often so large that congeneric identity is not obvious.

In a large number of families only one seedling type is known. For woody genera of Malesian Dicots this is ± 54%, but this percentage is evidently higher when also the herbaceous groups are taken into account, in which the Macaranga type (1) is the rule. This type is almost exclusive in families of the more advanced orders like Asterales, Campanulales and Lamiales, in which most species are herbaceous, and in the Rubiales (Grushvitskyi 1963). In the rather primitive Myristicaceae as far as is known to me only the Horsfieldia type/subtype (7a) is present.

Malesian Dicot families for which two or more seedling types have been recorded amount to almost 46% of the families when only woody genera are taken into account. In some families, like Leguminosae/Caesalpiniaceae, Leguminosae/Papilionaceae and Myrtaceae up to 5 and 6 seedling types are known. Table 8, p. 111, presents a list of quantitative seedling type distribution in genera of some Malesian Dicot families with woody representatives. Seedlings are not known from all genera in these families, but it is clear that in some families one seedling type is predominant (e.g. Euphorbiaceae), whereas in others several seeding types occur evenly distributed (e.g. Meliaceae).

At family level the seedlings are rarely so uniform that all can be recognised as being related. But sometimes morphological peculiarities are present in the majority of seedlings of one family which occur rarely in other families. In Bignoniaceae, as far as is known, always a more or less deep apical notch is present in the (para)cotyledons. In Burseraceae, lobed to palmate epigeal (para)cotyledons are present in a large number of genera (*Boswellia*, *Bursera*, *Canarium*, *Dacryodes*, *Garuga*, *Pachylobus*, *Protium* and *Santiria*), whereas in *Aucoumea*, *Crepidospermum*, *Scutinanthe* and *Tetragastris* the epigeal (para)cotyledons are entire (Guillaumin 1910). In *Haplolobus* the enclosed hypogeal cotyledons are plano-convex (Leenhouts and Widodo 1972). In Dipterocarpaceae the fleshy, more or less unequal cotyledons possess a typical two-winged shape in the genera *Anisoptera*, *Hopea* and *Neobalanocarpus*. Two-winged cotyledons are also present in *Shorea* and *Vatica*, where they may be either exposed, or remain in the fruit wall. In *Dryobalanops* the cotyledons are very unequal, both irregularly reniform and exposed. The cotyledons of *Dipterocarpus* are rather thin, much-folded, and remain in the fruit wall. A detailed study of cotyledon shapes in Dipterocarpaceae has been presented by Maury (1978).

Although the organs, and characters, of seedlings are limited in number, their diversity is so great that specific combinations of morphological characteristics may serve for the purpose of identification of seedlings. This has been demonstrated by Csapody (1968) and Muller (1978), who proved that it is possible to construct keys by means of which seedlings of almost entire regional floras can be identified; the first covering almost 1500, the second ± 1200 species. For seedlings in which the fruit wall and/or testa are persistent additional characters may be present in the shape, ornamentation or texture of these parts.

The conclusion is that the importance of seedlings for taxonomic considerations is variable, depending on the group under consideration, and that Léonard's hypotheses have no general validity. I do not see any objections against the occurrence of more than one seedling type within a genus.

Seedling ecology

Introduction

The seedlings of the various seedling types appear to have different preferences for types of vegetation, which is correlated with the amount and sort of light, the humidity, and changes in temperature at short intervals at soil level in the different biotopes. The Macaranga type (1) appears well adapted to open vegetations and to life on bare mineral soil. Herbaceous plants have almost always this type of seedling, as well as most pioneer shrubs and trees and many species from the secondary forest. In dense primary forests the Macaranga type is present too, but there it is less common.

Other seedling types seem much better adapted to the circumstances in the primary forest. To those categories belong all types which possess a persistent food reserve, like the Horsfieldia type (7), the Heliciopsis type (6), the Endertia type (11), the Sloanea type (2), etc. The majority of big-seeded trees, shrubs and climbers from the primary forest possess such seedlings, which do not thrive well in exposed places, but have on the contrary a much better chance to survive under dense cover than those of the former category. These types, which are in general much sturdier plantlets than Macaranga type seedlings, are more typical for woody plants.

The number of established seedlings in a vegetation is dependent on the supply of seeds which can germinate and establish in that environment. During the course of time such a stand of seedlings is subject to large losses, due to animal attack, suppression by other plants, and other environmental influences. Germination, establishment and speed of growth are specific, and may vary largely between the different seedling types and within one type amongst seedlings of different taxa. In some seedlings specific adaptations and protection devices are encountered.

Seedlings play a major role in the succession of vegetations. When they grow into maturity the plants may change the conditions in the environment to such an extent, especially in the early stages of the succession, that seedlings of the original composing species are not able to germinate. This results in the invasion of other seedlings, and consequently, on a long term, in a change of the composition of the vegetation.

Seedling type and vegetation

In each vegetation a characteristic combination of biotic and edaphic factors is present, which determines, to a large extent, the possibilities for seeds to germinate, and for seedlings to develop into maturity. Amongst these factors belong density of the vegetation, infra- and interspecific competition, cover of the mineral soil, humidity, temperature, light intensity and sort of light at soil level, and the variation therein over short and long time intervals.

In *open vegetations and on bare mineral soil* competition between the plants is of a low intensity. Fluctuations in humidity, temperature and light intensity are high. The amount of light is little influenced by the vegetation, and the complete spectrum of wavelengths can reach the soil. Species which are part of such vegetations, and which are able to germinate and establish, produce in the large majority of cases small seeds, in large quantities, which are easily dispersed over long distances. In general these seeds show a long viability and certain types of dormancy (see pp. 127–128) and sprouting of the seeds is mostly not simultaneous except after specific circumstances and calamities to the vegetation.

Seeds which germinate in such vegetations are mostly in the possession of only minor amounts of reserve food, which are either stored in the endosperm, or in the seedling tissues. Predominantly Macaranga type seedlings (1) are produced, with assimilating paracotyledons. Germination of the seed and initial establishment of the seedling is supported by the reserve food, which is ± exhausted after the seedling has produced the leaf-like, green paracotyledons. These organs are, for a prolonged period, the main food-producing parts, as photosynthetic activity of the hypocotyl is of minor importance, and in this initial phase the terminal bud remains at rest. When stem and leaf development takes place this means an enlargement of the photosynthetic area, because the paracotyledons remain persistent and functional till an often large and specific number of leaves has been formed.

In a closed vegetation, seeds which produce Macaranga type seedlings will as a rule not germinate. Seedlings of this type are usually small, and competition and overtopping by other plants will mostly result in their death. Almost all herbaceous plants produce this type of seedling, and it is also very common in woody representatives of secondary vegetations. In primary forest vegetations Macaranga type seedlings are usually much stouter, and less frequent.

In *closed primary vegetations* temperature, humidity and light conditions are very stable. The latter category requires special adaptations of the seedling to survive. Not only is the light intensity very low, in addition certain wavelengths of light are almost wanting, due to absorption by the canopy. Species from this type of vegetation produce in general relatively small amounts of seeds which are rather large. Special means for dispersal over long distances are usually not present, the seeds are often deposited under, or nearby, the parent plant. The viability of such seeds is usually short, dormancy is in general not present, and sprouting is ± simultaneous if conditions for germination are right.

Most seeds which germinate in this type of vegetation possess a more or less large amount of food, which is stored in the cotyledons, the endosperm, or more rarely the hypocotyl. A large variety of seedling types are produced: Sloanea type (2), Sterculia stipulata type (3), Ternstroemia type (4), Heliciopsis type (6), Horsfieldia type (7), Blumeodendron type (8), Coscinium type (10), Endertia type (11), Cynometra ramiflora type (12), Barringtonia type (13), Garcinia type (14) and Hodgsonia type (15). Some of these types are rare, whereas others are common. The food source is, in most cases, rather long persistent on the seedling, and supports the development of a strong root system and a stem with two or more developed leaves. If two opposite leaves are produced the cotyledons are soon dropped after the former have attained full size, after which the plant is dependent on the assimilates produced by these leaves. Further leaves are produced later, and the first leaves remain long persistent, e.g. in the Sloanea type (2) and occasionally in other types. In other cases the food source in the cotyledons or in the hypocotyl is much longer persistent, during which period a (much) larger number of leaves develop, in many cases preceded by a number of scales. Assimilates produced by these leaves form, in such seedlings, an additional food source which increases in importance, and finally replaces that of the cotyledons or the hypocotyl e.g. in the Horsfieldia type (7), Heliciopsis type (6) and the Barringtonia type (13). In the Blumeodendron type (8) and the Coscinium type (10) the cotyledons are dropped before a shoot is produced, the food is transferred first from these organs to the hypocotyl and this organ subsequently supports the development of a shoot and leaves.

Seeds of primary forest species germinate well under a closed canopy, and are able to withstand the low light conditions for a considerable period. In open places germination of such seeds is fair, but the adverse environmental conditions frequently affect the tender young parts, or even cause death. The seedlings are usually relatively large, which is of advantage in the case of covering by litter. They are rather well able to survive overtopping by other plants. For development in the sapling stage the plants are, to some extent, dependent on better light conditions, which occur through opening of the closed canopy.

Various *intermediate vegetation types* exist between the two mentioned. These represent succession stages. Each step in this succession changes the composition of the vegetation. From a condition with extensive fluctuations in light intensity, humidity and temperature a more stable situation results. In my experience, if a vegetation is hygrophilous and light-intercepting, then more seedlings occur which are dependent on their own food supply, and the proportion of Macaranga type seedlings is reduced. Although Macaranga type seedlings are also present in dense forest vegetations, and reversely seedlings with a food reserve occur in more open vegetations, an obvious preference of each category is present.

Specialised edaphic and climatic conditions may to some extent modify the above mentioned situation. *Lowland vegetations on limestone*, and *heath forest* on very poor soil, are usually low, much less light intercepting, and subject to larger fluctuations in humidity. In such vegetations Macaranga type seedlings occur in a higher percentage than in other lowland forest. In *monsoon forests* a large proportion of the trees is deciduous, and the light condition on the forest floor in the period they are leafless may facilitate development of Macaranga type seedlings, which are here more frequent than in dense forests in the everwet regions. A similar situation is present in the *temperate broad-leaved forests*; in the Netherlands a large percentage of the trees and shrubs are of the Macaranga type. Also in *beach forests* in the tropics a higher percentage of Macaranga type seedlings is encountered, although big seeds with a storage of food in the hypocotyl or the cotyledons are also common. The latter are able to germinate under high, as well as low, light conditions.

In *savanna regions* of tropical America Rizzini (1965) recorded predominant occurrence of epigeal seedlings for tree species, whereas hypogeal seedlings were more common in undershrubs. Taproot development was observed to precede shoot development by several months, especially in seedlings with a food reserve, obviously preventing loss of moisture in the vulnerable early stages of seedling development. Strong taproots prevail in this habitat, none was able to reach the ever moist soil strata in the first year, but in the second to third year most did. Die-back of the shoot took place regularly under natural circumstances, either from drought or from the regular fires.

Periodically inundated areas pose special problems for the seedlings to survive. In marsh forests regeneration by seedlings takes place during the irregular periods when the forest floor is dry. Fast development of a shoot with scales and leaves has been recorded for *Heritiera fomes* (Sterculiaceae, Heliciopsis type/subtype, 6a) and *Xylocarpus moluccensis* (Meliaceae, Horsfieldia type/subtype, 7a) by Troup (1921), and for *Mangifera gedebe* (Anacardiaceae, Heliciopsis type/subtype, 6a) where all leaves are developed. The striking length of the scaly stem part of the first two, 45 cm, was interpreted by Troup as an adaptation to the enviroment, which makes it possible for the seedling to continue assimilation when the area is flooded.

The *mangrove* is subject to daily inundation. All Rhizophoraceae genera from this vegetation have seedlings of the Rhizophora type (9), whereas non-mangrove genera of this family are of the Macaranga type (1). Fast development of the root system facilitates establishment of the seedling when it is washed ashore; sometimes the hypocotyl is 'planted' upright when it is dropped from the parent plant at low tide, and pierces the soft mud. In *Avicennia* (Verbenaceae, Ulbrich 1928, De La Mensbruge 1966) stiff hairs are present on the hypocotyl which secure anchorage to prevent the seedling being washed away. *Sonneratia* (Sonneratiaceae) produces numerous small seeds from which Macaranga type seedlings develop, which show no adaptations to the habitat.

Viability and dormancy

Some seeds retain their power of germination for a very long time, others lose their viability quickly. Species of the primary forest are notorious for their short viability. Many Dipterocarpaceae are known to lose their germination power within a week. Soepadmo and Eow (1976) recorded that sterilised *Durio zibethinus* (Bombacaceae) seeds in an airtight container remained viable for at least 32 days at 20°C. At 36°C they lost viability within one week. Unfortunately the retention of germination capability under natural circumstances, which is somewhere between these two figures, was not recorded. On the other hand, a long viability is recorded for many species from secondary vegetations. Such plants need a trigger for germination, or a germination block must be removed before they will germinate.

When seeds are disseminated they may arrive in a situation not suitable for germination. Seeds which possess dormancy may then remain resting in the soil until suitable conditions occur. In the soil of the primary forest numerous seeds of secondary forest plants are indeed present, waiting for the removal of the canopy to start germination. There must have been a selection pressure for a long viability in seeds with such characteristics. In former times secondary vegetations used to be restricted to relatively small areas, like landslips, windfall areas, and an occasional clearing (ladang). Seeds of pioneer species had often to wait a long time in the soil

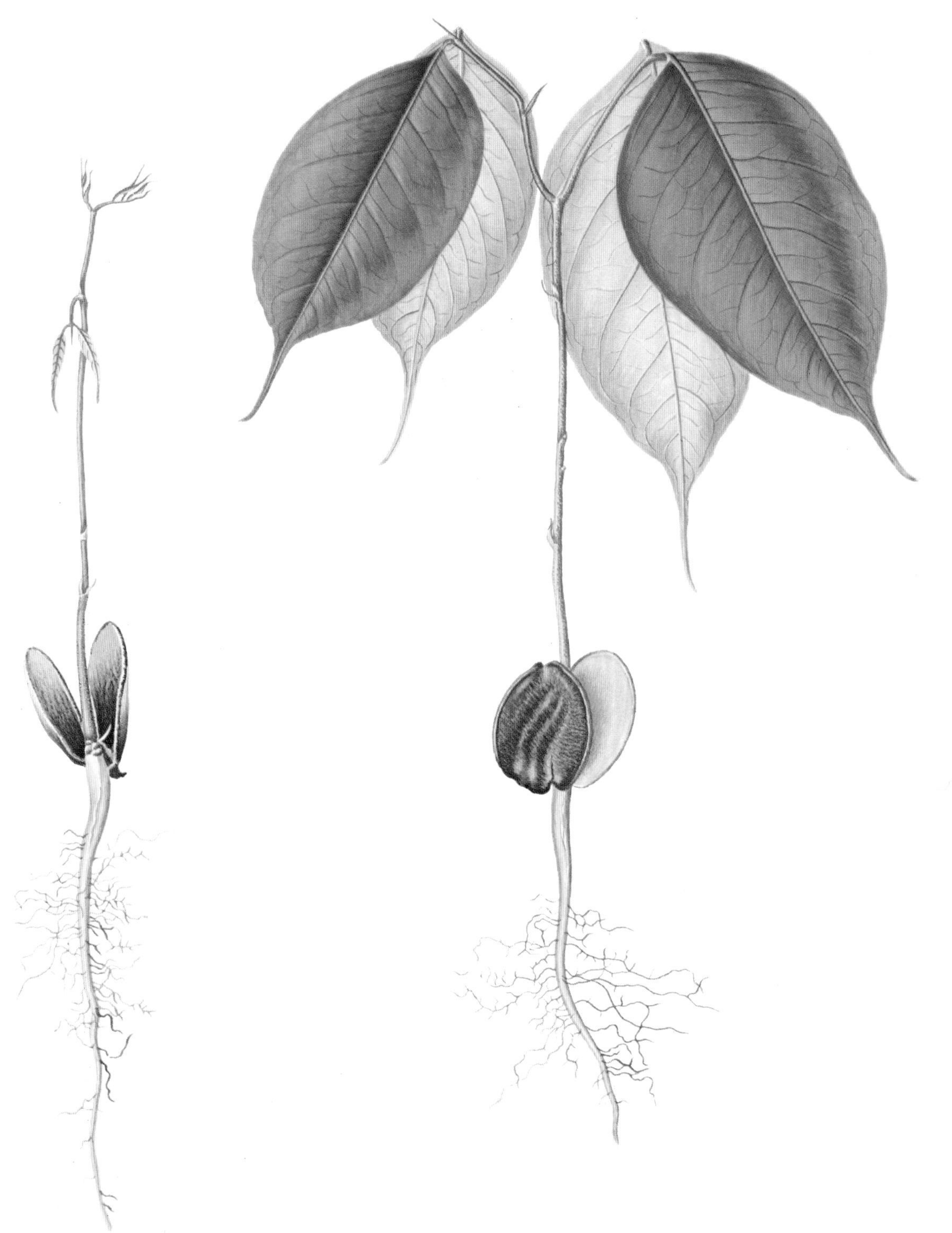

Plate 17. *Endertia spectabilis* Steen. & De Wit (Leguminosae/Caesalpiniaceae), De Vogel 1018.

Plate 18. a. *Aglaia sp*. (Meliaceae), De Vogel 932.
b. *Aglaia tomentosa* T. & B. (Meliaceae), De Vogel 804.

before they could germinate, and the species of which the seed survived longest had the best chance to reproduce this feature. A long viability is also found in many seeds from temperate regions, from monsoon areas, and in waterplants, in fact from all areas that are subject to a period with adverse conditions.

Ewart (1908) provided a long list, more than 175 pages, on 3000 tests on longevity of old seed collections, made by himself and based on literature. A small percentage of these tests concerns tropical seeds. He stated that the most resistant seeds show a pronounced decrease in germination power after 50 to 100 years. Further he speculated, based on known germination curves, that 'the probable extreme duration of vitality for any known seed may be set between 150 and 250 years (Leguminosae)'. Long-living seeds, which he named macrobiotic seeds, may last from 15 to over 100 years. Most of these are Leguminosae, while Malvaceae and Myrtaceae come next in importance. He attached a list including 181 species with macrobiotic seeds.

Dormancy is a state where a seed does not germinate under conditions which normally lead to germination. Sometimes, a more or less small proportion of the seeds of one crop may germinate, whereas the others remain dormant. This can be regarded as a positive feature for plants living in environments which are irregular in their succession of seasons, or for pioneer species. Sprouting of the seed is delayed until this is triggered off by an outward or inward agent. Seeds of some waterplants will not germinate, unless they have been dry for a certain period. Many plants of the temperate regions have seeds that must undergo a prolonged period of chilling, and in addition others must receive direct sunlight, otherwise they remain in rest. Fire is known to be the trigger for mass germination in several *Acacia* species, like *A. hebecladioides*, Leguminosae/Mimos. (Schnell 1971), *A. melanoxylon* (Villiers 1972), *Melaleuca leucadendron*, Myrtaceae, and many others.

Villiers (in Kozlowski 1972) discussed the recent literature on dormancy. He came to the conclusion that two types of dormancy exist. He reserved the term dormancy sensu stricto for: 'the state of arrested development whereby the organ or organism, by virtue of its structure or chemical composition, may possess one or more mechanisms preventing its own germination.' This is in contrast to the resting condition that is merely caused and 'maintained solely by unfavorable environmental conditions such as inadequate water supply', which he named quiescence. Secondary dormancy is a modified condition of dormancy that is imposed upon seeds when these are subjected to adverse conditions, by which a block against germination develops. This secondary dormancy is probably induced in the same way as normal dormancy, and it can be released by means of similar agents as normal dormancy is.

Several classifications have been made of the different types of dormancy. These are, more or less, all based on the same principles. Here the classification of Crocker (1916) is given, which is still largely in use.

1. The embryo is immature in the mature seed or fruit.
   This is known from e.g. *Buchanania* (Anacardiaceae), *Gironniera* (Ulmaceae) and others, where seemingly mature fruits are hollow, with only a very small embryo inside, which must develop into maturity before germination commences.
2. The seedcoat is impermeable for water.
   This is found in many Leguminosae seeds, and others. Gradual breakdown of the

testa results in germination which is spread over a long period. Economic treatment of this germination block is by immersing the seeds in boiling water, in sulphuric acid, or by mechanically damaging the seedcoat.

3. The seedcoat mechanically resists embryo growth.
According to Villiers (in Kozlowski 1972) this may not be a primary cause of dormancy; probably other factors are involved, and mechanical resistance usually occurs in addition to other factors.
4. Low permeability of the seedcoat to gasses, especially oxygen.
This may be one of the more important causes, but so far few investigations have been made.
5. Metabolic blocks in the physiology of the seed.
   a. Light-sensitivity.
   This has been proven to occur in seeds of many herbaceous plants (e.g. Van der Veen 1970). Germination of the seed may be stimulated or suppressed when seeds are subjected to different wavelengths of light, or even the photoperiodicity may have an effect. So far big-seeded primary forest species have not been checked on this feature. For a more detailed discussion on the mechanism of photo-sensitivity of seeds see next section.
   b. Requirement for chilling.
   Formation of gibberellins was induced by chilling in seeds of *Fagus sylvatica* and *Corylus avellana* (Frakland and Wareing 1962). Similar growth inhibitors and stimulators have been recorded in buds of mature plants. Comparable mechanisms with phytohormones possibly control the dormancy of seeds in e.g. monsoon climates.
6. Combinations of different causes of dormancy.
Indications have been found that often combinations of the above mentioned causes of dormancy play a role.

Light-sensitivity in seeds

Primary forest species normally produce seeds which are able to germinate in the forest under poor light conditions. Not only is the light intensity low, but in addition not all wavelengths of the light are transmitted in the same degree. Chloroplasts in the leaves absorb a high percentage of the red and the blue light, much more than far red, green and yellow light, thus altering the composition of the spectrum which eventually reaches the forest floor.

Stoutjesdijk (1972a) did some research on transmission of light in a number of vegetation types in temperate regions. In closed forest vegetations transmission of wavelengths between 400 and 700 nm is very low. More than 99% is absorbed by the canopy. Absorption of light over 700 nm is much lower, between 90 and 95%, depending on canopy density. The result is a 5–10 times higher proportion of far red light under the leaf-cover as compared with unfiltered light. In shrub vegetations transmission of light of both parts of the transmission curve is much higher: between 400 and 700 nm to 2.5%, in the far red light part of the curve increasing to 12.5 to 25%. In a germination experiment under a shrub vegetation he found inhibition of germination in some species which was absent in others. Seeds which require light for germination as well as those which require darkness can be strongly inhibited by light conditions under a canopy.

Transmission in tropical montane forests and tropical lowland forests have also been investigated (Stoutjesdijk 1972b, see Fig. 23). In these types of forest the ratio near-red/far-red is somewhat higher than in temperate forests. Still there is a large difference between the amounts of the different wavelengths of light that reach the forest floor. According to the curves published, the far red light portion of the curve is between 5–10 times higher than the visible light portion.

Seeds of many secondary forest species and weeds require direct sunlight for germination. Such seeds are named photoblastic seeds. Germination is regulated by a pigment, phytochrome, which is sensitive to different wavelengths of light. It can be present in two forms, a far red light absorbing form and a red light absorbing form. Increase of one form results in the decrease of its antagonist. When far red light is received the phytochrome equilibrium changes towards the red light absorbing form. The reverse occurs when red light is received. Phytochrome has a function in stimulation or inhibition of germination in seeds. The red light absorbing form of the pigment prevents germination, while the far red light absorbing form works as a trigger for sprouting.

The canopy of a closed forest vegetation transmits mainly the far red light. That

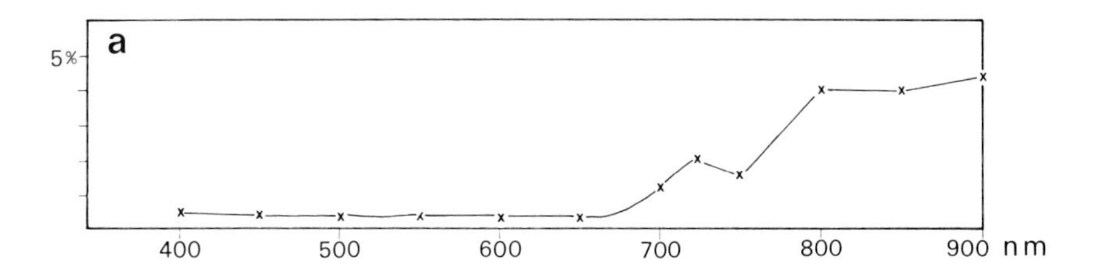

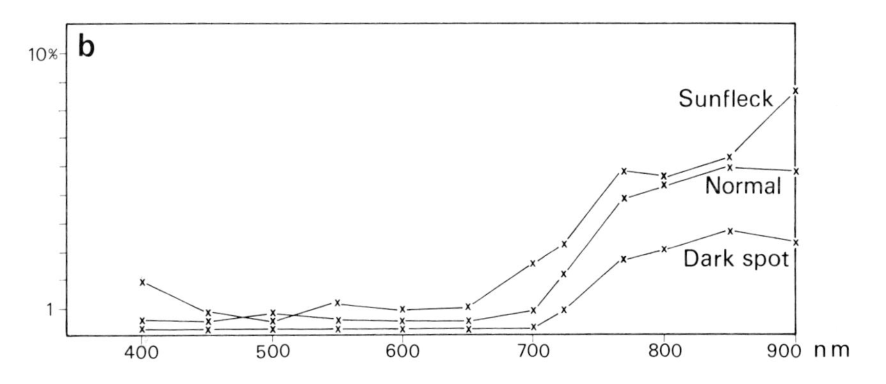

Fig. 23. Transmission curves of light in tropical forest vegetations, measured on the forest floor. – a. Pulau Peucang, W. Java, lowland forest slightly above sea level; b. Mt. Gedeh, Cibodas, W. Java, lower montane forest about 1450–1500 m altitude (after Stoutjesdijk 1972b).

causes the phytochrome in light-sensitive seeds to change into the red light absorbing form. Unless red light is received these seeds will not sprout. In forest vegetations this only takes place when the canopy is opened. The phytochrome will then turn into the far red light absorbing form, and germination of the seeds will commence.

Many of the secondary forest species, and also of other pioneer species like weeds etc. are clearly light-demanding, and will not germinate in darkness. Van der Veen (1970) drew attention to the fact that also seeds of species which are seemingly not sensitive for light and are able to germinate in light as well as in complete darkness, show, nevertheless, a certain reaction. If these seeds are put under a leaf cover or a canopy, or are subjected to far red light, they will not germinate unless red light is received. In a forest this situation occurs after canopy opening. Such seeds are then again able to germinate in light as well as in darkness.

Germination of seeds of primary forest species seems not to be influenced by the reception of either far red light or normal sunlight. To my knowledge no publications exist about experiments with primary forest seeds, but it is common knowledge that such seeds, especially the large ones, germinate well under a canopy as well as in the open.

In a compilation on germination of seeds and the influence of light, Van der Veen (1971) mentioned secondary dormancy caused by adverse light conditions in seeds which require red light for germination. When these seeds receive only far red light within a certain, prolonged period, they become entirely insensitive to light impulses, even if these are of the right sort. This insensitive situation continues for a very long period until the germination block slowly disappears. Essential for survival of the seed is that it remains in moist soil. This dormancy can be broken by means of damaging the seedcoat, strong drying of the testa and applying water again, or admittance of light.

Germination, establishment and speed of growth

A seed germinates when moisture, amount and sort of light, temperature and other factors are right. In germination, a certain sequence of steps is involved. The seedcoat absorbs water, it swells, and becomes more permeable for water, oxygen and carbon dioxide. Metabolism, which is in the resting seed extremely low, is stimulated by this process. Food reserves are mobilised by hydrolysis after the intake of water. These nutrients are transported to the regions of growth activity of the young plant. In seeds where the food is stored in the endosperm this is taken up by the growing plantlet, especially through the cotyledons or paracotyledons. In seedlings with foodstoring cotyledons export of food from the cotyledons takes place. The absorbed water also induces swelling of the seed contents, which results in opening of the seed. The radicle is then pushed out of the envelopments, and the early stages of germination are a fact.

In the stage of germination, establishment, and initial growth, the seedling is very vulnerable. Adverse conditions easily affect the young plant. This sometimes results in the entire loss of a crop of seedlings. A seedling has no means of active protection against adverse conditions. When such conditions occur the seedling is first hampered in its development, and finally dies.

Speed of growth of the seedling is in the earliest stages of germination mainly determined by the food contents of the seed and the genetic properties. Further development depends on the food reserve present in the seedling, and/or assimilates produced by the (para)cotyledons and the leaves. In addition it is often largely influenced by external factors, especially when the seedling has to rely partly or entirely on the food it produces through assimilation. The effect of light on seedling growth may be considerable, especially for plantlets with assimilating paracotyledons, but it may also have a large effect on seedlings with a long persistent food reservoir. Growth of *Kopsia arborea* (Apocynaceae) seedlings in the nursery was seen to be greatly influenced by the amount of light received. The plantlets were protected by a palm thatch roof and light from the east side was intercepted by trees. In the east side of the bed the seedlings were 2–3 times higher, those in the west side 4–5 times as big as those in the centre of the bed which received the least amount of light.

The genetic properties of a plant determine whether the seedling is a fast grower or not. *Terminalia copelandii* (Combretaceae) developed into a sapling 2.5 m high within two years. In that stage the sapling was in the possession of a number of sturdy lateral branches. *Mastixia trichotoma* (Cornaceae) produced in the same period only four leaves, and did not exceed 15 cm.

Seedling survival

Seedlings are vulnerable to external factors. A very high proportion dies in a young stage. Normally, the parent plant produces a more or less large crop of seeds. Immediately after germination the mortality of the seedlings is extremely high. In the remainder of the established seedlings this high mortality rate decreases gradually. The sounder the crop of seeds, the better the chance of the individual seedlings to survive. In an experiment with *Dalbergia latifolia* (Leguminosae/Papilionaceae) seedlings Eidmann (1933) demonstrated that the first germinated ones had a much better chance to survive than those which were somewhat retarded, the last germinated ones having a mortality seven times higher than those which germinated first. The first ones germinated were also for a much longer period resistant against adverse external factors.

Under natural conditions it is normal that a very high percentage of the seedlings dies in a juvenile stage. In the primary mixed tropical forests the continuation of a population of trees is through isolated saplings and pole trees. These are the remnants of often quite large numbers of established seedlings.

A very high death rate in seedling stands is quite natural. In monoecious species each plant of a population only has to produce one seedling in its entire life which grows into a mature seed-producing plant, to keep the population at more or less the same level. For dioecious species this number is two.

In most of the species the seedlings seem to 'take their chance'. Relatively large numbers of seedlings are produced during the life of the parent plant, and averagely only a very small proportion will reach maturity. Production of very large quantities of seeds in some species overcomes survival problems. This is especially the case with pioneer species and much less so in primary forest species. Sometimes a defence is present in the form of chemical compounds in the tissues of the plantlet which can be effective against predators. The majority of the seedlings have no special means of protection against adverse conditions, browsing, insect attack, damping off disease etc.

Large differences exist in the vulnerability and the possibilities of the seedlings to repair damage. Young seedlings of the Macaranga type are the most vulnerable. When the paracotyledons or cotyledons (either still covered or unfolded) are browsed off by rodents, usually the terminal bud is also bitten off, and the seedling is lost. During the present investigation this occurred with the collection of *Hydnocarpus polypetala* (Flacourtiaceae), of which almost all 80 specimens were thus killed in one night. Macaranga type seedlings are less vulnerable when they are further developed and have produced a stem with leaves. Browsing may then leave the lower portion of the stem intact, and sprouting from axillary buds may repair the damage.

Seedlings with exposed foodstoring or haustorial cotyledons are also very vulnerable. An attractive exposed food source is there present above the ground. This is often attacked by rodents, and may result in large losses in a stand of seedlings. In the Sloanea type/subtype (2a) and the Sterculia stipulata type (3) the cotyledons are relatively shortly persistent during the period they support the development of the first two opposite leaves. During this process the food reserve is used up, and when the cotyledons are dropped, damage to the top of the plant may leave the cotyledonary node unaffected, with the possibility of sprouting of the cotyledonary buds. The Endertia type (11) and the Cynometra ramiflora type (12) are in this respect slightly less vulnerable because the lower surface of the cotyledons remains covered.

In the Blumeodendron type (8), the Rhizophora type (9), the Coscinium type (10), the Ternstroemia type (4), the Barringtonia type (13), and the Garcinia type (14) the food reserve is stored in the hypocotyl, in the first three after transport from the cotyledons which are shortly persistent in a covered position. Attacks on the hypocotyl will mostly result in death of the seedling. However, in the nursery a specimen of *Durio zibethinus* (Bombacaceae, Blumeodendron type) of which the top of the hypocotyl was bitten off developed from the boundary of the cortex and stele a ring of accessory buds, several of which grew out into shoots. In the Barringtonia type and the Garcinia type the hypocotyl is protected by the envelopments, and situated at or below soil level, which makes it less vulnerable.

Damage to Blumeodendron type and Sloanea type seedlings may also occur when the cotyledons are not able to shed the envelopments, which remain clasping the hypocotyl tip. Wright (1904) recorded this 'suicidal method of germination' from several species of *Diospyros* (Ebenaceae, Sloanea type), which resulted mostly in death of the juvenile plant. The same was observed in several seedlings of *Mezzettia leptopoda* (Annonaceae, Fig. 38, Blumeodendron type) in the nursery. One or two weeks after the testa, endosperm and cotyledons normally would have been shed, accessory buds developed at non-specific places on the hypocotyl, of which one or a few grew into leafy shoots. One of these developed in the leader branch, and the entire seed body with the tip of the hypocotyl were dropped after some time.

Good protection of the food stored in endosperm or cotyledons is present in seedlings in which the enveloping fruit wall and/or testa remains persistent, viz. the Heliciopsis type (6), the Horsfieldia type (7) and the Hodgsonia type (15), especially in the typical subtypes of the first two where they are borne at or below soil level. Damage to the shoot leaves usually the lowermost, often small internodes intact, and sprouting of the cotyledonary buds or scale- or leaf buds results in replacement of the main axis. Damage to the cotyledons causes hampered development of the juvenile plant, but the seedling is usually able to survive. Dying back of the shoot through environmental circumstances occurs in several taxa (see also p. 44); this is replaced by a new shoot(s) from either the cotyledonary buds or leaf buds, which may take place several seasons in succession.

Special modifications of the morphology are known from several seedling types which provide a protected position for the plumule. In various families the hypocotyl of Macaranga type seedlings hardly develops, and the much elongated paracotyledonary petioles are connate in a tube resembling the hypocotyl (see also p. 37). The plumule is situated at the base of this tube, below soil level, well protected against damage. When it elongates it either travels up through the tube and appears between the paracotyledons, or it breaks through and emerges laterally. In three *Combretum* species, which show this feature, this has been interpreted as an adaptation to protect the seedlings against the regular fires in the savanna regions where they occur (Burtt 1972, Exell and Stace 1972). A similar case of protection of the plumule has been described in Horsfieldia type/subtype seedlings of *Marah* (Cucurbitaceae, Asa Gray 1877, Schlising 1969), where the fused cotyledonary petioles elongate and push the plumule deep into the soil. In the high altitude *Quercus semecarpifolia* (Troup 1921) the fused cotyledonary petioles protect the plumule until this starts elongating; die-back of the shoot occurs several years in succession.

The conditions in savanna regions are adverse for seedling development, and die-back of the shoot occurs regularly in dry spells and during fires. Rizzini (1965) recorded branching in the underground parts in several taxa, and a common occurrence of stunted growth of the shoot in seedlings and saplings. Hypogeal woody tubers are also produced.

Lignotubers, woody swellings commencing from the cotyledonary node, have been recorded in Australian *Eucalyptus* species exclusively from regions with an unstable climate, not from the moister areas (Kerr 1925). On the tubers large numbers of accessory buds are formed, which may develop into shoots if the main shoot is destroyed. Removal of the endbud, or checking root development was recorded to result in increase of the size of the tuber.

The seedling project

Collecting

Several collecting trips were made to areas with primary vegetation in Indonesia, to collect seeds for the study of seedlings. The seeds were grown in the Bogor Botanic Gardens at c. 250 m altitude, which excludes the possibility of survival of collections from above c. 1500 m.

In the everwet regions in Indonesia two fruiting seasons are present each year, one of which is rather pronounced, the other very modest. In monsoon areas fruiting is more or less confined to the wet period. Mast fruiting, a peak in the fruiting of many species, takes place once in a few years. The results of seed collecting can be very disappointing if it is not the real fruiting season.

Local variations in climate and ecological factors may influence the time of flowering and fruiting. Forests on similar soil, at the same altitude, some tens of kilometres apart, may have an entirely different rhythm of seasons. Forests on limestone may be in a sterile stage whereas the vegetation on nearby laterite is in full flower or fruit. Isolated trees appear to flower and fruit more profusely than the same species in undisturbed forest. Fruiting of upper storey species is usually confined to a period of a few weeks, whereas trees in the understoreys often have a less pronounced period, which is much longer.

Collecting was done in the customary manner, but for the living samples only mature or almost ripe fruits were taken, amounting were possible to 100 seeds. Specimens consisting of twigs, leaves and fruits are necessary for documentation. A sample was taken from the forest floor if seeds or fruits could not be collected from the tree; then the voucher of the fruits was completed, where possible with some fallen leaves. However, one should be aware of 'late fall' of unviable diaspores which sometimes occurs before the viable ones. Seeds or fruits from excrements or resting places of fruit-eating animals mostly germinate well, but identification is often difficult.

Damaged seeds may die and affect others; samples have also to be inspected for maturity, and apparent viability. Fruits collected from the tree may not be entirely mature, but the seeds may ripen when they are left in the fruit; a large proportion of near-mature seed eventually produced some seedlings. Differences in the vigour of seedlings in a seedcrop have been recorded by Eidmann (1933), see p. 131.

Collecting stations

The collecting stations are first in geographical, then in chronological order (Fig. 24):

Fig. 26. The work-shed in the nursery, where all descriptions and illustrations were made.

Fig. 27. View of the nursery. The beds are surrounded by bamboo (later cobbles) against erosion. The palm thatch roofs give protection against strong sunshine and downpours.

Fig. 28. View below the thatch roofs. Each collection is provided with its collection number on a bamboo label. Notice the difference in size of collections sampled and planted at about the same time.

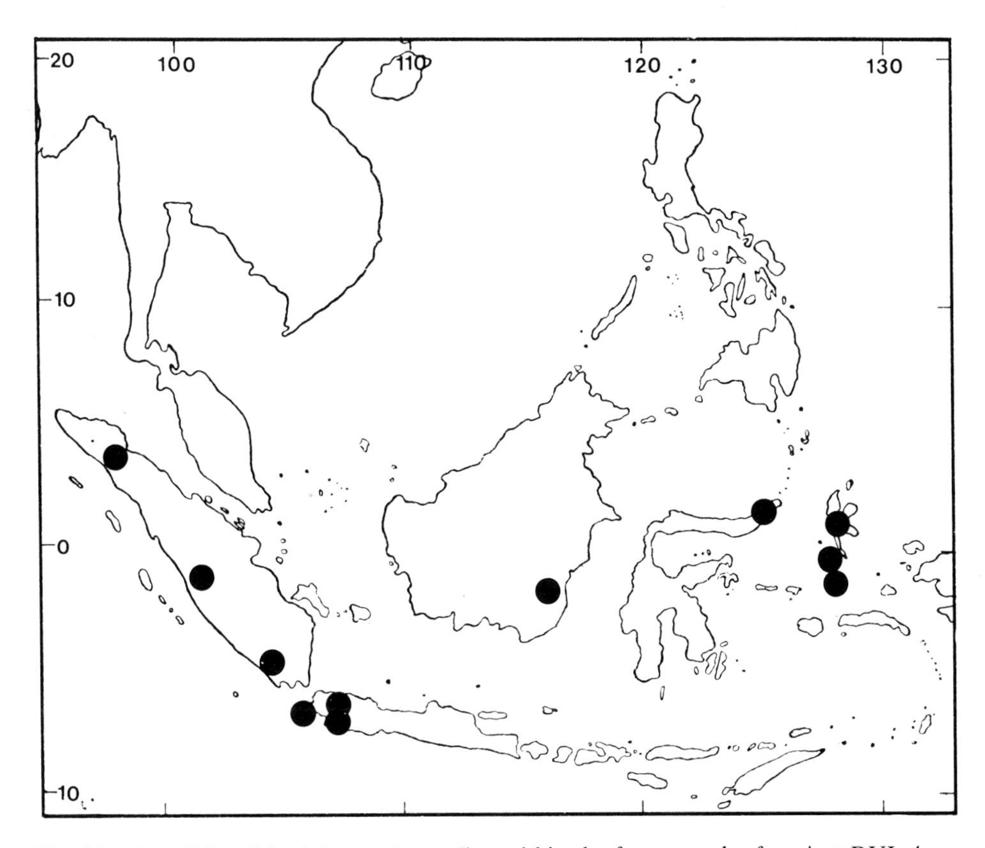

Fig. 24. Localities visited for seed sampling within the frame-work of project RUL 4.

South Sumatra, various localities along the Barisan Mountains between Seleman ·Enim, ± 3°58′ S, 103°48′ E, and Lake Ranau, ± 5°04′ S, 104°04′ E, March 7–21, 1972. Herbarium collections 1209–1348, 16 seed samples, of which 12 germinated.
North Sumatra, Sibolangit Botanic Gardens, 3°21′ N, 98°33′ E, July 9, 1972. Herbarium collections 4, 3 seed samples, of which 2 germinated. Gunung Leuser Nature Reserve, Ketambe, ± 3°34′ N, 97°46′ E, July 10–25, 1972. Herbarium collections 1412–1568, 59 seed samples, of which 14 germinated.
West Sumatra, various localities in the Barisan Mountains, Muaro, ± 0°36′ S, 100°59′ E, Air Sirah, 0°57′ S, 100°32′ E, between Tapa and Sungai Penuh, 2°03′ S, 101°18′ E, Rimbo Panti Nature Reserve, ± 0°20′ N, 100°04′ E, Febr. 21–March 21, 1974. Herbarium collections 2674–3012, 4 seed samples, of which 4 germinated.
West Java, Ujung Kulon Nature Reserve, mainland and Peucang Island, ± 6°44′ S, 105°15′ E, June 19–28, 1972. Herbarium collections 1363–1411, 40 seed samples, of which 29 germinated.
West Java, Ujung Kulon Nature Reserve, mainland and Peucang Island, ± 6°44′ S, 105°15′ E, Aug. 25–Sept. 3, 1972. Herbarium collections 1569–1584, 7 seed samples, of which 5 germinated.

Southeast Borneo, Meratus Mountains near Muara Uja, ± 1°50′ S, 115°41′ E, Nov. 1–Dec. 5, 1971. Herbarium collections 700–1199, 20 seed samples, of which 12 germinated.
Southeast Borneo, Meratus Mountains near Muara Uja, ± 1°50′ S, 115°41′ E, Oct. 10–Nov. 15, 1972. Herbarium collections 1585–2142, 30 seed samples, of which 21 germinated.
Southeast Borneo, Meratus Mountains near Muara Uja, ± 1°50′ S, 115°41′ E, March 3–21, 1973. Herbarium collections 2181–2397, 155 seed samples, of which 78 germinated.
North Celebes, various localities in Minahasa and Bolaang Mongondow: Mt. Lokan, Batu Angus Nature Reserve, Mt. Klabat, Mt. Soputan, Mt. Ambang, Oct. 1–Nov. 5, 1973. Herbarium collections 2431–2670, 110 seed samples, of which 74 germinated.
Northern Moluccas, various localities on the islands Halmahera, between 0°30′–1°10′ N and 127°30′–128°00′ E, Bacan, ± 0°45′ S, 127°30′–127°45′ E, and Obi, 1°20′–1°30′ S and 127°25′–127°45′ E, Sept. 17–Dec. 14, 1974. Herbarium collections 3014–4537, 191 seed samples, number that germinated not known.

From trees growing in the Bogor Botanic Gardens over 60 samples of seeds were obtained, of which ± 85% germinated. Seeds were also received from Mr. Dedy Darnaedi, Dr. J. Dransfield, Mr. R. Geesink, Dr. A. J. G. H. Kostermans, Mr. Sukasdi, Mr. Trimulio, Dr. J. F. Veldkamp, and Dr. W. J. J. O. de Wilde, from various localities. Most samples were Malesian, but some collections were from Thailand and Ceylon.

Voucher collections are marked by a note: SEEDS FOR CULT. TO BOGOR. The material will be deposited in the Herbaria: 1. BO, 2. L, 3. KEP, 4. K, 5. LAE, 6. CANB, 7. SING, 8. A. A herbarium of cultivated seedlings was made only in two sets, which bear the same collection number as the vouchers. They will be deposited in the herbaria of Bogor and Leiden. Seedlings in several stages of development were preserved where possible. Part of this material has recently been used for an anatomical study in which characters of (para)cotyledons, juvenile leaves, and leaves of the adult plant are compared.

Storage and transport

Seed collecting expeditions should not exceed one month actual collecting; if of longer duration, loss in viability in the first collected samples may exceed accessions of new collections. Samples can be stored in plastic bags, which facilitates inspection. Small holes in the bags are needed for air circulation. The moisture regime in the bags can be regulated by wrapping the seeds in moss, which also protects them from bruising and facilitates germination. Samples must be kept moist but not wet; free water is harmful.

To keep tags with collection numbers legible, the best method seems to write the collection number on the sticking side of transparent tape, and fold it with the adhesive sides together. As local material, bamboo chips c. 4 by 1 cm can be used, with the number inscribed with hard pencil. In addition the number must be written on the outside of the bag with waterproof felt-pen.

Seed samples must be cleaned of the fleshy parts. The complete fruits must first be stored. Many will open during storage, after which the fruit wall can be discarded, and the healthy seeds selected. In quite a number of taxa the seeds are dispersed while still enclosed in (part of) the fruit wall, which is shed during germination (e.g. *Canarium*, Figs. 51, 52), or (part of) the fruit wall remains persistent even after germination (e.g. *Eusideroxylon*, Fig. 102, *Barringtonia*, Fig. 13, p. 73). In such fruits the fleshy parts will decay by itself and can be brushed off; the persistent part must remain intact if the natural mode of germination is to be studied.

Fruit and seed characters may supply additional data for the construction of future keys to seedlings, but for practical reasons they have been left out of the descriptions. These characters can be found in taxonomic literature; in seedling literature they have been consequently included by Ng (1975, 1976).

Samples must be regularly checked for moisture and for dead, molded or germinating fruits and seeds. Seeds of plants from secondary vegetation can in general be kept longer than those from primary vegetations. Hardening of testas can be prevented by keeping the seeds moist. Grown-up seedlings collected in the forest survive well when several together are stored in long plastic bags, with the roots packed in moist moss. Leaves should be dry as water on the leaves will cause them to stick together and anaerobic rot will follow. Seedlings thus packed can survive for weeks, even if many bundles are stored on top of each other in a closed box. Airmailed seed samples must be kept under passenger conditions; this should be noted on the cover, and the recipient be notified, with flight number, by cable.

Nursery practice

The seeds were planted in a nursery in the Bogor Botanic Gardens which was made in a vacant lot and was shaded by some trees. The annual rainfall is c. 430 cm; the wettest month is January with 30–50 cm, the driest is August with 15–30 cm. In the four driest months 20 or more rainy days occur. Water was available nearby in an artificial pond and in the Ciliwung river.

The nursery is situated on the flat land near the river. The soil consists of reworked laterite, on top of the former river bed which is composed of big boulders, pebbles, and sand. Soil structure is fine-grained without stones, with a good crumb, and excellent drainage. Soil fertility is rather low, with a high deficiency of phosphorus (G. J. L. Muller, pers. comm.).

Soil processing started with removing the vegetation. The soil was dug up in three layers, each 15 cm deep, over the entire width of the field, in stretches 90 cm wide. Three adjacent trenches must be dug out, 45 cm, 30 cm, and 15 cm deep respectively. Of each step a 15 cm deep layer is dug out, crumbled and freed from root parts, and moved to the first trench, as shown in Fig. 25. This procedure is continued to fill trench 2, etc. until a required area has been covered, after which the remaining trenches are filled with the earth from the first ones. Manure or fertiliser was never applied to the soil; signs of mineral deficiency have never been observed in the growing plantlets. The beds were 90 cm wide, laid north–south, separated by paths 35 cm wide.

Protection from sun and rain, not sufficiently given by the trees overhead, was

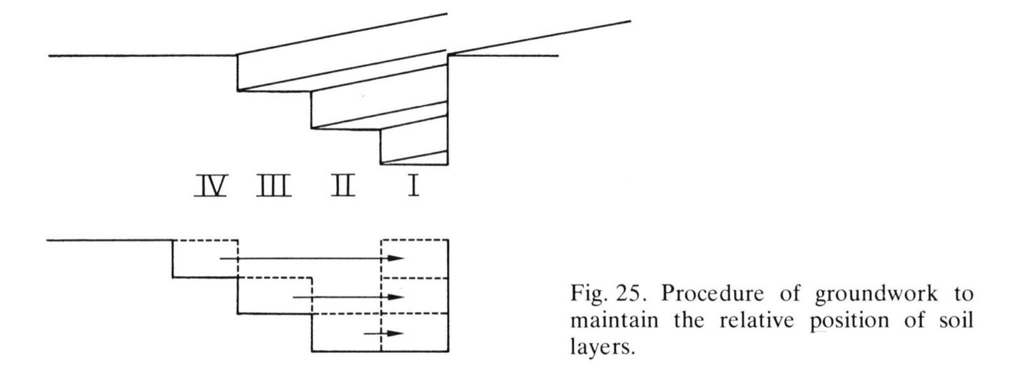

Fig. 25. Procedure of groundwork to maintain the relative position of soil layers.

provided by placing thatch roofs (atap) over the beds made of the leaflets of palms (Fig. 27). The roofs were tilting from 1 m height at the eastern side to 0.5 m at the western side; in the afternoon the shade of adjoining beds overlapped. The soil of the beds was kept in place with lengths of bamboo, later with rows of stones placed closely together (Fig. 28).

Weeding had to be done at least every two weeks. Watering was necessary after 5–7 days of drought, and then daily. Distances of 10 cm between the seedlings sufficed, in most cases, for two years' development. If longer periods of observation are foreseen, or with sturdy plants, 15 or 20 cm should be used. It is essential to plot all beds with the location of the samples and their number on a map, because labels will rot (bamboo) or disappear. Plastic labels are advised.

As far as possible 100 seeds were planted. Seeds were pressed halfway into the ground, thus achieving good contact with the soil, and still allowing the early processes of germination to be observed.

Hard-coatedness of seeds was never treated except in the cases of dry-stored seeds which were received, in which the seedcoat was cut; other methods used in commercial nurseries include immersion in boiling water, treatment with high concentrations of sulphuric acid, or, for small seeds, ponding with coarse sand. According to White (1908), the impermeable portion in small and medium-sized seeds is the cuticle, in large seeds the palissade cells.

Germination results of the c. 650 seed samples was over 60% for samples taken on expeditions, c. 85% for samples collected in the Gardens. In general the lowest germination percentages were in samples which were stored for long periods.

Occasional losses of seedlings occurred by damping off. In *Hydnocarpus polypetala* (Flacourtiaceae) in one night almost all hypocotyls were bitten off, presumably by rodents, in *Artocarpus lanceifolius* (Moraceae) almost all cotyledons were bitten off. In *Afzelia javanica* (Leguminosae/Caesalpiniaceae) the black seeds with the bright red, hard aril were once scattered all over the nursery by animals, but they were not damaged. Termites never attacked living plants in the nursery, but sometimes they damaged the hard remnants of the fruit walls, thus more or less exposing the normally hidden cotyledons or hypocotyls, and they heavily attacked bamboo constructions and thatch roofs.

Grown-up seedlings were gradually exposed to light when 50 cm high. Chlorosis as a result of too sudden exposure occurred in *Blumeodendron tokbrai* (Euphorbiaceae). Removal of the thatch roofs resulted in silting up of the soil after rain, and subsequently in erosion by the strong surface runoff.

Personnel in the nursery amounted to one botanist, one technical officer (later a second one joined for training), and two labourers, with occasional assistance from two other labourers for preparation of new beds. One draughtsman had a full-time job in making the illustrations, having a busy time shortly after each expedition when many collections started germinating. The nursery expanded over an area of c. 1000 m^2, including paths, a work-shed, etc., the effectively planted area covering 600 m^2. Empty places in the beds through loss of samples were gradually filled in with new incoming collections.

Additional data on nursery techniques in the tropics can be found in Barnard (1956) and Van Hall and Van de Koppel (1946).

Contents and limitations of the descriptive part

Notes to the descriptions

Due to the limitations in the knowledge of seedlings, it is impossible to select diagnostic characters, because of uncertainty on the possible variation between specimens of one population, of different populations, between species, and within one genus. In addition many genera, in Malesia possibly half of all woody genera, have not yet been described in the seedling stage. A reliable diagnosis of a seedling is only obtained if it is based on a combination of characters, as there are limited characters available in each seedling part. Therefore extensive descriptions are given, from which the essential characters can later be chosen for the construction of keys.

Changes in the morphology of the seedling may take place in relatively short intervals, so constant observation is required. Especially in the very juvenile stages the plantlets must be checked at least once a week. In certain stages temporary rests may occur in the development of the epigeal parts. Seedling growth may also be gradual or in flushes. Organs may be long persistent or not; their disappearance is often correlated with certain development stages.

To make comparison easy, each part is described in its main stages. For descriptions and illustrations, always one of the larger plants was selected. Since some variation is always present, several plants were checked for differences. All parts were described, of full-grown ones the measurements were taken, and the new situation was checked with the description previously made. Descriptions, all from living specimens, are based mostly on more than one seedling because replanted seedlings usually developed less well.

Notes to the illustrations

All illustrations were prepared from living specimens. With small collections the seedling was left in the soil, when more seedlings were present the one selected for illustration was lifted from the soil, in order to draw the root system. Retarded development after replanting did usually not permit using the same specimen again for illustration.

All parts were measured with measuring divider and drawn to scale. The leaves were mostly drawn flat to show the shape, in several illustrations the undersurface of one leaf is shown. Soil level is indicated by a horizontal line. A scale in centimetres, on which the soil level is indicated, is given for each illustration.

The line-drawings were made on thin typing paper, and later copied on thick drawing paper and inked. Also the watercolours, 33 in all, were first made on typing paper. To obtain a brilliant and natural looking shine of the leaves these parts were first filled in with white paint, covered with the required colours, after which the covering paint was washed away. With the Winsor and Newton paints used almost exact colour shades could be obtained. Unfortunately the high costs of printing did not allow to include all watercolours in this book.

The development of a seedling into a sapling is in most cases adequately documented by three or four illustrations. The first picture shows the mode of germination. Epigeal seedlings in a later stage are depicted with the (para)cotyledons extracted and fully developed, sometimes also with the first few leaves. Hypogeal seedlings in a later stage are illustrated with the shoot partly developed, showing the scale-leaves if present and preferably also some of the developed leaves. The third and fourth pictures show stages of the young sapling, usually with a relatively large number of leaves, and often with the (para)cotyledons and some of the lowest leaves already shed. In some cases it was necessary to provide more illustrations, or details are added, which give additional information. If possible the species possessing compound leaves in a later stage, or attaining a climbing habit, are also drawn in that stage.

The legend to the figures gives, besides the name of the species, the collection number (the abbreviation 'De V.' means De Vogel), dates of collecting and planting and (except at the beginning of the project) that of drawing. The latter information gives an idea about the rate of development of the seedling in the nursery, but it must be kept in mind that conditions and speed of growth in the forest may be different, and that seedlings from the same parent may differ much in vigour.

Presentation of results

The order of taxa is alphabetical. Names of subgenera are only provided where different seedling types occur within one genus.

Literature references are given under the genus, and can be found under the first species treated. The first reference is to literature on the adult plant: Flora Malesiana if the genus has been treated in that work, or, if not, a more local flora like Flora of Java, or the Tree Flora of Malaya. In the case where the genus is not treated there, the latest revision is given. The local floras do not always include the species of which the seedling is here described.

All further literature references are to seedling descriptions, arranged alphabetically according to author. The species considered are recorded for each reference. It has been tried to be as complete as possible, and also non-Malesian references have been included in order to give an overall picture on the state of knowledge on the seedlings of a genus. If a seedling of a certain species has been described more than once, all references are given.

The references are followed by the distribution of the genus in its entire area, this being considered from West to East, starting with Africa for pantropic areas. Distribution is not given in detail, except for the Malesian region. Partial areas indicated are: Africa (temperate and tropical North and South, Central), Europe, Middle East, India, Ceylon, Southeast Asia (including Burma, Thailand, and the entire Indo-Chinese peninsula), East Asia (including tropical and temperate China, Taiwan, the Ryukyus, and Japan), the Malesian area (with indication of the main islands and island groups), the Pacific (for which is indicated how far North, East and South the genus occurs), and America (temperate and tropical North and South, Central).

Information on the described seedling starts with the name of the species, followed by the number of the figure. Under *Development* the sequence of development of the parts is described by which the seedling emerges, supplemented by the occurrence of resting stages and other growth details. Under *Seedling* general information is given on the germination type and it is indicated to which type the seedling belongs. This is followed by a detailed morphological description. All parts of the seedling are described in the sequence from the root to the growing tip of the stem. Two stem parts have been described separately from the rest of the stem, viz. the hypocotyl and the first internode. These parts may have a special method of development, and are often different from other parts of the stem. Emergentia are described under the organs or parts on which they occur. Derived organs can be found under the part from which they are derived.

Reliability of identifications

In many families taxonomy is largely based on flowering material. Often fruits are under-represented in the herbarium, sometimes even of well-known and common species. In many cases this makes naming of fruiting plants difficult. Identification of the vouchers was done by Mr. Nedi, who compared the plants with the Bogor collections. For a final check the herbarium material was identified with the aid of a modern revision where existent. The most problematic families were Guttiferae, Myrtaceae, and Myrsinaceae.

If no voucher of the tree could be obtained it was tried to identify the collection by means of characters of the fruit, seed and seedling. Occasionally this was adequate to identify the material down to the species, sometimes the genus name could be found, but, more often, it was not even sufficient to determine the family. The description of the seedling was included in this book if no doubt existed with regard to the identity at genus or species level.

Plate 19. *Dysoxylum sp*. (Meliaceae), De Vogel 1028.

Plate 20. *Heliciopsis velutina* (Prain) Sleumer (Proteaceae), De Vogel 1074.

References

Adanson, M. 1763. Familles des plantes. Vol. I: cccxxv, 189 pp.; Vol. II: (24), 640 pp.
Agardh, C. A. 1829 – 1832. Lärobok i Botanik, etc. 2 Afd. Thomson, Malmö.
Allen, E. K. and O. N. Allen. 1965. Nonleguminous plant symbiosis. In: C. M. Gilmour and O. N. Allen (eds.), Microbiology and soil fertility: 77 – 106. Oregon State Univ. Press, Corvallis, Oregon.
Amann, G. 1954. Bäume und Sträucher des Waldes. 231 pp., illust. J. Neumann-Neudamm, Melsungen.
Arber, A. 1920. Waterplants. A study of aquatic Angiosperms. xvi, 436 pp. University Press, Cambridge.
Arber, A. 1925. Monocotyledons. A morphological study. xiv, 258 pp. University Press, Cambridge.
Baillon, M. H. 1858. Etude générale du groupe des Euphorbiacées. 684 pp. Masson, Paris.
Baillon, M. H. 1859. Monographie des Buxacées et des Stylocérées. 87, [2] pp., 3 pl. Masson, Paris.
Barnard, R. C. 1956. A manual of Malayan silviculture for inland lowland forests. Research Pamphlet no. 14: 199 pp. Forest Research Institute, Kepong, Malaya.
Bernhardi, J. J. 1832. Ueber die merkwürdigste Verschiedenheiten des entwickelten Pflanzenembryos und ihre Wert für Systematik. Linnaea 7: 561 – 613, pl. XIV.
Bokdam, J. 1977. Seedling morphology of some African Sapotaceae and its taxonomical significance. Med. Landbouwhogeschool Wageningen 77 (20): 84 pp., 23 fig., 2 tab.
Borbas, V. von. 1879. Keimung von Castanea und Quercus. Oest. Bot. Zeitschr. 29: 59 – 61.
Boyd, L. 1932. Monocotylous seedlings: Morphological studies in the post-seminal development of the embryo. Trans. Proc. Roy. Soc. Edinb. 31: 1 – 224.
Brandza, G. 1908. Recherches anatomiques sur la germination des Hypéricacées et des Guttifères. Ann. Sci. Nat. 9e sér VIII: 221 – 300, pl. V – XV.
Brenan, J. P. M. 1967. Leguminosae subfamily Caesalpinioideae. In: E. Milne-Redhead and R. M. Polhill (eds.), Flora of Tropical East Africa. 230 pp.
Buchenau, F. 1865. Morphologische Studien an deutschen Lentibularieen. Bot. Zeit. 23: 61 – 66, Taf. III.
Burger Hzn., D. 1972. Seedlings of some tropical trees and shrubs, mainly of South East Asia. 399 pp., 155 fig. Centre for Agricultural Publishing and Documentation (PUDOC), Wageningen.
Burkill, I. H. 1917a. Notes on Dipterocarps. 1. The seedling of Anisoptera costata Korth. J. Str. Br. R. As. Soc. no. 75: 43 – 48.
Burkill, I. H. 1917b. Notes on Dipterocarps. 2. Seedling and seed production of some species of Shorea. J. Str. Br. R. As. Soc. no. 76: 161 – 167.
Burkill, I. H. 1918. Notes on Dipterocarps. 3. Seedling of Shorea robusta Roxb. and the conditions under which it grows in pure forest. J. Str. Br. As. Soc. no. 79: 39 – 44.
Burkill, I. H. 1920a. Notes on Dipterocarps. 5. The embryo and seedling of Balanocarpus maximus King J. Str. R. As. Soc. no. 81: 3 – 4.
Burkill, I. H. 1920b. Notes on Dipterocarps. 4. On the embryo, seedling, and position of the flower in various species, and the embryo and seedling of Balanocarpus maximus King. J. Str. Br. R. As. Soc. no. 81: 49 – 76.

Burkill, I. H. 1922a. Notes on Dipterocarps. 7. On the fruits and germination of Isoptera borneensis. J. Str. Br. R. As. Soc. no. 86: 281 – 284.
Burkill, I. H. 1922b. Notes on Dipterocarps. 8. On some large fruited species, and in particular upon the effects of pressure on the interior of the fruit wall. J. Str. Br. R. As. Soc. no. 86: 285 – 291.
Burkill, I. H. 1923. Notes on Dipterocarps. 9. On the difference in the seedlings between Balanocarpus maximus King and B. heimii King. J. Mal. Br. R. As. Soc. 1: 218 – 222.
Burkill, I. H. 1925. Notes on Dipterocarps. 10. On Balanocarpus hemsleyanus King. J. Mal. Br. R. As. Soc. 3: 4 – 9.
Burkill, I. H. and F. W. Foxworthy. 1922. Notes on Dipterocarps. 6. On the genus Pachynocarpus. J. Str. Br. R. As. Soc. no. 86: 271 – 280.
Burtt, B. L. 1972. Plumular protection and some related aspects of seedling behaviour. Trans. Bot. Soc. Edinb. 41: 393 – 400.
Caesalpinus, A. 1583. De plantis libri XVI: (xxxviii), 621, (10) pp. Marescottum, Florentiae.
Camus, A. 1942. Soudure des cotylédons dans le genre Lithocarpus Blume. Bull. Mus. Nat. Hist. Nat. 14: 461 – 462.
Candolle, Alphonse L. P. P. de. 1846. Sur la durée relative de la faculté de germer dans des graines appartenant à diverses familles. Ann. Sci. Nat. sér. III, 6: 373 – 382.
Candolle, Augustin P. de. 1825. Mémoires sur la famille des Légumineuses [15 mémoires]. 515 pp., 2 tab., 70 pl.
Caspary, R. 1854. Ueber Samen, Keimung, Species und Nährpflanzen der Orobanchen. Flora 37: 577 – 588; 38: 593 – 603, pl. III.
Chamberlain, C. J. 1935. Gymnosperms, structure and evolution. xi, 484 pp., 397 fig. University of Chicago Press, Chicago.
Chapman, V. J. 1976. Mangrove vegetation. viii, 447 pp., 289 fig. Cramer, Vaduz.
Chibber, H. M. 1916. Studies in the germination of three Indian plants. Ann. Jard. Bot. Btzg 29: 52 – 56, pl. 8.
Compton, R. H. 1912. An investigation of the seedling structure in the Leguminosae. J. Linn. Soc. 41: 1 – 122.
Conde, L. F. and D. E. Stone. 1970. Seedling morphology in the Juglandaceae, the cotyledonary node. J. Arn. Arb. 51: 463 – 477.
Corner, E. J. H. 1958. Transference of function. J. Linn. Soc. (Bot.) 56: 33 – 40.
Cremers, G. 1973. Architecture de quelques lianes d'Afrique Tropicale. Candollea 28: 249 – 280.
Cremers, G. 1974. Architecture de quelques lianes d'Afrique Tropicale. Candollea 29: 57 – 110.
Crocker, W. 1916. Mechanisms of dormancy in seeds. Am. J. Bot. 3: 99 – 120.
Csapody, V. 1968. Keimlingsbestimmungsbuch der Dikotyledonen. 286 pp. Akademia Kiado, Budapest.
Darwin, C. R. 1880. The power of movement in plants. x, 592 pp. John Murray, London.
Dormer, K. J. 1945. On the absence of a plumule in some Leguminous seedlings. New Phytol. 44: 25 – 28.
Duke, J. A. 1965. Keys for the identification of seedlings of some prominent woody species in 8 forest types in Puerto Rico. Ann. Miss. Bot. Gard. 52: 315 – 350.
Duke, J. A. 1969. On tropical tree seedlings. 1. Seeds, seedlings, systems and systematics. Ann. Miss. Bot. Gard. 56: 125 – 161.
Eidmann, F. E. 1933. Kiemingsonderzoek bij een 55-tal wildhoutsoorten en groenbemesters. Med. Boschbouwproefstation 26: 156 pp. Landsdrukkerij, Batavia.
Engler, A. 1925. Guttiferae. In: A. Engler and K. Prantl (eds.), Die natürlichen Pflanzenfamilien 2e Aufl., 21: 166, fig. 70.

Engler, A. and E. Irmscher. 1919. Saxifragaceae-Saxifraga. Das Pflanzenreich IV, 117[II]. Pars generalis pp. 3 – 4.
Engler, A. and K. Krause. 1935. Loranthaceae. In: A. Engler and K. Prantl (eds.). Die natürlichen Pflanzenfamilien 2[e] Aufl., 16 b: 105 – 108, figs. 48, 52, 53.
Ewart, A. J. 1908. On the longevity of seeds. With an appendix by Miss Jean White. Proc. Roy. Soc. Victoria 21: 1 – 210, pl. I, II.
Exell, A. W. and C. A. Stace. 1972. Patterns of distribution in the Combretaceae. In: D. A. Valentine (ed.), Taxonomy, phytogeography and evolution: 307 – 323. Academic Press, London and New York.
Fahn, A. 1967. Plant anatomy: 264 – 269, 474 – 479. Pergamon Press, Oxford, London, Edinburgh, New York, Toronto, Sydney, Paris and Braunschweig.
Ferré, Y. de. 1952. Les formes de jeunesse des Abiétacées. Ontogénie-Phylogénie. 284 pp., 36 fig. Les Artisans de l'Imp. Douladoure, Toulouse.
Frakland, B. and P. F. Wareing. 1962. Changes in endogenous gibberellins in relation to chilling of dormant seeds. Nature (London) 194: 313.
Fritsch, K. 1904. Die Keimpflanzen der Gesneriaceen, mit besonderer Berücksichtigung von Streptocarpus, nebst vergleichenden Studien über die Morphologie dieser Familie. 188 pp. G. Fischer, Jena.
Gaertner, J. 1788 – 1807. De fructibus et seminibus plantarum. 1788 Vol. I: clxxxii, 384 pp., pl. 1 – 79, Stuttgardiae; 1791 Vol. II: lii, 520 pp., pl. 80 – 180, Tubingae; 1805 – 1807 Vol. III: 256 pp., pl. 181 – 225, Lipsiensis.
Gatin, C. L. 1906. Recherches anatomiques et chimniques sur la germination des palmiers. Ann. Sci. Nat. Bot. sér. IX, 3: 191 – 314.
Gilg, E. 1925. Dipterocarpaceae. In: A. Engler and K. Prantl (eds.), Die natürlichen Pflanzenfamilien 2e Aufl., 21: 246 – 247, 251, fig. 109e.
Gill, A. M. and P. B. Tomlinson. 1969. Studies on the growth of Red Mangrove (Rhizophora mangle L.). 1. Habit and general morphology. Biotropica 1: 1 – 9.
Gillis, W. T. 1971. The systematics and ecology of poison-ivy and the poison-oaks (Toxicodendron, Anacardiaceae). Rhodora 73: 172.
Gray, A. 1877. The germination of the genus Megarrhiza Torr. Am. J. Sci. 14: 21 – 24.
Grushvitskyi, I. V. 1963. Subterranean germination and the function of the cotyledons. Bot. Zhurnal 48 (6): 906 – 915 [Russian].
Guillaumin, A. 1910. Recherches sur la structure et le développement sur Burséracées, application à la systématique. 100 pp. Thèse, Faculté des Sciences de Paris, sér. A. no. 620. Masson, Paris.
Hall, C. J. J. van, and C. van de Koppel. 1946. De landbouw in den Indischen Archipel. 4 Vol. W. van Hoeve, 's Gravenhage.
Hallé, F. and R. A. A. Oldeman. 1970. Essay sur l'architecture et la dynamique de croissance des arbres tropicaux. Coll. Monogr. Bot. Biol. Veget. 6: 178 pp.
Hallé, N. 1970. Rubiacées. Flore du Gabon 2e partie 17: 219.
Hara, H. 1946. Taxonomic studies of useful plants in Japan. III. Phaseolus angularis, P. Ricciardianus and P. hirtus. Misc. Rep. Res. Inst. Nat. Res. 10: 1 – 12.
Hara, H. and S. Kurosawa. 1963. Notes on the Rubia cordifolia group. Sci. Rep. Tôhoku Univ. Ser. IV (Biol.) 29: 257 – 259.
Heckel, E. 1899. Sur le processus germinatif dans la graine de Ximenia americana L. et sur la nature des écailles radiciformes propres a cette espèce. Rev. Gén. Bot. 11: 401 – 408.
Heckel, E. 1901. Sur le processus germinatif dans les genres Onguekoa et Strombosia de la famille des Olacacées. Ann. Mus. Colon. Marseille 8: 17 – 27.
Henderson, M. R. 1949. The genus Eugenia (Myrtaceae) in Malaya. Gard. Bull. 12: 1 – 16.

Hickel, R. 1914. Graines et plantules des arbres et arbustes indigènes et communément cultivés en France. II. Angiospermes. 347 pp., 85 + 2 pl. Protat Frères, Versailles.
Hill, A. W. 1906. Morphology and seedling structure of the geophilous species of Peperomia. Ann. Bot. 20: 395 – 427, pl. XXIX-XXX.
Hill, T. G. and E. de Fraine. 1908. On the seedling structure of Gymnosperms I. Ann. Bot. 22: 689 – 712, pl. 35.
Hill, T. G. and E. de Fraine. 1909a. On the seedling structure of Gymnosperms II. Ann. Bot. 23: 189 – 227, pl. 15.
Hill, T. G. and E. de Fraine. 1909b. On the seedling structure of Gymnosperms III. Ann. Bot. 23: 433 – 458, pl. 30.
Hill, T. G. and E. de Fraine. 1910. On the seedling structure of Gymnosperms IV. Ann Bot. 24: 319 – 333, pl. 22 – 23.
Hill, T. G. and E. de Fraine. 1913. A consideration of the facts relating to the structure of seedlings. Ann. Bot. 27: 257 – 272.
Hou, D. 1958. Rhizophoraceae. Flora Malesiana I, 5: 431, 443 – 444, fig. 6, 9, 10, 23.
Hou, D. 1971. Chromosome numbers of Trigonobalanus verticillata Forman (Fagaceae). Acta Bot. Neerl. 20: 543 – 549, 4 fig., 2 pl.
Hou, D. 1972. Germination, seedling, and chromosome number of Scyphostegia borneensis Stapf (Scyphostegiaceae). Blumea 20: 88 – 92, 2 pl.
Hou, D. 1978. Florae Malesianae praecursores LVI. Anacardiaceae. Blumea 24: 10, fig. 19, h.
Irmisch, T. 1859. Ueber Lathyrus tuberosus und einige andere Papilionaceen. Bot. Zeit. 17: 57 – 63, 65 – 72, 77 – 85, Taf. III.
Irmisch, T. 1861. Ueber Polygonum amphibium, Lysimachia vulgaris, Comarum palustre, Menyanthes trifoliata. Bot. Zeit. 19: 105 – 109, 113 – 115, 121 – 123, Taf. IV.
Irmisch, T. 1865. Ueber einige Ranunculaceen. Bot. Zeit. 23: 29 – 32, 37 – 39, 45 – 48, Taf. II.
Jackson, B. D. 1928. A glossary of botanical terms with their derivation and accent. 481 pp. Duckworth and Co., London; Hafner, New York.
Jacobs, M. 1966. The study of seedlings. Fl. Mal. Bull. no. 21: 1416 – 1421.
Jarrett, F. M. 1959. Studies in Artocarpus and allied genera. I. General considerations. J. Arn. Arb. 40: 1 – 29.
Kalkman, C. 1972. Mossen en vaatplanten. Bouw, levenscyclus en verwantschappen van de Cormophyta. 304 pp., illust. Oosthoek, Utrecht.
Keng, H. 1967. Observations of Ancistrocladus tectorius. Gard. Bull. 22: 117 – 119, fig. 3 – 5, pl. 1.
Kerr, L. R. 1925. The lignotubers of Eucalypt seedlings. Proc. Roy. Soc. Vict., n.s. 37: 79 – 96.
Klebs, G. 1885. Beiträge zur Morphologie und Biologie der Keimung. Unters. Bot. Inst. Tübingen I: 536 – 635.
Knaap-van Meeuwen, M. S. 1970. A revision of four genera of the tribe Leguminosae--Caesalpinioideae-Cynometreae in Indo-Malesia and the Pacific. Blumea 18: 4 – 7.
Kostermans, A. J. G. H. 1951. The genus Teijsmanniodendron Koorders (Verbenaceae). Reinwardtia 1: 75 – 106, fig. 2.
Kozlowski, T. T. 1971. Growth and development of trees. Vol. I. Seed germination, ontogeny and shoot growth. xii, 443 pp. Vol. II. Cambial growth, root growth, and reproductive growth. xiv, 514 pp. Academic Press, New York, London.
Kozlowski, T. T. (ed.) 1972. Seed biology. Vol. I. Importance, development, and germination. xiii, 430 pp. Vol. II. Germination control, metabolism, and pathology. xi, 447 pp. Vol. III. Insects, and seed collecting, storage, testing and certification. xi, 422 pp. Academic Press, New York, London.

Kuijt, J. 1969. The biology of parasitic flowering plants. 246 pp. University of California Press, Berkeley, Los Angeles.
Lamprecht, H. 1948a. The genetic basis of evolution. Agric. Hort. Genet. 6: 83 – 86.
Lamprecht, H. 1948b. Zur Lösung des Artproblems. Agric. Hort. Genet. 6: 87 – 141.
Leenhouts, P. W. 1956. Burseraceae. Flora Malesiana I, 5: 209 – 296.
Leenhouts, P. W. 1959. Revision of the Burseraceae of the Malaysian area in a wider sense. Xa. Canarium Stickm. Blumea 9: 275 – 475.
Leenhouts, P. W. 1966. Canarium sect. Africanarium nov. sect. (Burserac.). Blumea 13: 396.
Leenhouts, P. W. and S. H. Widodo. 1972. Some notes on the seedling of Haplolobus (Burseraceae). Blumea 20: 311 – 314.
Leeuwenberg, A. J. M. 1965. The Loganiaceae of Africa. VII. Strychnos II. Acta Bot. Neerl. 14: 227, fig. 2.
Léonard, J. 1957. Genera des Cynometreae et des Amherstieae africaines (Leguminosae-Caesalpinioideae). Esai de blastogénie appliquée à la systématique. Mém. Acad. Roy. Belg., Classe des Sci. 30 (2): 1 – 312.
Link, D. H. F. 1807. Grundlehren der Anatomie und Physiologie der Pflanzen: 234 – 244. Danckwerte, Göttingen.
Lovell, P. H. and K. G. Moore. 1970. A comparative study of cotyledons as assimilating organs. J. Exp. Bot. 21 (69): 1017 – 1030.
Lovell, P. H. and K. G. Moore. 1971. A comparative study of the role of the cotyledon in seedling development. J. Exp. Bot. 22 (70): 153 – 162.
Lubbock, J. 1892. A contribution to our knowledge of seedlings. Vol. I: viii, 608 pp.; vol. II: 646 pp. Kegan Paul, Trench, Trübner and Co., London.
Maheshwari, P. 1950. An introduction to the embryology of Angiosperms. x, 453 pp. McGraw-Hill, New York, Toronto, London.
Malpighi, M. 1687. Opera omnia. Vol. I. Anatome plantarum. 170 pp., 142 icones; Vol. II. Medico-anatomica continens. 379 pp., icones. Van Der Aa, Lugduni Batavorum.
Malpighi, M. 1697. Opera posthuma: 110, 187, 10 pp., 19 tab. Gallet, Amstelodami.
Marshall, P. E. and T. T. Kozlowski. 1973. The role of cotyledons in growth and development of woody angiosperms. Can. J. Bot. 52: 239 – 245.
Marshall, P. E. and T. T. Kozlowski. 1974. Photosynthetic activity of cotyledons and foliage leaves of young angiosperms. Can. J. Bot. 52: 2023 – 2032.
Marshall, P. E. and T. T. Kozlowski. 1976. Compositional changes in cotyledons of woody angiosperms. Can. J. Bot. 54: 2473 – 2477.
Marshall, P. E. and T. T. Kozlowski. 1977. Changes in structure and function of epigeous cotyledons of woody angiosperms during early seedling growth. Can. J. Bot. 55: 208 – 215.
Maury, G. 1968. Germinations anomales chez les Dipterocarpacées de Malaisie. Bull. Soc. Hist. Nat. Toulouse 104: 187 – 202.
Maury, G. 1970. Différents types de polyembryonie chez quelques Dipterocarpacées asiatiques. Bull. Soc. Hist. Nat. Toulouse 106: 282 – 288.
Maury, G. 1978. Dipterocarpacées du fruit à la plantule. Thèse. Vol. I: 432 pp.; Vol. II: 344 pp., illus. Impr. de l'Université Paul Sabatier, Toulouse.
Mensbruge, G. de la. 1966. La germination et les plantules des essences arborées de la forêt dense humide de la Côte d'Ivore. 389 pp. Publication 26 du Centre Technique Forestier Tropical, Nogent-sur-Marne.
Meijer, W. 1968. The study of seedlings of Bornean trees. Bot. Bull. Herb. For. Dep. Sandakan 11: 112, 16 pl.
Meijer Drees, E. 1941. Germination and seedlings of some Acacia species. Tectona 34: 1 – 45.
Mirbel, M. 1809. Nouvelles recherches sur les charactères anatomiques et physiologiques qui distinguent les plantes monocotylédones des plantes dicotylédones. Ann. du Muséum XIII: 54 – 86, pl. 2 – 8.

Muller, F. M. 1978. Seedlings of the North-Western European lowland. A flora of seedlings. 654 pp., 403 pl. W. Junk, The Hague/Centre for Agricultural Publishing and Documentation (PUDOC), Wageningen.

Ng, F. S. P. 1975. The fruits, seeds and seedlings of Malayan trees I-XI. The Malays. For. 38: 33 – 99.

Ng, F. S. P. 1976. The fruits, seeds and seedlings of Malayan trees XII-XV. The Malays. For. 39: 110 – 146.

Nooteboom, H. P. 1972. Simaroubaceae. Flora Malesiana I, 6: 969, fig. 15.

Ohashi, H. 1973. Desmodium and its allied genera. Ginkgoana 1: 26 – 30, fig. 4 – 5.

Pancho, J. V. and E. A. Bardenas. 1972. Classification of weed seedlings and seeds. Lecture note 6c: 12 pp. mimeogr. Second Weed Science Training Course 1972, Bogor.

Paijens, J. P. D. W. 1967. A monograph of the genus Barringtonia (Lecythidaceae). Blumea 15: 165 – 169, 187, 197, 230, fig. 1.

Pennington, T. D. and B. T. Styles. 1975. A generic monograph of the Meliaceae. Blumea 22: 419 – 540.

Pierre, L. 1879 – 1907. Flore forestière de la Cochinchine I: pl. 68; II: pl. 94; III: pl. 205, 218, 221, 225; IV: pl. 244, 263; V: pl. 253, 361. Doin, Paris.

Pilger, R. 1935. Santalaceae. In: A. Engler and K. Prantl (eds.), Die natürlichen Pflanzenfamilien 2e Aufl., 16 b: 52 – 53, fig. 27.

Planchon, J. E. and J. Triana. 1860 – 1862. Mémoire sur la famille des Guttifères. Ann. Sci. Nat. sér. IV, 13: 306 – 376, pl. 16; 14: 226 – 367, pl. 15 – 18; 15: 240 – 319, 16: 263 – 308.

Richard, A. 1819. Nouveaux élémens de botanique et de physiologie végétale. xv, 410 pp., 8tab. Béchet, Paris.

Richard, L. C. M. 1808. Démonstrations botaniques ou Analyse du fruit considéré en général: 48 – 50. Gabon, Paris.

Rizzini, C. T. 1965. Experimental studies on seedling development of Cerrado woody plants. Ann. Miss. Bot. Gard. 52: 410 – 426.

Roeper, J. A. C. 1824. Enumeratio Euphorbiarum quae in Germania et Pannonia gignuntur. viii, 68 pp., 3 tab. Rosenbusch, Goettingae.

Rooden, J. van, L. M. A. Akkermans and R. van der Veen. 1970. A study on photoblastism in seeds of some tropical weeds. Acta Bot. Neerl. 19: 257 – 264.

Roxburgh, W. 1795 – 1820. Plants of the coast of Coromandel 3: pl. 212, 213 [1811], p. 67, pl. 270 [1820].

Ruggles Gates, R. 1951. Epigeal germination in the Leguminosae. Bot. Gaz. 113: 151 – 157.

Rumphius, G. E. 1741. Herbarium Amboinense I: p. 16, Tab. 2; II: p. 150, Tab. 47; III: p. 65, Tab. 36, p. 111, Tab. 72. Changuion, Catuffe und Uytwerf, Amsterdam; Gosse, Neaulme, Moetjens and van Dole, 's Hage; Neaulmue, Utrecht.

Sargeant, E. 1902. The origin of the seed-leaf in Monocotyledons. New Phytol. 1: 107 – 113, fig.

Sargeant, E. 1903. A theory of the origin of Monocotyledons, founded on the structure of their seedlings. Ann. Bot. 17: 1 – 92, 7 pl., 10 text-fig.

Sastrapradja, D., S. Sastrapradja, S. R. Aminah and I. Lubis. 1975. Species differentiation in Javanese Mucuna with particular reference to seedling morphology. Ann. Bog. 6: 57 – 68.

Sastrapradja, S., D. Sastrapradja, S. R. Aminah, I. Lubis, and S. Idris. 1972. Comparative seedling morphology of Mucuna pruriens group. Ann. Bog. 5: 131 – 136.

Sastrapradja, S., I. Lubis, S. H. A. Lubis, and D. Sastrapradja. 1976. Studies in the Javanese species of Canavallia. II. Variation in nature and seedling characters. Ann. Bog. 6: 97 – 110.

Schlising, R. A. 1969. Seedling morphology in Marah (Cucurbitaceae) related to the Californian Mediterranean climate. Am. J. Bot. 56: 552 – 560.

Schnell, R. 1971. Introduction à la phytogéographie des pays tropicaux. Les problèmes généraux. Vol. 2: 607. Gauthier-Villars, Paris.

Schopmeyer, C. S. 1974. Seeds of woody plants in the United States. Agriculture Handbook 450: 883 pp. Forest Service, U.S. Department of Agriculture, Washington, D.C.

Sleumer, H. 1935. Olacaceae. In: A. Engler and K. Prantl (eds.), Die natürlichen Pflanzenfamilien 2ᵉ Aufl., 16b: 8 – 9.

Sleumer, H. 1942. Icacinaceae. In: A. Engler and K. Prantl (eds.), Die natürlichen Pflanzenfamilien 2ᵉ Aufl., 20 b: 339, 391, fig. 116.

Sleumer, H. 1971. Icacinaceae. Flora Malesiana I, 7: fig. 28 on p.58.

Soegeng Reksodihardjo, W. 1959. Kostermansia Soegeng, a new genus in Bombacaceae (Durioneae). Reinwardtia 5: 1 – 9, fig. 4.

Soepadmo, E. and B. K. Eow. 1976. The reproductive biology of Durio zibethinus Murr. Gard. Bull. 29: 25 – 33.

Stebbins, G. L. 1971. Adaptive radiation of reproductive characteristics in Angiosperms. II. Seed and seedlings. Ann. Rev. Ecol. Syst. II: 237 – 260.

Steenis, C. G. G. J. van. 1972. The mountain flora of Java. viii, 90 pp., 26 fig., 72 photogr., 57 pl. Brill, Leiden.

Stoutjesdijk, P. 1972a. Spectral transmission curves of some types of leaf canopies with a note on seed germination. Acta Bot. Neerl. 21: 185 – 191.

Stoutjesdijk, P. 1972b. A note on the spectral transmission of light by tropical rainforest. Acta Bot. Neerl. 21: 346 – 350.

Swamy, B. G. L. 1949. Further contributions to the morphology of the Degeneriaceae. J. Arn. Arb. 30: 10 – 39.

Takhtajan, A. 1959. Die Evolution der Angiospermen. 344 pp. Gustav Fischer, Jena.

Tantra, I. G. M. 1976. A revision of the genus Sterculia L. in Malesia (Repisi marga Sterculia L. di Malesia) (Sterculiaceae). Communication Lembaga Penelitian Hutan 102: ii, 1 – 194. Bogor.

Taylor, C. J. 1960. Synecology and silviculture in Ghana. Thomas Nelson and Sons, London.

Thomson, T. 1858. On the structure of the seeds of Barringtonia and Careya. J. Proc. Linn. Soc. Bot. 2: 47 – 53.

Troup, R. S. 1921. The silviculture of Indian trees. Vol. I: lvii, 1 – 336. iii; Vol. II: xi, 337 – 783, iv; Vol. III: xii, 785 – 1195; 490 fig. Clarendon Press, Oxford.

Ulbrich, E. 1928. Biologie der Früchten und Samen (Karpobiologie). Biologische Studienbücher VI: 27 – 30, Abb. 4. Springer, Berlin.

Vassilczenko, I. T. 1936. Ueber die Bedeutung der Morphologie der Keimung des Samen für die Pflanzensystematik und die Entstehungsgeschichte der Pflanzen. Acta Inst. Bot. Acad. Sci. U.S.S.R. ser. I, 3: 7 – 66 [Russian, summary in German].

Vassilczenko, I. T. 1937. Morphologie der Keimung der Leguminosen in Zusammenhang mit ihrer Systematik und Phylogenie. Acta Inst. Bot. Acad. Sci. U.S.S.R. ser. I, 4: 347 – 425 [Russian, summary in German].

Vassilczenko, I. T. 1941. The morphology of germination in representatives of the fam. Chenopodiaceae in connection with their systematics. Acta Inst. Bot. Acad. Sci. U.S.S.R. ser. I, 5: 331 – 357 [Russian, summary in English].

Vassilczenko, I. T. 1946. On the question of the evolutional importance of morphological peculiarities of the seedlings of flowering plants. Volumes of scientific works carried out in Leningrad in the course of the three years of the national war. The U.S.S.R. Academy of Science, V. L. Komarov Botanical Institute, Leningrad [Russian].

Vassilczenko, I. T. 1947. Morphology of germination of Labiatae and its systematical signification. Acta Inst. Bot. Acad. Sci. U.S.S.R. ser. I, 6: 72 – 104 [Russian, summary in English].
Veen, R. van der. 1970. The importance of the red-far red antagonism in photoblastic seeds. Acta Bot. Neerl. 19: 809 – 812.
Veen, R. van der. 1971. Zaadkieming in licht. Gorteria 5: 225 – 227.
Villiers, T. A. 1972. Seed dormancy. In: T. T. Kozlowski, Seed biology. Vol II: 219 – 281. Academic Press, New York, London.
Vogel, E. F. de. 1978. Germination and seedlings. In: D. Hou, Anacardiaceae. Flora Malesiana I, 8: 400 – 401.
Vogel, E. F. de. 1979. Seedlings of Dicotyledons. Structure, development, types. Thesis. xiv, 203 pp, 72 pl. Centre for Agricultural Publishing and Documentation (PUDOC), Wageningen (text of p. 1 – 149 of the present book).
Voorhoeve, A. G. 1965. Liberian hight forest trees. Belmontia I, 8: 1 – 416.
Warming, E. 1877. Om Rhizophora mangle L. Bot. Not. 77: 14.
Weberling, F. and P. W. Leenhouts. 1966. Systematisch-morphologische Studien an Terebinthales-Familien (Burseraceae, Simaroubaceae, Meliaceae, Anacardiaceae, Sapindaceae). Abh. Akad. Wiss. Lit., Math.-Naturw. Klasse, Jahrg. 1965, 10: 536 – 542, 570 – 571.
White, J. 1908. The occurrence of an impermeable cuticle on the exterior of certain seeds. In: A. J. Ewart, On the longevity of seeds. Proc. Roy. Soc. Victoria 21: 203 – 210, pl. I, II.
Wieland, G. R. 1911. A study of some American fossil Cycads. V. Further notes on seed structure. Am. J. Sci. IV, 32: 133 – 155.
Wieland, G. R. 1916. American fossil Cycads. II. Taxonomy. Carnegie Institute, Washington, Publ. 34.
Winkler, A. 1874. Ueber die Keimblätter der deutschen Dicotylen. Verh. bot. Ver. Brandenb. 16: 6 – 21, 54 – 56, Taf. 2.
Winkler, A. 1876. Nachträge und Berichtigungen zur Uebersicht der Keimblätter der deutschen Dikotylen. Verh. bot. Ver. Brandenb. 18: 105 – 108.
Winkler, A. 1884. Die Keimblätter der deutschen Dikotylen. Verh. bot. Ver. Brandenb. 26: 30 – 41, Taf. 1.
Wright, H. 1904. The genus Diospyros in Ceylon: its morphology, anatomy and taxonomy. Ann. Roy. Bot. Gard. Peradeniya 2: 1 – 106, 133 – 210, pl. I-XX.
Zimmermann, W. 1930. Die Phylogenie der Pflanzen. 452 pp. Gustav Fischer, Jena.

Descriptions and illustrations

ALANGIACEAE

ALANGIUM

Fl. Java 2: 161; Ng, Mal. Forester 38 (1975) 38 (*A. ebenaceum*).
Monograph: Bloembergen, Bull. Jard. Bot. Buit. III, 16 (1939) 139–235.
Genus in tropical Africa, through the Indian Ocean, India, and Ceylon, S.E. Asia, tropical and temperate E. Asia up to Japan, Taiwan, all over Malesia, N.E. Australia, the Solomons, in the Pacific as far as Fiji, including New Caledonia.

Alangium javanicum (K. & V.) Wang. – Fig. 29.

Development: The taproot and hypocotyl emerge from one pole of the fruit, carrying the paracotyledons enclosed by endosperm and generally also by the fruit wall. After a brief resting stage while the paracotyledons are upright, they spread, shedding the fruit wall and endosperm, although they often have difficulty in freeing themselves completely. Then follows a second resting stage.
Seedling epigeal, phanerocotylar. Macaranga type.
Taproot slender, fibrous, cream-coloured, with rather many long, slender, shortly branched, cream-coloured sideroots.
Hypocotyl strongly enlarging, terete, slender, to 7 cm long, green, with scattered minute, simple, appressed hyaline hairs.
Paracotyledons 2, opposite, foliaceous, in the 11th–13th leaf stage dropped; simple, spreading, without stipules, short-petiolate, herbaceous, green. Petiole to 4 by 2 mm, flattened, slightly hairy like the hypocotyl. Blade ovate, to 4.3 by 3 cm; base obtuse, slightly abruptly decurrent; top obtuse; margin entire; glabrous; main nerves 3, above raised, below rather inconspicuous, ending free.
Internodes: First one terete, elongating to 1.2 cm long, otherwise like the hypocotyl; next ones much shorter, to 0.5 cm, higher ones slightly longer, the 10th one to 2 cm long, 13th one to 5 cm long. Stem slightly zig-zag, rather slender, soon with many warty, brown, elevated lenticels.
Leaves all spirally arranged, simple, without stipules, herbaceous, green. Petiole in cross-section semi-orbicular, at the base thickened, above slightly channelled; to 5 by 2 mm, in higher ones slightly larger, the 10th one to 20 by 2 mm; hairy as the hypocotyl. Blade oblong, higher ones obovate-oblong to obovate-lanceolate, 4th one to 12 by 3.7 cm, each next one slightly larger, 10th one to 14 by 4.3 cm, 13th one to 18 by 7.5 cm, higher ones (often also lower ones) usually much smaller, the axil often with a developed branch; base wedge-shaped, in the lower ones and often in the higher ones slightly obtuse; top cuspidate to caudate; margin entire; midrib below slightly hairy, glabrescent; nerves pinnate, above sunken, below very prominent, lateral nerves joined to an undulating marginal nerve.

Specimens: 1453 from N. Sumatra. Alluvial flat, primary forest on deep clay, low altitude.
Growth details: In nursery germination good, ± simultaneous, growth vigorous.
Remarks: Two species described, similar in germination pattern and general mor-

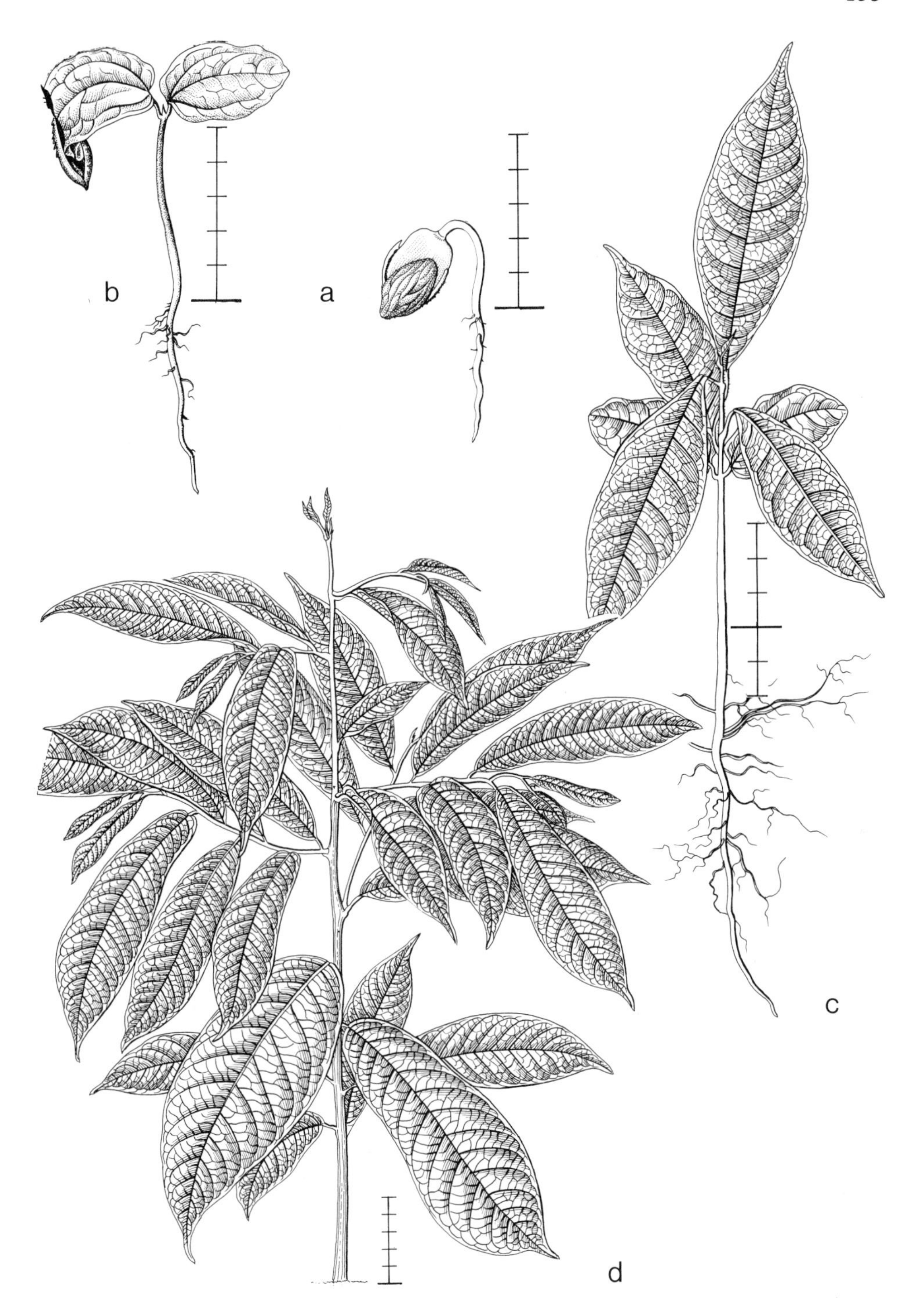

Fig. 29. *Alangium javanicum.* De V. 1453, collected 14-vii-1972, planted 27-vii-1972. a, b, and c date?; d. 17-ix-1973.

phology. *A. ebenaceum* differs in details only.

ANACARDIACEAE

GLUTA

Fl. Males. I, 8: 446; Burger, Seedlings (1972) 36 (*G. renghas*); Troup, Silviculture 1 (1921) 244 (*Melanorrhoea usitata* = *Gluta usitata*).
Genus in Madagascar, W. India, throughout S.E. Asia up to Hainan, Malesia except in the Philippines.

Gluta macrocarpa (Engl.) Ding Hou – Fig. 30.

Development: The fruit wall splits longitudinally at one side, the taproot emerges and the cotyledonary petioles enlarge, pushing the plumular bud free from the envelopments, by which the shoot is now able to elongate. No resting stages occur.
Seedling hypogeal, cryptocotylar, with hyaline exudate which turns black when dry. Heliciopsis type and subtype.
Taproot long, rather sturdy, hard, dark brown, with many long, rather slender, dark brown sideroots; these much branched, short, brown, white tipped.
Hypocotyl not enlarging.
Cotyledons 2, secund, succulent, ineffective in the stage with 6–8 developed leaves, but longer persistent; simple, without stipules, petiolate. The whole body irregularly roundish in outline, top-shaped, flattened above and below, to 3.5 by 3 by 2 cm, the bluish grey, dull fruit wall persistent, above with an erect, often eccentrically placed, ± terete, at its top abruptly swollen stalk up to 1.5 cm long. Petiole slightly elongating, not exceeding the fruit wall, to 2 by 5 mm, at the upper side prolonged in a ridge at the upper margin of the cotyledon, at the base distinctly widened and clasping the stem, glabrous, green turning brown.
Internodes: First one terete, elongating to 2.5 cm, green, glabrous, rather soon with many small, slightly warty, brownish lenticels; next ones like the first one, irregular in length, in the lower part 0.5 to 4 cm long, in the leafy part below to 0.5 cm long, higher up gradually longer, 10th one to 3.5 cm long. Stem rather slender, straight, where damaged turning black.
Leaves all spirally arranged, simple, spreading, without stipules, shortly petiolate, herbaceous, dull reddish green turning green, glabrous, the first ± 4 much to slightly reduced. Petiole in the lower leaves to 7 by 2 mm, in 20th leaf to 7 by 3.5 mm, in cross-section semi-orbicular, channelled above, distinctly narrowed at the tip. Blade in the lower developed leaves oblong, to 15 by 5.5 cm; in the higher ones obovate-lanceolate, to 18 by 5.5 cm in the 6th developed leaf, in 20th one to 25 by 7 cm, in 25th one to 35 by 9 cm; base ± acute, in higher ones narrowed, tip slightly rounded to emarginate; top caudate, in higher ones cuspidate; margin entire, slightly undulating; nerves pinnate, midrib slightly prominent above, distinctly so below, lateral nerves sunken above, prominent below, at the margin curved back at the tip.

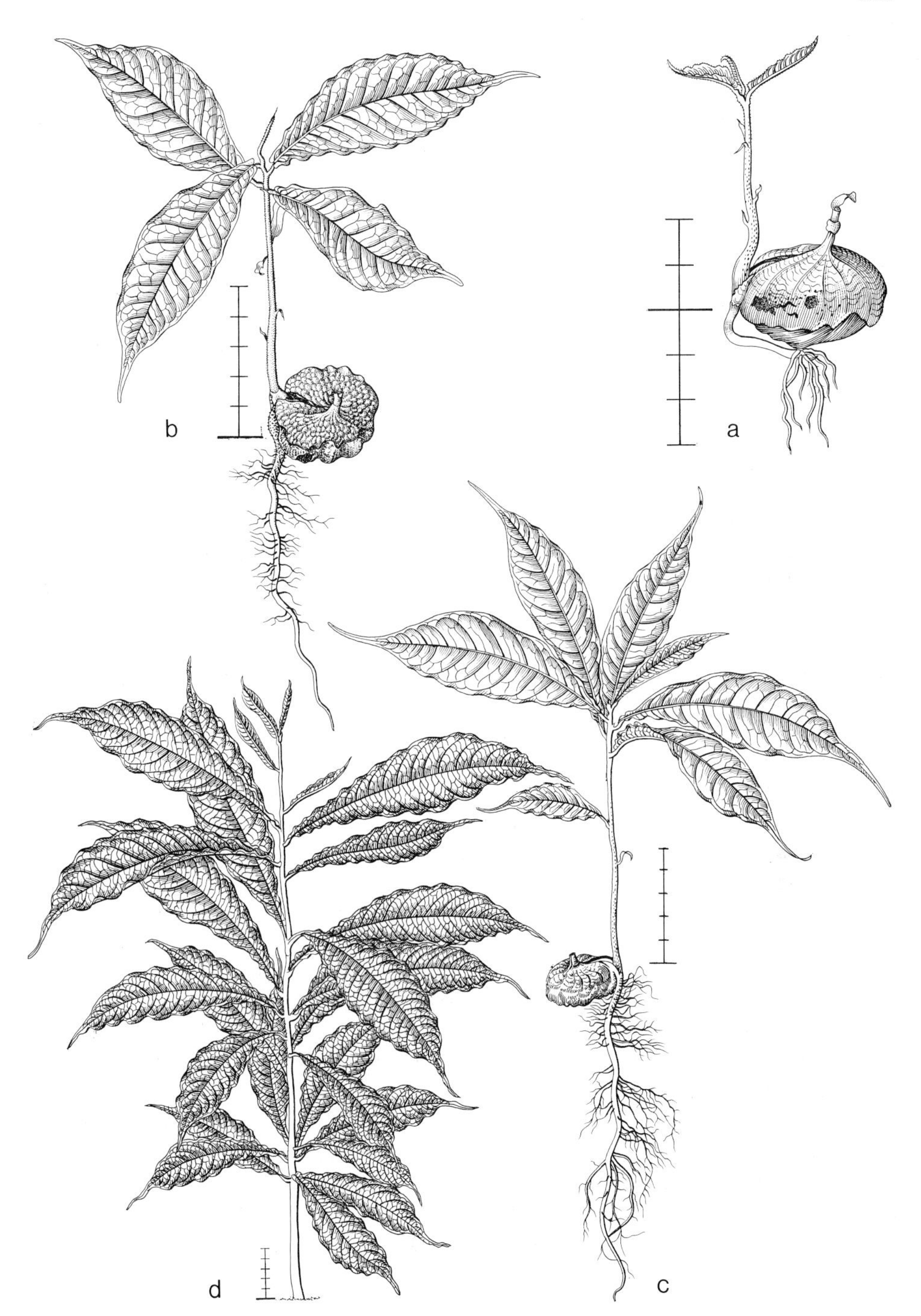

Fig. 30. *Gluta macrocarpa.* De V. 2332, collected 14-iii-1973, planted 22-iii-1973. a. 11-iv-1973; b. 8-v-1973; c. 4-vi-1973; d. 14-iii-1974.

Specimens: 2332 from S.E. Borneo, low hills, primary forest on deep clay, low altitude.
Growth details: In nursery germination good, ± simultaneous.
Remarks: Three species described, similar in germination pattern and general morphology, differing in details only.

KOORDERSIODENDRON

Fl. Males. I, 8: 486; Meijer, Bot. Bull. Sandakan 11 (1968) 112 (*K. pinnatum*).
Genus in Malesia: Borneo, Philippines, Celebes, Moluccas, and New Guinea.

Koordersiodendron pinnatum Merr. – Fig. 31.

Development: The taproot and hypocotyl emerge from one pole of the fruit, the elongating hypocotyl carrying the enclosed cotyledons. The cotyledonary petioles elongate carrying the fruit away from the plumule, which is now free to elongate. When two leaves have developed, a resting stage follows. In 8 months plant to 45 cm high with 15 leaves.
Seedling epigeal, cryptocotylar. Heliciopsis type, Koordersiodendron subtype.
Taproot rather long and slender, flexuous, brownish, with rather many long, slender, branched, brownish sideroots.
Hypocotyl strongly elongating, terete, slender, to 6 cm long, rather densely covered with minute, simple, white to pale brownish hairs; lenticels scattered, white turning brownish.
Cotyledons 2, secund, succulent, in the 2nd–3rd leaf stage dropped; simple, without stipules, petiolate. The whole body ellipsoid, to 3 by 1.7 by 1.3 cm, the dark brown, cracking, fibrous fruit wall persistent. Petiole flattened, exceeding the testa to 10 mm, 2.5 mm wide, slightly hairy like the hypocotyl. Blade auriculate at the base, these sometimes exceeding the fruit wall.
Internodes: First one strongly elongating, terete, to 12 cm long, greenish turning reddish brown, indument and lenticels like the hypocotyl; next ones like the first one, rather irregular in length, 1–2 cm long. Stem slender, ± straight.
First two leaves opposite, compound, without stipules, petiolate, leaflets petiolulate, herbaceous, reddish green when young, turning green. Petiole to 25 by 1.5 mm, terete, at the base slightly thickened, hairy like the hypocotyl. Rhachis up to 5.5 cm long, like the petiole. Blade imparipinnate, (3–) 4 (–5) jugate with terminal leaflet, elliptic in outline. Petiolules to 5 by 0.5 mm, ± terete, like the petiole, that of the terminal leaflet up to 1.3 by 0.5 mm. Leaflets semi-opposite; at the base obtuse, the paired ones unequally so; at the top acuminate; margin entire; slightly hairy like the hypocotyl, glabrescent; lower ones ovate-oblong, 3–4 by 1.5 cm, higher ones more ovate-lanceolate, to 6.5 by 1.5 cm; nerves pinnate, slightly sunken above, prominent below, especially the midrib, side-nerves ± free ending.
Next leaves spirally arranged with gradually increasing length and number of leaflets; 7th one with 4–5 (–6) pairs, to 24 by 14 cm, 14th leaf 9–10 pairs, 20th one with 10–11 pairs, to 38 by 20 cm, further as the first leaves.

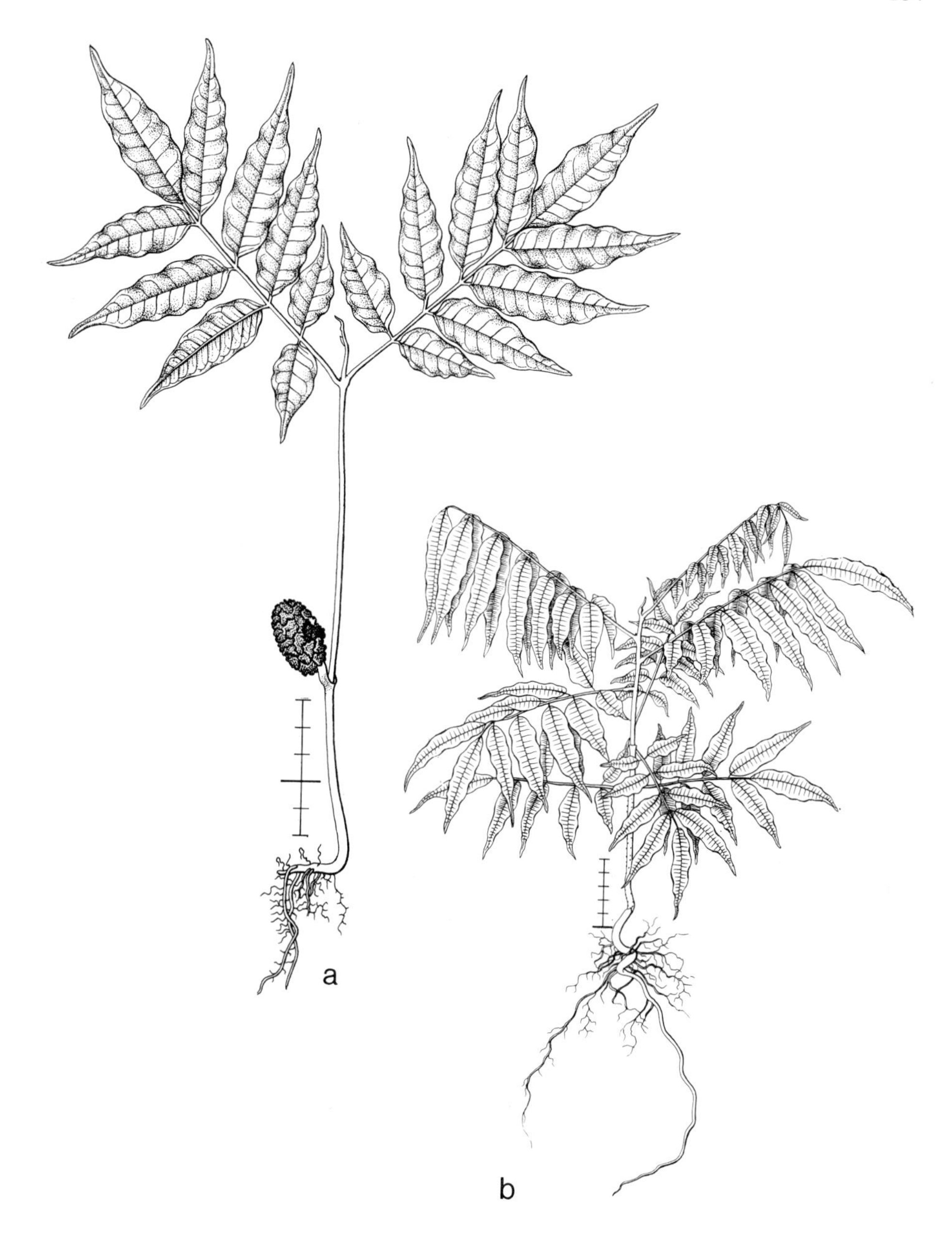

Fig. 31. *Koordersiodendron pinnatum.* De V. 719a, collected 11-ix-1971, planted 7-xii-1971. a and b date?.

Specimens: 719a from S.E. Borneo. Alluvial flat, primary forest on deep clay, low altitude.
Growth details: In nursery germination fair.

MANGIFERA

Fl. Males. I, 8: 422; Duke, Ann. Miss. Bot. Gard. 52 (1965) 340 (*M. indica*); Lubbock, On Seedlings 1 (1892) 374 (*M. indica*); Pierre, Flore Forestière Cochinchine 5 (1879–1899) pl. 361 (*M. indica*); Troup, Silviculture 1 (1921) 237 (*M. indica*).
Genus in India and Ceylon, throughout S.E. Asia, S.W. China, Malesia as far as the Solomons.

Mangifera gedebe Miq. – Fig. 32.

Development: The big reniform fruit opens at one auricle. By elongation of the cotyledonary petioles, the developing taproot and shoot are pushed out of the fruit. The petioles are recurved into the hollow in the fruit margin. A resting stage follows after development of each flush of leaves.
Seedling hypogeal, cryptocotylar. Heliciopsis type and subtype.
Taproot long, slender, sturdy, reddish brown, with numerous short, close-set, dark brown, shortly branched sideroots.
Hypocotyl slightly enlarging to 1 cm, clearly different from root and stem, in cross-section elliptic, slightly tapering to the base, pale reddish green, with many small, round, white, warty lenticels, glabrous.
Cotyledons 2, secund, succulent, shed in the 14th–16th leaf stage; simple, without stipules, petiolate. The whole body reniform, to 8 by 6 by 4 cm, the dark grey-brown hard endocarp and fruit wall persistent. Petioles conspicuously protruding, curved back to the convex margin, to 3 by 0.5 cm, in cross-section semi-orbicular, above broadly and deeply channelled, green, sometimes reddish above, with white lenticels, glabrous.
Internodes: First one strongly elongating, to 20 cm long, ± elliptic in cross-section, reddish green, shining, smooth, glabrous; next ones as first one, irregular in length, to 10 cm long but often much smaller, further as first internode. Stem long, slender, straight.
Leaves all spirally arranged, simple, spreading, without stipules, petiolate, herbaceous, dark purple when young, turning (reddish) green, glabrous. Petiole to 10 by 2 mm, swollen, terete, the higher ones shorter; curved, distinctly constricted at the junction with the blade. Blade lanceolate, in higher ones linear, to 20 by 5.3 cm in the lowest leaves, ultimately up to 43 by 5.5 cm; base narrowed over a long stretch into the petiole; top acuminate, its tip distinctly apiculate; margin entire; nerves pinnate, midrib distinctly raised above and below, sidenerves less so, ending in anastomoses.

Specimens: Bogor Botanic Gardens VI.D.5, from S. Sumatra.
Growth details: In nursery germination fair, ± simultaneous.

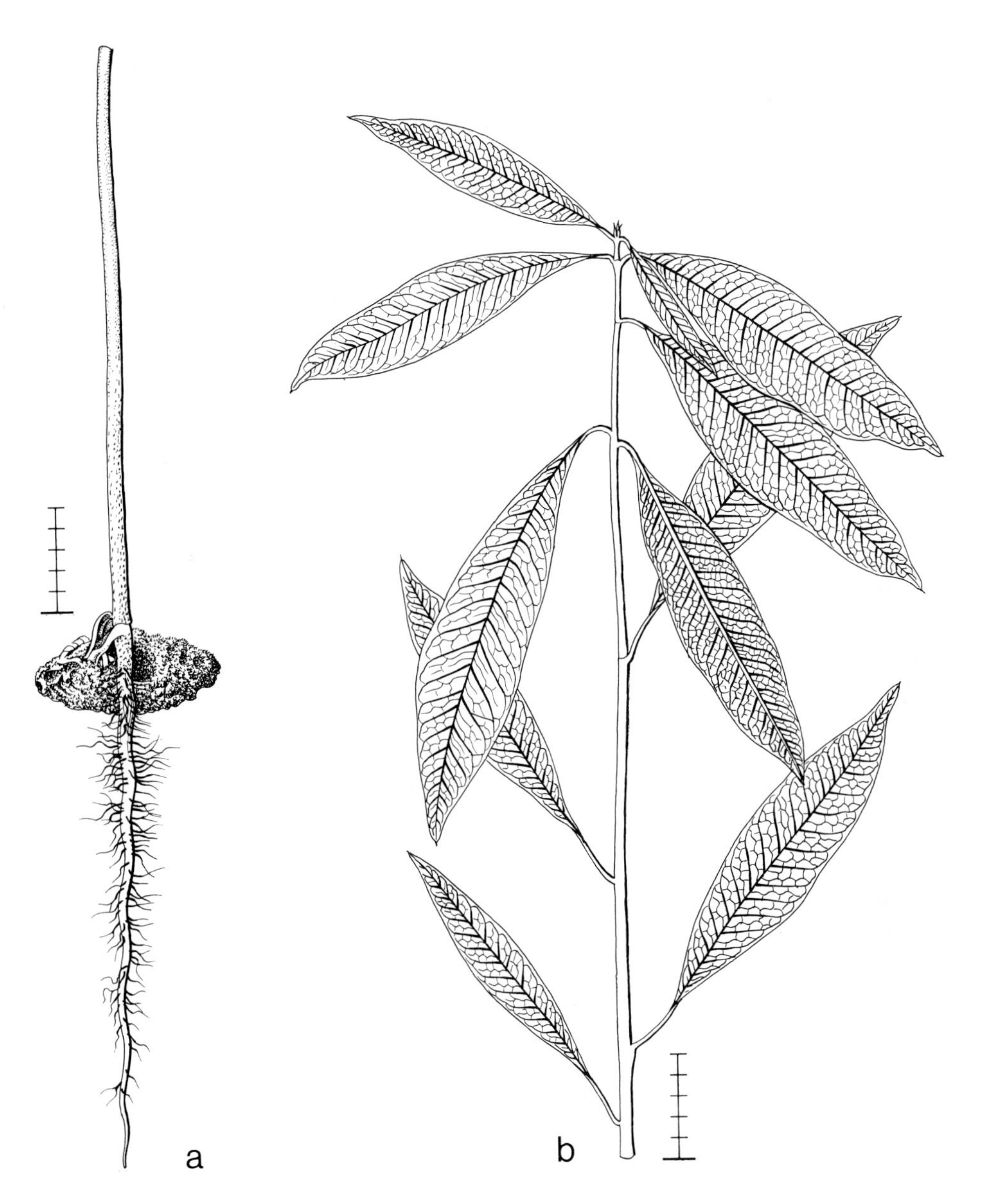

Fig. 32. *Mangifera gedebe.* Bogor Bot. Gard. VI. D. 5, collected and planted 10-i-1973. a. 11-iv-1973.

Remarks: Two species described, similar in germination pattern and general morphology. In *M. indica* the first one to three pairs of leaves are opposite (Troup), or all leaves are spirally arranged, the first up to six scale-like (Lubbock). Polyembryony is recorded to occur.

MELANOCHYLA

Fl. Males. I, 8: 490.
Genus in W. Malesia: Malay Peninsula, Borneo, Sumatra, Java.

Melanochyla fulvinervis (Bl.) Ding Hou – Fig. 33.

Development: The taproot emerges from one pole of the fruit. The cotyledonary petioles elongate and spread distinctly at their upper margin, splitting and separating the fruit wall, thus enabling the shoot to emerge. No resting stages occur.
Seedling hypogeal, cryptocotylar. Heliciopsis type and subtype.
Taproot long, coarse, fibrous, dark brown, with rather few, sturdy, few-branched, dark brown sideroots, the latter often densely covered by short, red root hairs.
Hypocotyl not enlarging.
Cotyledons 2, secund, succulent, ineffective after the 3rd–4th leaf stage, but often longer persistent; simple, without stipules, petiolate, red where exposed. The whole body ellipsoid, to 25 by 1.5 by 1.5 cm, the dark brown fruit wall persistent, gaping above; fruit wall coarse, with drops of shining black resin. Petiole for more than halfway connate to the blade, forming a wing-like structure, distinctly laterally bent, free part to 4 by 3 mm, in cross-section ± triangular, over its entire length with vague, irregular cross-ridges.
Internodes: First one terete, slightly elongating, to 2 mm long, greenish turning brownish, rather densely covered by minute, simple, brownish-hyaline hairs, slowly glabrescent; next ones as first one, in the leafless part of the stem to 2 cm long, in general smaller, in the leafy part lower ones to 8 mm long, often smaller, 12th to 4 cm long. Stem slender, ± straight, often with black spots.
Leaves all spirally arranged, simple, without stipules, petiolate, subcoriaceous, red when young, turning green; first ± 10 ones much reduced, scale-like, sessile, narrowly triangular, to 3 by 0.5 mm, soon shrivelling but persistent, developed leaves often intercalated with scales. Petiole in developed leaves in cross-section semi-orbicular, lower ones to 7 by 1.5 mm, in 12th one to 12 by 3 mm, hairy as first internode. Blade oblong to ovate-oblong, the lower ones to 8.3 by 3.7 cm, 12th one to 18 by 7 cm; base more or less narrowed, in higher ones obtuse with the margins turned upright; top cuspidate, its tip rounded; margin entire; above and below on the midrib hairy as first internode when very young, soon glabrescent; nerves pinnate, slightly prominent above, much so below, especially the midrib, lateral nerves curved at their tip, free ending.

Specimens: 2349 from S.E. Borneo, low hills, primary forest on deep clay, low altitude.

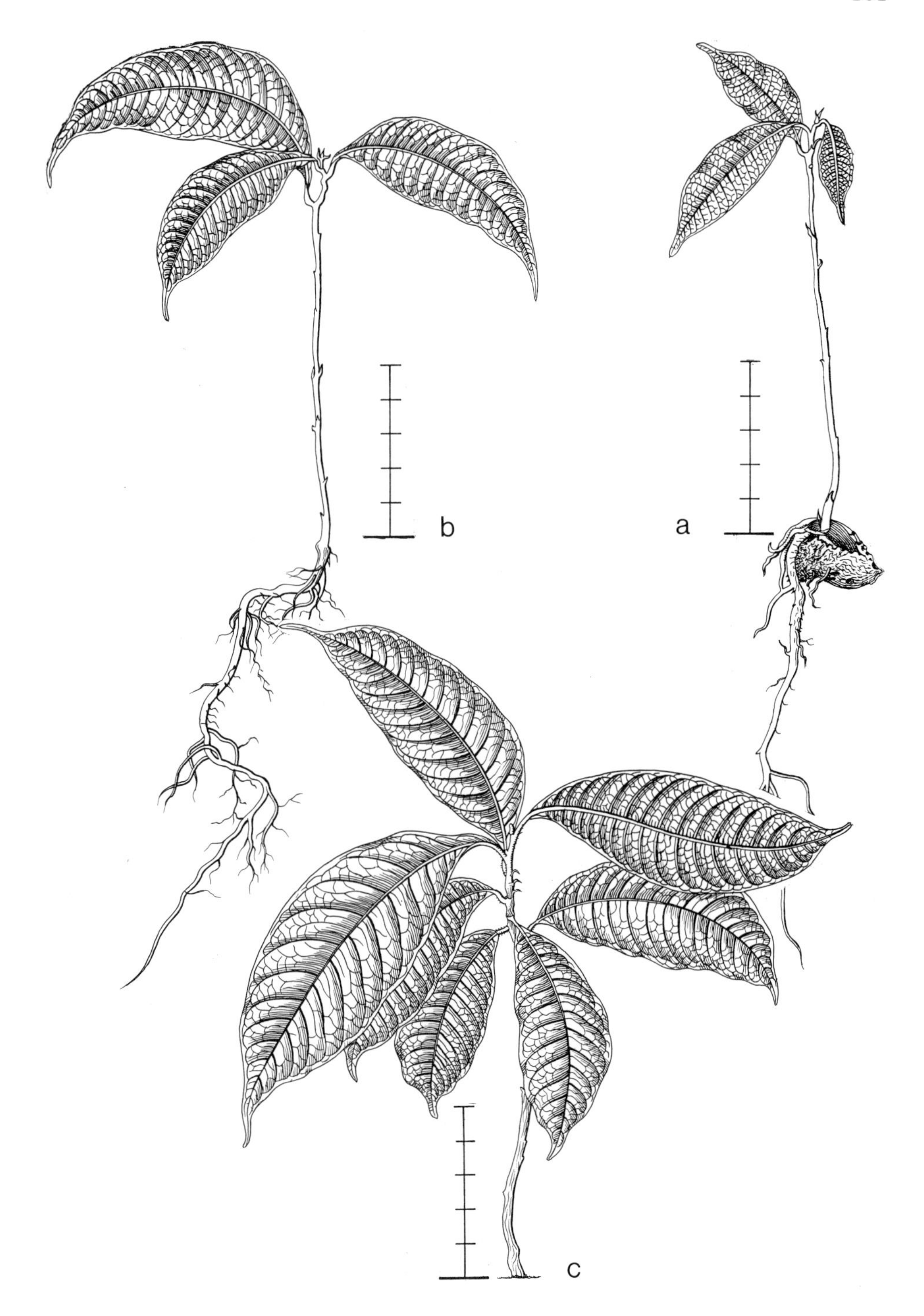

Fig. 33. *Melanochyla fulvinervis*. De V. 2349, collected 15-iii-1973, planted 22-iii-1973 . a. 14-iv-1973; b. 4-vi-1973; c. 13-xi-1973.

Growth details: In nursery germination good, ± simultaneous. Many seedlings germinated around parent tree in forest.

PARISHIA

Fl. Males. I, 8: 541.
Genus in S.E. Asia and throughout W. Malesia, including the Philippines.

Parishia insignis Hook. f. – Fig. 34.

Development: The taproot and hypocotyl emerge from the top of the fruit between the wing-like enlarged calyx-lobes, and curve between the wings to reach the soil. When the root is established, the curved hypocotyl elongates, and the spreading cotyledons are withdrawn from the fruit. A short resting stage occurs with the cotyledons and 2 developed leaves.
Seedling epigeal, phanerocotylar. Sloanea type and subtype.
Taproot long, slender, flexuous, fibrous, dark brown, with many rather short, few-branched, brown sideroots.
Hypocotyl strongly enlarging, terete, slender, to 3.2 cm long, clearly distinguishable from taproot and stem, slightly swollen, slightly tapering to the top, greenish brown, hypogeal part brownish, rather densely covered with minute, hyaline, simple hairs.
Cotyledons 2, opposite, succulent, in the 3rd leaf stage dropped; simple, sessile, swollen, without stipules, green, glabrous. Blade ovate in outline, inside slightly convex, outside distinctly concave, to 12 by 7 by 5 mm; base slightly narrowed; top caudate; margin entire.
Internodes: First one strongly enlarging, in cross-section slightly elliptic, to 3.5 cm long, slender, brownish green, hairy as the hypocotyl; next internodes as first one, variable in length, before the 15th one not exceeding 2.5 cm in length. Stem slender, in the beginning slightly twisted, later ± straight.
First two leaves opposite, in general compound, without stipules, petiolate, herbaceous, green. Petiole terete, to 7.5 by 0.7 mm, swollen at the base, hairy as the hypocotyl. Blade imparipinnate (1–) 2 (–3)-foliolate. Leaflets elliptic, opposite, ± sessile, to 2.4 by 1.2 cm, the lateral ones usually smaller; base acute, often oblique; top cuspidate; margin entire; slightly hairy as the hypocotyl, especially on the margin and the nerves; nerves pinnate, above slightly raised, below much so, lateral nerves ending in anastomoses.
Next leaves spirally arranged, compound, the lowest 4–8 3-foliolate, sometimes intercalated with unifoliolate or 2-foliolate ones, higher ones 5-foliolate. Petiole as that of first leaf, in 12th one to 5 by 1.5 cm. Rhachis as petiole, in 12th one to 5.5 cm long. Petiolules short, in 12th leaf those of the lateral leaflets to 4 by 2 mm, that of the terminal one to 2 by 2 mm. Blade of leaflets in lowest leaves ovate, in 12th leaf 7 by 4 cm, in higher ones ovate-oblong, to 10 by 4.3 cm. Blade of terminal leaflet 10.5 by 4.7 cm, further as preceding leaves.

Specimens: 1863 from S.E. Borneo, mountain ridge, primary forest on limestone

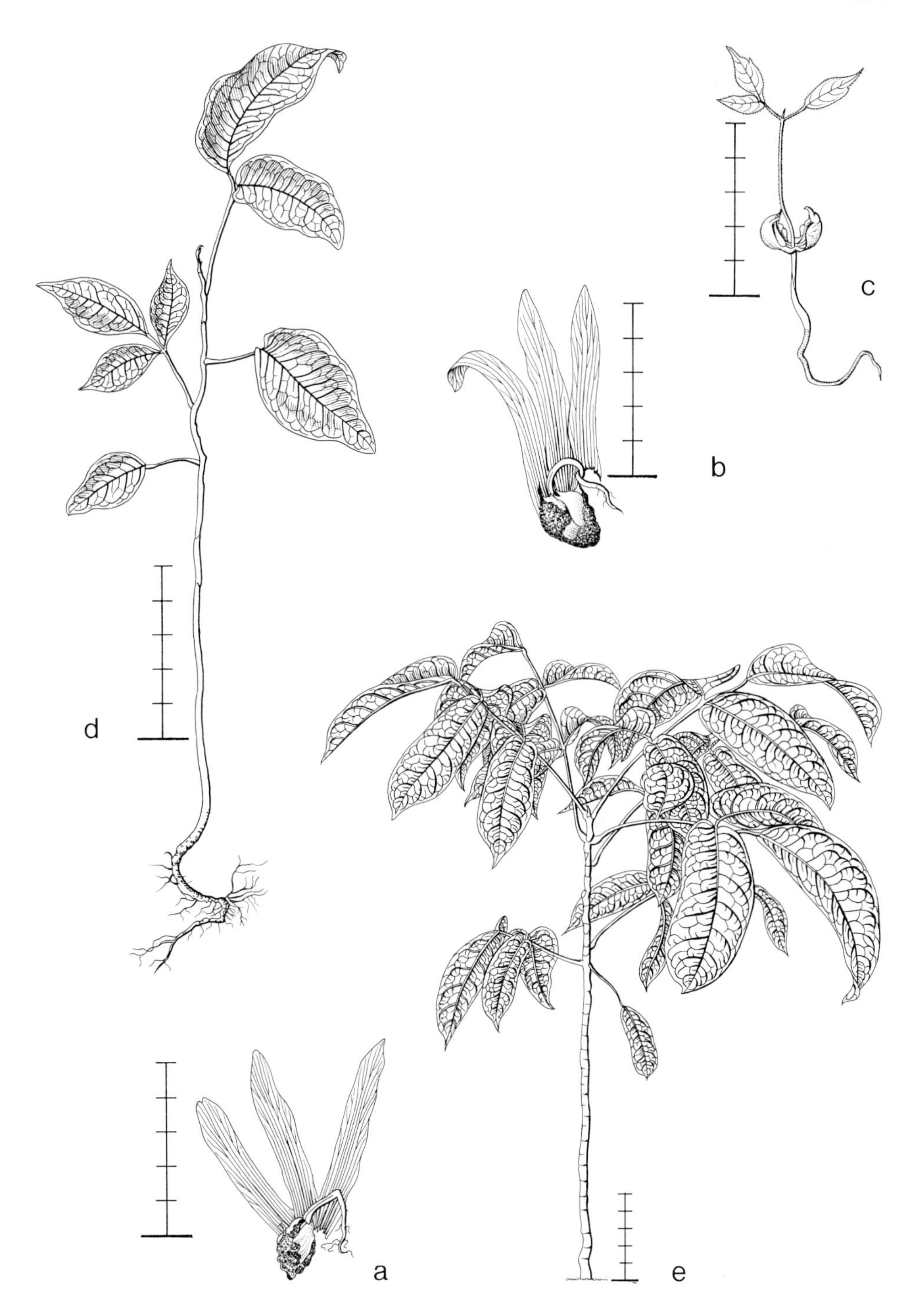

Fig. 34. *Parishia insignis.* De V. 1863, collected 26-x-1972, germinated seeds planted 17-xi-1972. a, b, and c date?; d. 15-viii-1973; e. 6-vii-1974.

rock, medium altitude.
Growth details: In nursery germination poor. Between the moment of germination, and the root entering the soil, the seedling is liable to be affected by adverse conditions. Drying of the root often occurs.

SPONDIAS

Fl. Males. I, 8: 479; Duke, Ann. Miss. Bot. Gard. 52 (1965) 340 (*S. purpurea*); Mensbruge, Germination (1966) 233 (*S. mombin*); Troup, Silviculture 1 (1921) 247 (*S. mangifera*).
Genus in Indo-Malesia and American tropics. *S. mombin* is a native of tropical America, commonly cultivated in Java.

Spondias mombin L. – Fig. 35.

Development: Taproot and hypocotyl both emerging from the blunt end of the fruit, piercing the fruit wall. The hypocotyl first curves, then withdraws the cotyledons from the envelopments. A short resting stage occurs with two cotyledons spread, and a short one with the cotyledons and the first two leaves developed.
Seedling epigeal, phanerocotylar. Sloanea type and subtype.
Taproot long, slender, flexuous, white turning cream-coloured, sideroots many, branched, white turning cream-coloured.
Hypocotyl strongly enlarging, terete, to 4.5 cm long, sometimes slightly angular, reddish, glabrous.
Cotyledons 2, opposite, slightly succulent, in the 2nd–4th leaf stage dropped; simple, without stipules, sessile, fleshy, green, glabrous. Blade linear, to 35 by 3 mm; top acute; margin entire, above flat, below concave, 3–5-nerved, main nerve sunken above, all nerves either inconspicuous to very prominent below, varying in different specimens.
Internodes: First one in cross-section rather angular, elongating to 1.7 cm long, greenish, sometimes tinged red, glabrous, later with whitish, warty lenticels; next ones as first one, the second one to 1.5 cm long, further ones not exceeding 0.8 cm. Stem slender, ± straight.
First two leaves opposite, compound, without stipules, petiolate, leaflets petiolulate, herbaceous, green, glabrous. Petiole to 10 by 1.5 mm, in cross-section semi-orbicular, above with a longitudinal ridge in the centre, reddish, especially above. Blade trifoliolate, sometimes two of the leaflets partly fused, ± triangular in outline, to 3.5 by 3.5 cm; petiolules as petiole, to 1.2 by 0.8 mm, that of the median leaflet to 3.5 mm long. Lateral leaflets ovate, to 2 by 1.2 cm; base rounded to slightly acute, often inequal; top acute to acuminate; margin rather regularly dentate, teeth acute; nerves pinnate, slightly raised above, midrib prominent below, lateral nerves much less so, joined into a clear marginal nerve close to the margin. Median leaflet to 3.5 by 1.8 cm; base acute, equal, further as the lateral leaflets.
Next leaves spirally arranged, the lower 3–5 5-foliolate; blade to 6.5 by 6 cm in outline, leaflets larger than those of the first leaf; further leaves 7-foliolate, not seen

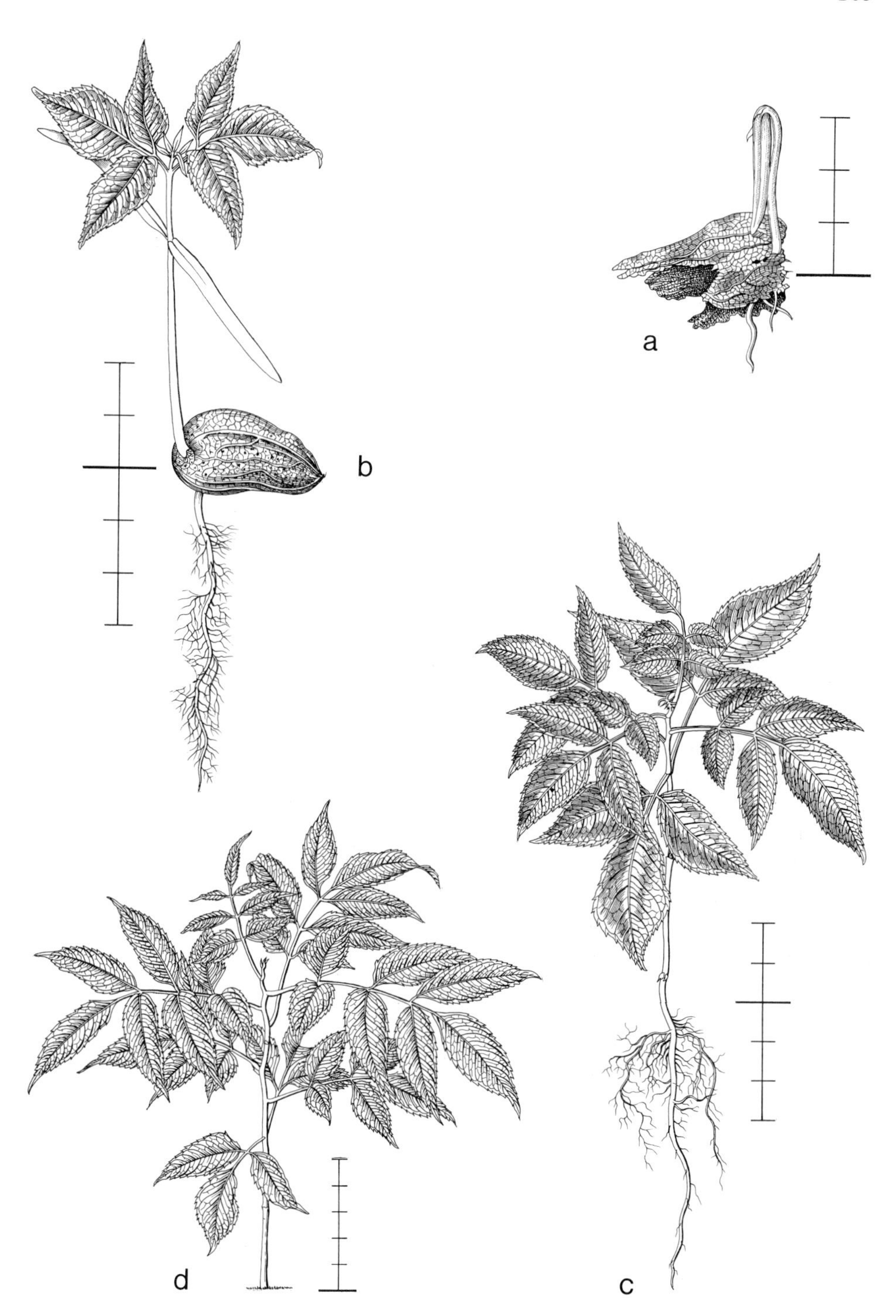

Fig. 35. *Spondias mombin.* Bogor Bot. Gard., without number, collected and planted 10-i-1973. a. 18-vii-1973; b. 19-vii-1973; c. 31-viii-1973; d. 19-vii-1974.

beyond 10th leaf stage.

Specimens: Bogor Botanic Gardens, without collection number.
Growth details: In nursery germination fair, ± simultaneous. The seedlings looked healthy at the start, but at the 5th–10th leaf stage for no apparent reason dropped their leaves and died.
Remarks: Three species described, similar in germination pattern and general morphology, differing in details only.

SWINTONIA

Fl. Males. I, 8: 440.
Genus throughout S.E. Asia and W. Malesia.

Swintonia cf. *schwenkii* (T. & B.) T. & B. ex Hook. f. – Fig. 36.

Development: The fruit wall and testa split lengthwise along one side, the taproot and hypocotyl emerge from just below the wing-like enlarged petals curving along the side. The cotyledonary petioles elongate, bringing the plumule free from the envelopments. No resting stage occurs.
Seedling ± epigeal, cryptocotylar. Heliciopsis type, Koordersiodendron subtype.
Taproot long, slender, fibrous, dark brownish red, with many long, slender, branched, brownish red sideroots.
Hypocotyl epigeal, enlarging to 2 cm long, terete, brownish, glabrous.
Cotyledons 2, secund, succulent, in the 2nd–3rd leaf stage dropped; simple, without stipules, petiolate, dark red where exposed. The whole cotyledonary body 2.3 by 1.3 by 1.3 cm, the dark brown, 5-winged fruit wall persistent. Petiole ⊥ on the blade, in cross-section semi-orbicular, to 5 by 2 mm, above channelled, dark red, glabrous. Wings of the fruit ± equal in size, to 4 by 0.8 cm.
Internodes: First one strongly elongating, terete, slender, to 9 cm long, reddish green, glabrous; next ones as first one, to 2.3 cm long, but in general not exceeding 1.5 cm. Stem slender, slightly zig-zag, growing in flushes with 2–5 leaves and some scales at the base, rather soon branched.
First two leaves opposite, simple, without stipules, petiolate, subcoriaceous, reddish yellow when young, turning pale green, pale greyish green below, glabrous. Petiole in cross-section semi-orbicular, to 13 by 1 mm, base distinctly swollen, pale green, glabrous. Blade oblong, to 12 by 4.3 cm; base acute; top cuspidate; margin entire; nerves pinnate, midrib above slightly raised, below prominent, lateral nerves above hardly raised, below rather prominent, shortly decurrent along the midnerve, almost at the tip curved and ascending in the slightly thickened margin.
Next leaves spirally arranged, simple, in each flush two developed ones with some scales. Scales ± 2 in each flush, sessile, much reduced, triangular, soon dropped. Petiole as those of first leaves, gradually becoming larger, in 4th leaf to 15 by 1 mm, in 10th leaf to 28 by 1.5 mm, at the base distinctly swollen. Blade oblong, in 5th leaf to 10.5 by 3.2 cm, in 10th leaf to 18 by 6 cm, further as first ones.

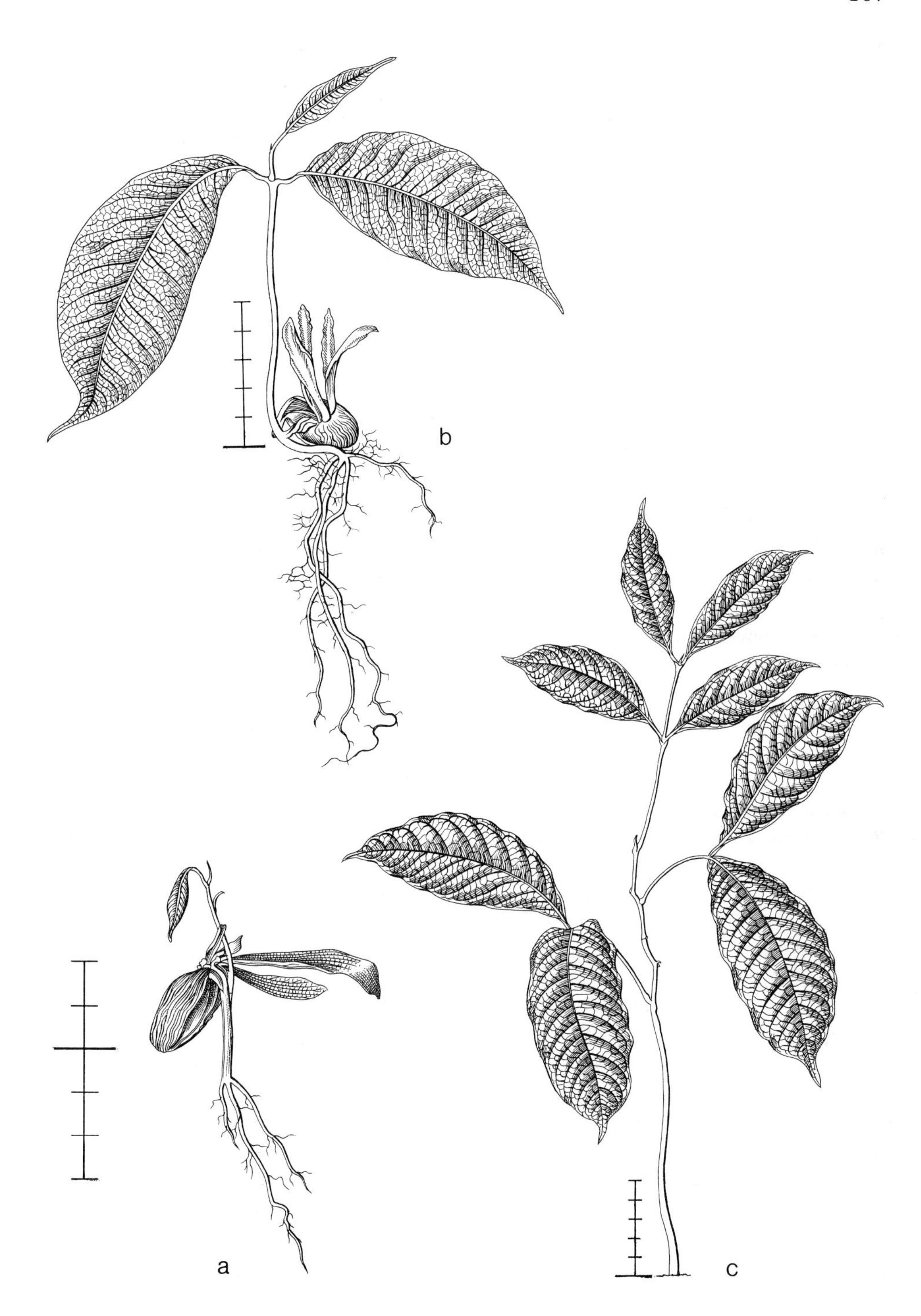

Fig. 36. *Swintonia* cf. *schwenkii.* De V. 2383, collected 16-iii-1973, planted 22-iii-1973. a. 18-iv-1973; b. 23-iv-1973; c. 19-xi-1973.

Specimens: 2383 from S.E. Borneo, primary forest on deep clay, medium altitude.
Growth details: In the forest germination abundant. Germinating seeds difficult to transport, in general dying soon.

ANNONACEAE

CYATHOCALYX

Fl. Java 1: 113.
Genus in India and Ceylon, throughout S.E. Asia, all over Malesia, in the Pacific as far as Santa Cruz.

Cyathocalyx sumatranus Scheff. – Fig. 37.

Development: The taproot and hypocotyl emerge from one blunt pole of the seed, carrying the enclosed paracotyledons up on the curved hypocotyl. A resting stage occurs with the paracotyledons enclosed in the testa on top of the straight or still curved hypocotyl. After some time the two valves with the enclosed ruminate endosperm are shed by spreading of the paracotyledons. Then follows a second temporary rest.
Seedling epigeal, phanerocotylar. Macaranga type.
Taproot sturdy, ± fleshy, brownish cream, with rather many quite long, hardly branched sideroots.
Hypocotyl strongly enlarging, slender, in cross-section slightly elliptic, to 13 cm long, towards the top ± 4-angular, dark green with many light green, linear lenticels, glabrous.
Paracotyledons 2, opposite, foliaceous, when dropped not known but not before the 9th leaf stage; for a considerable time hidden in the hard, ± ellipsoid, to 3.5 by 2 by 1.5 cm big, light brown, two-valved testa; afterwards expanding, simple, without stipules, shortly petiolate, herbaceous, green, glabrous. Petiole flattened, to 4 by 3 mm, in the centre with a longitudinal ridge formed by the descending midnerve. Blade ovate to ovate-oblong, to 7.5 by 4.1 cm; base obtuse, descending slightly along the petiole; top acute; margin entire; nerves 5, main nerves above prominent, especially the midrib, below much more so, ending free near the top of the blade, the marginal lateral nerves rather inconspicuous, the more central lateral nerves descending shortly along the midnerve into the petiole.
Internodes: First one shortly elongating, terete, to 7 mm long, slightly swollen towards the top, herbaceous, green, slightly hairy with minute, appressed, hyaline hairs; next ones as first one, irregular in length, 1–2 cm long, 8th one to 2.7 cm long. Stem rather slender, slightly zig-zag.
Leaves all spirally arranged, simple, petiolate, without stipules, herbaceous, green. Petiole in cross-section semi-orbicular, above slightly channelled, in first ones to 6 by 1 mm, in 8th one to 14 by 1.5 mm. Blade obovate-lanceolate, in first leaf to 12.5 by 3.5 cm, 5th one to 22 by 5.5 cm, but often smaller; base wedge-shaped; top

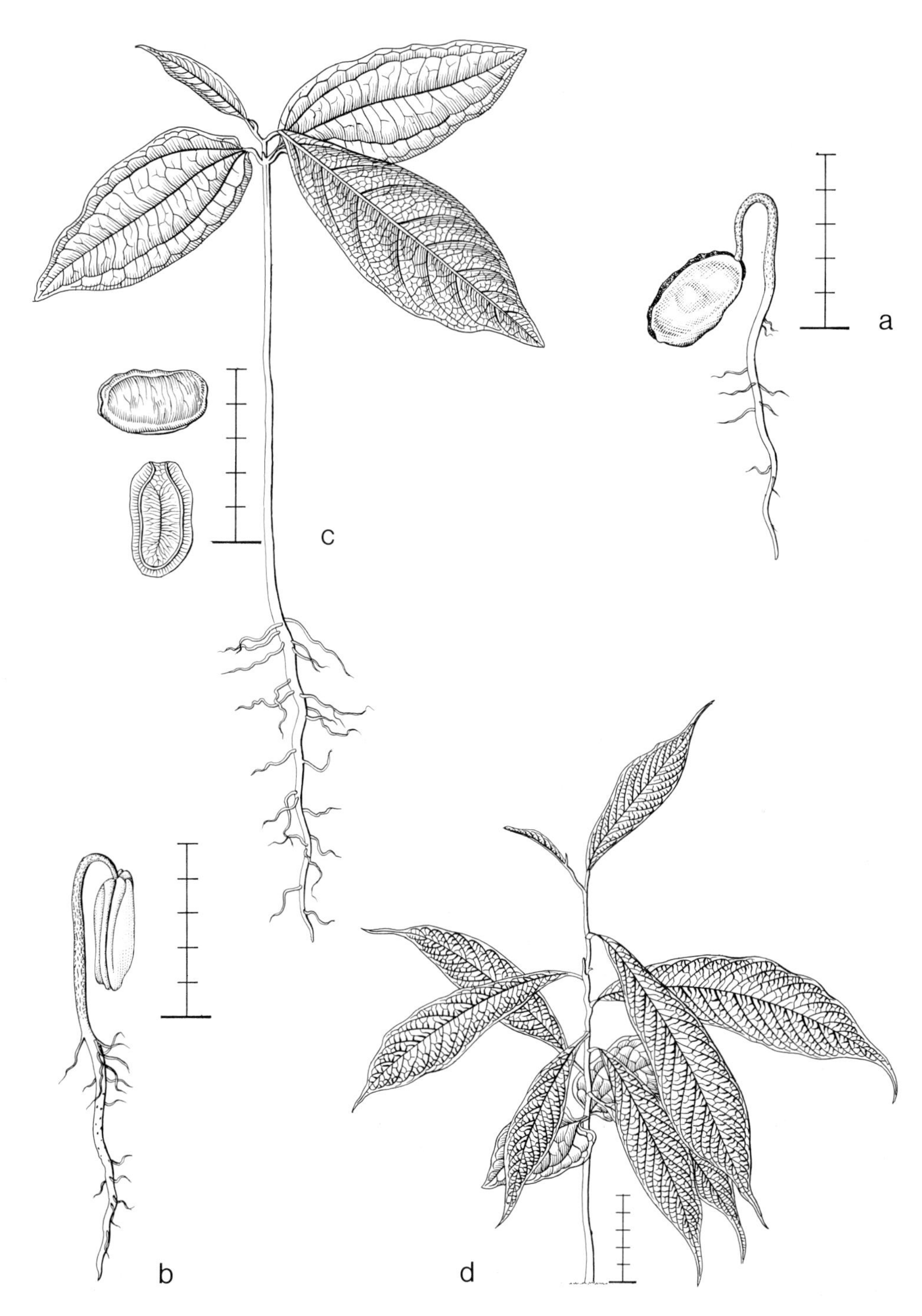

Fig. 37. *Cyathocalyx sumatranus.* De V. 1526, collected 23-vii-1972, planted 27-vii-1972. a, b, and c date?; d. 13-viii-1973.

acuminate to caudate, margin entire; on the base of the margin and below on the nerves slightly hairy with hairs as on the first internode; nerves pinnate, midrib sunken but above slightly prominent, sidenerves above sunken, at their tip curved to the next one, all nerves below very prominent, forming an undulating blade.

Specimens: 1526 from N. Sumatra, primary hill forest on deep clay, low altitude.
Growth details: In forest frequently scattered seedlings. In nursery germination and establishment fair, easy in cultivation.

MEZZETTIA

Tree Fl. Malaya 1: 77.
Genus in W. Malesia: Malay Peninsula, Sumatra, and Borneo.

Mezzettia leptopoda Oliv. – Fig. 38.

Development: The hypocotyl and taproot emerge from one pole of the seed. After establishment of the taproot, the hypocotyl becomes erect, carrying the cotyledons up. After a long resting period, the cotyledons, still enclosed in the hard testa and endosperm, are shed.
Seedling epigeal, cryptocotylar. Blumeodendron type.
Taproot long, sturdy, distinctly swollen, hard fleshy; sideroots very few, short, not branched, in the young plant restricted to the upper part and collar of the taproot.
2Hypocotyl strongly enlarging, terete, sturdy, to 10 cm long, green, glabrous.
Cotyledons 2, opposite, on top of the hypocotyl, flat, serving as sucker organ, dropped with the start of the development of the shoot; without stipules, sessile. Whole body ellipsoid in outline, to 4 by 3 by 2.5 cm, one side flattened, the other convex, along the margin with a deep groove, the woody, thick, brown valves of the seed persistent.
Internodes: First one elongating, terete, to 1.5 cm long, green, glabrous; next internodes as first one, hardly becoming larger, ± the first 20 to 2 cm long, the next 15 to 3 cm long, higher ones to 4 cm long. Stem slender, in the beginning erect, straight, later drooping.
Leaves alternate, simple, when young involute and covering a younger leaf, without stipules, petiolate, herbaceous, green, glabrous. Petiole in cross-section semi-orbicular, in the first leaf to 4 by 1.5 mm, in 20th leaf to 5 by 2 mm, jointed at the base. Blade in the first leaf somewhat irregularly formed, ovate-oblong, to 8 by 2.7 cm, but often much smaller, higher ones lanceolate, 10th one to 16.5 by 3.5 cm, 20th one to 21 by 4.6 cm, base acute to slightly narrowed; top caudate; margin entire; nerves pinnate, midrib above slightly raised, below prominent, lateral nerves above slightly sunken, below slightly raised, all curved at their tip to the next one.

Specimens: 1412 from N. Sumatra, alluvial flat, primary forest on deep clay, low altitude.
Growth details: In nursery germination good, over a long stretch of time.

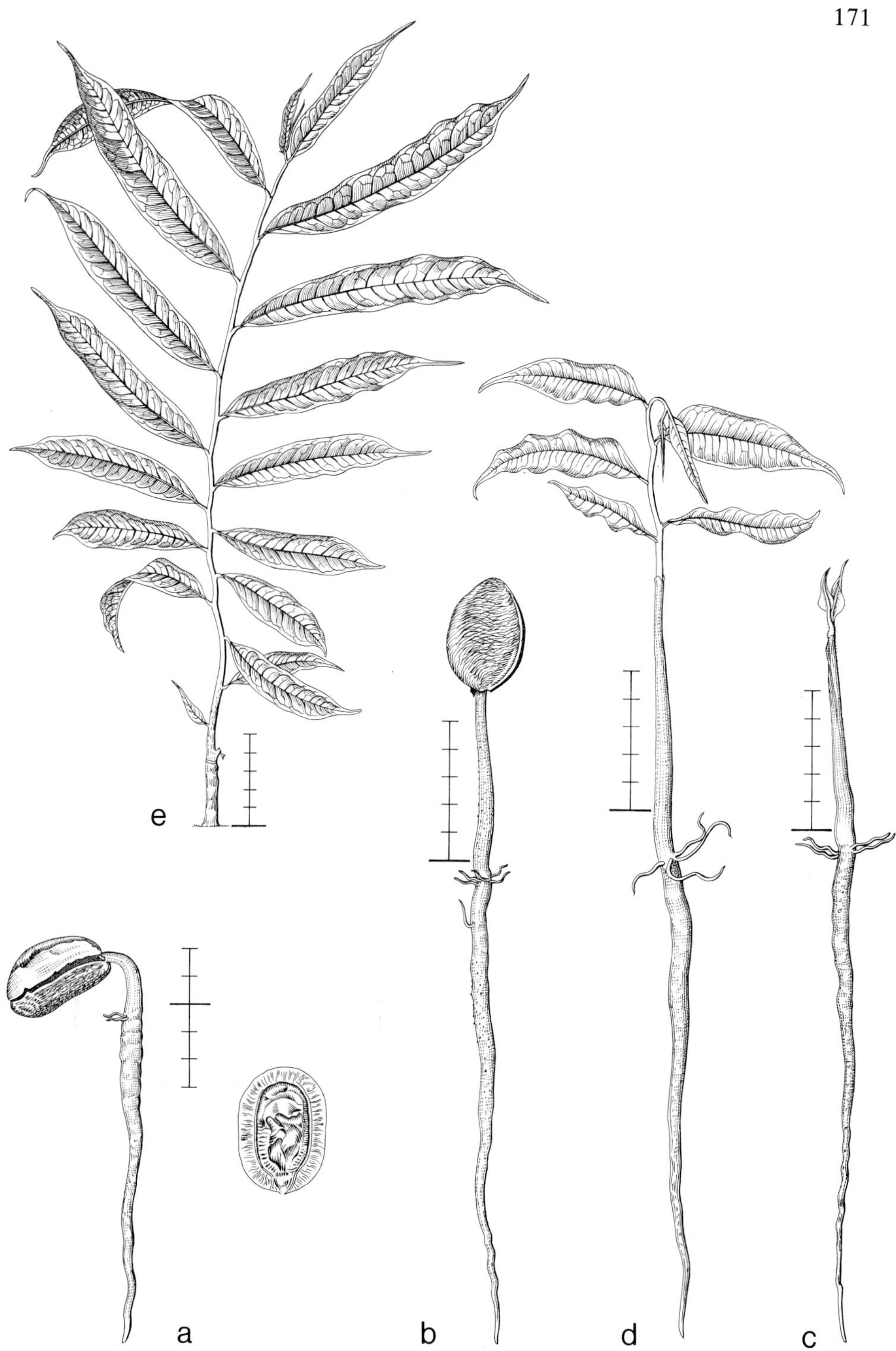

Fig. 38. *Mezzettia leptopoda.* De V. 1412, collected 10-vi-1972, planted 27-vii-1972. a, b, c, and d date?; e. 7-viii-1973.

MEZZETTIOPSIS

Ridley, Sarawak Mus. J. 1 (1913) 73–98.
Genus endemic in Borneo.

Mezzettiopsis creaghii Ridl. – Fig. 39.

Development: The taproot and hypocotyl emerge from the margin of the seed, the hypocotyl looped with the enclosed cotyledons at soil level, sometimes later becoming erect with the cotyledons on top. After a resting period, the cotyledonary petioles are detached and the cotyledons are shed, after which the shoot starts developing.
Seedling 'epigeal', in general the seed remaining at soil level at the top of the curved epigeal hypocotyl, cryptocotylar. Blumeodendron type.
Taproot long, slender, flexuous, brownish cream, with rather few long, slender, few-branched, creamy-brown sideroots.
Hypocotyl strongly elongating, terete, to 7 cm long, with two longitudinal grooves, dark bluish to reddish green, glabrous, curved in the resting stage, distinctly warty with brownish lenticels.
Cotyledons 2, normally remaining at soil level on top of the curved hypocotyl, shed when the shoot starts developing; without stipules, minutely petiolate. Whole body in cross-section semi-globular, to 1.2 by 1 by 0.5 cm, the pale brownish testa persistent, usually flattened on one side, along the margin with a distinct, wide groove, surface slightly globulate to undulate. Petioles hardly exceeding the testa, in cross-section elliptic, to 2 by 1 mm.
Internodes: First one terete, elongating to 1 cm, rather densely hairy with short, appressed, hyaline hairs; next ones as first one, 5th one to 1.5 cm long, rarely to 4.5 cm. Stem slender, rather straight, often soon branched.
First two leaves opposite, simple, conduplicate-induplicate when young, without stipules, shortly petiolate, herbaceous, green. Petiole in cross-section semi-orbicular, to 3 by 1 mm, swollen at the base, hairy as first internode. Blade ovate-oblong, to 5.5 by 2.2 cm; base rounded; top acuminate to cuspidate; margin entire; below on the nerves slightly hairy as the hypocotyl; nerves pinnate, main nerve above sunken, below prominent, sidenerves above slightly sunken, below slightly prominent, marginal nerve inconspicuous, widely undulating.
Next leaves spirally arranged. Petiole in 3rd leaf to 2 by 1 mm, in 8th leaf to 2 by 1 mm. Blade in 3rd leaf to 7 by 3.8 cm, in 8th leaf to 8 by 3.2 cm; nervation as in first leaves, but lateral nerves free ending.

Specimens: 2247 from S.E. Borneo, primary hill forest on deep clay, low altitude.
Growth details: In nursery germination fair, rather simultaneous, long delayed.

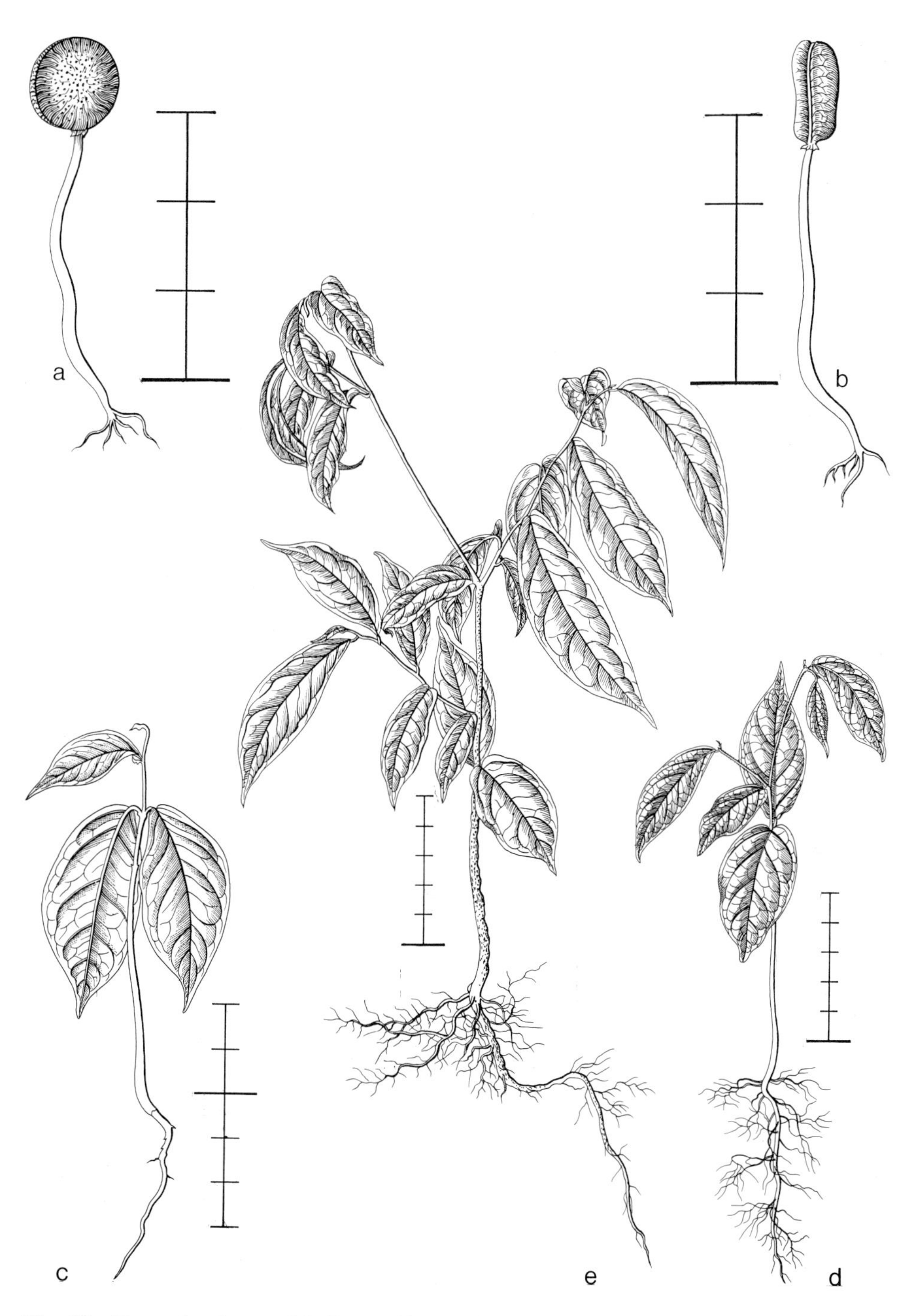

Fig. 39. *Mezzettiopsis creaghii.* De V. 2247, collected 9-iii-1973, planted 22-iii-1973. a, b, and c. 25-vi-1973; d. 5-xi-1973; e. 2-vii-1974.

POLYALTHIA

Tree Fl. Malaya 1: 85; Grushvitskyi, Bot. Zhurn. 48 (1963) 906–915 (*P. suberosa*).
Genus in Madagascar, India, and Ceylon, throughout S.E. Asia, all over Malesia, N.E. Australia, in the Pacific as far as Tonga Arch., including the W. Carolines, and New Caledonia.

Polyalthia lateriflora King var. *elongata* Boerl. – Fig. 40.

Development: The short hypocotyl and taproot emerge from one pole of the seed. The cotyledonary petioles elongate, bringing the plumule free from the envelopments, enabling the shoot to elongate. No resting stage occurs.
Seedling hypogeal, cryptocotylar. Horsfieldia type and subtype.
Taproot long, sturdy, hard and tough, creamy-brown turning dark grey-brown, with rather few, sturdy, tough, creamy-brown, white tipped sideroots.
Hypocotyl somewhat elongating, subterranean, terete, to 2 cm long, dark brown, in texture not differing from the root, later with small, somewhat elevated, paler lenticels, glabrous.
Cotyledons 2, secund, flat but somewhat fleshy, in the 2nd–3rd leaf stage shed; irregularly formed, without stipules, petiolate. Whole body ellipsoid, to 2.7 by 1.3 by 1.4 cm, circumcised with a longitudinal groove, the pale creamy-brown testa persistent. Petiole in cross-section semi-orbicular, to 10 by 3.5 mm, distinctly curved, above flat, dark brown, glabrous.
Internodes: First one strongly elongating, terete, to 10 cm long, green, turning dark brownish grey, rather densely covered with minute, simple, brown to hyaline hairs and with scattered, warty lenticels; next internodes as first one, second one to 1 cm long, third one to 5 cm long, 5th one up to 15 cm long. Stem rather slender, slightly zig-zag.
Leaves all spirally arranged, the lowest ones sometimes reduced or semi-opposite, simple, without stipules, shortly petiolate, subcoriaceous, reddish when young, turning green. Petiole in cross-section semi-orbicular, in the lower leaves to 5 by 2 mm, in 3rd leaf to 7 by 2 mm, brownish green, hairy as the first internode. Blade oblong to obovate-oblong, in the first leaf to 11 by 5 cm, but often much smaller, in 4th leaf to 10 by 5 cm; base acute; top cuspidate; margin entire; below slightly hairy on the nerves; nerves pinnate, above sunken, below prominent, sidenerves ending in anastomoses; pellucid with minute hyaline dots.

Specimens: 2487 from N. Celebes, primary forest on shallow clayey sand on rock, medium altitude; 2592, from N. Celebes, primary forest on bare rock, low altitude.
Growth details: In nursery germination good, rather long delayed, ± simultaneous.
Remarks: Two species described. In *P. suberosa* the seeds are lifted above soil level by the elongation of the hypocotyl, and the cotyledons are dropped together with the seedcoat (Blumeodendron type).

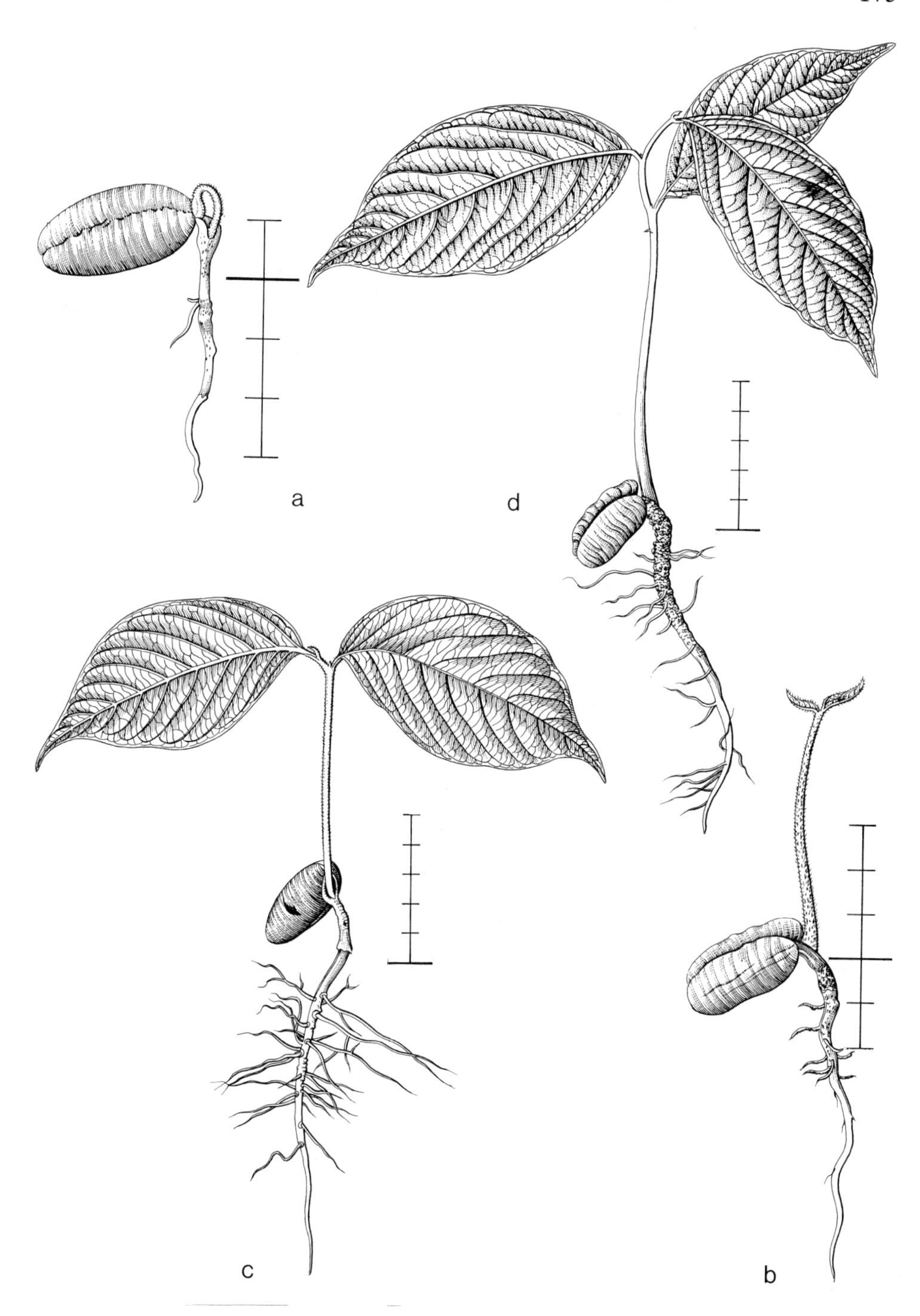

Fig. 40. *Polyalthia lateriflora* var. *elongata*. De V. 2487, collected 8-x-1973, planted 7-xi-1973 (a, c); De V. 2592, collected 22-x-1973, planted 7-xi-1973 (b, d). a and b. 12-xii-1973; c and d. 22-i-1974.

PSEUDUVARIA

Fl. Java 1: 111.
Genus in S.E. Asia, throughout Malesia including the Philippines and New Guinea.

Pseuduvaria reticulata (Bl.) Merr. – Fig. 41.

Development: The taproot and hypocotyl emerge from one pole of the seed, which is carried up by the hypocotyl becoming erect. The cotyledonary petioles push the plumule free from the envelopments, enabling the shoot to emerge. No resting stage occurs.
Seedling epigeal, cryptocotylar. Horsfieldia type, Pseuduvaria subtype.
Taproot rather sturdy, slightly fleshy, hard, brown to creamy-brown, with few, branched, creamy-white sideroots.
Hypocotyl strongly enlarging, to 4 cm long, slender, terete, brownish, rather densely covered with minute, simple, patent, brownish hyaline hairs.
Cotyledons 2, secund, thin but somewhat fleshy, in 2nd–5th leaf stage dropped; without stipules, petiolate, remaining in the ruminate endosperm. Whole body ellipsoid, ± 13 by 8 by 6 mm, with a groove along the margin, the rather soft, greyish, wrinkled testa persistent. Petioles exceeding the testa for ± 2 mm, in cross-section flattened, hairy as the hypocotyl.
Internodes: First one strongly elongating, terete, to 3.5 cm long, green, hairy as the hypocotyl; next internodes much shorter, the lowest ones to 1.3 cm long, 5th one to 2.5 cm long, further as the first internode. Stem rather slender, erect, slightly zig-zag.
Leaves all spirally arranged, simple, without stipules, petiolate, subcoriaceous, green. Petiole short, terete, in the first leaves to 3 by 0.8 mm, in 10th leaf to 5 by 1 mm, brownish by dense covering of hairs as on the hypocotyl. Blade ± ovate, higher ones oblong, first one to 3 by 2 cm, but often somewhat reduced in size, 10th one to 10 by 3.8 cm; base rounded; top acute to acuminate, in higher ones cuspidate; margin entire; above on the nerves and along the margin, and below especially on the nerves hairy as the hypocotyl; nerves pinnate, above sunken, below prominent, lateral nerves free ending.

Specimens: 1415 from N. Sumatra, planted on deep clay, and 1537 from N. Sumatra, alluvial flat near river, primary forest on deep clay, all low altitude.
Growth details: In nursery germination fair to poor, ± simultaneous.

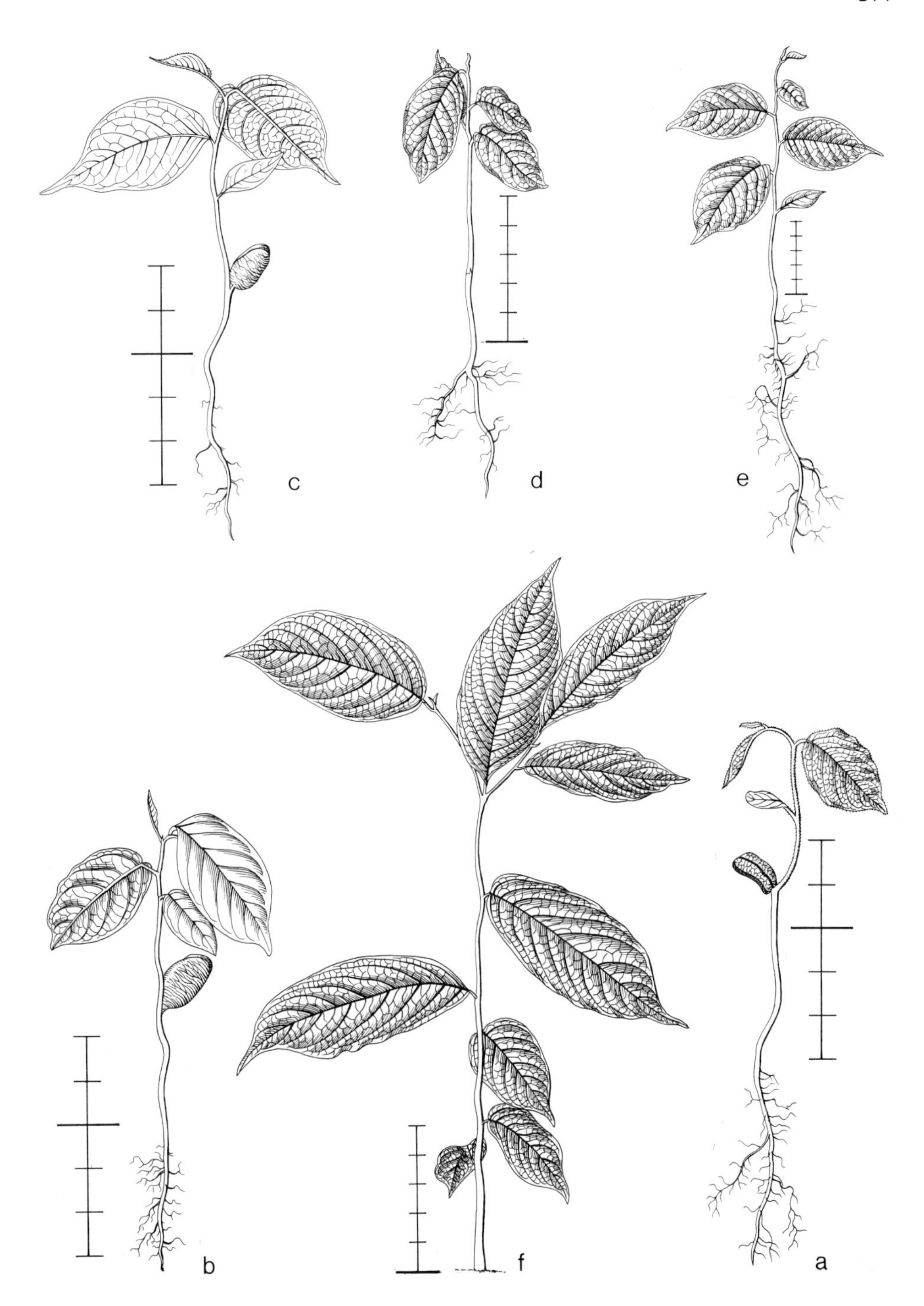

Fig. 41. *Pseuduvaria reticulata*. De V. 1415, collected 9-vii-1972, planted 27-vii-1972 (a, b, c, and d); De V. 1537, collected 21-vii-1972, planted 27-vii-1972 (e, f). a, b, and c date?; d. 2-viii-1973; e and f date?.

STELECHOCARPUS

Fl. Java 1: 102.
Genus in S.E. Asia, throughout Malesia as far as the Solomons, except in the Philippines.

Stelechocarpus burahol (Bl.) Hook. f. & Thoms. – Fig. 42.

Development: The hypocotyl and taproot emerge from one pole of the seed which remains at soil level. The cotyledonary petioles elongate, pushing the plumule free from the envelopments. The shoot is then enabled to emerge. No resting stage occurs.
Seedling hypogeal, cryptocotylar. Horsfieldia type and subtype.
Taproot long, greyish, fusiform, distinctly swollen in the upper part, tapering to base and top, lower part fibrous, slender, along the whole length with small, unbranched, thin, greyish sideroots.
Hypocotyl hypogeal, strongly elongating, in cross-section terete, to 6 cm long, lengthwise grooved, with dark brown, warty lenticels, glabrous, greyish.
Cotyledons 2, secund, thin but somewhat fleshy, in the 1st–3rd leaf stage shed, remaining in the ruminate endosperm, without stipules, petiolate. Whole body ± ellipsoid, to 3 by 2 by 1.5 cm, distinctly flattened below, the brownish testa persistent, with a central ridge, circumcised along the margin with a groove. Petioles hardly exceeding the testa.
Internodes: First one strongly elongating, terete, to 2.5 cm long, green when young, turning greyish green, covered with rather close-set, simple, hyaline hairs; next internodes as first one, those in the scaly part of the stem 3–6 cm long, in the leafy part much smaller, in general not exceeding 1–2 cm, sometimes to 5 cm long. Stem rather slender, straight.
Leaves all spirally arranged, simple, without stipules, petiolate, herbaceous, reddish when young, turning green, glabrous, the lower 1–2 abortive, scale-like. Petiole in cross-section semi-orbicular, in the lower ones to 5 by 1.5 mm, in 10th one to 8 by 1.5 mm, above with a longitudinal ridge formed by the descending midnerve. Blade oblong, the lower ones 4.5 by 2 cm, in the 4th–5th leaf to 10 by 3.3 cm, in 10th one to 13.5 by 5 cm; base acute to narrowed; top cuspidate; margin entire; nerves pinnate, midrib above and below prominent, lateral nerves in the lower leaves rather inconspicuous, above slightly sunken, below slightly prominent, forming a much undulating marginal nerve.

Specimens: Dransfield 2542 from W. Java, secondary forest on deep clay, low altitude.
Growth details: In nursery germination good, over a stretch of time, long delayed. Seedlings slow growing.

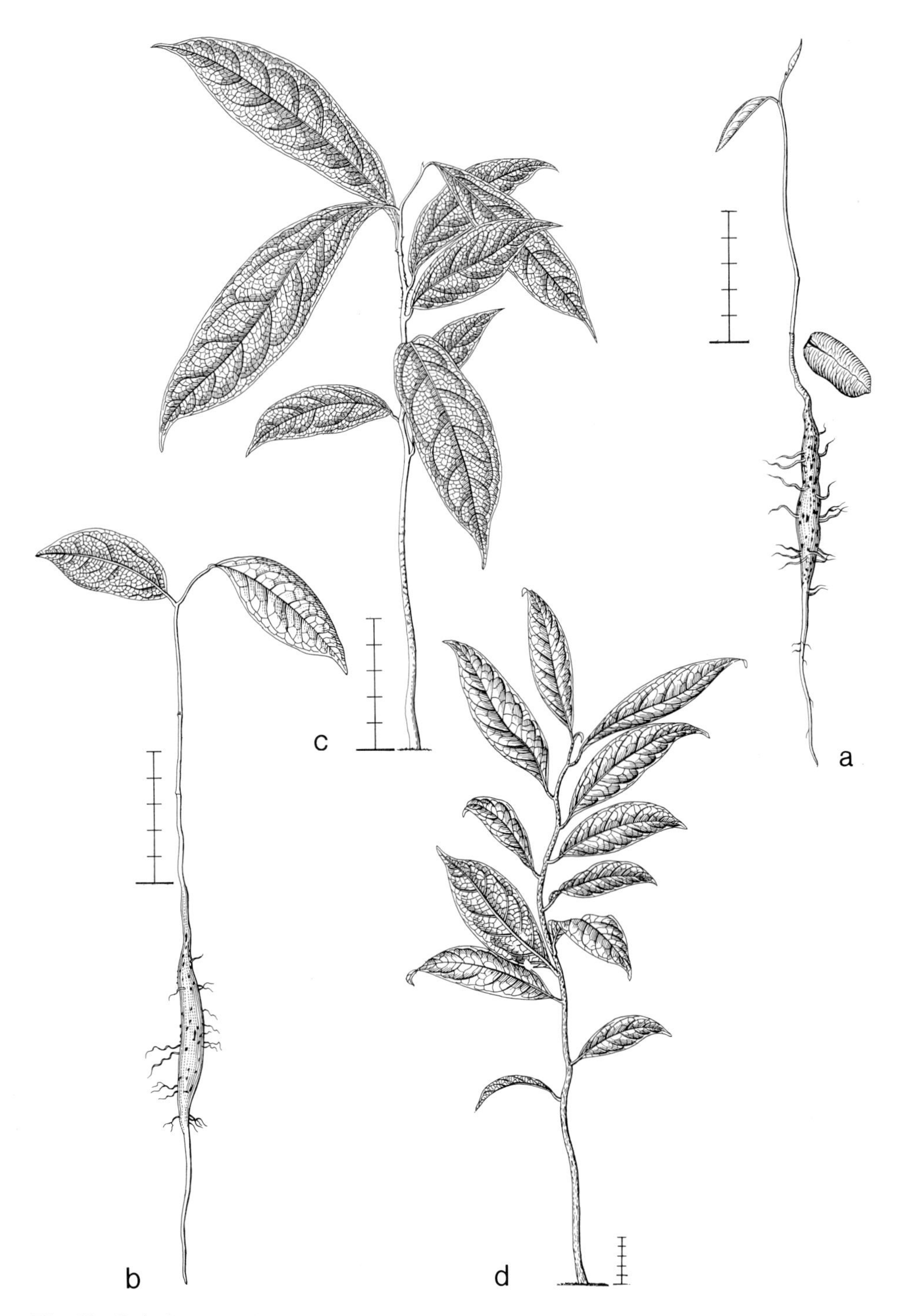

Fig. 42. *Stelechocarpus burahol.* Dransfield 2542, collected 17-vi-1972, planted 20-vi-1972. a, b, c, and d date?.

TRIVALVARIA

Fl. Java 1: 112.
Genus in S.E. Asia and W. Malesia, including Borneo and W. Java.

Trivalvaria macrophylla (Bl.) Miq. – Fig. 43.

Development: The hypocotyl and taproot emerge from the seed, which is carried up by the hypocotyl becoming erect. After a resting period, with the paracotyledons upright, enclosed in testa and endosperm, the paracotyledons expand and spread, shedding the testa and endosperm. Then follows a second temporary rest.
Seedling epigeal, phanerocotylar. Macaranga type.
Taproot long, sturdy, hard, creamy-white, turning blackish, with few, rather short, very slender, cream-coloured, few-branched sideroots.
Hypocotyl strongly elongating, terete, to 5 cm long, dark green, below paler, glabrous.
Paracotyledons 2, opposite, foliaceous, dropped in the 8th–12th leaf stage; simple, without stipules, shortly petiolate, coriaceous, green, glabrous. Petiole in cross-section semi-orbicular, to 0.5 by 3 mm. Blade ovate, but often irregularly formed, to 3.5 by 2 cm; base slightly retuse; top slightly acute; margin entire; nerves pinnate, midrib above slightly prominent at the base, below raised, sidenerves hardly raised, curved at the tip.
Internodes: First one elongating to 7 mm long, terete, green turning brownish black, rather densely hairy with short, simple, appressed, hyaline hairs; next internodes as first ones, hardly gaining in length, 10th one to 8 mm long. Stem sturdy, slightly zig-zag.
Leaves all spirally arranged, simple, without stipules, shortly petiolate, coriaceous, green, paler green below. Petiole in cross-section semi-orbicular, in the first leaf to 1 by 1 mm, in the 10th leaf to 4 by 2 mm, below hairy as first internode. Blade elliptic, first one to 3.2 by 2.3 cm, 10th one to 10.5 by 5.8 cm; base ± acute; top acuminate; margin entire; below on the nerve hairy as first internode; nerves pinnate, midrib above slightly sunken, below prominent, lateral nerves above slightly raised, more so below, ending in anastomoses.

Specimens: 1406 from W. Java, primary forest on coral sand and coral rock, low altitude.
Growth details: In nursery germination good, ± simultaneous. Plants very slow growing.

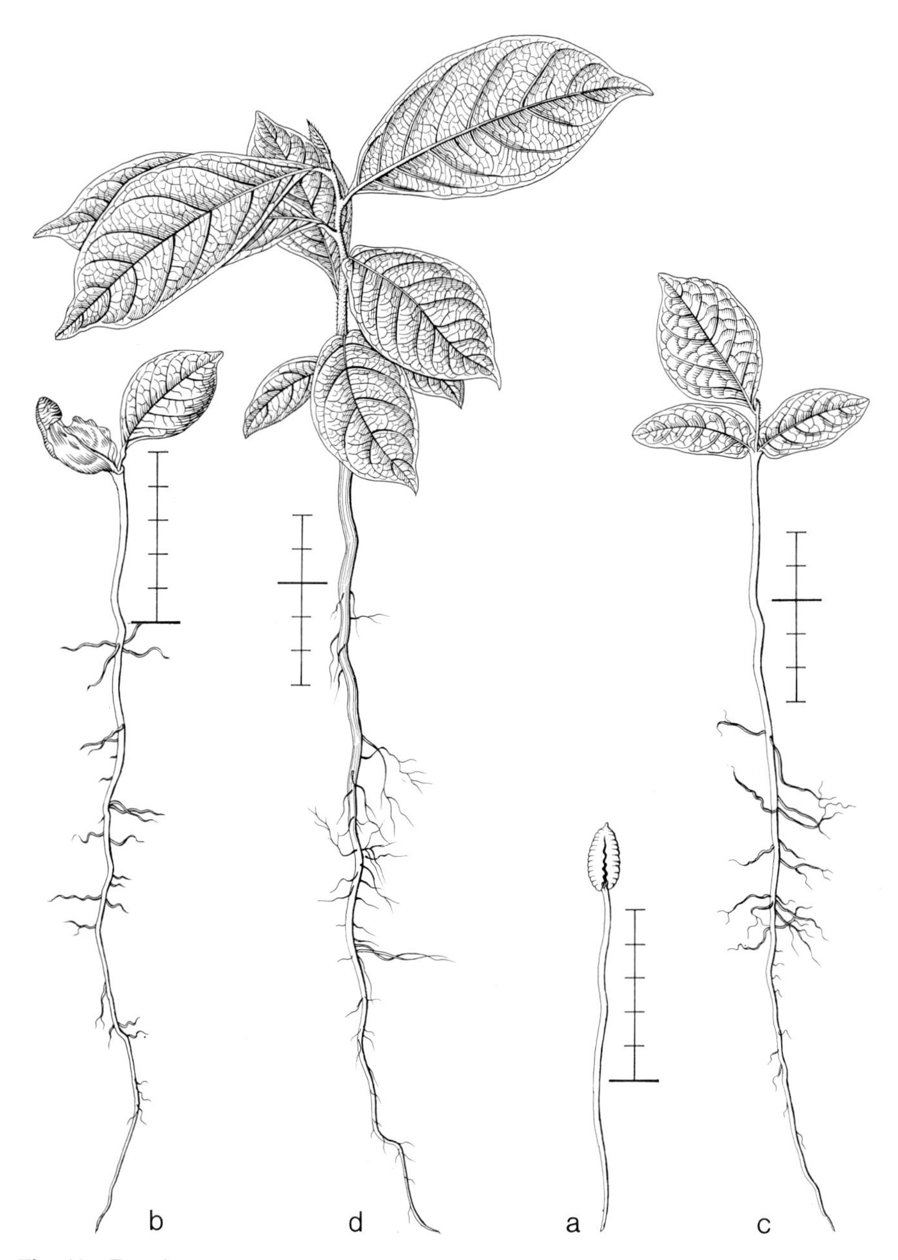

Fig. 43. *Trivalvaria macrophylla*. De V. 1406, collected 26-vi-1972, planted 2-vii-1972. a, b, and c date?; d. 2-viii-1973.

APOCYNACEAE

CERBERA

Fl. Java 2: 232.
Genus in Madagascar, throughout the Indian Ocean, India, and Ceylon, S.E. and E. Asia, all over Malesia, N.E. Australia, in the Pacific as far as the Marquesas and Tuamotu Arch., including the Marianas and New Caledonia.

Cerbera manghas L. – Fig. 44.

Development: The taproot emerges from the obtuse pole of the fruit. The endocarp valves open slightly by secondary growth of the lower internodes, thus presenting an opening through which the shoot may emerge. No resting stage occurs.
Seedling hypogeal, cryptocotylar. Heliciopsis type and subtype.
Taproot strong, stout, thick fleshy, greyish cream, when young often not able to emerge from the fruit wall, with few, short, hardly branched sideroots, these often interwoven with the fibres of the endocarp.
Hypocotyl not enlarging, formed as a strong cross-ridge between shoot and root, clasped in the fruit wall.
Cotyledons 2, secund, succulent, when shed not known, simple obovate, to 2.5 by 2 by 0.4 cm; without stipules, petiolate, creamy-white. The whole body 5.5 by 4.5 by 4 cm, the persistent endocarp spongy-fibrous, disintegrating to a rather smooth body 4.5 by 3 by 2 cm, with some persistent bases of fibres. Petioles flattened, to 8 by 4 mm, creamy-white.
Internodes: First one strongly elongating to 6 cm long, ± terete, glabrous, with many, warty, brownish lenticels, green, turning silvery green; next ones as first one, the lower irregular in length to 3.5 cm long, those in the leafy part more regular, below to 1 cm long, higher in the stem to 2 cm long. Stem rather sturdy, straight.
Leaves all spirally arranged, simple, spreading, without stipules, petiolate, herbaceous, green, glabrous, with white latex, the lower ones very reduced, gradually passing into the normal leaves. Petiole in cross-section semi-orbicular, to 15 by 1.5 mm in the lower developed leaves, in the 35th leaf to 20 by 2.5 mm, slightly tinged red. Blade in the lowest leaves oblong, to 9.5 by 3.3 cm, in the 30th leaf lanceolate to 16 by 4 cm; base acute, slightly decurrent; top acuminate to cuspidate; margin entire; nerves pinnate, midrib slightly raised above, much prominent below, lateral nerves hardly raised, fine, joined to an inconspicuous marginal nerve.

Specimens: Dransfield 2539 from W. Java, seashore, on coral sand, low altitude.
Growth details: In nursery germination fair, not simultaneous, over a long stretch of time.

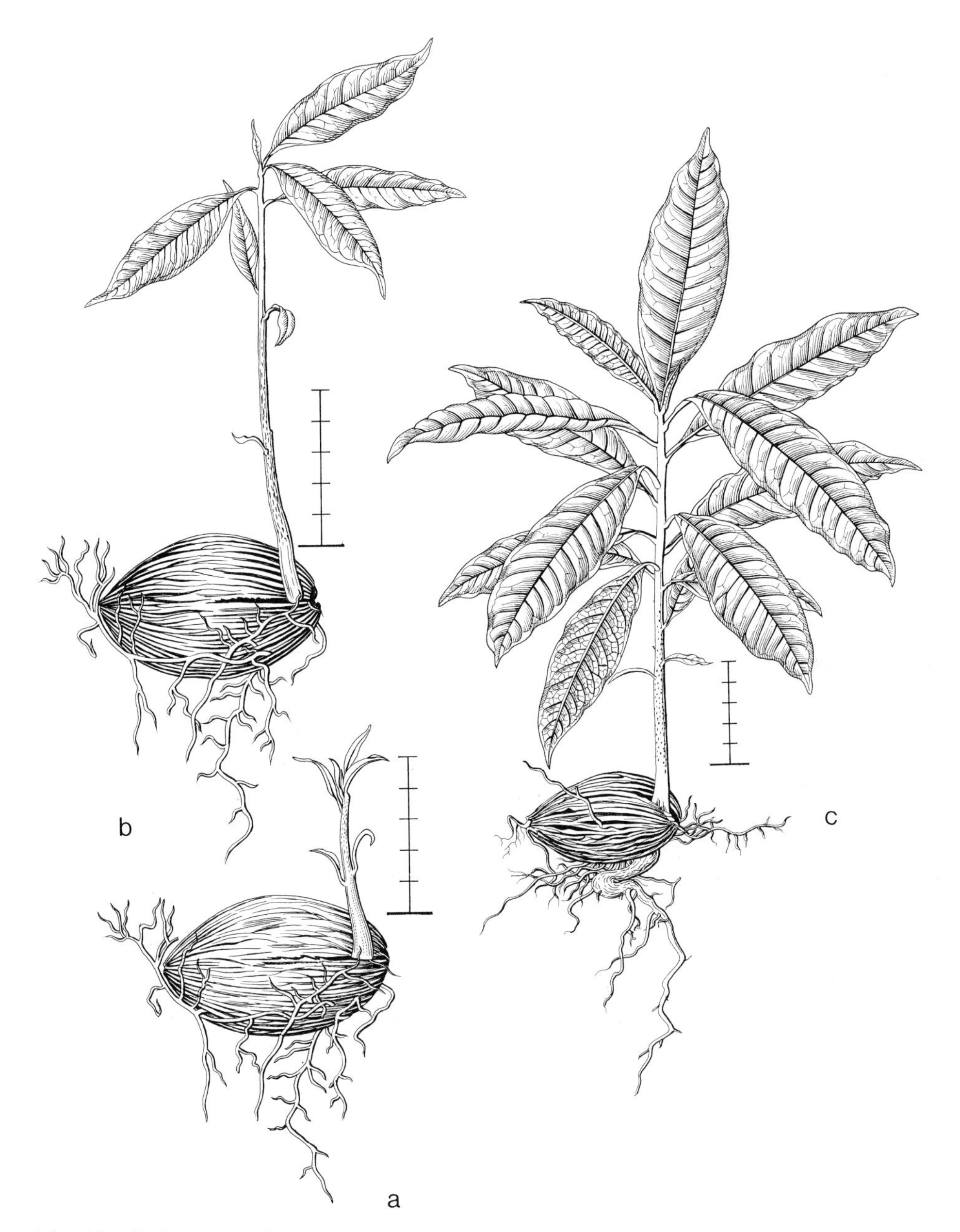

Fig. 44. *Cerbera manghas.* Dransfield 2539, collected 16-vi-1972, planted 20-vi-1972. a, b, and c date?.

KOPSIA

Fl. Java 2: 232.
Revision: Markgraf, Blumea 20 (1973) 416.
Genus in S.E. Asia, throughout Malesia to the New Hebrides.

Kopsia arborea Bl. – Fig. 45.

Development: The taproot emerges from one pole of the fruit, the fruit wall splitting at that place. The cotyledonary petioles elongate, thus bringing the plumule free from the envelopments. The shoot is then able to develop unhampered. Resting stages do not occur.
Seedling hypogeal, cryptocotylar. Heliciopsis type and subtype.
Taproot sturdy, hard fleshy, creamy-white, with rather few, long, unbranched, later much-branched sideroots.
Hypocotyl slightly elongating, clearly distinguishable from the root, ± terete, thickened, to 5 mm long, warty, cream-coloured, glabrous.
Cotyledons 2, secund, succulent, shed in the 6th leaf stage. The whole body ellipsoid, ± 2 by 1 cm, the bluish black fruit wall, of which the soft outer wall slowly disintegrates, slightly splitting but persistent. Petioles rather flat, to 5 by 3 mm, partly hidden in the fruit wall.
Internodes: First one strongly elongating, angular when young, terete when older, to 5.5 cm long, green, creamy-white at the base, glabrous; next ones as first one, irregular in length, to 2(–3) cm long in the lower ones, 15th one to 3 cm long, 20th one to 6 cm long. Stem rather slender, straight, with a cross-ridge at the nodes.
Leaves decussate, sometimes minute to abortive or inequal in size, simple, without stipules, shortly petiolate, herbaceous, green, glabrous, with white latex. Petiole in cross-section semi-orbicular, to 5 by 0.8 mm in the lower ones, in 7th pair to 5 by 1.5 mm, in 20th pair to 10 by 3 mm. Blade ovate-oblong, in the lower leaves to 4.5 by 1.6 cm, but often smaller or irregularly formed, in 7th pair to 8 by 3.8 cm, in 20th pair to 19 by 6 cm, base decurrent; top obtuse to acute, in higher ones acuminate with obtuse tip; margin entire; nerves pinnate, rather inconspicuous above, midrib prominent below, lateral nerves not prominent, dark green, joined into an inconspicuous marginal nerve.

Specimens: 1363 from W. Java, much depleted primary forest on limestone rock, low altitude.
Growth details: In nursery germination good. Seedlings retarded in growth when put under shade.

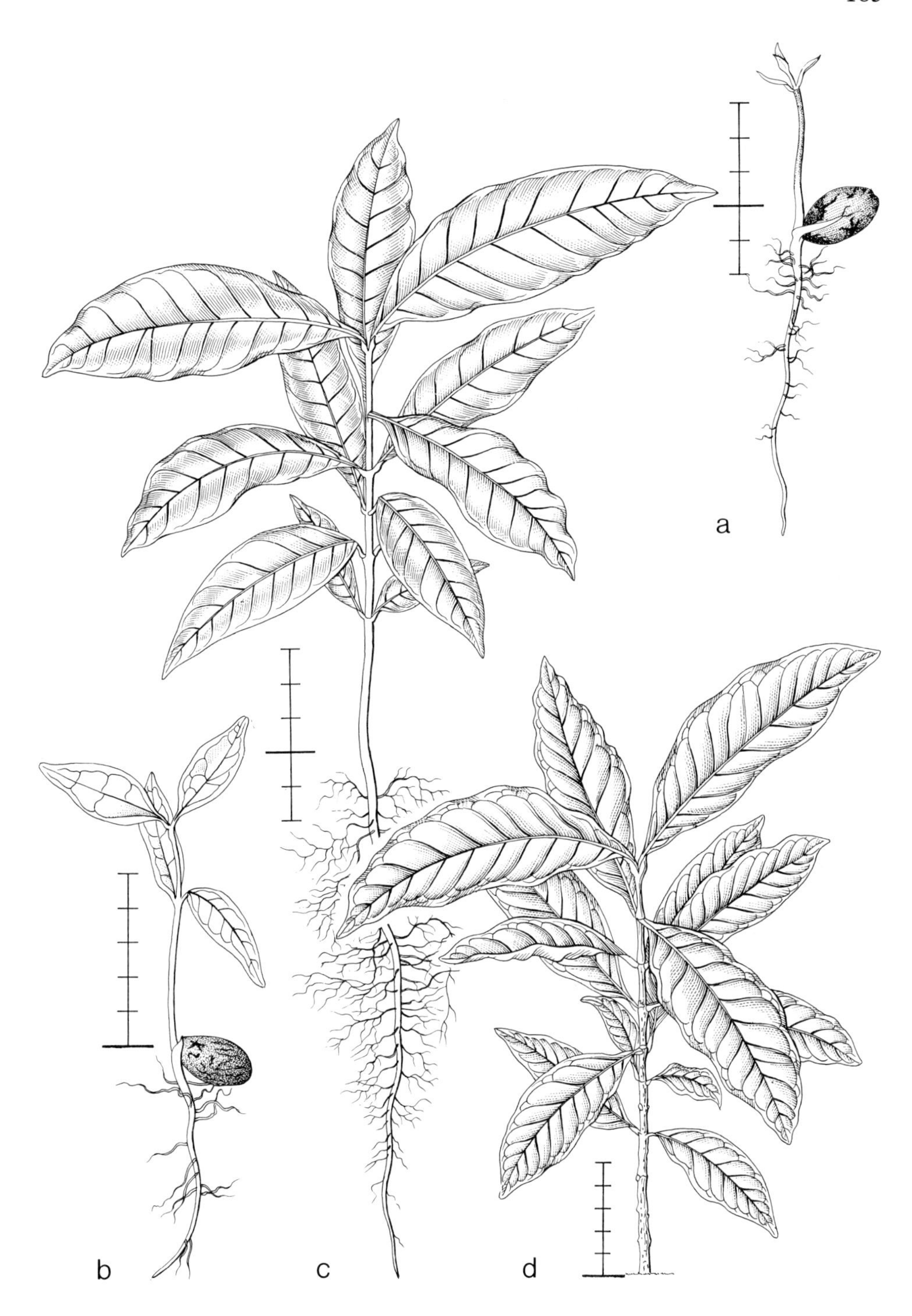

Fig. 45. *Kopsia arborea.* De V. 1363, collected 21-vi-1972, planted 2-vii-1972. a, b, and c date?; d. 30-vii-1973.

OCHROSIA

Fl. Java 2: 232.
Genus in S.E. and E. Asia up to Japan, throughout Malesia, Australia, in the Pacific as far as Tonga Arch., including the Bonins, Central Polynesia and New Caledonia.

Ochrosia acuminata Val. – Fig. 46.

Development: The taproot emerges from the acute pole of the fibrous endocarp, but is often entangled in the fibres and not able to become free. The hypocotyl emerges from the same pole, carrying the paracotyledons free. A short resting stage occurs with developed paracotyledons.
Seedling epigeal, phanerocotylar. Macaranga type.
Taproot rather short, sturdy, creamy-brown, with few, short, later much-branched creamy-white sideroots.
Hypocotyl strongly elongating, terete, just below the paracotyledons in cross-section ovate, longitudinally with two ridges, to 11 cm long, distinctly tapering to the top, green, glabrous.
Paracotyledons 2, opposite, foliaceous, in the 8th–12th leaf stage dropped; simple, almost sessile, without stipules, herbaceous, green, glabrous. Blade ovate, to 4.6 by 2.4 cm; base obtuse; abruptly narrowed into a 1 mm long, broad petiole; top obtuse; margin entire; nerves pinnate, midrib abruptly prominent above, slightly so below, lateral nerves above slightly sunken, below slightly prominent, rather inconspicuous.
Internodes: First one terete with a groove from between the leaves, to 2 cm long, green, glabrous; next internodes as first one, second one to 1 cm long, higher ones gradually larger, 6th one to 2 cm long, 10th one to 2.5 cm long. Stem straight, rather slender.
Leaves all decussate to whorled, simple, without stipules, petiolate, herbaceous, green, glabrous. Petiole in cross-section semi-orbicular, in first leaves to 10 by 1 mm, in 4th pair to 12 by 1.5 mm, in 10th pair to 12 by 1.5 mm, above with raised midrib. Blade lanceolate, in first pair to 5.5 by 1.6 cm, in 4th pair to 9 by 2.5 cm, in 10th pair to 12.5 by 3.3 cm; base narrowed into the petiole; top acuminate with acute tip; margin entire; nerves pinnate, midrib above raised, below prominent, lateral nerves above sunken, below slightly raised and dark green, ending in anastomoses or curved to the next one; when damaged from the nerves abundant white latex.

Specimens: 2431 from N. Celebes, primary forest on deep clay, medium altitude.
Growth details: In nursery germination fair, ± simultaneous.
Remarks: Two species cultivated, similar in germination pattern and general morphology. *O. oppositifolia* differs in details only.
According to Fosberg and Sachet, Micronesica 10 (1974) 251–256 and Markgraf, Blumea 25 (1979) 241, *Ochrosia* has to be split up into two groups; plants of which the fruit has a spiny endocarp belong to the genus *Neisosperma*.

Fig. 46. *Ochrosia acuminata.* De V. 2431, collected 3-x-1973, planted 7-xi-1973. a. 27-xi-1973, abnormal germination; b. 27-xi-1973, normal germination; c. 28-iii-1974.

WILLUGHBEIA

Tree Fl. Malaya 2: 5.
Revision: Markgraf, Blumea 20 (1973) 410.
Genus in E. India and Ceylon, throughout S.E. Asia and W. Malesia, including Borneo and Java.

Willughbeia coriacea Wall. – Fig. 47.

Development: The taproot emerges from one pole of the seed. The cotyledons remain secund at soil level, and spread slightly, by which the testa is shed. The shoot emerges from the opening. No resting stage occurs.
Seedling hypogeal, phanerocotylar. Sloanea type, Palaquium subtype.
Taproot long, slender, flexuous, yellow, with many long, hardly branched, slender, yellow sideroots.
Hypocotyl not enlarging.
Cotyledons 2, turned secund, succulent, in the 6th leaf stage shed; without stipules, petiolate, hard fleshy, dark purplish red, subterranean parts red, glabrous. Petiole terete to flattened, to 4 by 2 mm, often with 2 lateral grooves on the lower side. Blade elliptic in outline, to 3.5 by 1.5 by 0.5 cm; base and top rounded; margin entire; outside slightly concave, inside flat; nerves not visible.
Internodes: First one strongly elongating, to 8 cm long, terete, dark green, glabrous; next internodes as first one, second one to 2.5 cm long, 5th one to 4 cm long. Stem slender, straight.
Leaves all decussate, not folded when young, simple, without stipules, shortly petiolate, subcoriaceous, shining reddish green when young, turning dull green, glabrous, with some white latex. Petiole in cross-section semi-orbicular, in first leaves to 10 by 1.5 mm, in 8th pair to 10 by 3 mm. Blade oblong, in second pair to 11.5 by 4.5 cm, in 8th pair to 21 by 8.5 cm; base acute, slightly narrowed; top cuspidate; margin entire; nerves pinnate, midrib above raised, below prominent, lateral nerves above slightly sunken, below raised, at the tip curved back to the next one.

Specimens: 2236 from S.E. Borneo, primary hill forest on deep clay, low altitude.
Growth details: In nursery germination good, ± simultaneous.

BOMBACACEAE

DURIO *sect.* BOSCHIA

Tree Fl. Malaya 1: 106; Burger, Seedlings (1972) 56 (*D. zibethinus*); Meijer, Bot. Bull. Sandakan 11 (1968) 112 (*D. graveolens*); Soepadmo and Eow, Gard. Bull. 29 (1976) 25–33 (*D. zibethinus*).
Monograph: Kostermans, Reinwardtia 4 (1958) 357.
Genus in Ceylon, S.E. Asia, throughout Malesia, except the N. Philippines, the Les-

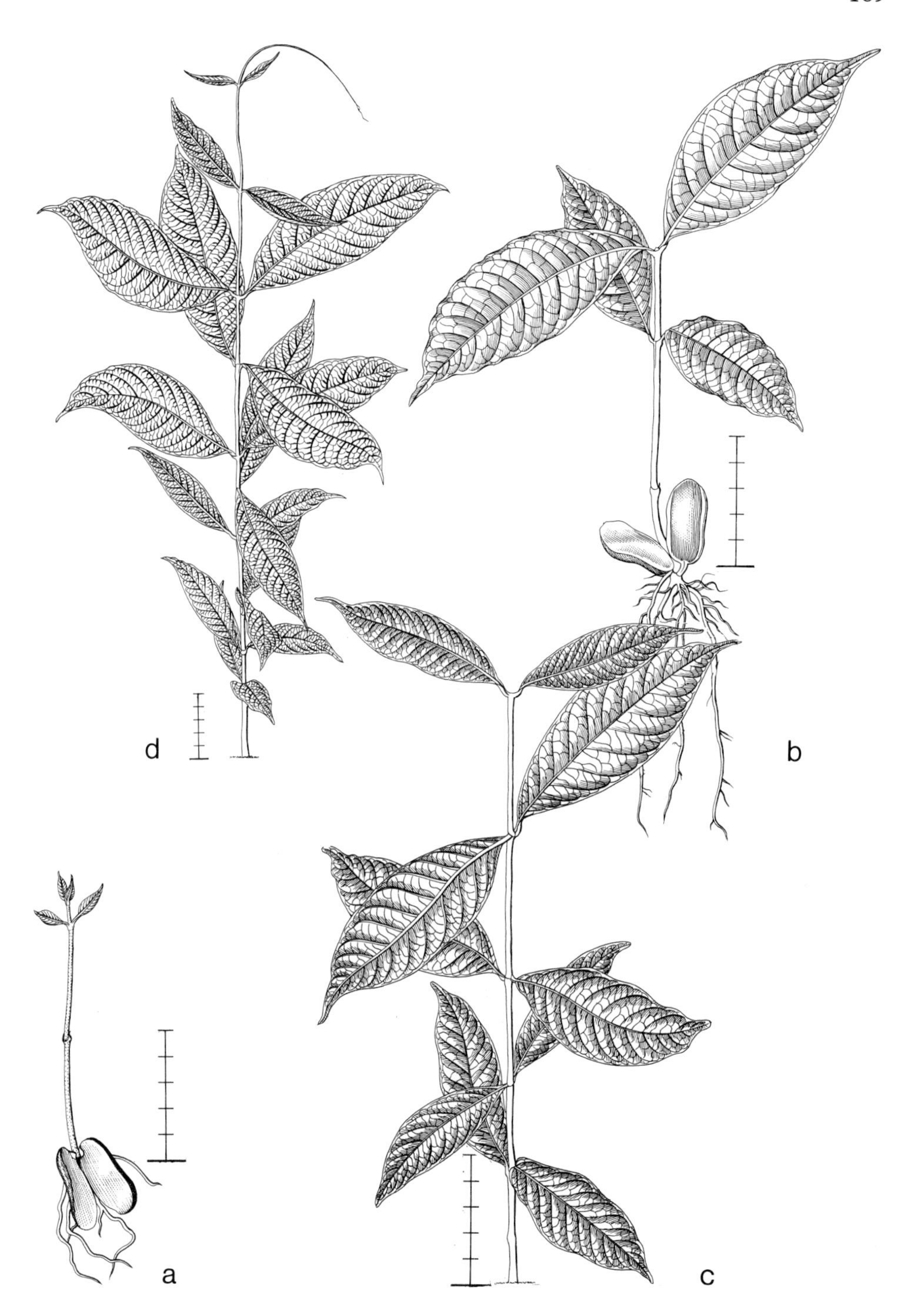

Fig. 47. *Willughbeia coriacea.* De V. 2236, collected 8-iii-1973, planted 22-iii-1973. a. 7-iv-1973; b. 2-v-1973; c. 29-iv-1974; d. 16-vii-1974.

ser Sunda Islands, and New Guinea east of Geelvinkbaai.

Durio excelsus (Korth.) Bakh. – Fig. 48.

Development: The hypocotyl emerges from one pole of the seed, after which the taproot starts elongating. The cotyledonary petioles elongate considerably, carrying the enclosed cotyledons free from the plumule, enabling the shoot to emerge from between them. No resting stage occurs.
Seedling epigeal, cryptocotylar. Horsfieldia type, Pseuduvaria subtype.
Taproot long, sturdy, fleshy, brownish, with rather few, few-branched, brownish sideroots.
Hypocotyl strongly enlarging, fusiform, terete, distinctly tapering into the root, at the top much thicker than the first internode, to 3.5 cm long, creamy-brown turning brownish grey, rather smooth, with small, slightly elevated, brown flakes, almost glabrous.
Cotyledons 2, secund, succulent, in the 3rd–4th leaf stage dropped, simple, without stipules, petiolate. Whole body ellipsoid, to 2.3 by 1.6 by 1.6 cm, the blackish testa persistent. Petioles strongly elongating, in cross-section semi-orbicular, to 18 by 3 mm but often smaller, with small flakes as on the hypocotyl, brownish grey, glabrous.
Internodes: First one slightly elongating, terete, to 1.5 cm long, brownish grey, densely covered with brownish, peltate scales; next internodes as first one, to 2 cm long, in general shorter. Stem rather straight, slender.
Leaves all spirally arranged, simple, conduplicate-induplicate when young, stipulate, petiolate, subcoriaceous, above green, below (bluish-) green. First ± 5 leaves much reduced, scale-like, narrowly triangular, to 2 by 0.5 cm, greyish brown and hairy as first internode. Stipules 2 at the base of each leaf, subulate, dropping soon, erect but twisted, lanceolate-triangular, to 5 by 1 mm, above with a fleshy, longitudinal ridge, with minute simple, hyaline hairs, below hairy as the first internode. Petiole terete, in 3rd leaf to 18 by 1 mm, the upper half distinctly swollen, coloured and hairy as the first internode. Blade ovate-oblong to oblong, in 3rd leaf to 6.3 by 2.8 cm; base rounded; top cuspidate; margin entire; above glabrous, below densely covered with minute, silvery-hyaline, simple hairs, and few scattered stellate or peltate scales, especially on the nerves; nerves pinnate, midrib above slightly sunken, below much prominent, lateral nerves above rather inconspicuous, below rather prominent, free ending.

Specimens: 2324, from S.E. Borneo, primary hill forest on deep clay, low altitude.
Growth details: In nursery germination poor. The root has difficulties in establishing, often dying off, and being replaced by a secondary root. Finally, all seedlings obtained died.
Remarks: Nine species cultivated for the project, one of which had been described previously; one other was described in literature. In seedlings of *D. excelsus*, and some of *D. zibethinus*, the mode of germination is as described above. In *D. acutifolius, D. dulcis, D. griffithii, D. kuteiensis, D. oxleyanus*, some seedlings of *D. zibethinus* (Soepadmo and Eow 1976), *D. sp.* De V. 2360, and *D. sp.* De Wilde 13924 the mode of germination is as described under *D. oxleyanus* (Fig. 49).

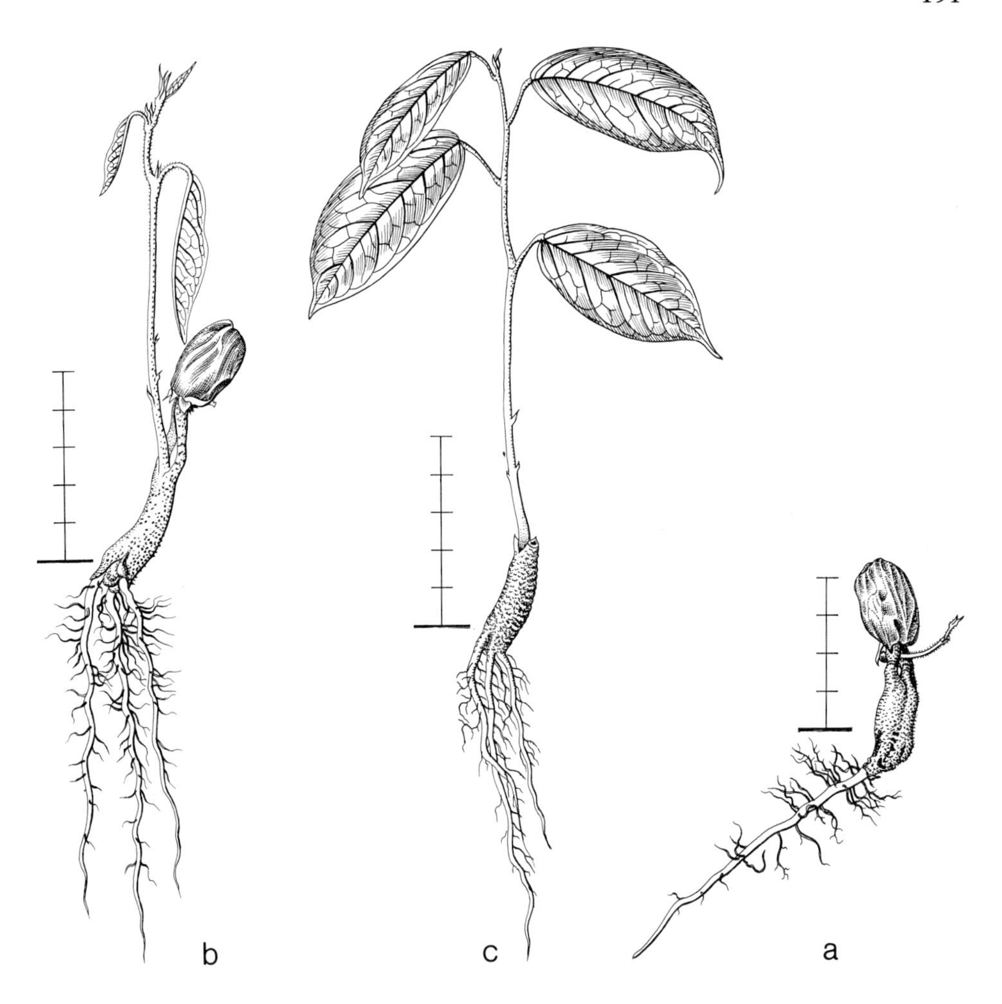

Fig. 48. *Durio excelsus.* De V. 2324, collected 14-iii-1973, planted 22-iii-1973. a and b. 7-v-1973; c. 24-vii-1973.

DURIO *sect.* (EU)DURIO

For references and genus distribution see *Durio* sect. *Boschia.*

Durio oxleyanus Griff. – Fig. 49.

Development: The hypocotyl emerges from one pole of the seed, after which the taproot starts elongating. The hypocotyl elongates, first curves, then becomes erect, carrying the enclosed cotyledons on top. After a resting period the cotyledons in the testa are pushed off by the developing shoot.
Seedling epigeal, cryptocotylar. Blumeodendron type.
Taproot long, sturdy, hard fleshy, brown, with rather few sturdy, much-branched, creamy-brown sideroots, especially around the collar.
Hypocotyl strongly enlarging, fusiform, terete but slightly angular in cross-section, distinctly tapering to the top, to 15 cm long, greyish brown, densely covered with brownish, stellate hairs, to the top with more peltate scales.
Cotyledons 2, opposite, succulent, on top of the hypocotyl, pushed off by the developing shoot; simple, without stipules, petiolate. Whole body ellipsoid, to 5 by 3.3 by 3 cm, the reddish brown testa persistent. Petioles strongly elongating, in cross-section semi-orbicular, to 1.5 by 0.2 cm, often smaller, hairy as the hypocotyl.
Internodes: First one strongly elongating, terete, to 12 cm long, greenish to brownish grey, densely covered with hyaline to brownish, peltate scales; next internodes as first one, irregular in length, 3–12 cm long. Stem very slender, straight.
Leaves all spirally arranged, simple, conduplicate-induplicate when young, stipulate, petiolate, subcoriaceous, above green, below pale (bluish-)green. First 1–2 leaves much reduced, scale-like, dropping soon, hairy as first internode. Stipules 2 at the base of each leaf, subulate, dropping soon, erect, linear, to 10 by 1.5 mm, above with a fleshy succulent longitudinal ridge, with minute, simple, hyaline hairs, below hairy as first internode. Petiole terete, in lower ones to 3.5 by 1.5 cm, the upper 1 cm distinctly swollen, the basal part slightly so, colour and hairs as in the first internode, ultimately to 4.5 cm long. Blade ovate-lanceolate, in lower leaves to 17 by 4.3 cm, ultimately to 21 by 7 cm; base rounded to slightly emarginate; top cuspidate to caudate; margin entire; above glabrous, below densely hairy with stellate, hyaline, thin-branched hairs, especially on the nerves and along the margin with brownish peltate scales; nerves pinnate, midrib above sunken, below very prominent, lateral nerves above hardly prominent, below prominent, at their tip curved and forming an undulating, irregular marginal nerve.

Specimens: 2325 from S.E. Borneo, primary hill forest on deep clay, low altitude.
Growth details: In nursery germination good, ± simultaneous.
Remarks: See *D. excelsus*.

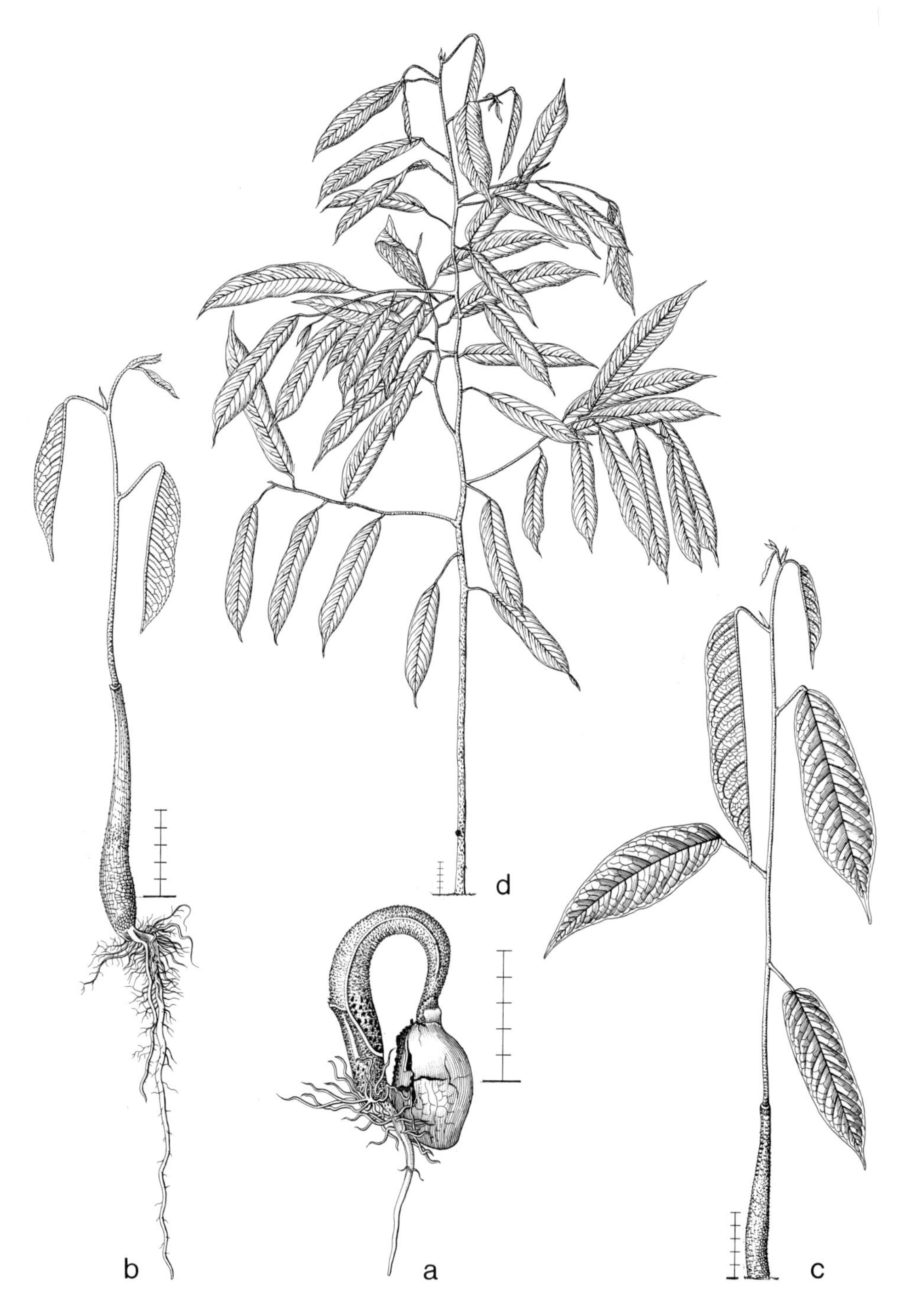

Fig. 49. *Durio oxleyanus*. De V. 2325, collected 14-iii-1973, planted 22-iii-1973. a. 11-iv-1973; b. 9-v-1973; c. 4-vi-1974; d. 29-vii-1974.

NEESIA

Fl. Java 1: 420.
Monograph: Soepadmo, Reinwardtia 5 (1961) 481.
Genus in S.E. Asia and W. Malesia, including Borneo and Java.

Neesia altissima (Bl.) Bl. – Fig. 50.

Development: The hypocotyl and taproot emerge from one pole of the seed, and establish, carrying the paracotyledons enclosed in endosperm and testa on top. After a short resting stage the unfolding paracotyledons free themselves from the testa, the endosperm often adhering to one of the paracotyledons for some time. After a second resting stage with expanded paracotyledons, the shoot starts its growth.
Seedling epigeal, phanerocotylar. Macaranga type.
Taproot long, slender, fibrous, brownish, especially in the upper part with many, rather short, hardly branched sideroots.
Hypocotyl first slowly developing, with the freeing of the paracotyledons strongly elongating, terete, to 7 cm long, pale green, densely hairy with simple, rather long, soft, hyaline hairs.
Paracotyledons 2, opposite, foliaceous, in the 6th–12th leaf stage dropped, at first enveloped by the red-brown testa and creamy-white endosperm, when freeing often the tips held together for some time by the adhering endosperm, spreading, without stipules, shortly petiolate, herbaceous, above green, below light green. Petiole in cross-section semi-orbicular, to 4 by 2.5 mm, above broadly channelled, hairy as the hypocotyl. Blade elliptic, to 6 by 4.5 cm; base retuse, top broadly rounded; margin entire, above with distant, along the margin the close-set, simple, rather stiff, hyaline hairs; nerves pinnate, midrib above slightly raised, below much so, lateral nerves below prominent, ± free ending.
Internodes: First one elongating to 7 cm long, terete, green, hairy as the hypocotyl; next internodes as first one, gradually becoming larger, 6th one to 1.5 cm long, 10th one to 2.5 cm long. Stem thick, rather slender, straight.
Leaves all spirally arranged, simple, ± conduplicate-induplicate when young, stipulate, petiolate, herbaceous, slightly reddish when young, turning green. Stipules 2 at the base of each petiole, persistent, sessile, ovate to ovate-oblong, 6th one to 6 by 3 mm, 20th one to 15 by 5 mm, below and above on the midnerve densely hairy with bundles of long, simple, stiff hyaline hairs, and few, smaller stellate ones. Petiole terete, in the 5th leaf to 7 by 2 mm, gradually becoming slightly larger, densely hairy as the stipules. Blade obovate-oblong, 5th one to 7 by 3 cm, 20th one to 20 by 7.5 cm; base slightly obtuse; top acuminate; margin entire, rather densely hairy as stipules, especially on the nerves; nerves pinnate, above hardly raised, below much prominent, especially the midrib, lateral nerves free ending.

Specimens: Dransfield 2531, from W. Java, low altitude.
Growth details: In nursery germination good, ± simultaneous.

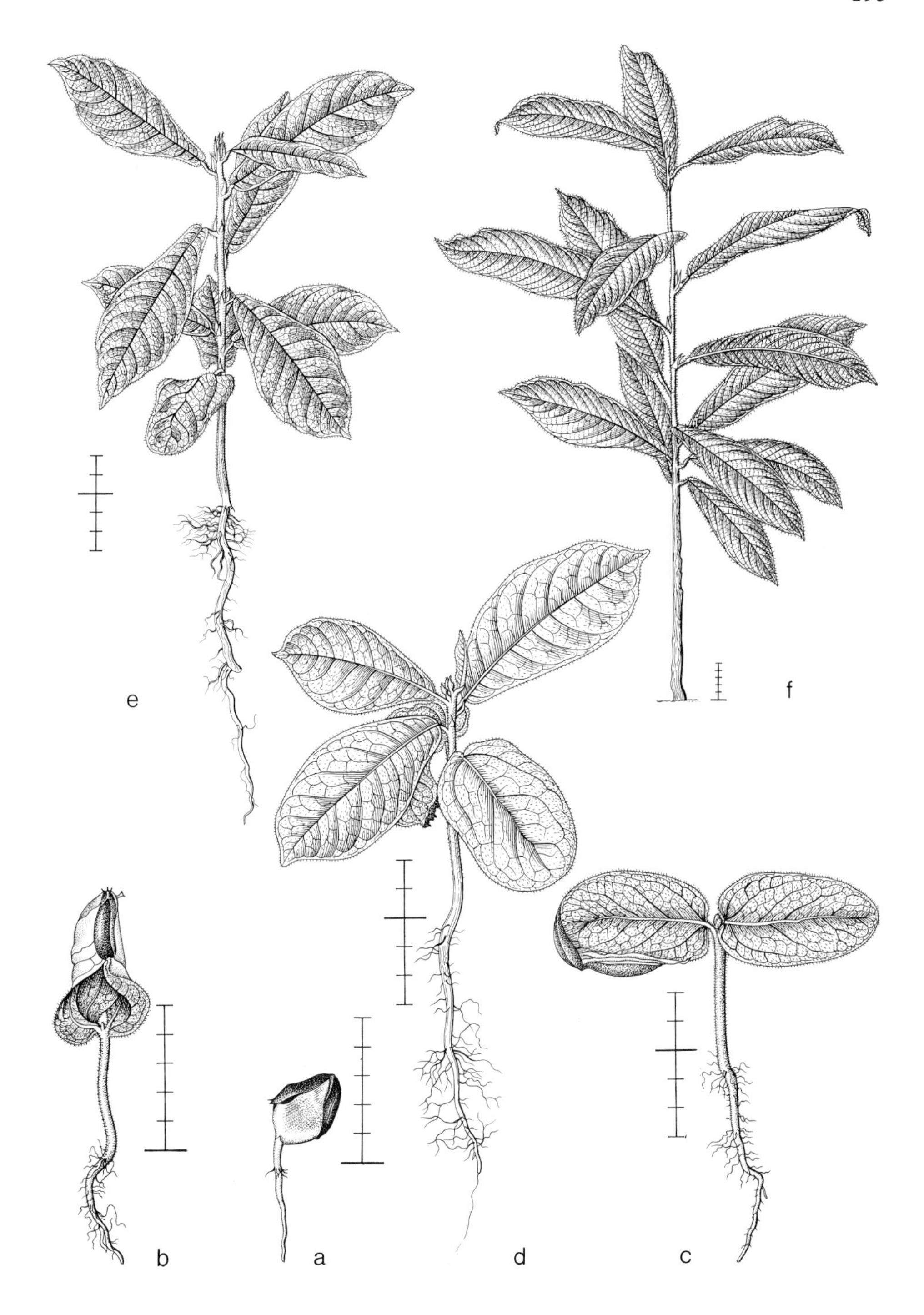

Fig. 50. *Neesia altissima.* Dransfield 2531, collected 15-vi-1972, planted 20-vi-1972. a, b, c, d, and e date?; f. 15-viii-1974.

BURSERACEAE

CANARIUM *sect.* PIMELA

Fl. Males. I, 5: 249; Guillaumin, Structure et le développement des Burséracées (1910) 219 (*C. commune = vulgare*, *C. moluccanum = indicum*, *C. occidentale = schweinfurthii*, *C. rufum = littorale*); Lubbock, On Seedlings 1 (1892) 333 (*C. strictum*); Mensbruge, Germination (1966) 220 (*C. schweinfurthii*); Meijer, Bot. Bull. Sandakan 11 (1968) 112 (*C. littorale*, *C. megalanthum*, *C. pseudodecumanum*); Ng, Mal. Forester 38 (1975) 44 (*C. littorale*, *C. megalanthum*, *C. patentinervium*, *C. pseudo-sumatranum*, *C. reniforme*); Voorhoeve, Liberian high forest trees (1965) 75 (*C. schweinfurthii*); Weberling and Leenhouts, Abh. Math. Naturw. Klasse Mainz 10 (1966) 536 (*C. asperum*, *C. indicum*, *C. kipella*, *C. oleosum*, *C. schweinfurthii*, *C. subulatum*, *C. vulgare*).
Monograph: Leenhouts, Blumea 9 (1959) 275–475.
Genus in Tropical Africa, India and Ceylon, throughout S.E. Asia, all over Malesia, N. Australia, in the Pacific as far as Tonga, including New Caledonia.

Canarium hirsutum Willd. – Fig. 51.

Development: A triangular valve is detached from the stone; taproot and hypocotyl emerge and establish. The paracotyledons are withdrawn from the stone and testa by the first bent, then becoming erect, hypocotyl. A short resting stage occurs with expanded paracotyledons.
Seedling epigeal, phanerocotylar. Macaranga type.
Taproot long, slender, fibrous, brown, with some bundles of short, slender, unbranched sideroots.
Hypocotyl strongly elongating, terete, to 6.5 cm long, pale green, densely covered with short, simple, hyaline hairs.
Paracotyledons 2, opposite, foliaceous, in the 3rd–5th leaf stage dropped; compound, without stipules, petiolate, herbaceous, green. Petiole in cross-section semi-orbicular, to 3 by 2.5 mm, hairy as the hypocotyl. Petiolules as petiole, to 1 by 1 mm. Blade palmate, triangular in outline, to 5 by 6 cm. Leaflets 3, ovate-lanceolate, to 5 by 1.3 cm; base and top acute; margin entire; along the margin and at the base of the main nerve slightly hairy; median one symmetric, lateral ones asymmetric, especially at the base; nerves 3, the midrib below prominent, above slightly so, lateral nerves less conspicuous, looping, joined into a rather inconspicuous marginal nerve.
Leaves all spirally arranged, with pseudostipules, petiolate, herbaceous, green, gradually becoming larger. Lower ± 7 leaves simple. Petiole terete, in 5th leaf to 45 by 1.5 mm, hairy as the hypocotyl, at the base swollen, at the top slightly so. Pseudostipules 2, on the petiole just above the basal thickening, subulate, those in 5th leaf to 5 by 0.5 mm, green, hairy as hypocotyl. Blade oblong to obovate-oblong, in the 5th leaf to 11.5 by 5.5 cm; base truncate; top acuminate; margin regularly serrate; below on the nerves and above scattered over the blade rather sparsely hairy with quite long, stiff, simple, hyaline hairs; nerves pinnate, midrib above and below raised, lateral nerves above slightly sunken, below prominent, very regular, at the tip

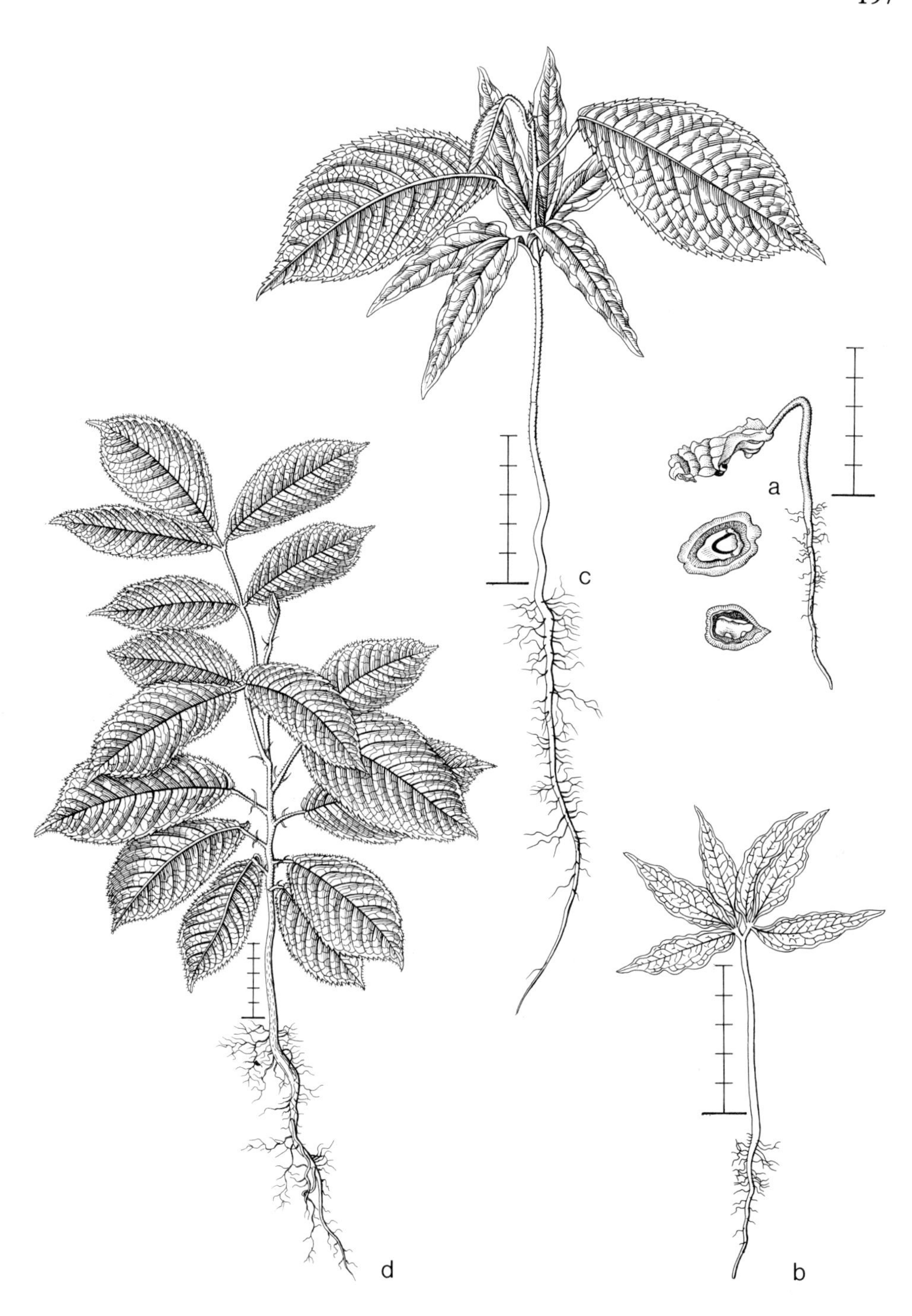

Fig. 51. *Canarium hirsutum.* De V. 1397, collected 23-vi-1972, planted 2-vii-1972. a, b, c, and d date?.

curved to form a much undulating marginal nerve, veins ending free in the teeth. Next leaves as first ones, compound, pseudostipulate, without stipellae; beyond the 7th leaf 3-foliolate, gradually gaining in number of leaflets, in 30th leaf with up to 13 leaflets. Petiole terete, in second compound leaf to 6.5 cm long, in 6th one to 15 cm long, in 20th one to 10 cm long, at the base distinctly swollen, and there flattened above. Rhachis as the petiole, in second compound leaf to 1.7 cm long, in 6th one to 4.5 cm long, in 20th one to 24 cm long. Blade imparipinnate with terminal leaflet. Leaflets petiolulate, the lateral ones opposite. Petiolules terete, in the second compound leaf to 2 by 1 mm, in 6th one to 4 by 1.5 mm, in 20th one to 8 by 3 mm. Blade of lateral leaflets elliptic to oblong, sometimes more obovate-oblong, the basal ones in each leaf the smaller, in second compound leaf to 8.5 by 3.5 cm, in 20th one to 23 by 8 cm; base rounded, slightly retuse, often somewhat asymmetric; top abruptly acuminate; margin regularly serrate, further as in simple leaves. Terminal leaflet more obovate-oblong, in second compound leaf to 12 by 5.5 cm, in 20th one to 22 by 8.3 cm, further as lateral leaflets.

Specimens: 1397 and Dransfield 2526 from W. Java, Udjung Kulon Nature Reserve, alluvial flat, primary forest on coral sand, low altitude.
Growth details: In nursery germination good, over a rather long stretch of time.
Remarks: Seedlings in subgenus *Canarium*, sect. *Pimela* all have leaf-like, thin, palmate, three-lobed paracotyledons that are long persistent. The first leaves are mostly all spirally arranged (*C. asperum*, *C. hirsutum*, *C. kipella*, *C. oleosum*, *C. strictum*, *C. subulatum*). *C. pseudo-sumatranum* is different in that the first two leaves are opposite; in this it is similar to the seedlings of sect. *Canarium*, and *C. schweinfurthii* (subg. *Africanarium*).
Seedlings in subgenus *Canarium*, sect. *Canarium* all have rather short-lived, relatively thin but food-storing, palmate cotyledons. The first two leaves are always opposite. Three-lobed cotyledons are most common (*C. indicum*, *C. littorale*, *C. patentinervium*, *C. reniforme*, *C. vulgare*). Five-lobed cotyledons are present in *C. megalanthum*, where the lobes are long and very slender, the midlobe sometimes with two branches, and in *C. pseudodecumanum*, where they are short and stunted. The only species described of subgenus *Africanarium* has leaf-like 5-lobed paracotyledons that are long persistent, with the first two leaves opposite. The systematic position as subgenus was partly based on the seedling characters, then thought unique in *Canarium*, but also present in *C. pseudo-sumatranum*, sect. *Pimela*.

CANARIUM *sect.* CANARIUM

For references and genus distribution see *Canarium* sect. *Pimela.*

Canarium littorale Bl. var. *tomentosum* Leenh. – Fig. 52.

Development: A small, triangular, lateral valve becomes loose from the stone. The taproot establishes, the hypocotyl becomes erect, withdrawing the cotyledons from the stone. A short resting stage occurs with the cotyledons and 2 developed leaves.

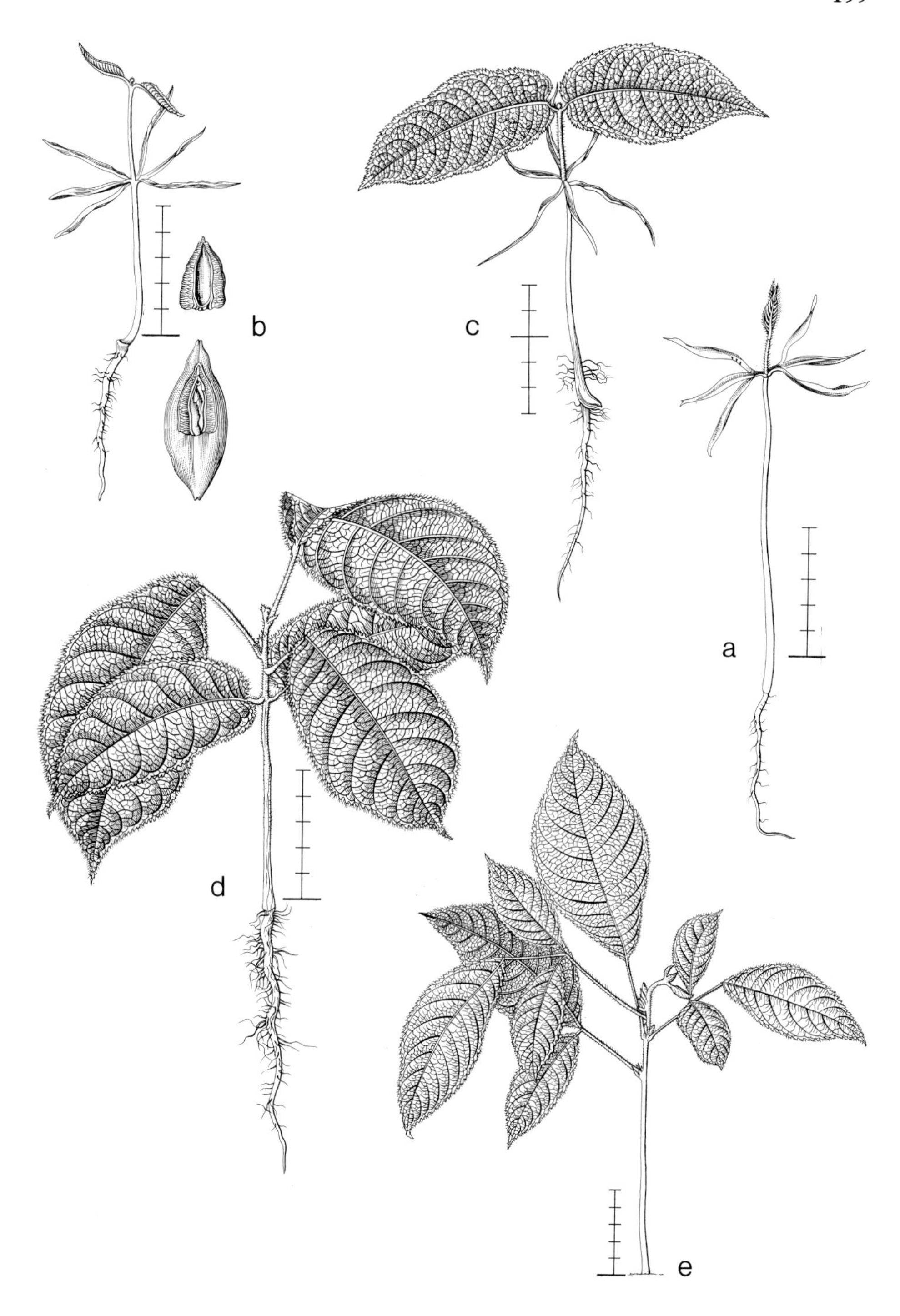

Fig. 52. *Canarium littorale* var. *tomentosum*. De V. 2299, collected 12-iii-1973, planted 22-iii-1973. a. 1-xii-1973; b. 27-v-1973; c. 23-vii-1973; d. 7-iii-1974; e. 29-vii-1974.

Seedling epigeal, phanerocotylar. Sloanea type and subtype.
Taproot sturdy, fleshy, brown, with many, paler lenticels, lateral roots many, short, unbranched. Later sometimes additional roots develop from the base of the hypocotyl.
Hypocotyl strongly elongating, terete, to 7 cm long, at its base with a fleshy, ovate, lateral outgrowth, dull brownish, densely covered by minute, simple, inconspicuous brown hairs.
Cotyledons 2, opposite, succulent, in the 2nd–3rd leaf stage dropped; compound, ± equal, without stipules, ± sessile, hard fleshy, dark blackish green, glabrous. Blade palmate, triangular in outline, to 3.5 by 8 cm. Leaflets 3, linear, ± subulate, slightly flattened and swollen, to 50 by 4 by 2 mm, the midlobe to 3.5 cm long; top rounded; margin entire; nerves not visible.
Internodes: First one strongly elongating, to 3 cm long, terete, green, densely covered with patent, rather long, simple, reddish, later brown-hyaline hairs; next internodes as first one, 5th one to 1 cm long. Stem rather slender, straight.
First two leaves opposite, simple, without stipules, shortly petiolate, herbaceous, green. Petiole terete, to 5 by 1.5 mm, hairy as the hypocotyl. Blade ovate-oblong, to 8.5 by 4 cm; base exsculptate; top cuspidate; margin double serrate, each tooth with small lower top and a larger upper one, on both sides with scattered hairs as on the first internode, slightly glabrescent; nerves pinnate, above slightly raised, below much so, lateral nerves distinctly curved at the end towards the next one, each tooth provided with a vein. Next leaves spirally arranged, simple, with pseudostipules. The lowest ± 5 leaves simple. Petiole as that in the first leaves, at base and tip slightly swollen, in 7th leaf to 5.5 cm long. Pseudostipules 2 on the swollen base of the petiole, persistent, erect, sessile, narrowly triangular to spathulate, in 7th leaf to 8 by 2 mm; top acute; margin towards the tip with some coarse teeth; outside sparsely hairy as the hypocotyl. Blade elliptic, in 3rd leaf to 6.5 by 3.5 cm, in 7th one to 16.5 by 8.5 cm; base acute to obtuse; top acuminate; margin serrate; further as first leaves. Subsequent leaves compound. Pseudostipules as those in simple leaves, spathulate, in 3rd compound leaf to 10 by 5 mm, with large irregular teeth. Petiole as in first leaves, in 3rd compound one to 5.5 cm long, swollen at the base only. Rhachis as the petiole, in second compound leaf to 1.5 cm long, in 5th one to 6 cm long. Blade imparipinnate with terminal leaflet, the lower 3–4 ones (2–) 3-foliolate, next ones 5-foliolate. Leaflets without stipellae, petiolulate, the lateral ones in opposite pairs. Petiolules terete, as petiole, in second compound leaf to 5 by 1 mm, in 5th compound one to 10 by 1.5 mm, those of the terminal leaflets much smaller. Blade of lateral leaflets elliptic, in first compound leaf to 4.7 by 2.5 cm, in 5th one to 11.5 by 5 cm, but the basal ones in each leaf much smaller. Blade of terminal leaflet elliptic, in first compound leaf to 9.5 by 5.5 cm, in 5th one to 8.5 by 5 cm, further as the blade of simple leaves.

Specimens: 2299 from S.E. Borneo, primary hill forest on deep clay, low altitude.
Growth details: In nursery germination fair, rather long delayed, over a long stretch of time.
Remarks: See *C. hirsutum*.

DACRYODES *sect.* TENUIPYRENA

Fl. Males. I, 5: 219; Duke, Ann. Miss. Bot. Gard. 52 (1965) 337 (*D. excelsa*); Mensbruge, Germination (1966) 220 (*C. klaineana*); Ng, Mal. Forester 38 (1975) 55 (*D. costata, D. kingii, D. rostrata*); Voorhoeve, Liberian high forest trees (1965) 80 (*D. klaineana*).
Genus pantropic; tropical Africa, S.E. Asia and throughout Malesia, with the exception of Central and E. Java, most of Celebes, the Moluccas, and the Lesser Sunda Islands.

Dacryodes rostrata (Bl.) H. J. Lam f. *cuspidata* H. J. Lam – Fig. 53.

Development: By slight unfolding of the cotyledons the endocarp valves open. The taproot and hypocotyl emerge, and establish. The cotyledons expand, thus shedding the envelopments, and are carried above the soil by the hypocotyl becoming erect. No resting stage occurs.
Seedling epigeal, phanerocotylar. Sloanea type and subtype.
Taproot long, sturdy, fibrous, brownish, sideroots many in the upper part, slender, brownish, much-branched.
Hypocotyl strongly elongating, terete, to 5.5 cm long, reddish brown, with minute, close-set, simple hyaline hairs, glabrescent.
Cotyledons 2, opposite, slightly succulent, ineffective in the 3rd–4th leaf stage, but finally shed beyond 8th leaf stage; more or less appressed along the stem and hypocotyl, sessile, without stipules, coriaceous, greenish, glabrous. Blade palmatifid, in outline ± orbicular, 10 by 10 cm, palmately nerved, nerves not raised, pale yellow. Lobes to 13, ovate-lanceolate, the lateral ones much smaller, each one undulating and curved in an irregular way; top acute, its tip rounded; margin entire.
Internodes: First one strongly elongating, to 8 cm long, terete, dull green, hairy as the hypocotyl; next internodes as first one, irregular in length, 0.5–1.5 cm long. Stem slender, slightly zig-zag.
Leaves all spirally arranged, without stipules, petiolate, subcoriaceous, slightly reddish green when young, turning green. First 6–9 leaves simple. Petiole terete, or slightly flattened above, to 4 by 0.1 cm, top slightly thickened, base distinctly so, hairy as hypocotyl. Blade ovate-oblong, to 13.5 by 6.5 cm; base acute to rounded; top caudate; margin entire; on the nerves and along the margin with scattered hairs as on the hypocotyl; nerves pinnate, midrib above slightly prominent, below much so, lateral nerves above sunken, below prominent, free ending. Next leaves compound. Petiole above terete to slightly flattened, in first compound leaf to 7 by 0.1 cm, but in general much smaller, at the base distinctly swollen, further as in simple leaves. Rhachis as the petiole, in first compound leaf to 3.5 cm long, in general shorter. Blade imparipinnate, 3-foliolate, beyond the 4th compound one often with 4 leaflets, higher leaves not seen. Leaflets without stipellae, petiolulate. Petiolule of lateral leaflets in first compound leaf to 8 by 1 mm, in 4th one to 1.5 cm long, at base and tip slightly swollen; that of terminal leaflet swollen, to 3 by 1.5 mm. Blade of leaflets as those of simple leaves, that of the lateral ones generally smaller than that of terminal one, in first compound leaf the former 7.5 by 3.5 cm, the latter 8.5 by 3.5

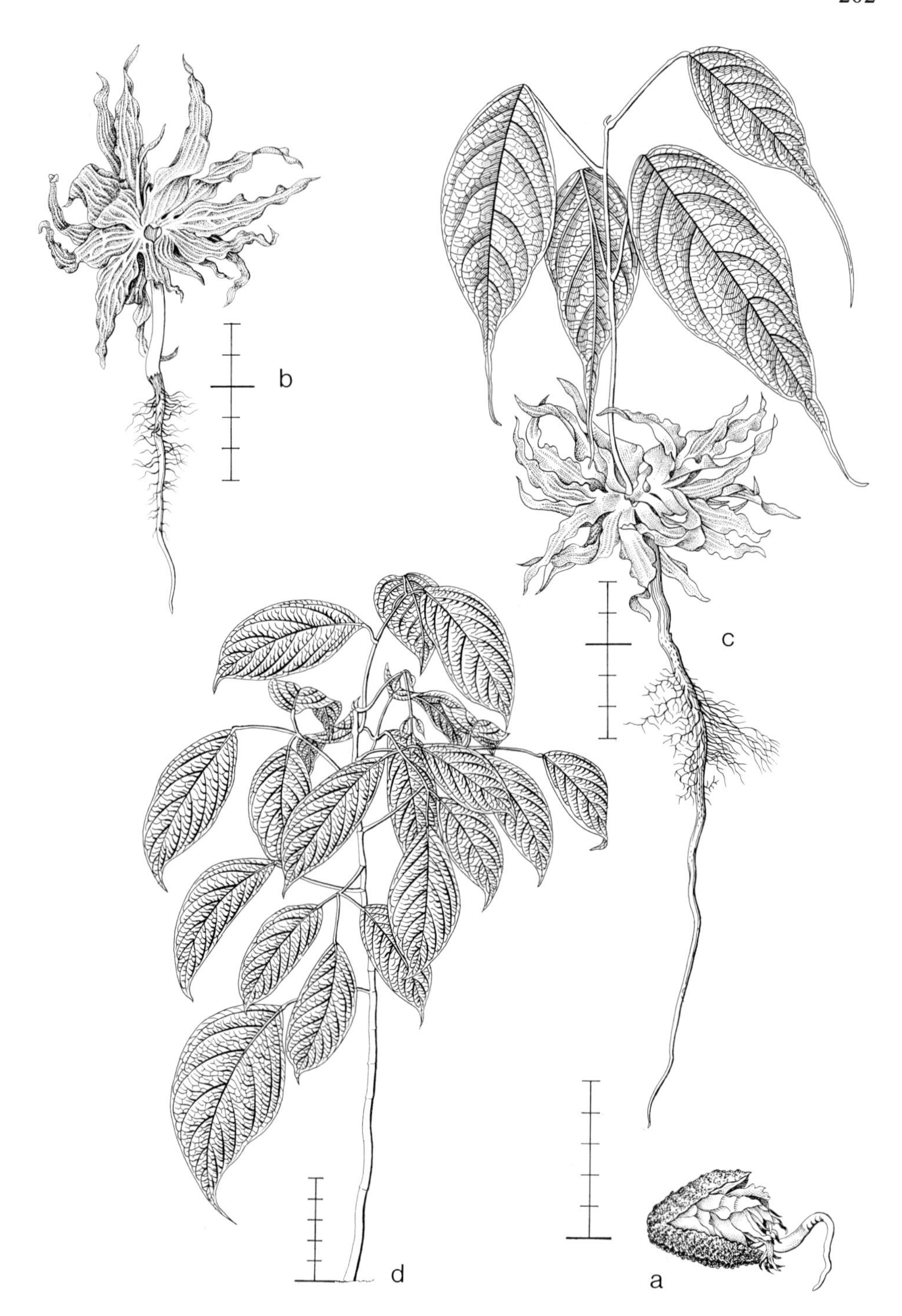

Fig. 53. *Dacryodes rostrata* f. *cuspidata.* De V. 2274, collected 11-iii-1973, planted 22-iii-1973. a and b. 7-iv-1973; c. 4-v-1973; d. 22-vii-1974.

cm; further as the first leaves.

Specimens: 2274 from S.E. Borneo, primary hill forest on deep clay, low altitude.
Growth details: In nursery germination fair, rather simultaneous.
Remarks: Most species described have the small, succulent, relatively short-lived, palmate, 5-lobed cotyledons free and opposite above the soil on an epigeal hypocotyl. The first leaves are opposite (*D. costata*, *D. kingii*, *D. klaineana*, *D. rugosa*). *D. rostrata* has the cotyledons succulent but flattened, with up to 13 lobes, long persistent, although ineffective beyond the 3rd–4th leaf stage. The first two leaves are recorded to be opposite (Ng), or all are spirally arranged. *D. excelsa* is different in the secund, hypogeal, palmate cotyledons, the extremely long first internode, and the opposite first two leaves that are compound.

Dacryodes rugosa H. J. Lam – Fig. 54.

Development: The taproot and hypocotyl emerge from one pole of the fruit. The cotyledons spread, and are withdrawn from testa and endocarp by the hypocotyl becoming erect. A short resting stage occurs with the cotyledons and two developed leaves.
Seedling epigeal, phanerocotylar. Sloanea type and subtype.
Taproot long, slender, fibrous, creamy-brown, with many scattered, short, short-branched, creamy-brown sideroots.
Hypocotyl strongly elongating, terete, to 4.5 cm long, reddish brown, densely covered with minute, simple, hyaline hairs.
Cotyledons 2, opposite, succulent, in the 3rd–4th leaf stages dropped; compound, ± equal, without stipules, petiolate, hard fleshy, more or less dark red, rather densely hairy as the hypocotyl. Petiole in cross-section semi-orbicular, slightly flattened, to 2 by 1.5 mm. Blade palmate, to 2.7 by 1.7 cm in outline. Lobes 5, ± fusiform, distinctly swollen, to 17 by 3 by 3 mm, the midlobe the longer, irregularly sculptured with depressions; top more or less rounded, narrowed at the base; nerves not visible.
Internodes: First one strongly elongating, terete, to 6.5 cm long, pale orange-red when young, turning greenish, densely covered with minute, simple, hyaline hairs, glabrescent; next internodes as the first one, the second one to 15 mm long, higher ones gradually becoming larger, irregular in length, the 6th one to 2.5 cm long, 15th one to 6 cm long. Stem slender, distinctly zig-zag, especially in the lower part.
First two leaves opposite, simple, without stipules, petiolate, subcoriaceous, reddish when young, turning green. Petiole terete, to 20 by 1 mm, at base and top abruptly swollen, densely hairy as the hypocotyl. Blade ovate-oblong, to 7 by 2.8 cm; base acute to almost obtuse; top caudate, its tip clavate; margin entire; below slightly hairy as the hypocotyl, especially on the nerves; nerves pinnate, midrib above raised, below distinctly so, lateral nerves above hardly raised, below prominent, ending in an irregular, much undulating marginal nerve. Next leaves spirally arranged. The lowest 8–12 leaves simple, the first one much smaller to as large as the opposite ones, gradually the higher ones larger, 7th one to 12 by 4.5 cm, its petiole to 5.5 cm long. Subsequent leaves spirally arranged, compound. Petiole as those of the first leaves, in lowest compound one to 3.5 cm long, in 6th compound one to 6 cm long. Rhachis as

Fig. 54. *Dacryodes rugosa.* De V. 2318, collected 13-iii-1973, planted 22-iii-1973. a. 7-v-1973; b. 24-vii-1973; c. 10-iii-1974; d. 29-vii-1974.

the petiole, in first compound one 2.5 cm long, in 6th one to 4 cm long. Blade ± triangular in outline, imparipinnate with terminal leaflet, sometimes intercalated with paripinnate ones; the lower (3–)5–8 trifoliolate, subsequent ones 4–5-foliolate. Leaflets in opposite pairs, petiolulate, without stipellae. Petiolule of lateral leaflets as the petiole, in lowest compound leaf to 1 cm long, in 6th one to 1.5 cm long, distinctly swollen at base and top; that of the terminal leaflet swollen, to 2 mm long. Blade of lateral leaflets smaller than that of terminal one, the former in first compound leaf to 9 by 3.3 cm, the latter 10.5 by 4.8 cm, in 6th one the lateral ones to 10.5 by 4.5 cm, the terminal one to 12.5 by 5.8 cm; further as simple leaves.

Specimens: 2318 and 2344 from S.E. Borneo, alluvial flat, primary forest on deep clay, low altitude.
Growth details: In nursery germination good, ± simultaneous.
Remarks: See *D. rostrata*.

SANTIRIA

Fl. Males. I, 5: 229; Ng, Mal. Forester 38 (1975) 60 (*S. laevigata, S. rubiginosa*).
Genus in tropical W. Africa, and all over Malesia but not in Java and the Lesser Sunda Islands.

Santiria tomentosa Bl. – Fig. 55.

Development: A valve detaches from the fruit wall, enabling the taproot and hypocotyl to emerge. By the hypocotyl becoming erect the cotyledons are carried above the soil, in the meantime spreading, by which the envelopments are shed. A short resting stage occurs with 2 developed leaves. A second prolonged resting stage occurs with 6 developed leaves.
Seedling epigeal, phanerocotylar. Sloanea type and subtype.
Taproot long, slender, flexuous, brownish, with many short, slender, unbranched, brownish sideroots.
Hypocotyl strongly elongating, terete, to 6 cm long, often shorter, light greenish brown turning dark brown, densely hairy with minute, simple, yellowish hairs; in the lower part clearly separated from the root, slightly warty.
Cotyledons 2, opposite, succulent, in the 2nd–3rd leaf stage dropped; compound, without stipules, almost sessile, hard fleshy, brown, glabrous. Petiole flattened. Blade palmatipartite, to 2 by 1 cm. Lobes 5, irregularly curved, irregularly spindle-shaped with flattened sides, in the centre swollen; base slightly narrowed; top acute; the central one symmetric, ± oblong, the lateral ones more irregular, more obovate-oblong.
Internodes: First one strongly elongating, distinctly narrower than the hypocotyl, to 6 cm long, green, brownish by dense covering of hairs as on the hypocotyl; next internodes as first one, 0.5–1 cm long. Stem slender, straight.
First two leaves opposite, simple, without stipules, petiolate, herbaceous, green, sparsely hairy as the hypocotyl, especially on the nerves. Petiole terete, to 13 by 1

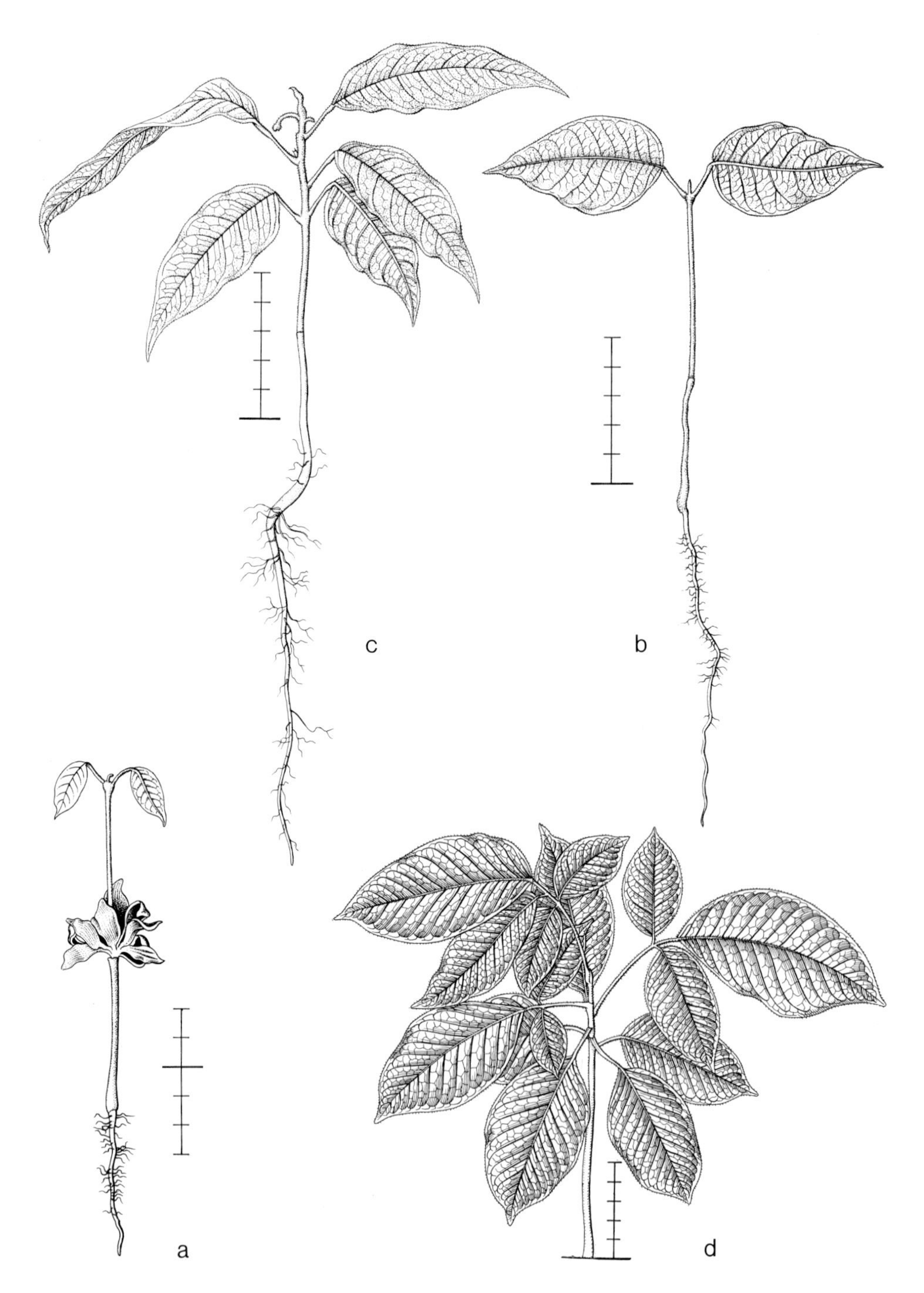

Fig. 55. *Santiria tomentosa.* De V. 747, collected 12-xi-1971, planted 7-xii-1971. a, b, and c date?; d. 10-vii-1973.

mm, at base and top thickened, hairy as the first internode. Blade ovate, to 6.5 by 4 cm; base obtuse; top acuminate; margin entire, slightly irregular; nerves pinnate, midrib above slightly raised, below much so, lateral nerves below raised, forming a vague, undulating marginal nerve. Next leaves spirally arranged. The lowest ± 16 leaves simple. Petiole to 35 by 1.5 mm. Blade ovate-oblong, irregular in size, to 11.5 by 5.5 cm in the highest ones, the lower ones smaller, further as the first leaves. Subsequent leaves spirally arranged, compound. Petiole as those of the lower leaves. Rhachis as petiole. Blade imparipinnate, with a big terminal leaflet and much smaller lateral leaflets, the lower ones 2–3-foliolate, gradually with more leaflets. Leaflets ± sessile, when an even number is present usually opposite. Lateral leaflets in 2nd compound leaf ovate, to 7 by 3.8 cm. Terminal leaflet ovate, to 12 by 6.7 cm; base obtuse, more so in the lateral leaflets; top acuminate; further as the first leaves.

Specimens: 747 from S.E. Borneo, primary hill forest on deep clay, low altitude.
Growth details: In nursery germination poor. When transplanted, the seedlings are liable to die. Development slow: in 8 months the seedling to 13 cm high, with 6 leaves.
Remarks: Three species described. *S. laevigata* and *S. tomentosa* are similar in germination pattern and general morphology, with the succulent, palmate cotyledons on a long epigeal hypocotyl, the first two leaves opposite and simple. *S. rubiginosa* differs in the hypogeal, secund cotyledons. The first two leaves are opposite and compound.

CAPPARIDACEAE

CAPPARIS

Fl. Males. I, 6: 69; Duke, Ann. Miss. Bot. Gard. 52 (1965) 332 (*C. coccolobifolia*, *C. cynophallophora*); ibid. 56 (1969) 148 (*C. pittieri*); Lubbock, On Seedlings 1 (1892) 182 (*C. sp.*).
Genus pantropic and in the subtropics; Africa, S. Europe, Near East, throughout the Indian Ocean, India and Ceylon, throughout S.E. Asia and S. China, all over Malesia, N. Australia, in the Pacific as far as Tuamotu Arch., including the Marianas, Hawaii, Norfolk and New Caledonia, tropical and subtropical America.

Capparis pubiflora DC. – Fig. 56.

Development: The hypocotyl and taproot emerge from the seed, which is carried up by the first which becomes erect. After a resting stage, the testa is shed by spreading of the paracotyledons. A long resting stage occurs with developed paracotyledons only. The plantlet grows very slow, in flushes.
Seedling epigeal, phanerocotylar. Macaranga type.
Taproot absent to much reduced and conical, additional roots long, thin, much-branched, white, one taking over the function of the main root.

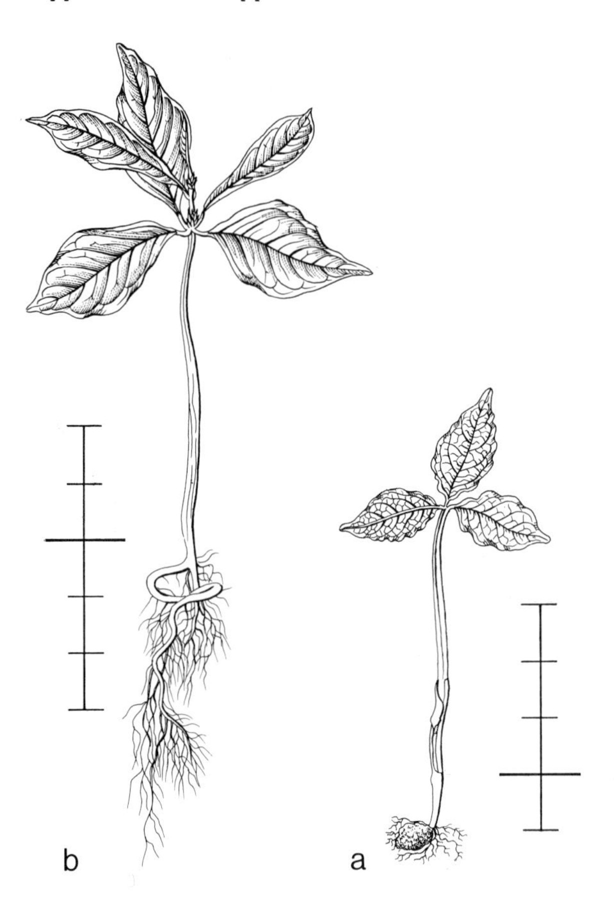

Fig. 56. *Capparis pubiflora.* De V. 1531, collected 21-vii-1972, planted 23-vii-1972. a date?; b. 9-viii-1973.

Hypocotyl strongly elongating, in cross-section 3-angular, to 5.5 cm long, green, glabrous, slightly tapering to the top, on the angles with a longitudinal, sharp ridge.
Paracotyledons 3, in one whorl, foliaceous, in the 2nd–4th leaf stage dropped; simple, stipulate, shortly petiolate, herbaceous, green, glabrous. Stipules 2 at the base of each paracotyledon, subulate, to 0.5 by 0.05 mm, reddish green. Petiole in cross-section semi-orbicular, to 2 by 0.5 mm, reddish green. Blade ovate, to 3.2 by 1.6 cm; base slightly narrowed; top acute; margin entire; nerves pinnate, above all depressed, below prominent, marginal nerve not conspicuous.
Internodes: First one elongating to 5 mm, terete, green, glabrous; lowest internodes of each next flush crowded, small, to 0.2 mm, and higher ones more elongating; those in first flush to 2 mm long. Stem slender, straight, developing in flushes.
First two leaves subopposite, simple, stipulate, petiolate, herbaceous, green, glabrous. Stipules 2 at the base of each petiole, subulate, persistent, to 0.5 by 0.05 mm, green, soon turning dark brown. Petiole in cross-section semi-orbicular, to 3 by 0.5 mm, above slightly channelled. Blade oblong to lanceolate, to 3.5 by 1 cm; base acute

to slightly narrowed; top cuspidate; margin entire; nerves pinnate, above all sunken, below prominent, lateral nerves curved near the end, forming an undulating, marginal nerve.
Next leaves spirally arranged, in first two flushes lower up to 7 ones abortive, present as slightly elevated warts, when young provided with caducous stipules as in the first leaves, which form a bundle on top of the endbud, when the shoot is developed soon dropped; at most 2 developed leaves present in each flush. Blade in leaves of second flush to 3.8 by 1.3 cm.

Specimens: 1531 from N. Sumatra, alluvial flat, primary forest on deep clay, low altitude.
Growth details: In nursery germination poor. Seedlings very slowly developing; all died before the third flush developed.
Remarks: Five species described, similar in germination pattern and general morphology; *C. pubiflora* differs in having three paracotyledons, the other species have two.

CELASTRACEAE

BHESA

Fl. Males. I, 6: 280.
Genus in India and Ceylon, throughout S.E. Asia and Malesia except Java, Celebes, and the Moluccas, as far east as Louisiades.

Bhesa robusta (Roxb.) Ding Hou – Fig. 57.

Development: The hypocotyl and root emerge from the sharp pole of the seed. After establishment the hypocotyl becomes erect, carrying the enclosed cotyledons up. After a resting stage the testa is shed by spreading of the cotyledons. A second short resting stage occurs with the cotyledons and two developed leaves.
Seedling epigeal, phanerocotylar. Sloanea type and subtype.
Taproot not developing. Additional roots many, from the base of the hypocotyl, slender, little branched, white, one developing into the main root.
Hypocotyl strongly enlarging, terete, to 8 cm long, green, glabrous.
Cotyledons 2, opposite, rather succulent, in the 2nd leaf stage dropped; simple, without stipules, sessile, tender fleshy, reddish cream turning greenish cream, glabrous, above with 5, greenish cream, slightly depressed nerves.
Internodes: First one hardly elongating, in cross-section elliptic, to 4 mm long, green, glabrous; next internodes as first one, rather irregular in length, 0.1–2 cm long. Stem first compressed later slender, straight.
First two leaves opposite, simple, stipulate, almost sessile, herbaceous, shining reddish turning green, glabrous. Stipules 4, between the petioles, rather soon caducous, narrowly triangular, to 3 by 1 mm, pale green. Petiole in cross-section semi-orbicular,

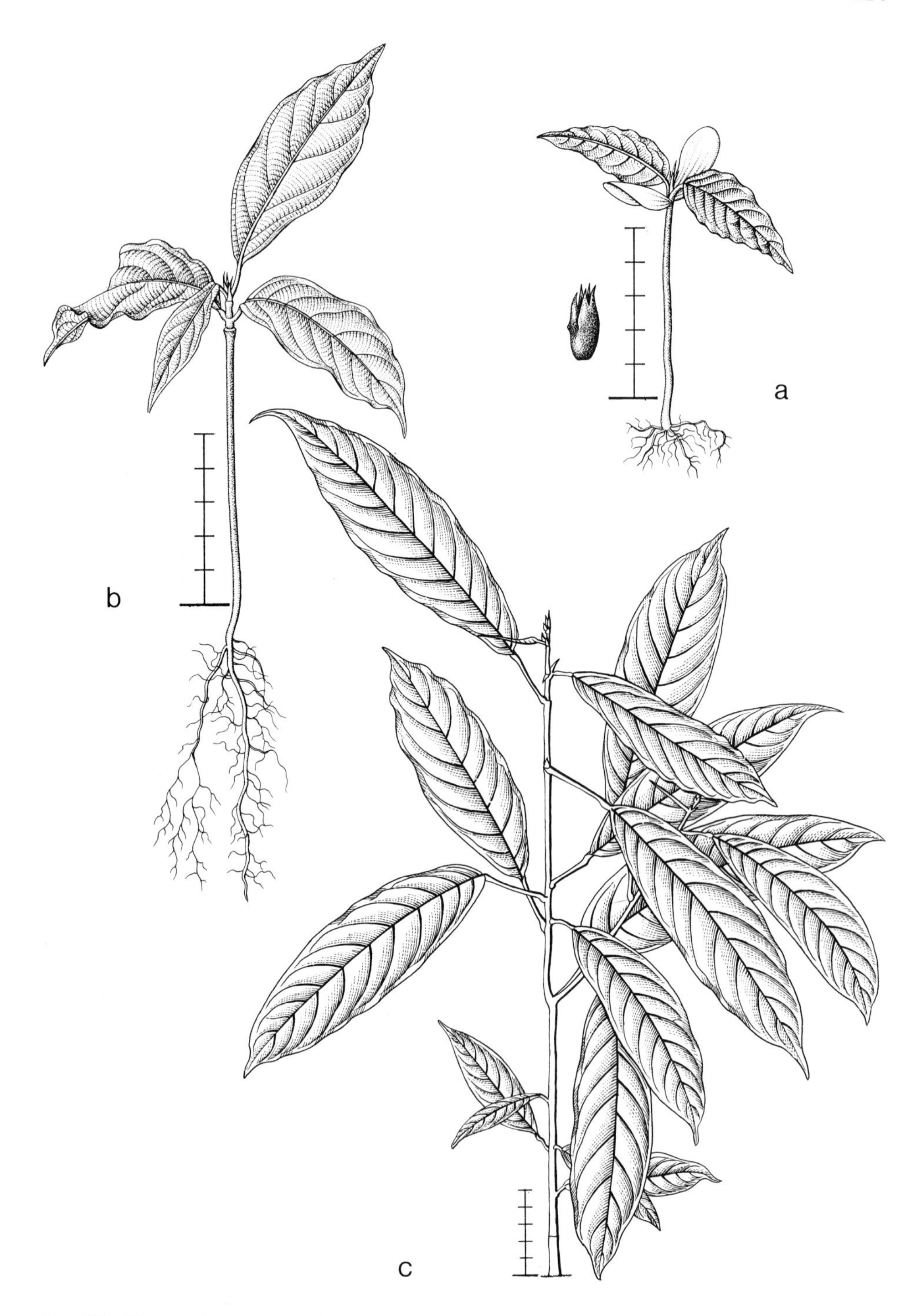

Fig. 57. *Bhesa robusta.* Dransfield 2392, collected 30-i-1972, planted 5-ii-1972. a date?; b. 17-v-1972; c date?.

to 2 by 2 mm. Blade ovate-oblong, to 6 by 2.5 cm; base narrowed into the petiole; top cuspidate; margin slightly undulating; nerves pinnate, above slightly prominent, below more so, lateral nerves free ending, veins numerous, not branched, parallel. *Next leaves* spirally arranged, gradually the higher ones larger. Stipules 2 at the base of each petiole, to 5 by 3 mm. Petiole in the 6th leaf to 25 by 1 mm, in 12th leaf to 50 by 2 mm, terete, above channelled, at base and tip distinctly swollen. Blade in 6th leaf to 12 by 3 cm, in 12th leaf to 25 by 8 cm, the higher ones gradually larger.

Specimens: Dransfield 2392, Malay Peninsula, primary hill forest, low altitude.
Growth details: In nursery germination good, ± simultaneous.

LOPHOPETALUM

Fl. Males. I, 6: 262; Meijer, Bot. Bull. Sandakan 11 (1968) 112 (*L. beccarianum*).
Genus in W. India, throughout S.E. Asia and Malesia, but not in E. Java and the Lesser Sunda Islands.

Lophopetalum javanicum (Zoll.) Turcz. – Fig. 58.

Development: The hypocotyl and taproot emerge from the margin at the side of the seed. By the hypocotyl becoming erect, the enclosed cotyledons are carried free from the soil. At the opposite margin the cotyledons spread, forming a gaping slit, from which the shoot appears. A short resting stage occurs with 2 developed leaves.
Seedling epigeal, cryptocotylar. Endertia type and subtype.
Taproot rather long, sturdy, cream-coloured, later turning orange-yellow, with few, short, thin, unbranched, later much-branched sideroots.
Hypocotyl strongly enlarging, piercing the testa at the margin in the centre of the seed, terete, to 2 cm long, cream-coloured turning grey-green, glabrous.
Cotyledons 2, opposite, flat but succulent, in the 3rd–5th leaf stage dropped; simple, sessile, without stipules, at the upper margin somewhat spreading, there rupturing the margin of the seedcoat, covered by the brown testa which forms a papery wing around the entire margin of the seed, hard fleshy, green where exposed, glabrous. Blade much broader than long, its form visible by a slight elevation and darker brown colour in the seed, to 4 by 1 cm, at the sides rounded, nerves not visible.
Internodes: First one strongly elongating, terete, to 4.5(–7) cm long, tapering to the top, and there somewhat angular and with two longitudinal grooves, green, glabrous; next internodes as first one, distinctly angular, red when young, turning green, second one to 7 mm long, gradually becoming larger, 6th one to 1.5 cm long. Stem slender, rather straight.
First two leaves opposite, simple, stipulate, shortly petiolate, herbaceous, reddish green turning green and dull green below, glabrous. Stipules caducous, leaving two warts at the base of the petiole. Petiole in cross-section semi-orbicular, to 2 by 1 mm. Blade very irregular in size, oblong, to 6 by 2.8 cm but in general much smaller; base narrowed into the petiole; top acuminate; margin entire; nerves pinnate, midrib above and below prominent, lateral nerves above sunken, below prominent, ending

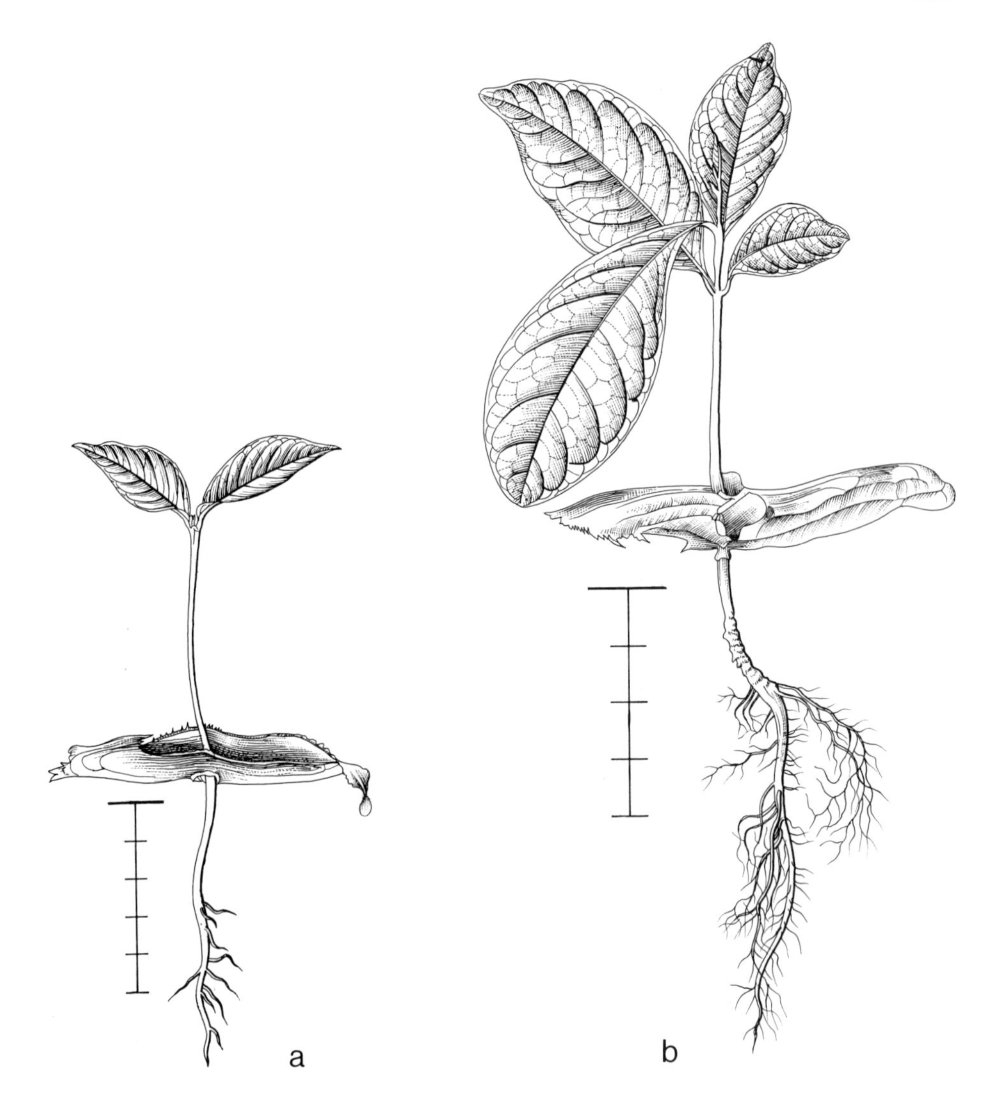

Fig. 58. *Lophopetalum javanicum*. De V. 2454, collected 6-x-1973, planted 7-xi-1973. a. 11-xii-1973; b. 2-iv-1974.

in anastomoses.
Next leaves spirally arranged. Stipules 2 at the base of each leaf, triangular, to 0.5 by 0.5 mm, reddish cream. Petiole in cross-section semi-orbicular, in lower leaves to 3 by 1 mm, in 9th leaf to 4 by 1.5 mm, at the base distinctly jointed, bright red turning green, glabrous. Blade oblong to obovate oblong, those in first leaves to 6 by 2.8 cm, but often much smaller, in 9th one to 9 by 4 cm; base narrowed into the petiole; top acuminate, its tip rounded, apiculate below the tip; margin entire; further as first leaves.

Specimens: 2454 from N. Celebes. Primary forest on deep clay, low altitude.
Growth details: In nursery germination fair, ± simultaneous.
Remarks: Two species described, similar in germination pattern and general morphology. *L. beccarianum* differs in details only.

SALACIA

Fl. Males. I, 6: 404; Mensbruge, Germination (1966) 281 (*S. bipindensis*).
Genus pantropic; tropical Africa, India, Ceylon, S.E. Asia, and S. part E. Asia, all over Malesia, N.E. Australia, in the Pacific as far east as Fiji, including the W. Carolines and New Caledonia, tropical Central and S. America.

Salacia sp. – Fig. 59.

Development: The root emerges from one pole of the seed, the slightly elongating cotyledonary petioles bring the plumule free, enabling the shoot to emerge. A short resting stage occurs after development of each flush.
Seedling hypogeal, cryptocotylar. Horsfieldia type and subtype.
Taproot long, sturdy, hard, brownish cream, with many short, thick, unbranched, creamy-white sideroots.
Hypocotyl not enlarging.
Cotyledons 2, fused into one body, when shed not known. Whole body ± ellipsoid, irregularly deformed, to 4 by 2 by 2 cm; the brownish testa persistent. Petioles very short, hardly exceeding the seedcoat.
Internodes: First one shortly elongating, terete, to 1 cm long, green to creamy-white, warty by many thick, elevated, lengthwise split lenticels; next internodes as first one, irregular in length. Stem rather sturdy, straight.
Leaves all spirally arranged, simple, involute when young, stipulate, shortly petiolate, herbaceous, green, glabrous. First flush with up to 16 lower leaves very reduced, higher ones developed. Reduced leaves scale-like, triangular, to 2 by 1 mm, often abortive, the higher ones sessile, gradually passing into the normal leaves. Developed leaves 4–8. Stipules broadly triangular, to 0.3 by 0.8 mm; top acute; margin slightly dentate; persistent, but shrivelling soon. Petiole in cross-section semi-orbicular, to 3 by 1.5 mm, above with a longitudinal ridge formed by the descending midnerve, at the base distinctly jointed. Blade in leaves of first flush oblong, to 15 by 6 cm; base obtuse; top cuspidate; margin entire; nerves pinnate,

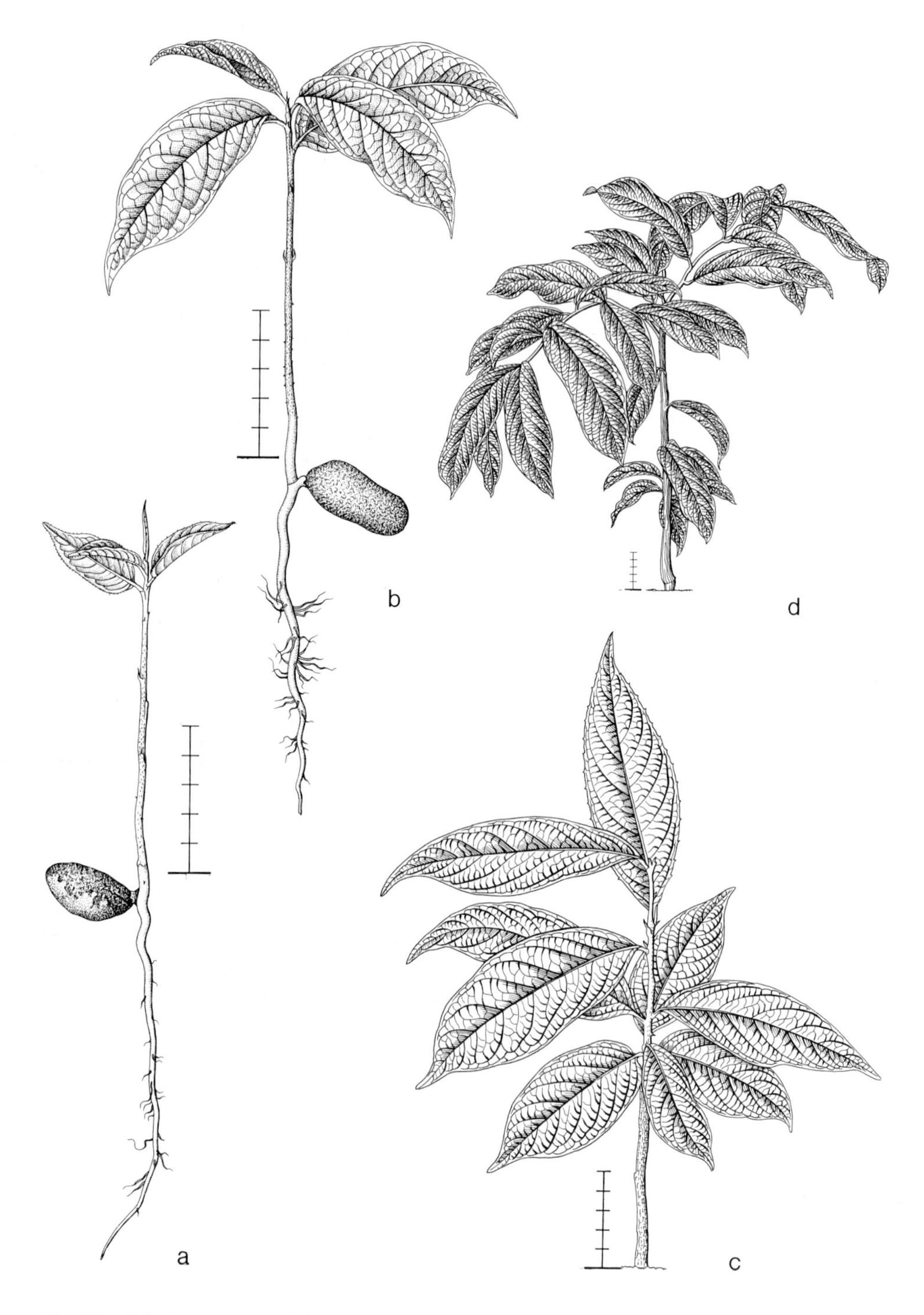

Fig. 59. *Salacia sp.* Dransfield 2538, collected 15-vi-1972, planted 20-vi-1972. a date?; b. 5-ix-1972; c. 27-i-1973; d. 19-viii-1974.

midrib above slightly prominent, below more so, lateral nerves above slightly prominent, descending along the midnerve, ending in anastomoses. Next flushes with 1–3 scales and 2 developed leaves, all spirally arranged, as those of first flush, those on the lateral branches opposite. Petiole in 10th leaf in cross-section semi-orbicular, to 5 by 2.5 mm. Blade oblong, in 10th leaf to 17 by 5.8 cm, base acute to obtuse; top acuminate; further as in the first leaves.

Specimens: Dransfield 2538 from W. Java, primary hill forest, low altitude.
Growth details: In nursery germination good, ± simultaneous.
Remarks: Two species described, similar in germination pattern and general morphology. *S. bipindensis* differs in details only.

SIPHONODON

Fl. Males. I, 6: 394.
Genus in E. India, S.E. Asia, all over Malesia.

Siphonodon celastrinus Griff. – Fig. 60.

Development: The hypocotyl and taproot emerge from the disintegrating fruit. By the hypocotyl becoming erect, the paracotyledons are withdrawn from the fruit. A short resting stage occurs with developed paracotyledons.
Seedling epigeal, phanerocotylar. Macaranga type.
Taproot long, slender, fibrous, creamy-yellow with many slender, branched, creamy-yellow sideroots.
Hypocotyl strongly enlarging, slender, in cross-section quadrangular, to 4.5 cm long, green, glabrous.
Paracotyledons 2, opposite, foliaceous, shed between 8th and 20th leaf stage; without stipules, almost sessile, herbaceous, green, glabrous. Blade elliptic, oblique, to 2 by 1.4 cm; base and top obtuse; margin entire; nerves pinnate, with footnerves, midrib above prominent, below ± sunken, other nerves rather inconspicuous, forming an intricate pattern.
Internodes: First one in cross-section quadrangular, to 1 cm long, distinctly winged along the margins, green, glabrous; next internodes as the first one, lower ones to 4–7 mm long, 10th one to 1 cm long, 15th one to 1.5 cm long. Stem rather slender, straight.
Leaves all spirally arranged, simple, ± sessile, stipulate, herbaceous, green, glabrous. Stipules triangular, to 0.3 by 0.3 mm, herbaceous, green. Blade in lower ones oblong, to 3 by 1.1 cm, the higher ones lanceolate, 5th one 6 by 1.9 cm, 15th one to 13 by 3.3 cm; base slightly narrowed; top acute; margin serrate, teeth ± 0.5 mm; nerves pinnate, midrib above prominent, below more so, lateral nerves above sunken, below raised, veins ending free in the teeth.

Specimens: 1408 from W. Java, primary forest on coral sand, low altitude, and 2368 from S.E. Borneo, hill Dipterocarp forest on deep clay, low altitude.

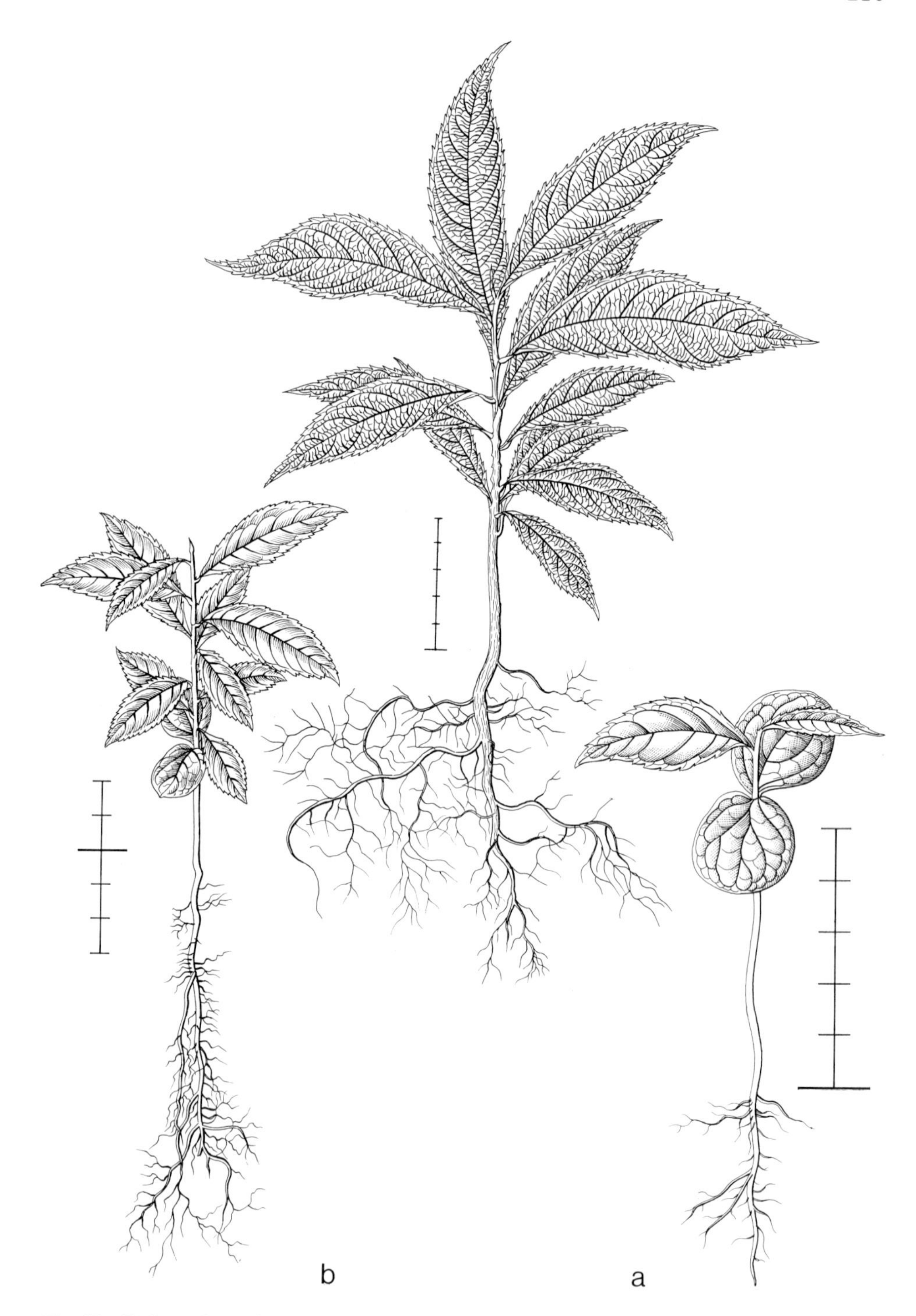

Fig. 60. *Siphonodon celastrinus.* De V. 1408, collected 26-vi-1972, planted 2-viii-1972. a, b, and c date?.

Growth details: In nursery germination ± simultaneous.

CONNARACEAE

CONNARUS

Fl. Males. I, 5: 525; Duke, Ann. Miss. Bot. Gard. 56 (1969) 148 (*C. panamensis*). *Genus* pantropic; tropical Africa, India, throughout S.E. and E. Asia, Malesia, N.E. Australia, in the Pacific as far as Fiji, including the W. Carolines, tropical S. and Central America.

Connarus grandis Jack – Fig. 61.

Development: The taproot emerges from one pole of the seed, the testa and usually the gaping fruit wall remain persistent. The testa splits along the margins of the cotyledons, which spread, after which the shoot starts developing. A resting stage occurs with 4 developed leaves.
Seedling hypogeal, cryptocotylar. Heliciopsis type and subtype.
Taproot slender, flexuous, brownish, with many long, shortly branched, brownish sideroots.
Hypocotyl hardly enlarging, terete, to 5 by 5 mm, pale brownish, glabrous.
Cotyledons 2, secund when the fruit wall is persistent, when the fruit wall is artificially removed more or less secund to opposite, succulent, after the 4th leaf stage shed; remaining at soil level, when free from the fruit wall the upper one slightly raised to enable the plumule to develop, simple, sessile, swollen, without stipules, ± irregularly elliptic, to 4 by 2.5 by 0.8 cm, the exposed part turning green, glabrous; the persistent thin testa splitting along the margins of the cotyledons in a regular way, shining blackish brown.
Internodes: First one elongating to 2.5 cm long, in cross-section elliptic, with 2 longitudinal ridges, especially below, green, slightly hairy with minute, simple, appressed, hyaline hairs, later developing many rusty-brown, warty lenticels; next internodes as the first one, irregular in length, second one to 5 mm long, 5th one to 2 cm long, 15th one to 4.5 cm. Stem slender, straight, erect.
Leaves spirally arranged. First 4–6 abortive, present as brownish warts, the lowest 2 (sub)opposite. Next up to 14 ones unifoliolate, the lower developed ones often subopposite, conduplicate-induplicate when very young, petiolate, without stipules, herbaceous, green, below hairy as first internode, glabrescent. Petiole terete, to 35 by 1.5 mm, jointed and distinctly thickened at base and top. Blade ovate-oblong, lowest 4 to 14.5 by 5 cm, 10th one to 20 by 9.5 cm; base rounded, top caudate; margin entire; nerves pinnate, above flat, the midrib above slightly prominent, below very prominent, lateral nerves curved at the tip, forming a widely undulating marginal nerve. Subsequent leaves imparipinnate, 3-foliolate, but the lowest 1–2 often with one developed lateral leaflet only. Petiole as those of the unifoliolate leaf, in 20th leaf to 8.5 by 0.25 cm, swollen and jointed at the base. Rhachis as petiole, in

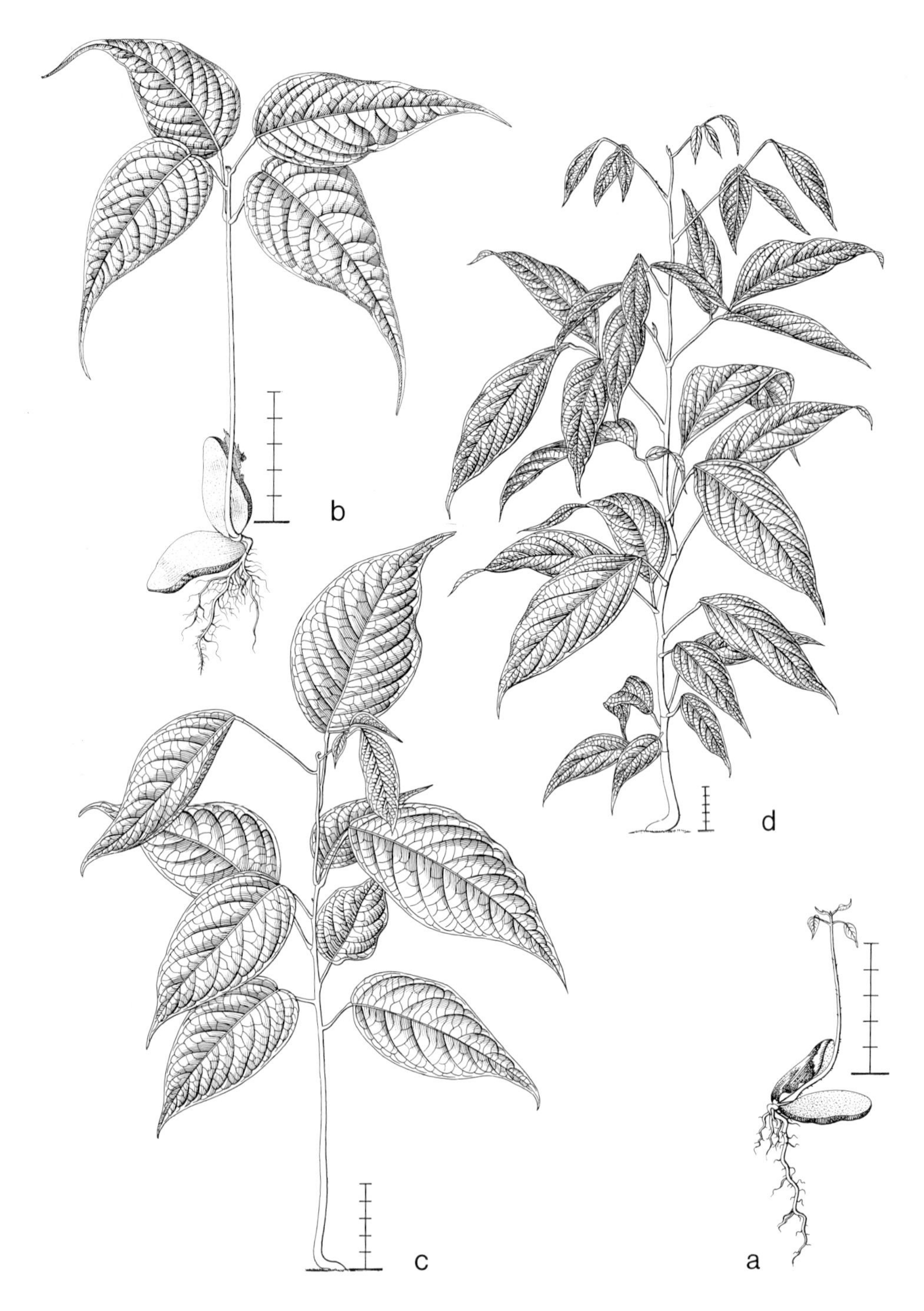

Fig. 61. *Connarus grandis*. De V. 1718, collected 20-x-1972, planted 17-xi-1972. a and b date?; c. 15-viii-1973; d. 3-xii-1973.

20th leaf to 3.5 by 2 mm. Leaflets petiolulate, without stipules. Petiolule swollen, terete, in 20th leaf to 8 by 2 mm in the lateral leaflets, to 7 by 3.5 mm in the terminal leaflet. Blade of leaflets ovate-oblong, in 20th leaf the lateral leaflets to 15 by 6 cm, terminal leaflet to 20.5 by 10 cm; base more acute to slightly narrowed; further as simple leaves.

Specimens: 1718 from S.E. Borneo, alluvial flat, primary forest on deep clay, medium altitude.
Growth details: In nursery germination good, ± simultaneous, but germination poor if the seeds are not freed from the pod.
Remarks: Three species described. *C. panamensis* differs in that the cotyledons become free from the fruit wall and are borne on an epigeal hypocotyl, and the first two leaves are opposite, borne on an elongated first internode, cataphylls being absent.

Connarus odoratus Hook. f. – Fig. 62.

Development: The taproot emerges from one pole of the gaping fruit, which splits along the upper margin. The cotyledons spread slightly in the fruit wall, enabling the shoot to emerge from the slit. No resting stage occurs.
Seedling hypogeal, cryptocotylar. Heliciopsis type and subtype.
Taproot long, slender, fibrous, brownish, with many rather long, little branched, brownish sideroots.
Hypocotyl little enlarging, terete, to 4 mm long.
Cotyledons 2, secund, succulent, in the 2nd–4th leaf stage shed; simple, without stipules, sessile, slightly spreading but remaining within the dark brown fruit wall. Whole body ± ellipsoid, to 3.5 by 1.8 by 2 cm, at its free tip with a rounded stalk 4 by 4 by 4 mm big, from this outgrowth along the upper side towards the other pole with a gaping slit. Cotyledonary blades loose in the fruit wall, elliptic in outline, to 2.7 by 1.5 by 0.3 cm, obliquely attached at the base, outside concave, inside flat; base and top rounded; margin entire; reddish cream, on the outside the bluish black, persistent, shining testa.
Internodes: First one strongly elongating, terete, to 9 cm long, green turning greyish brown, rather densely covered with short, stellate to tree-like, brown hairs, glabrescent; next internodes as the first one, irregular in length, between the lower leaves to 1.5 cm long, the 10th one to 2.5 cm long. Stem slender, slightly zig-zag.
First two leaves opposite, simple, spreading, without stipules, petiolate, subcoriaceous, green, when very young brownish and whitish by indument as on first internode, soon glabrescent. Petiole terete, to 22 by 1 mm, base and top distinctly swollen, hairy as the first internode. Blade long cordate, to 8 by 4.3 cm; base broadly auriculate; top caudate; margin entire; nerves pinnate, above sunken, below prominent, lateral nerves at the tip curved to the next one, forming an undulating marginal nerve.
Next leaves spirally arranged, lowest 6–10 simple. Petiole as those of the first leaves, ultimately to 5 cm long. Blade oblong, ultimately to 13.5 by 6 cm; base acute; further as the first leaves. Subsequent leaves compound. Petiole as those of the first leaves, in 3rd compound leaf to 7.5 cm long, but in general much smaller, swollen only at the

Fig. 62. *Connarus odoratus*. De V. 2343, collected 15-iii-1973, planted 22-iii-1973. a and b. 9-v-1973; c. 25-vii-1973; d. 14-iii-1974.

base. Rhachis as the petiole, in 5th compound leaf to 10 cm long. Blade in 3 lowest compound leaves trifoliolate, next 2–3 5-foliolate, subsequent ones 7-foliolate, imparipinnate with terminal leaflet. Lateral leaflets (sub)opposite, petiolulate. Petiolule terete, jointed at the base, wrinkled, in 5th leaf to 5 by 2 mm. Blade of leaflets as those of simple leaves; lateral ones in lowest compound leaf to 6 by 2.5 cm, in 5th one to 13.5 by 5 cm, lowest leaflets in each leaf smaller; terminal one in first compound leaf to 7 by 3.3 cm, in 5th one to 17 by 7 cm, further as simple leaves.

Specimens: 2343 from S.E. Borneo, alluvial flat, primary forest on deep clay, low altitude.
Growth details: In nursery germination good, ± simultaneous.
Remarks: See *C. grandis*.

ROUREA

Fl. Males. I, 5: 510; Duke, Ann. Miss. Bot. Gard. 52 (1965) 332 (*R. surinamensis*).
Genus pantropic; tropical Africa, including Madagascar, India and Ceylon, throughout S.E. and E. Asia, Malesia, N.E. Australia, in the Pacific as far as Samoa and Tonga Is., including New Caledonia, Central and tropical S. America.

Rourea minor (Gaertn.) Leenh. – Fig. 63.

Development: The taproot emerges from the acute pole of the seed, the testa splitting lengthwise. The bases of the cotyledons are pushed aside by secondary thickening of the cotyledonary node, by which an opening is made through which the plumule emerges.
Seedling hypogeal, cryptocotylar. Horsfieldia type and subtype.
Taproot long, slender, fibrous, brownish, when older with numerous long, fine, branched, brownish sideroots.
Hypocotyl not enlarging.
Cotyledons 2, secund, succulent, when shed not known, between 7th and 15th leaf stage; almost sessile, simple, without stipules. The whole body lengthwise partially splitting, spreading at the sharp pole, ± ovoid, to 13 by 7 by 7 mm, the blunt pole abruptly constricted into a more or less pronounced point, the brownish testa persistent.
Internodes: First one strongly elongating, slender, ± terete, to 5 cm long, base brownish, towards the top shining green, glabrous; next internodes as the first one, the lower ones to 5 mm long, 10th one to 15 mm long, in the erect part finally to 2.5 cm long; in the climbing leafless part, which usually develops from a lateral shoot after the death of the endbud, to 6 cm long, but usually much smaller. Stem erect, at first slightly zig-zag, later straight, climbing.
First two leaves opposite, often abortive or scale-like, simple, without stipules, petiolate, subcoriaceous, greenish red when young, turning green, glabrous. Petiole terete, to 10 by 0.4 mm, abruptly thickened at base and top for a length to 1 mm, at the top distinctly jointed. Blade ovate-oblong, to 4 by 1.7 cm; base obtuse to retuse;

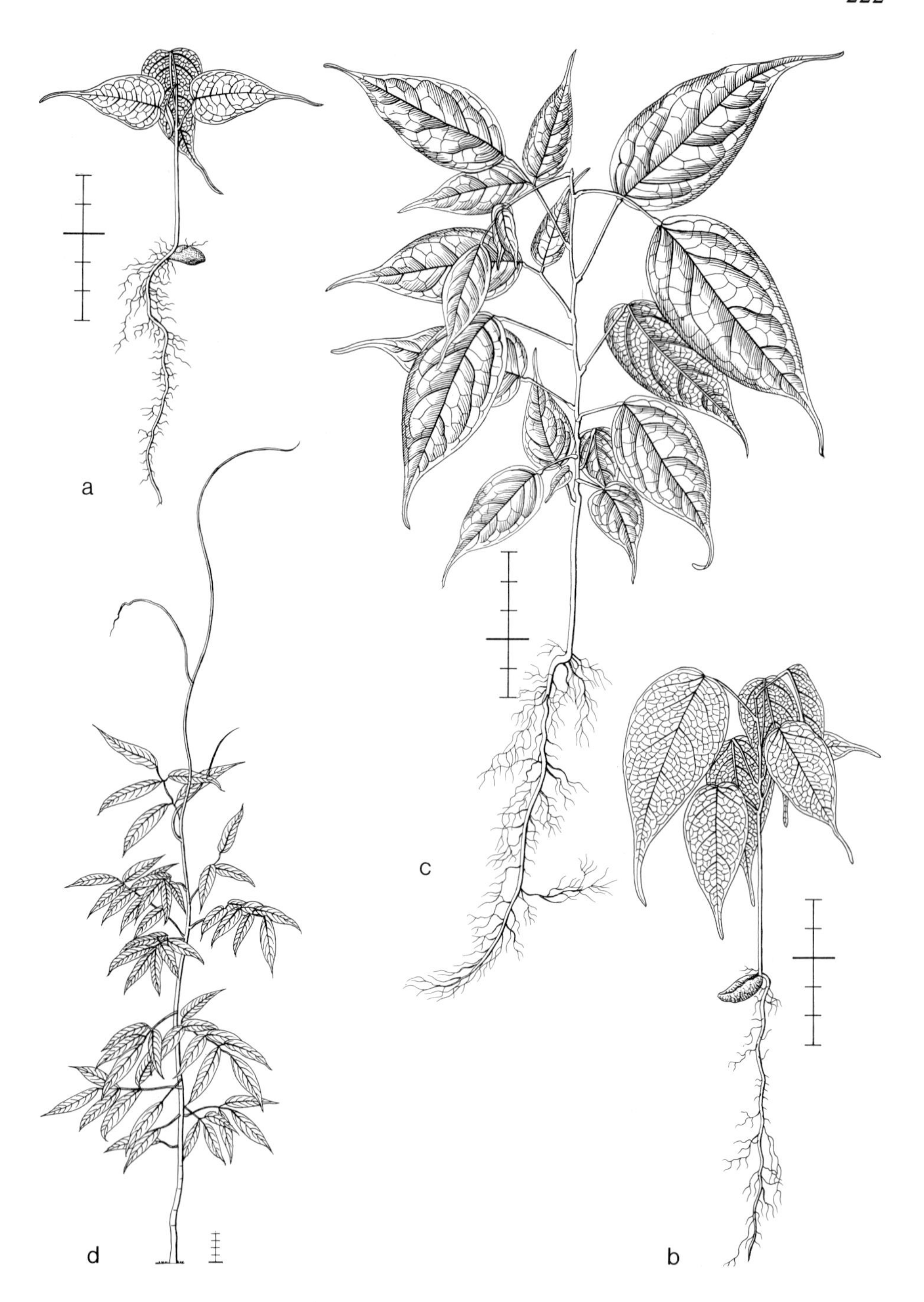

Fig. 63. *Rourea minor.* Dransfield 2523, collected 14-iv-1972, planted 20-vi-1972. a, b, and c date?; d. 15-viii-1974.

top caudate; margin entire; nerves pinnate, midrib below prominent, sidenerves slightly so, forming a vague marginal nerve.
Next 10–12 leaves spirally arranged, simple, without stipules, irregular in size, as first ones. Petiole as those of the first leaves, to 3.5 by 1 mm. Blade ultimately to 10 by 3.5 cm.
Subsequent leaves imparipinnate, 3-foliolate, sometimes intercalated with unifoliolate leaves, ultimately with 4 pairs of leaflets and top leaflet, without stipules and stipellae. Later on the climbing part of the stem all leaves reduced to scales. Petiole terete, thickened at the base only, finally to 7 cm long. Rhachis terete, in 3-foliolate leaves to 17 by 0.8 mm, in 9-foliolate leaves to 12 cm long, straight. Lateral leaflets (semi-)opposite, petiolulate. Petiolules terete, to 5 by 1 mm, wrinkled. Blade of lateral leaflets ovate-oblong, to 7.5 by 3.3 cm, those in the higher, more compound leaves more slender, obovate-lanceolate, to 10 by 3 cm. Terminal leaflet ovate-oblong, in 3-foliolate leaves to 10 by 4 cm, in 9-foliolate leaves to 10 by 3.5 cm, further as the first leaves.

Specimens: Dransfield 2523 from W. Java, coastal forest on coral sand, low altitude.
Growth details: In nursery the germination percentage was high.
Remarks: Two species described, and one more (*R. sp.*, De V. 2098) cultivated for the project, all similar in germination pattern and general morphology. They differ in details only.

CORNACEAE

MASTIXIA

Fl. Java 2: 159.
Monograph: Matthew, Blumea 23 (1976) 51–93.
Genus in India and Ceylon, throughout S.E. Asia, all over Malesia and the Solomon Islands.

Mastixia trichotoma Bl. var. *maingayi* (Clarke) Danser – Fig. 64.

Development: The taproot and hypocotyl emerge from one pole of the fruit. By the hypocotyl becoming erect, the paracotyledons enclosed in the endocarp are lifted from the soil. After a long resting period the two unequal valves of the endocarp are shed, and the paracotyledons expand, after which a second temporary rest follows.
Seedling epigeal, phanerocotylar. Macaranga type.
Taproot sturdy, woody, creamy-white, with few strong, unbranched sideroots as the taproot.
Hypocotyl strongly elongating, terete, to 6 cm long, hard herbaceous, dark bluish purple turning pale green, with many white lenticels, glabrous.
Paracotyledons 2, opposite, foliaceous, in 5th–7th leaf stage shed; during the resting stage hidden in the brownish, cracking seedcoat and stone; simple, spreading,

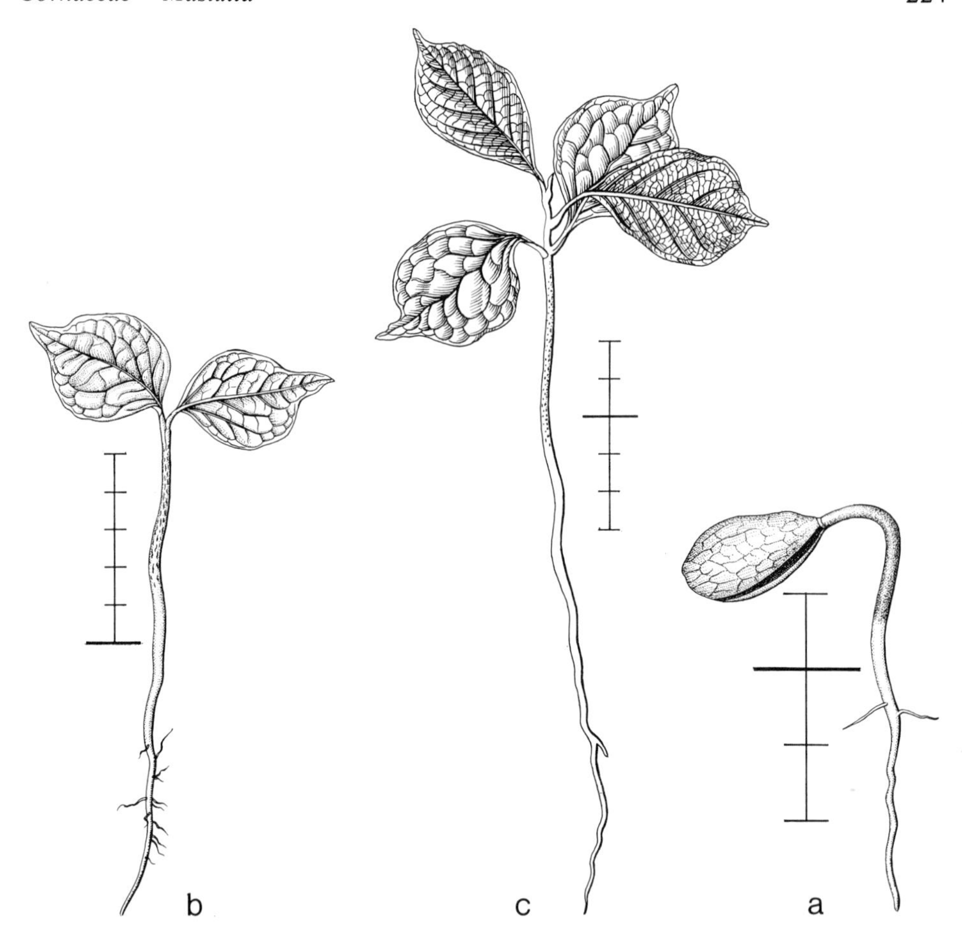

Fig. 64. *Mastixia trichotoma* var. *maingayi*. De V. 1249, collected 11-iii-1972, planted 23-iii-1972. a, b, and c date?.

without stipules, shortly petiolate, herbaceous, green, glabrous. Petiole in cross-section semi-orbicular, to 3 by 2 mm. Blade ovate, to 4.5 by 3.2 cm; base obtuse to narrowed; top acuminate; margin entire; vaguely 3-nerved, midrib above sunken, below prominent, other nerves less so, ending in anastomoses.

Internodes: First one hardly elongating, terete, to 3 mm long, green, densely hairy with appressed, simple, hyaline hairs; next internodes as first one, the lower ones short, the higher ones slightly more elongated, 4th one to 5 mm long. Stem short and stunted.

Leaves all spirally arranged, crowded, simple, without stipules, petiolate, herbaceous, reddish when young, turning green. Petiole in cross-section semi-orbicular, with a longitudinal groove above, to 13 by 1.5 mm, hairy as the first internode, soon glabrescent. Blade ± obovate-oblong, 4th one to 7 by 3 cm, higher ones not seen; base acute; top cuspidate; margin at the top with some scattered, inconspicuous teeth; above on the midnerve, below on the nerves and veins hairy as the first internode, slightly glabrescent; nerves pinnate, above distinctly sunken, below very prominent, lateral nerves especially towards the top of the leaf forming an inconspicuous, undulating marginal nerve; teeth provided with a vein.

Specimens: 1249 from S. Sumatra, much disturbed hill forest on deep clay, low altitude.
Growth details: In nursery germination fair. The plantlets establish quite well, but are very slow growing and remain much retarded in growth. After a long time of mere survival finally all died.

CUCURBITACEAE

HODGSONIA

Fl. Java 1: 304.
Monograph: Chakravarty, Rec. Bot. Surv. Ind. 17 (1959) 27.
Genus in N.E. India, throughout S.E. Asia, W. Malesia except the Philippines, and the Lesser Sunda Islands.

Hodgsonia macrocarpa (Bl.) Cogn. – Fig. 65.

Development: The creeping stem emerges from a slit at the obtuse end of the seed, developing additional roots on the nodes.
Seedling hypogeal, cryptocotylar. Hodgsonia type.
Main root absent; additional roots from the lower nodes, slender and much-branched, creamy-white.
Hypocotyl not developing, remaining in the seed.
Cotyledons 2, enclosed in the thick, hard, brown testa. The whole body reniform, to 7 by 5.5 by 3 cm, at one margin with a thick-walled slit, very long adhering.
Internodes: First emerging one short, to 0.6 cm long, angular, whitish, with minute, scattered, hyaline hairs; next internodes as first one, the third and later ones much longer, to 20 cm long, green spotted pale green; nodes thickened. Stem slender, flexuous, never erect.
Leaves all spirally arranged, the first ± 6 scale-like. Developed ones simple, without stipules, shallow to deep 3-lobed, petiolate, herbaceous, green, pale green below. Scales much reduced, sessile; top acute to acuminate; margin entire; hairs as first internode; the upper ones slightly 3-lobed, abruptly passing in the normal leaves. Petiole in cross-section semi-orbicular, above more or less flattened and vaguely to distinctly channelled, to 6 by 0.4 cm, hairy as the first internode. Blade ± rhomboid in outline, to 19 by 18 cm; base ± narrowed; top of lobes acute to acuminate; margin of lobes entire; nerves palmate, main nerves 3, above sunken, below much prominent, lateral nerves ± forming a right angle with the main nerves, at the tip forming a clear marginal nerve; tendrils reduced at the lower scales, fully developed from the 6th leaf onward, terete, to 30 cm long, on the left side of the leaf, branched, one branch up to

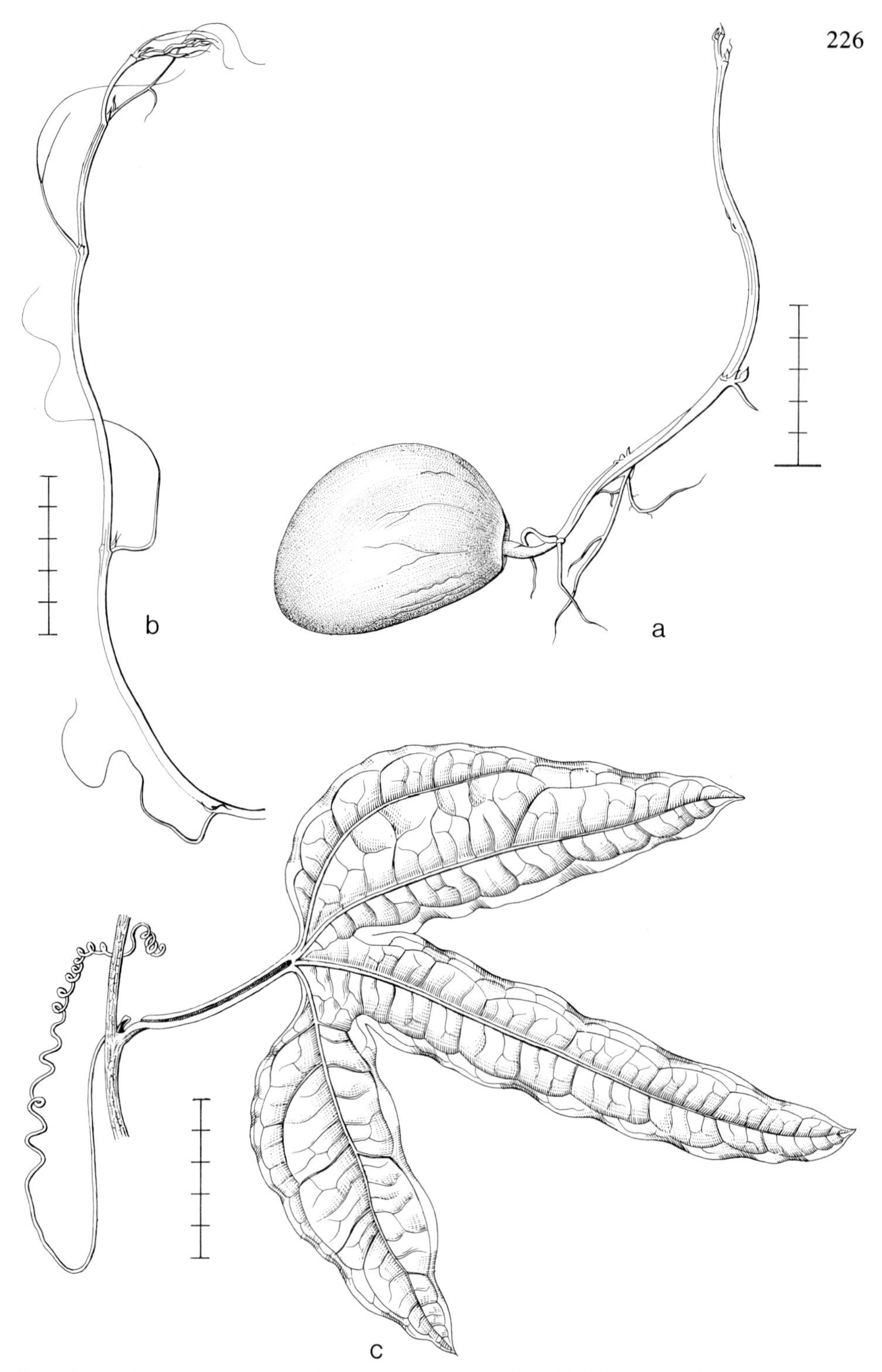

Fig. 65. *Hodgsonia macrocarpa*. De V. 1539, collected 22-vii-1972, planted 27-vii-1972. a, b, and c date?.

8 cm long, the other one much shorter, both branches coiled either in opposite or in the same direction, hairy as first internode.
Axillary buds naked, hairy as the stem, pale green, at the side opposite the tendril with an ovate, swollen and succulent, sessile, ant-frequented, dark green scale with minute, paler green, sunken pits.

Specimens: 1539 from N. Sumatra, alluvial flat, primary forest on deep clay, low altitude.
Growth details: In nursery germination poor.

DIPTEROCARPACEAE

DIPTEROCARPUS

Fl. Males. (in press); Burkill, J. Str. Br. R. As. Soc. no. 81 (1920) 51 (*D. crinitus*, *D. grandiflorus*, *D. kerrii*, *D. scortechinii*); Gilg, Nat. Pflanzenfam. ed. 2, 21 (1925) 251 (*D. retusus*); Maury, Bull. Soc. Hist. Nat. Toulouse 104 (1968) 187–202 (*D. baudii*, *D. costulata*, abnormal germ.); Meijer, Bot. Bull. Sandakan 11 (1968) 112 (*D. gracilis*, *D. stellatus*); Pierre, Flore Forestière Cochinchine 3 (1879–1899) pl. 215 (*D. intricatus*), pl. 218 (*D. tuberculatus*), pl. 221 (*D. punctulatus*); Troup, Silviculture 1 (1921) 37 (*D. indicus*).
Genus in India and Ceylon, throughout S.E. Asia and W. Malesia, including Java and the Philippines.

Dipterocarpus cf. *caudiferus* Merr. – Fig. 66.

Development: The taproot and hypocotyl pierce the tip of the fruit between the wings which represent two enlarged calyx lobes, curving and entering the soil. Meanwhile the cotyledonary petioles elongate, carrying the plumule free from the fruit wall, after which the shoot starts elongating. No resting stage occurs.
Seedling epigeal, but the enclosed cotyledons remain resting on the soil, cryptocotylar. Heliciopsis type, Koordersiodendron subtype.
Taproot long, sturdy, woody, dark reddish brown, with many long, very slender, much-branched, creamy-brown sideroots.
Hypocotyl terete, epigeal, elongating to 3.5 cm long, dark red-brown, especially in its upper part densely hairy with small, simple, hyaline hairs.
Cotyledons 2, secund, succulent, in the 4th leaf stage shed; simple, without stipules, long petiolate, covered by the dark brown, 2-winged fruit wall. Cotyledonary body smooth, the lower part rounded, the part enclosed by the wings conical, to 4 by 2.5 cm. Fruit wings 2, sessile, lanceolate, to 16 by 3.5 cm; base slightly narrowed; top rounded; margin entire; 3-nerved; 3 additional wings much smaller. Petiole in cross-section semi-orbicular, to 40 by 4 mm, above channelled, densely hairy as the hypocotyl.
Internodes: First one strongly elongating, terete, to 11 cm long, green turning greyish

Fig. 66. *Dipterocarpus* cf. *caudiferus.* De V. 2354, seedlings collected 15-iii-1973, planted 22-iii-1973. a. 17-iv-1973; b. 25-vii-1973; c. 8-iv-1974.

green, densely hairy with rather long, patent, hyaline bundles of hairs, glabrescent; next internodes as the first one, the lower ones to 1 cm long, the 10th one to 3 cm long. Stem often rather twisted (also in natural stands), slender.
First two leaves opposite, simple, when young pleated and hidden in the stipules, stipulate, petiolate, subcoriaceous, green. Stipules 2, interpetiolar, linear, to 25 by 3 mm, herbaceous, reddish pink, hairy as the first internode, shrivelling soon but rather long persistent. Petiole terete, to 10 by 2 mm, hairy as the first internode. Blade obovate, to 7 by 3.5 cm; base acute; top cuspidate; margin slightly undulating; along the margin and on the nerves hairy as the hypocotyl; nerves pinnate, above sunken, below prominent, lateral nerves free ending, veins ⊥ to the sidenerves, parallel.
Next leaves spirally arranged. Stipule 1, splitting longitudinally, when young covering the leaf, 7th one to 60 by 6 mm, on lateral branch in 20th leaf stage to 11 mm long, as those of first leaves. Petiole as in first leaves, but base and especially the top distinctly swollen, in 7th leaf to 25 by 1.5 mm, in 15th one to 30 by 2 mm, further as in first leaves. Blade (obovate-)lanceolate, in 7th leaf to 16 by 5 cm, in 15th one to 26 by 7 cm, further as in first leaves, surface regularly undulating.

Specimens: 2354 from S.E. Borneo, alluvial flat, primary forest on deep clay, low altitude.
Growth details: In nursery germination fair, ± simultaneous. In forest germination fairly abundant.
Remarks: Most species described show a distinct epigeal hypocotyl combined with long cotyledonary petioles (*D. baudii*, *D. caudiferus*, *D. costulata*, *D. crinitus*, *D. gracilis*, *D. grandiflorus*, *D. indicus*, *D. kerrii*, *D. palembanicus*, *D. retusus*, and *D. scortechinii*). In *D. tuberculatus* and *D. punctulatus* the hypocotyl is very short, and is pushed out of the fruit by elongation of the cotyledonary petioles which become very long, together with the undeveloped plumule.

Dipterocarpus palembanicus Sloot. – Fig. 67.

Development: The taproot and hypocotyl pierce the tip of the fruit between the wings which represent two enlarged calyx lobes, curve, and enter the soil. Meanwhile the cotyledonary petioles elongate, carrying the plumule free from the fruit wall, after which the shoot starts elongating. No resting stage occurs.
Seedling epigeal, but the enclosed cotyledons remain resting on the soil, cryptocotylar. Heliciopsis type, Koordersiodendron subtype.
Taproot long, sturdy, woody, dark brown, with many few-branched, creamy-white sideroots.
Hypocotyl terete, epigeal, elongating to 4.5 cm long, dark brown, especially in its upper part densely hairy with bundles of rather long, stiff, simple, brownish hyaline hairs and bundles of much shorter hyaline ones.
Cotyledons 2, secund, succulent, in the 4th–5th leaf stage shed, simple, without stipules, long petiolate, covered by the dark brown, 2-winged persistent fruit wall. Cotyledonary body distinctly 5-ridged, to 3.5 by 3 by 3 cm including the much elevated, sharp ridges. Fruit wings 2, sessile, lanceolate, to 10 by 2.3 cm; base slightly narrowed; top rounded; margin entire; 5-nerved, of which 3 central ones more

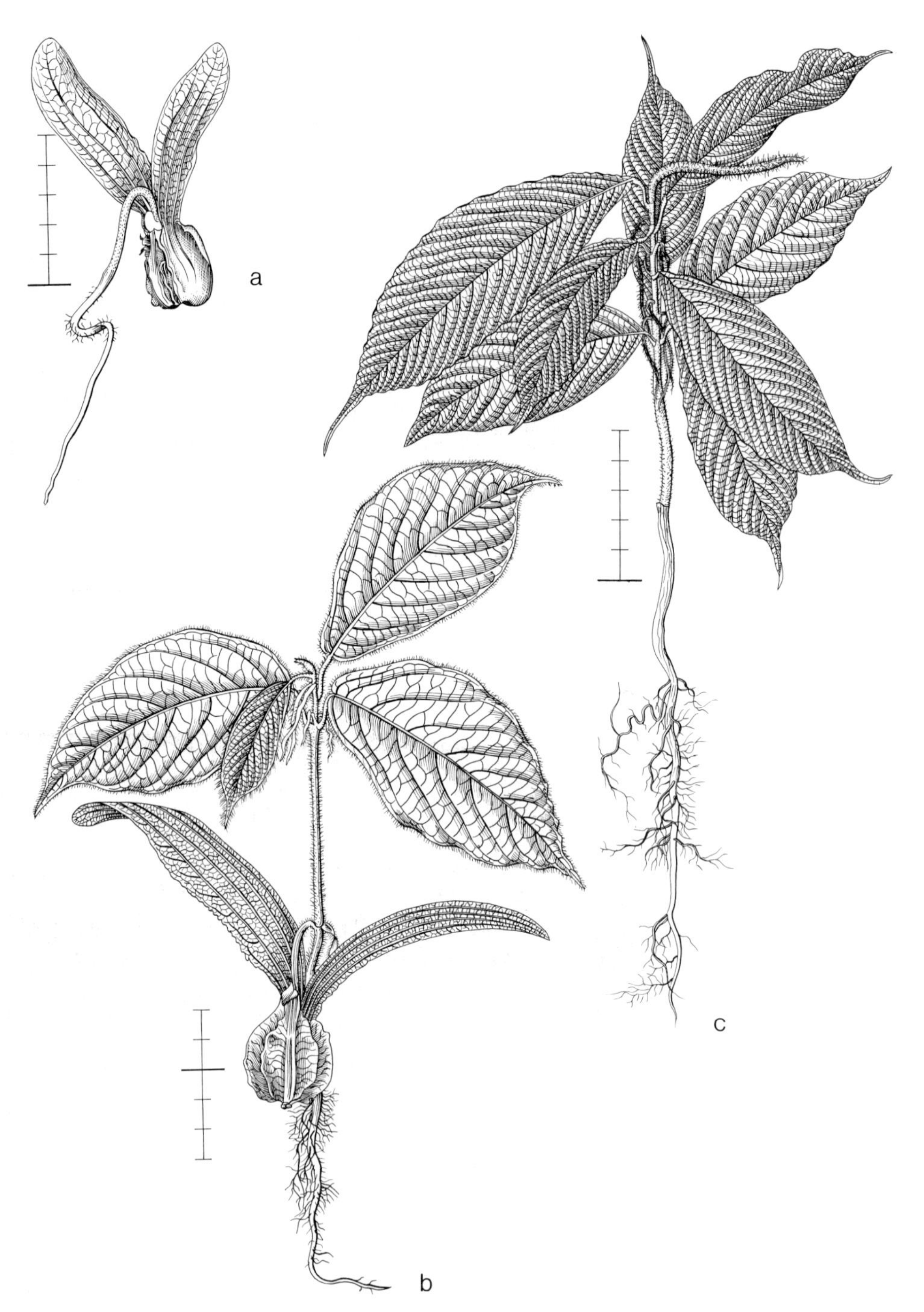

Fig. 67. *Dipterocarpus palembanicus*. De V. 2308, collected 12-iii-1973, planted 22-iii-1973. a. 31-iii-1973; b. 6-v-1973; c date?.

prominent; 3 additional wings much smaller. Petiole in cross-section semi-orbicular, to 3 by 0.2 cm, channelled above, densely hairy as the hypocotyl.
Internodes: First one strongly elongating, terete, to 6 cm long, green, densely hairy with rather long hairs on the hypocotyl; next 2 internodes as the first one, very short, to 3 mm long; the higher ones gradually larger, irregular in length, 5th one to 2 cm long, 10th one to 4 cm long. Stem straight, slender.
First two leaves opposite, simple, spreading, stipulate, petiolate, subcoriaceous, green. Stipules 2, when young connate and covering the endbud, splitting, interpetiolar, one at each side, sessile, linear, to 20 by 2 mm; top acute; margin entire; pale reddish, when the shoot starts further growth soon shrivelling but rather long persistent, hairy as the hypocotyl with simple and stellate hairs. Petiole terete, to 15 by 1.5 mm; base slightly swollen; top more so. Blade ± elliptic, to 10 by 5.5 cm; base slightly exsculptate; top cuspidate; margin entire; on the nerves and the margin on both sides with appressed long hairs as on the hypocotyl, above glabrescent; nerves pinnate, above sunken, below very prominent, lateral nerves very regular, free ending.
Next leaves all spirally arranged. Stipules two at the base of each petiole, connate when young, splitting longitudinally, in 5th leaf to 40 by 2.5 mm, in 8th leaf to 6 cm long, further as those of first leaf. Petiole in 5th leaf to 20 by 1.5 mm, in 10th leaf to 2.5 cm long. Blade obovate-oblong, in 5th leaf to 13 by 4.2 cm, in 10th leaf to 17.5 by 6.5 cm; base obtuse; top caudate; below on the nerves and above on the blade densely covered with long, simple, appressed hyaline hairs, above soon glabrescent; further as first leaves.

Specimens: 2308 from S.E. Borneo, primary hill forest on deep clay, low altitude.
Growth details: In nursery germination good, ± simultaneous.
Remarks: See *D.* cf. *caudiferus*.

DRYOBALANOPS

Fl. Males. (in press); Burkill, J. Str. Br. R. As. Soc. no. 81 (1920) 56 (*D. aromatica*); Meijer, Bot. Bull. Sandakan 11 (1968) 112 (*D. lanceolata*).
Genus in W. Malesia: Sumatra and Borneo.

Dryobalanops cf. *lanceolata* Burck – Fig. 68.

Development: The taproot and hypocotyl emerge from the apex of the fruit between the wings, curve, and enter the soil between the wings. After establishment of the root the cotyledons and shoot are pulled out of the fruit wall. No resting stage occurs.
Seedling epigeal, phanerocotylar. Sloanea type and subtype.
Taproot strong, slender, fibrous, creamy-brown, with many small, short, few-branched, creamy-white sideroots.
Hypocotyl enlarging to 6 cm long, terete, above green, below brown, glabrous, with some small, green, warty lenticels.
Cotyledons 2, opposite, succulent, in the 4th–6th leaf stage dropped; simple, distinctly unequal, without stipules, petiolate, hard fleshy, reddish, glabrous. Petiole

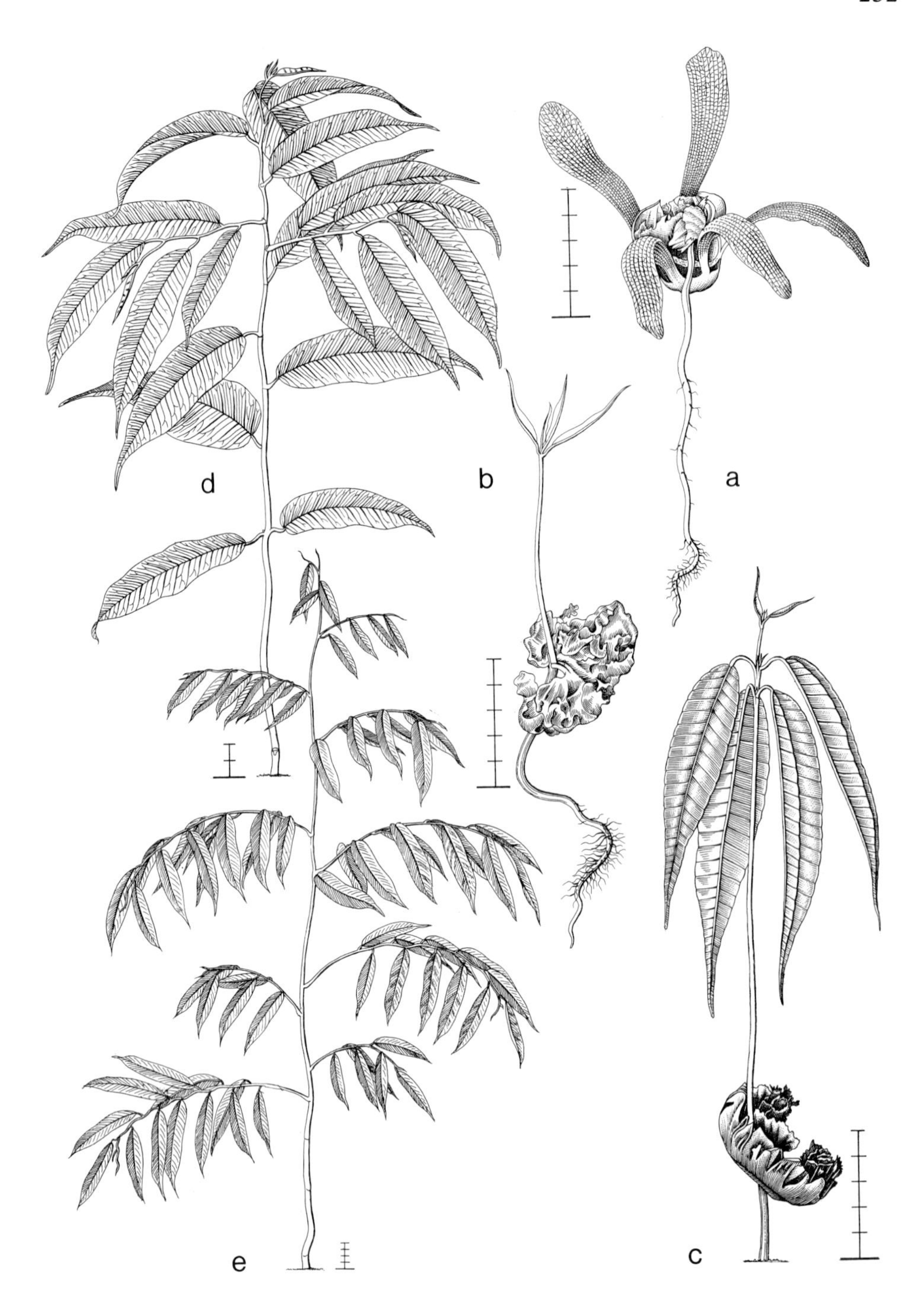

Fig. 68 *Dryobalanops* cf. *lanceolata.* Bogor Bot. Gard., without number, collected and planted i-1973. a, b, and c date?; d. 27-vii-1973; e. 10-vii-1974.

flattened, in cross-section elliptic, above channelled, that of the larger cotyledon to 8 by 4 mm, that of the smaller one 5 by 3 mm. Blade of the larger one ± reniform, irregularly lobed and incised, to 3 by 7 cm; base distinctly retuse with irregular auricles 1.5 by 1.5 cm big; top and margin irregularly lobed and incised; upper surface with irregular warts of different size, below rather smooth; blade of the smaller one ± reniform, slightly lobed, to 0.8 by 2 cm; base slightly retuse; top and margin irregularly lobed; surface slightly warty on both sides.
Internodes: First one strongly elongating, terete, to 22 cm long, reddish when young, turning green, glabrous; second one short, in cross-section elliptic, above the lower leaves with a longitudinal groove; next internodes terete, as first one, third one to 1 cm long, higher ones irregular in length, ranging from 2–16 cm long. Stem slender, straight.
First 4 leaves decussate, developed in one flush, conduplicate-induplicate when young, simple, stipulate, petiolate, subcoriaceous, dull reddish purple when young, turning green, with a distinct smell of turpentine when crushed, glabrous. Stipules 2 with each leaf, dropping soon, narrowly triangular, to 5 by 1 mm; top acute; margin entire. Petiole ± terete, to 10 by 2 mm, the lower 2 mm distinctly constricted, above channelled lengthwise. Blade ovate-lanceolate, to 17 by 4.8 cm; base obtuse, rounded; top caudate; margin entire; nerves pinnate, midrib above slightly sunken, below prominent, lateral nerves numerous, rather inconspicuous, parallel, in an angle of ± 70–80° with the main nerve, at the tip forming a hardly undulating, inconspicuous marginal nerve near the margin.
Next leaves spirally arranged. Blade ultimately to 17 by 5 cm, further as the first leaves.

Specimens: Bogor Botanic Gardens, without number. Origin unknown.
Growth details: In nursery germination good, ± simultaneous.
Remarks: The two species described are similar in germination pattern and general morphology; *D. aromatica* differs in details only.

HOPEA

Fl. Males. (in press); Burkill, J. Str. Br. R. As. Soc. no. 81 (1920) 58, 66 (*H. curtisii*, *H. micrantha*, *H. zeylanica*, the latter two under *Balanocarpus*); Pierre, Flore Forestière Cochinchine 4 (1879–1899) pl. 244 (*H. odorata*); Troup, Silviculture 1 (1921) 47 (*H. parviflora*).
Genus in India and Ceylon, throughout S.E. Asia, North up to Taiwan, all over Malesia except for the Lesser Sunda Islands.

Hopea dryobalanoides Miq. – Fig. 69.

Development: The taproot and hypocotyl emerge from the acute apex of the fruit between the wings, curving and entering the soil. After establishment of the root, the cotyledons expand and are withdrawn from the fruit wall. A short resting stage occurs with the cotyledons and two developed leaves.

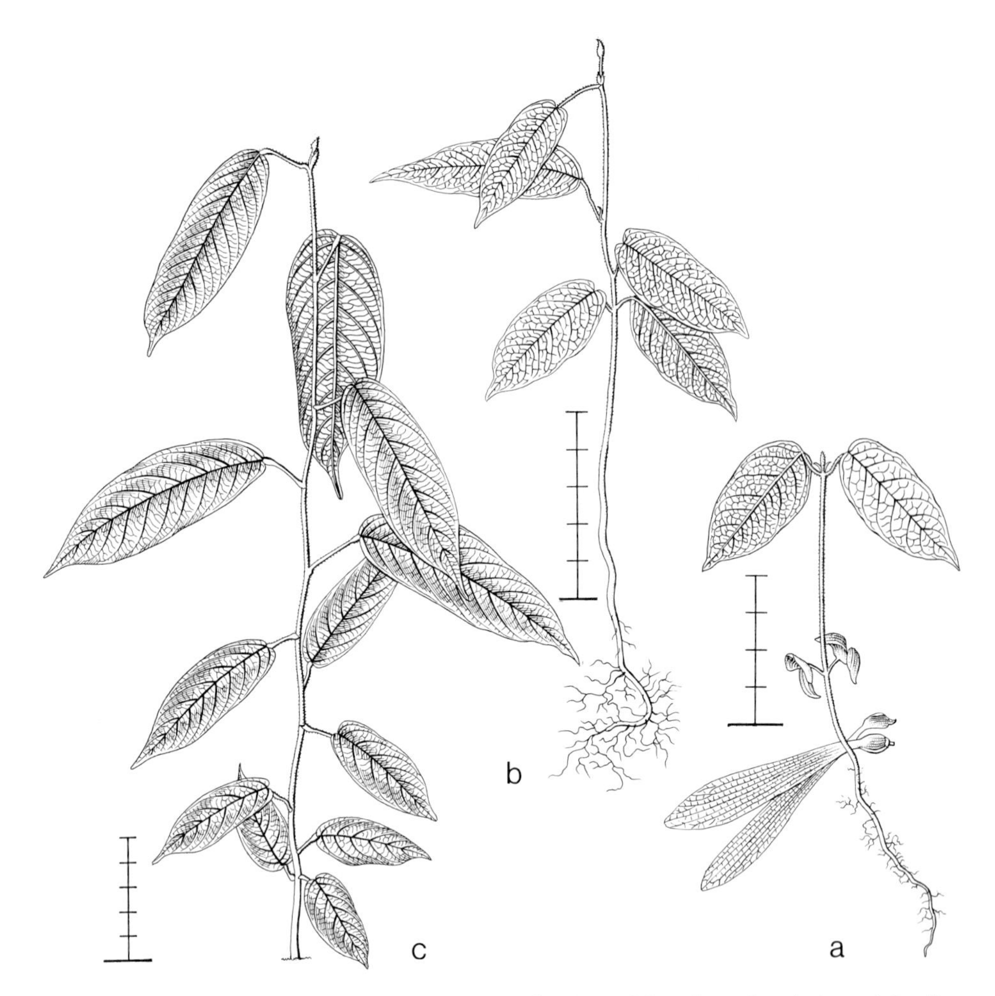

Fig. 69. *Hopea dryobalanoides.* Bogor Bot. Gard. VII. B. 23, collected and planted date?. a, b, and c date?.

Seedling epigeal, phanerocotylar. Sloanea type and subtype.
Taproot long, slender, fibrous, brown, with many rather short, branched creamy-white sideroots.
Hypocotyl strongly elongating, terete, to 5 cm long, the lower part brown, the slightly swollen green part below the cotyledons to 5 mm long, clearly separated by an undulating margin, hairy throughout with stiff, patent, hyaline, simple hairs.
Cotyledons 2, opposite, succulent, in the 2nd–3rd leaf stage shed; unequal, without stipules, petiolate, hard fleshy, pale green, with minute white spots. Petiole in cross-section ± elliptic, to 5 by 1 mm, hairy as the hypocotyl. Blade two-winged, much broader than long; top retuse; wings ± ovate, their tip obtuse; the one enclosed in the fruit slightly bigger, its base slightly retuse, below concave, the wings grooved towards the top, above smooth, with along the adaxial margin of the wings a deep groove; the other with base obtuse, below concave, smooth, along the margin with a slight groove, above convex at the base, at the upper part concave, there grooved as the lower side in the enclosed cotyledon.
Internodes: First one strongly elongating, terete, to 3.5 cm long, green, slightly hairy as the hypocotyl; next internodes as the first one, slightly irregular in size, 2–4 cm long, in higher ones much larger, especially these latter rather densely covered with minute, hyaline, stellate hairs. Stem slender, straight.
First two leaves opposite, when young conduplicate-induplicate, simple, stipulate, petiolate, subcoriaceous, above shining green, below brownish green, later turning green. Stipules sessile, dropping soon, narrowly triangular, to 2 by 0.5 mm; top acute, later rounded; margin entire; pale yellowish green, hairy as the higher internodes. Petiole terete, to 7 by 0.7 mm, hairy as the hypocotyl. Blade ± elliptic, to 3.5 by 1.9 cm; base retuse; top obtuse to slightly acuminate; margin entire; below slightly hairy on the nerves as in the higher internodes; nerves pinnate, above not raised, below prominent, especially the midrib, yellowish, sidenerves ± free ending.
Next leaves spirally arranged. Stipules more elliptic, with rounded top, especially the later ones, 15th one to 13 by 7 mm. Petiole distinctly swollen at the top, in 15th leaf to 20 by 1.5 mm. Blade ovate, gradually becoming larger, in 15th leaf to 14 by 5.7 cm, ultimately to 17 by 5.5 cm; base obtuse; top cuspidate, further as the first leaves.

Specimens: Bogor Botanic Gardens VII. B. 23, from Bangka.
Growth details: In nursery germination fair, ± simultaneous.
Remarks: All species described show the same germination pattern and general morphology, they differ in details only.

SHOREA

Fl. Males. (in press); Burkill, J. Str. Br. R. As. Soc. no. 76 (1917) 160 (*S. bracteolata*, *S. gibbosa*, *S. leprosula*, *S. macroptera*, *S. rigida*); ibid. no. 79 (1918) 39 (*S. robusta*); ibid. no. 81 (1920) 3 (*S. maxima*, under *Balanocarpus*), 67 (*S. bracteolata*, *S. curtisii*, *S. gratissima*, *S. macroptera*, *S. pauciflora*, *S. sentilata*, *S. sericea*); ibid. no. 86 (1922) 281 (*S. sumatrana*, as *Isoptera borneensis*), 285 (*S. singkawang*, as *S. thistletoni*); J. Mal. Br. R. As. Soc. 1 (1923) 218 (*S. maxima*, under *Balanocarpus*); ibid. 3 (1925) 4

(*S. hemsleyana*, under *Balanocarpus*); Meijer, Bot. Bull. Sandakan 11 (1968) 112 (*S. mecistopterix*, *S. xanthophylla*); Roxburgh, Plants of Coromandel 3 (1819) pl. 212 (*S. robusta*); Troup, Silviculture 1 (1921) 55 (*S. robusta*).
Genus in India and Ceylon, throughout S.E. Asia, all over W. Malesia, including the Philippines, in E. Malesia in Celebes and the Moluccas.

Shorea laevis Ridl. – Fig. 70.

Development: The taproot and hypocotyl pierce the tip of the fruit between the wings, curve, and enter the soil. After establishment of the root the cotyledons are withdrawn from fruit wall and testa by unfolding of the cotyledons, and carried above the soil by the hypocotyl becoming erect. No resting stage occurs.
Seedling epigeal, phanerocotylar. Sloanea type and subtype.
Taproot long, slender, fibrous, brownish, with many long, slender, shortly branched, brownish cream sideroots.
Hypocotyl strongly elongating, terete, to 4 cm long, brownish, glabrous.
Cotyledons 2, opposite, succulent, in the 3rd–4th leaf stage dropped; inequal in ornamentation, without stipules, petiolate, hard fleshy, green turning orange-yellow, glabrous. Petiole in cross-section flattened, above convex, to 5 by 1.2 mm. Blade 2-winged, much wider than long; the bigger one to 7 by 20 mm, top broadly emarginate, wings sessile, ovate, to 10 by 7 mm, their tip acute, margin entire, below convex, faintly longitudinally grooved, inside flat, smooth, at the base with a longitudinal groove; the smaller one to 5 by 17 mm, top broadly emarginate, wings sessile, ± ovate with a ± straight base, to 8 by 5 mm, their tip acute, margin entire, below flat, smooth, the upper side convex, smooth, the swollen margin faintly longitudinally grooved.
Internodes: First one strongly elongating, terete, to 2.5 cm long, green, glabrous; next internodes as first one, to 1 cm long. Stem slender, erect.
First two leaves opposite, simple, stipulate, petiolate, herbaceous, green, glabrous. Stipules sessile, soon dropping, linear, to 3 by 0.5 mm; top acute; margin entire. Petiole terete, to 10 by 0.5 mm, faintly thickened in the upper part. Blade often reduced in size, ovate-lanceolate, to 5.5 by 1.2 cm; base broadly rounded; top caudate, its tip rounded; margin entire; nerves pinnate, above and below hardly raised, pale yellow, lateral nerves ending free.
Next leaves spirally arranged. Petiole terete, in 10th one to 6 mm long, in the upper part distinctly thickened and curved. Blade ovate-oblong, in 5th leaf to 5.5 by 2.2 cm; base obtuse to truncate; top caudate, its tip rounded, further as in the first leaves.

Specimens: 2375 from S.E. Borneo, primary forest on deep clay, medium altitude.
Growth details: In nursery germination poor.
Remarks: In almost all species the seedling represents the Sloanea type and subtype, with the fleshy cotyledons exposed on top of an epigeal hypocotyl. The cotyledons are 2-winged, with a moustache-like form, and often slightly unequal in size and ornamentation (*S. bracteolata*, *S. curtisii*, *S. gibbosa*, *S. gratissima*, *S. hemsleyana*, *S. laevis*, *S. leprosula*, *S. macroptera*, *S. maxima*, *S. mecistopterix*, *S. pauciflora*, *S. rigida*, *S. sentilata*, *S. sericea*, *S. singkawang*, *S. sumatrana*, *S. xanthophylla*), or they

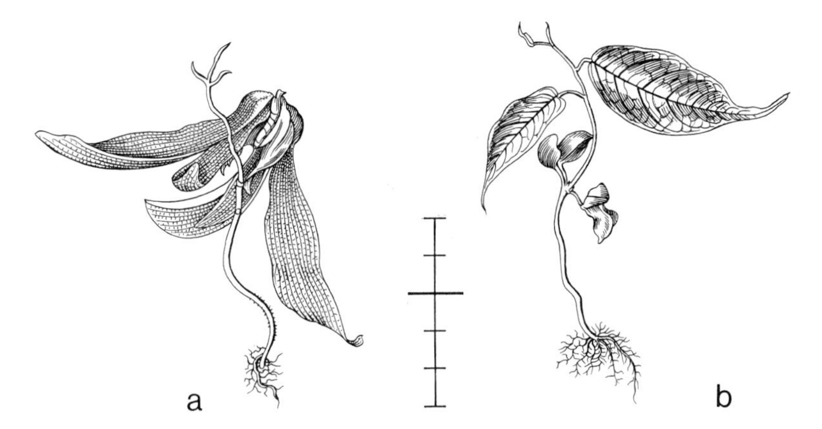

Fig. 70. *Shorea laevis*. De V. 2375, collected 16-iii-1973, planted 22-iii-1973. a. 18-iv-1973; b. 12-v-1973.

have an irregularly square form, broad, and folded, especially along the top margin (*S.* cf. *lamellata*). *S. robusta* seedlings differ in that the cotyledons are never freed from the fruit wall. The cotyledonary petioles elongate considerably, bringing the plumule and radicle out of the seed (Heliciopsis type and subtype).

Shorea cf. *lamellata* Foxw. – Fig. 71.

Development: The taproot and hypocotyl pierce the apex of the fruit between the wings, curve, and enter the soil. After establishment of the root the cotyledons spread and are withdrawn from fruit wall and testa by the hypocotyl becoming erect. No resting stage occurs.
Seedling epigeal, phanerocotylar. Sloanea type and subtype.
Taproot long, slender, fibrous, creamy-brown, with many long, slender, shortly branched, creamy-white sideroots.
Hypocotyl strongly elongating, terete, to 5.5 cm long, green turning brownish, densely hairy with minute, simple, hyaline hairs.
Cotyledons 2, opposite, succulent, in the 3rd leaf stage dropped; more or less equal, without stipules, petiolate, hard fleshy, yellowish pink, glabrous. Petiole in cross-section flattened, to 5 by 2 mm. Blade much wider than long, to 1.5 by 3.5 cm, irregularly folded, especially along the top margin; base obtuse, slightly narrowed into the petiole; top margin obtuse, irregularly incised, undulating, generally in the centre with one or two, more or less deep incisions; lateral margin entire; nerves not visible.
Internodes: First one strongly elongating, terete, to 5.5 cm long, green, hairy as the hypocotyl; next internodes as first one, lowest ones to 4 cm long, 10th one to 8 cm long. Stem slender, straight.
First two leaves opposite, simple, conduplicate-induplicate when young, stipulate, petiolate, herbaceous, green. Stipules 2, interpetiolarly connate, subulate, linear, to 5 by 0.5 mm; top acute; margin entire; hairy as the hypocotyl. Petiole terete, to 5 by 2

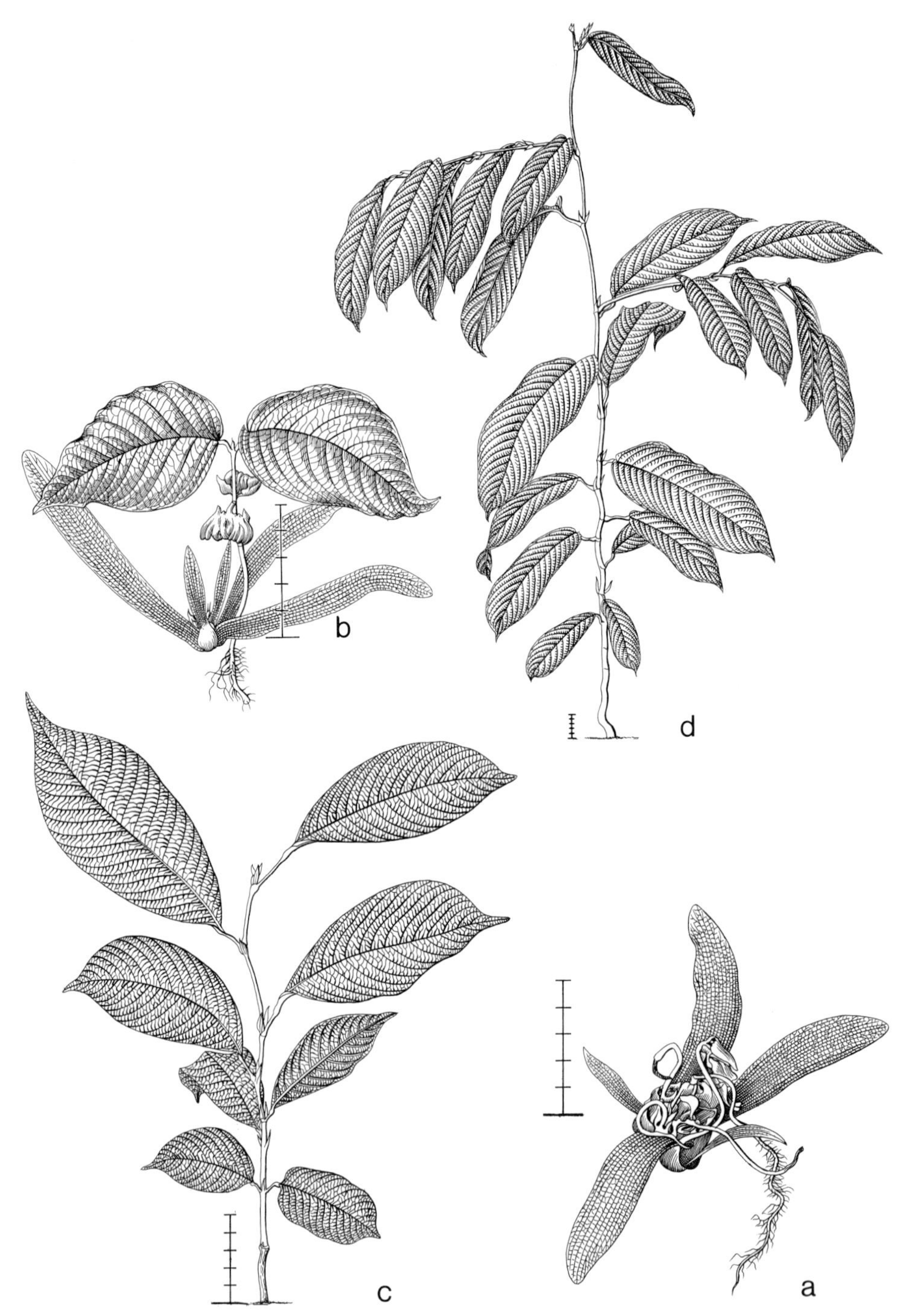

Fig. 71. *Shorea* cf. *lamellata*. De V. 2372, collected 16-iii-1973, planted 22-iii-1973. a. 17-iv-1973, fruit polyembryonic, with only one seedling established, in the others the root dried up; b. 18-iv-1973; c. 17-xi-1973; d. 6-viii-1974.

mm, thickened at the top, hairy as the hypocotyl. Blade ovate, to 9 by 5.5 cm; base obtuse; top acuminate, its tip acute; margin entire; at the base slightly hairy as the hypocotyl; nerves pinnate, midrib above slightly raised, below much so, lateral nerves less so, very regular, at their tip either anastomosing or curved to the next one. *Next leaves* spirally arranged, as the first ones. Stipules 2 at the base of each petiole, sessile, in the lower leaves lanceolate, to 8 by 2 mm, higher ones more ovate, in 20th leaf to 19 by 10 mm, slightly asymmetric; top rounded; margin entire; when young along the margin hairy as the hypocotyl; few-nerved. Petiole terete, in lowest leaves to 10 by 1.2 mm, in 10th one to 3 cm long, in 20th one to 4.5 cm long, at the top distinctly swollen, at the base less so. Blade elliptic to slightly obovate-oblong, in the lowest leaves to 9.5 by 5.2 cm, higher ones more oblong, in 10th one to 25 by 11 cm, in 20th one to 38 by 16 cm; base acute to obtuse; top cuspidate; margin entire; when young hairy as the hypocotyl, soon glabrescent; lateral nerves very regular, above deeply sunken, giving the leaf an undulating surface, free ending.

Specimens: 2372 from S.E. Borneo, primary hill forest on deep clay, low altitude.
Growth details: In nursery germination good, ± simultaneous. The fruits show a rather high percentage of polyembryony, with up to 3 seedlings in a fruit. In general only one of these succeeds in establishing, the roots of the other seedlings dry up in the process of reaching the soil.
Remarks: See *S. laevis*.

Shorea sp. – Fig. 72.

Development: The hypocotyl and taproot emerge from the apex of the fruit between the wings, form a curve, and enter the soil between the wings. After establishment of the root the cotyledons spread and are pulled free from the fruit wall. Sometimes a valve still adheres the cotyledons. No resting stage occurs.
Seedling epigeal, phanerocotylar. Sloanea type and subtype.
Taproot sturdy, hard, brownish, with numerous rather small, thin, creamy-white, branched sideroots.
Hypocotyl enlarging to 7 cm long, terete, with a vague, longitudinal ridge below each cotyledon, creamy-white turning dull brown, with close-set, appressed, simple hairs forming a felt-like indument.
Cotyledons 2, opposite, succulent, not known when dropped, between the 4th and 10th leaf stage; unequal in ornamentation, without stipules, petiolate, creamy-white when young, turning green. Petiole flattened, in cross-section elliptic, to 2 by 0.4 cm, at the base channelled above and there hairy as the hypocotyl. Blade 2-winged, slightly wider than long, to 4 by 5 cm; top deeply retuse; wings obliquely ovate, turned downwards, to 4 by 2 by 2 cm, with sharp edges and flat sides, at the outer side convex, at the top rounded, at the end acute, sometimes irregularly cut off, succulent, creamy-white turning green, glabrous, with minute, white warts; one cotyledon outside entirely convex, without incision in the centre, faintly longitudinally ribbed, inside with flat sides, with a broad depression on the edge; the other cotyledon with a broad incision in the centre above, there with flat sides, with a longitudinal, broad depression, wings outside faintly longitudinally ribbed, further as the other one.

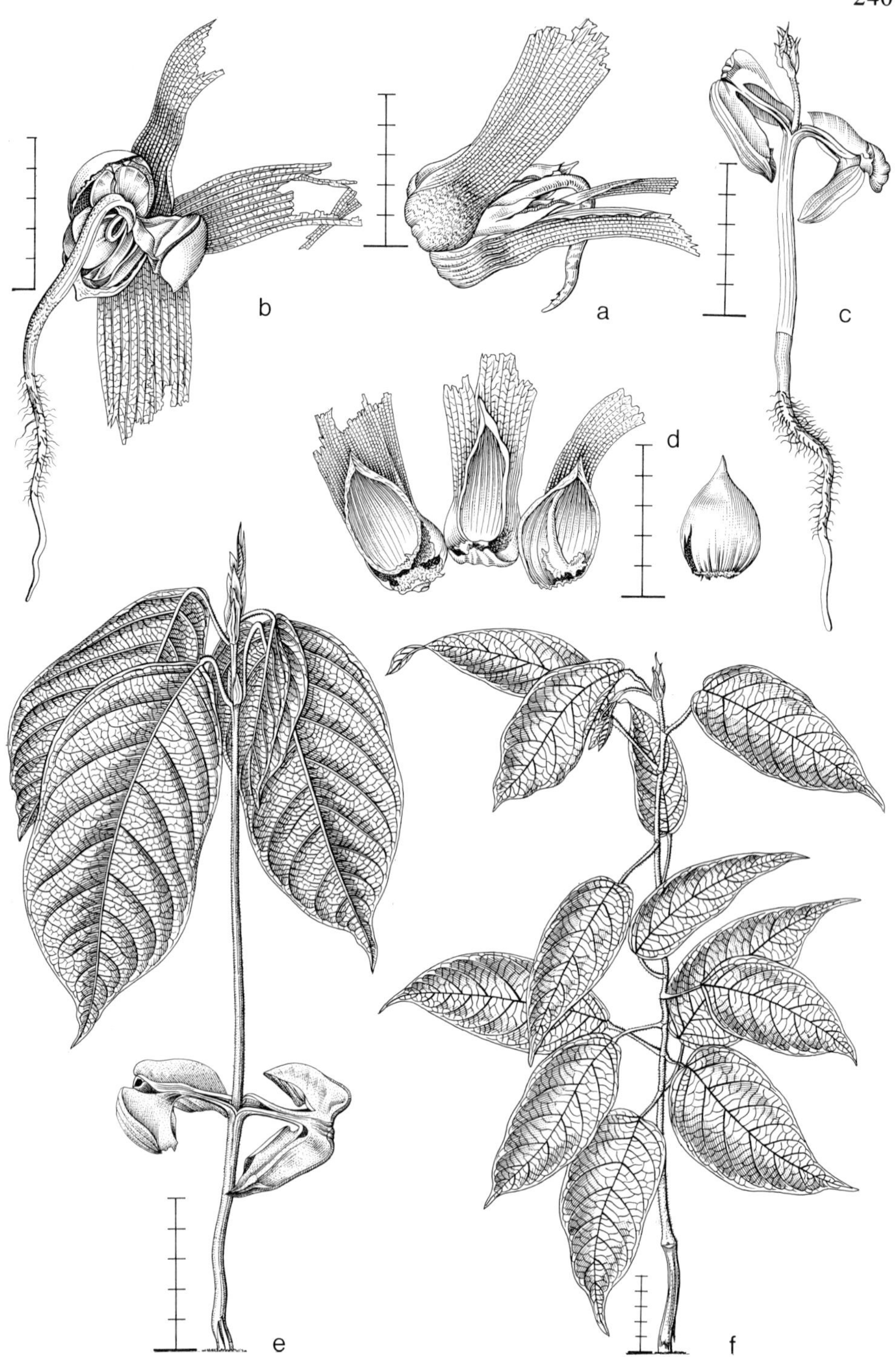

Fig. 72. *Shorea sp.* Bogor Bot. Gard. VII. C. 1, collected and planted i-1973. a, b, c, and d date?; e. 28-viii-1973; f date?.

Internodes: First one strongly elongating, to 15 cm long, terete, green, with few scattered, scale-like to stellate hairs; next internodes as first one, second one to 2.5 cm long, higher ones gradually larger, 10th one to 5.5 cm long. Stem straight, slender.
First two leaves opposite, conduplicate-induplicate when young, unfolding in a quite early stage, stipulate, petiolate, stiff herbaceous, bright orange-red when young, turning green. Stipules in general interpetiolate connate, sometimes partly so to entirely free, dropping when the leaf is fully developed, simple, sessile, ovate-oblong, to 21 by 8 mm; top obtuse; margin entire; hairy with rather close-set, simple, minute, hyaline hairs; greenish, often reddish when young; with up to 9 ± parallel nerves. Petiole terete, to 3.5 by 0.2 cm, at the top distinctly thickened, covered with rather distant stellate hairs. Blade ± oblong, to 20 by 9 cm; base obtuse to slightly retuse; top cuspidate; margin entire; below and above on the midnerve with minute, simple, hyaline hairs, when young on both sides with minute, globose, hyaline glands; nerves pinnate, below much prominent, lateral nerves ending free.
Next leaves spirally arranged, the lowest two sometimes subopposite. Stipules 2 at the base of each leaf, soon dropping, ovate, but those in the lower leaves and on the lateral branches more narrow, in 18th leaf to 2 by 1.2 cm, below more or less densely hairy with appressed, small stellate scales, further as the first leaves. Petiole as those of first leaves, in 10th leaf to 50 by 1.5 mm, in higher ones those with a lateral shoot in the axil much smaller. Blade as in first leaves, oblong, in 5th one to 13 by 6.3 cm, in 10th one to 17 by 7 cm; base rounded; top acuminate to slightly cuspidate; margin entire; further as those of first leaves.

Specimens: Bogor Botanic Gardens VII.C.1, from Borneo.
Growth details: In nursery germination good, ± simultaneous.
Remarks: See *S. laevis*.

VATICA

Fl. Males. (in press); Burkill, J. Str. Br. R. As. Soc. no. 81 (1920) 61 (*V. nitens*); Meijer, Bot. Bull. Sandakan 11 (1968) 112 (*V. papuana*).
Genus in India and Ceylon, throughout S.E. Asia, all over Malesia except the Lesser Sunda Islands.

Vatica venulosa Bl. – Fig. 73.

Development: The hypocotyl and taproot emerge from the apex of the fruit between the wings. The cotyledonary petioles elongate, bringing the plumule free from the fruit, enabling the shoot to emerge. A short resting stage occurs with 2 developed leaves.
Seedling hypogeal, cryptocotylar. Heliciopsis type, Koordersiodendron subtype.
Taproot long, slender, flexuous, brownish cream, with many much-branched, long, slender, creamy-white sideroots.
Hypocotyl terete, enlarging to 1 cm long, green, glabrous.
Cotyledons 2, secund, succulent, in the 4th leaf stage shed; without stipules, petiol-

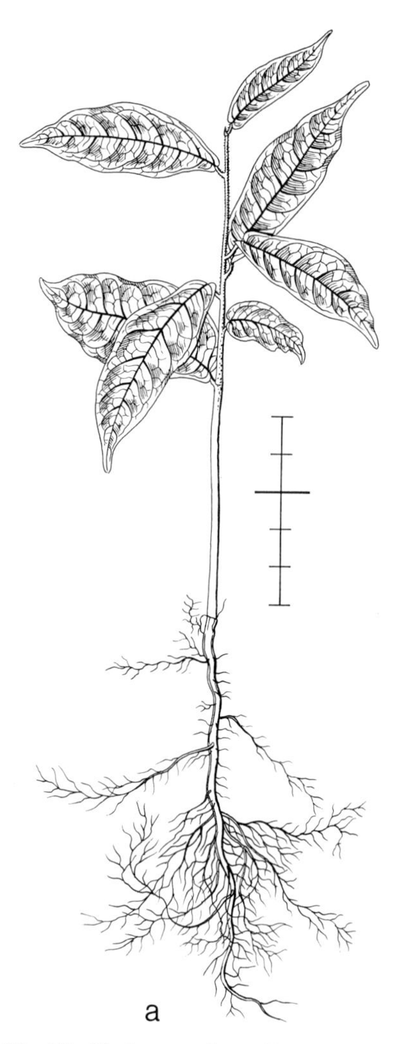

Fig. 73. *Vatica venulosa.* Bogor Bot. Gard. VII. B. 48a, collected and planted i-1973. a. 5-viii-1973, the cotyledons are already shed.

ate, covered by the persistent, 5-winged, brownish fruit wall. Whole body obrhomboid, to 1.3 by 1.3 by 1.3 cm, enclosed by the free, curved, wing-like enlarged sepals. Petiole in cross-section semi-orbicular, to 1 cm long, distinctly curved, channelled above, green, densely covered with minute, simple, hyaline hairs.
Internodes: First one more or less strongly elongating, terete, to 7 cm long, but in general much shorter, greenish, rather sparsely covered with brownish, appressed, simple hairs; next internodes as the first one, irregular in length, the lower ones to 3.5 cm long, but in general not exceeding 1.5 cm, beyond the 5th–8th one much smaller, in the 10th one to 0.5 cm long. Stem slender, straight.

First two leaves in general opposite, simple, conduplicate-induplicate when young, stipulate, shortly petiolate, subcoriaceous, pinkish red when young, turning green. Stipules dropping soon, not described. Petiole terete, to 4 by 1.5 mm, slightly hairy as the first internode. Blade ovate, to 5 by 3 cm, but often much smaller; base slightly auriculate, auricles rounded, to 1 by 4 mm big; top ± acute to acuminate; margin entire; below slightly hairy as the first internode; nerves pinnate, midrib above and below raised, lateral nerves below slightly raised, at their tip curved, and forming an undulating marginal nerve.
Next leaves spirally arranged. Stipules 2 at the base of each leaf, dropping soon, sessile, triangular, in 4th leaf 2.5 by 1 mm, in 15th leaf to 2 by 1 mm, green, slightly hairy as the hypocotyl. Petiole terete, up to the 15th leaf to 4 by 1.5 mm. Blade ovate-oblong, in 4th leaf to 6.5 by 2.6 cm, but in general much smaller, in 10th leaf to 10 by 4 cm; in lower leaves base slightly auriculate, in higher ones obtuse; top acuminate, its tip rounded; further as the first leaves.

Specimens: Bogor Botanic Gardens VII. B. 48a, from Bangka.
Growth details: In nursery germination fair, ± simultaneous.
Remarks: All species described are similar in germination pattern and general morphology. The epigeal part of the hypocotyl may be more or less distinct. They differ in details only.

EBENACEAE

DIOSPYROS

Fl. Java 2: 184; Csapody, Keimlingsbestimmungsbuch der Dikotyledonen (1968) 131 (*D. virginiana*); Hickel, Graines et Plantules (1914) 315 (*D. lotus*, *D. virginiana*); Lubbock, On Seedlings 2 (1892) 203 (*D. embryopteris*, *D. sp.*); Mensbruge, Germination (1966) 313 (*D. gabinensis*, *D. mannii*, *D. mespiliformis*, *D. sanzaminika*, *D. xanthochlamys*); Ng, Mal. Forester 39 (1976) 115 (*D. argentea*, *D. confertiflora*, *D. diepenhorstii*, *D. ismailii*, *D. maingayi*, *D. pendula*, *D. pilosanthera* var. *chikusensis*, *D. sumatrana*, *D. trengganensis*); Schopmeyer, Agric. Handbook 450 (1974) 373 (*D. virginiana*); Troup, Silviculture 2 (1921) 646 (*D. embryopteris* = ? *D. sp.*, *non embryopteris*, Kostermans, pers. comm. and also the first leaves are mistaken for the cotyledons, *D. melanoxylon*); Wright, Ann. R. Bot. Gard. Peradeniya 2 (1904) 1–106, 133–210 (*D. acuta*, *D. affinis*, *D. attenuata*, *D. crumenata*, *D. ebenum*, *D. embryopteris*, *D. gardneri*, *D. hirsuta*, *D. insignis*, *D. melanoxylon*, *D. montana*, *D. oocarpa*, *D. ovalifolia*, *D. pruriens*, *D. quaesita*, *D. sylvatica*, *D. thwaitesii*, *D. toposia*).
Genus pantropic; tropical Africa, throughout the Indian Ocean, India and Ceylon, S.E. and tropical E. Asia, all over Malesia, N. and E. Australia, in the Pacific as far as Samoa and Tonga Islands, including the W. Carolines and Hawaii, Central and tropical S. America.

Diospyros curraniopsis Bakh. – Fig. 74.

Development: The taproot and hypocotyl pierce the testa. By elongating of the hypocotyl the enclosed cotyledons are carried above the soil. After a resting period testa and endosperm are shed by expanding of the cotyledons. A second resting stage occurs with the cotyledons and with 2 leaves developed. When the seeds are by accident not freed from the fruit wall, the latter is carried up by several elongating hypocotyls.
Seedling epigeal, phanerocotylar. Sloanea type and subtype.
Taproot long, slender, fibrous, black, with rather many thin, quite short, unbranched, later much-branched, black sideroots.
Hypocotyl strongly enlarging, slender, terete, to 11 cm long, bluish grey-green, turning darker grey-green, in the 2-leaved stage already distinctly longitudinally grooved, glabrous.
Cotyledons 2, opposite, slightly succulent, in the 2nd–3rd leaf stage dropped; soon irregularly rolled, sessile, without stipules, more or less fleshy, pale green, glabrous, during germination and the first resting stage covered by the shining brown testa. Whole body reniform in outline, thickened at the back, to 17 by 10 by 6 mm, with a clear suture along its margin. Blade elliptic, flat, with recurved margins when older, to 15 by 8 mm; base and top rounded; margin entire; main nerves 5, parallel, rather inconspicuous.
Internodes: First one slightly elongating, to 5 mm long, in cross-section elliptic, with scattered, appressed, long black hairs; next internodes as the first one, second one to 2 cm long, higher ones irregular in length. Stem very thin and slender, ± straight, beyond 6th–10th leaf often with most of the leaves entirely reduced.
First two leaves opposite, simple, without stipules, shortly petiolate, herbaceous, green. Petiole in cross-section semi-orbicular, to 4 by 1 mm, hairy as the first internode. Blade oblong, to 7 by 3 cm, but in general smaller; base acute to slightly acuminate; top cuspidate; margin entire; above shining, below with few, scattered, appressed, black hairs; nerves pinnate, above sunken, below prominent, especially the midnerve, lateral nerves ending in anastomoses.
Next leaves spirally arranged. Petiole in cross-section semi-orbicular, in 3rd leaf to 3 by 1.5 mm, in 8th leaf to 5 by 1.5 mm. Blade more lanceolate, in 3rd leaf to 6 by 2 cm, in 5th–8th leaf to 11 by 2.8 cm; base narrowed into the petiole; top acuminate to cuspidate; further as the first leaves.

Specimens: 2216 from S.E. Borneo, primary forest on deep clay, low altitude.
Growth details: In nursery germination good, ± simultaneous.
Remarks: A good character for *Diospyros* seedlings seems the colour of the root system: where recorded this is always black. The hypocotyl also often turns black. In most big-seeded species the hypocotyl remains more or less short, either hypogeal or above the ground, and the cotyledons remain in the seed at soil level until they are shed (Blumeodendron type). Only then the shoot starts developing, the first leaves being either developed and opposite (*D. macrophylla*), or all leaves are spirally arranged, in this case the lowest leaves being either reduced and scale-like, or fully developed (*D. embryopteris* = *D. sp.*, *D. insignis*, *D. macrophylla*, *D. maingayi*, *D.*

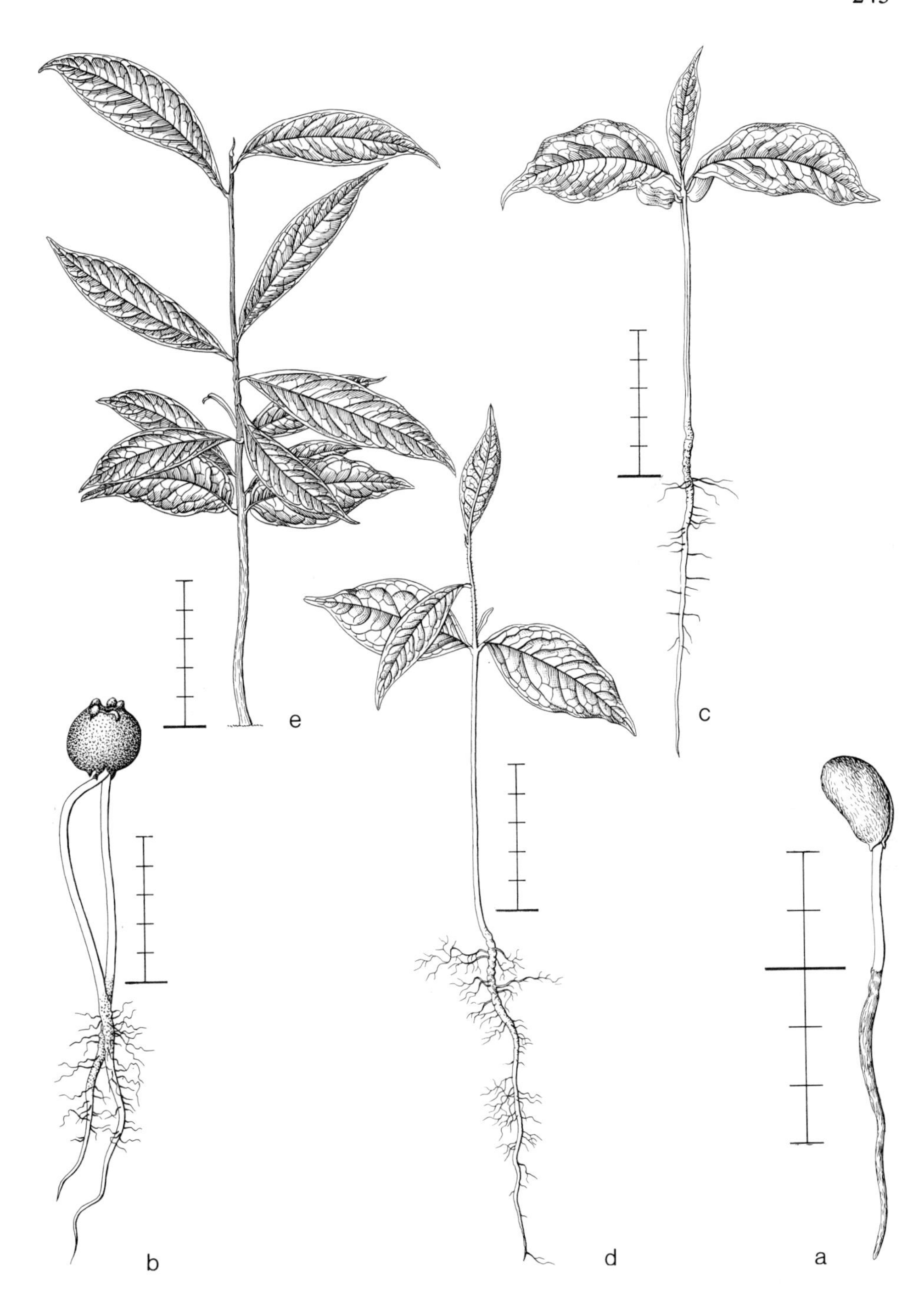

Fig. 74. *Diospyros curraniopsis*. De V. 2216, collected 7-iii-1973, planted 22-iii-1973. a. 5-iv-1973; b. (abnormal germination) and c. 1-v-1973; d. 17-vii-1973; e. 4-iii-1974.

oocarpa, *D. pendula*, *D. toposia*). The small-seeded species can be divided into two groups: one with long-lasting green paracotyledons (Macaranga type), and a second one with non-photosynthetic, soon dropped cotyledons (Sloanea type). The seedlings of the Macaranga type all have the paracotyledons on a long hypocotyl; they are easily extracted from the seed and endosperm. The leaves are all spirally arranged, except in *D. embryopteris*, where, according to Lubbock, the first two are opposite, which is according to Ng also the case in *D. confertiflora* (*D. confertiflora*, *D. discocalyx*, *D. ebenum*, *D. embryopteris*, *D. gabonensis*, *D. gardneri*, *D. ismailii*, *D. montana*, *D. sumatrana*, *D. sylvatica*, *D. virginiana*, *D. xanthochlamys*).
The seedlings of the Sloanea type all have the pale or whitish, short-lived cotyledons on a long hypocotyl. The epicotyl is rather short, and bears two opposite leaves. Wright mentions specifically for *D. ovalifolia*, *D. pruriens*, and also for other species a tendency for the cotyledons to remain in the seed because of difficulties in freeing. Also in *D. sanzaminika* the seeds are carried above the soil but the cotyledons are not freed. Wright comments on this peculiar behaviour, and the waste of endosperm if the cotyledons succeed in freeing early, and attributes the rarity of some species of the genus *Diospyros* in Ceylon to this 'curious suicidal mode of germination' (*D. affinis*, *D. argentea*, *D. attenuata*, *D. curraniopsis*, *D. crumenata*, *D. diepenhorstii*, *D. hirsuta*, *D. mannii*, *D. melanoxylon*, *D. moonii*, *D. ovalifolia*, *D. pilosanthera* var. *chikusensis*, *D. pruriens*, *D. quaesita*, *D. sanzaminika*, *D. thwaitesii*, *D. trengganensis*).

Diospyros macrophylla Bl. – Fig. 75.

Development: The taproot and hypocotyl emerge from one pole of the seed. In the first resting stage the cotyledons remain at soil level, covered by the testa, and the opaque hard endosperm which disintegrates, becomes gelatinous and turns bluish green. Then the cotyledons are shed and the shoot starts developing. A second short resting stage occurs with two developed leaves.
Seedling hypogeal, cryptocotylar. Blumeodendron type.
Taproot sturdy, hard fleshy, black, with rather few, short, few-branched, black sideroots.
Hypocotyl not easy to distinguish from the taproot, ± 5 cm long, its lower part subterranean, ± smooth, the short epigeal part with cross-cracks, glabrous, blackish.
Cotyledons 2, on top of the hypocotyl, slightly succulent, shed before the shoot is withdrawn from the seed; sessile, base obtuse, top obtuse, margin entire, coriaceous, glabrous, whitish, just before being shed strongly undulating and rupturing the testa along the margin of the seed. Whole body reniform in outline, thickened at the back, to 4.5 by 1.3 cm, the dark brownish, leathery testa persistent.
Internodes: First one strongly elongating, flattened when young, turning ± terete, to 12 cm long, green, densely covered with short, stiff, patent, blackish hairs; next internodes as the first one, irregular in length, the lower ones 1–2 cm long, gradually becoming larger, the 10th one to 3.5 cm long. Stem rather slender, ± straight.
First two leaves opposite, simple, conduplicate-induplicate when young, without stipules, petiolate, herbaceous, green. Petiole in cross-section semi-orbicular, to 4 by 2 mm, hairy as the hypocotyl. Blade ± elliptic, to 8 by 4.2 cm; base obtuse; top

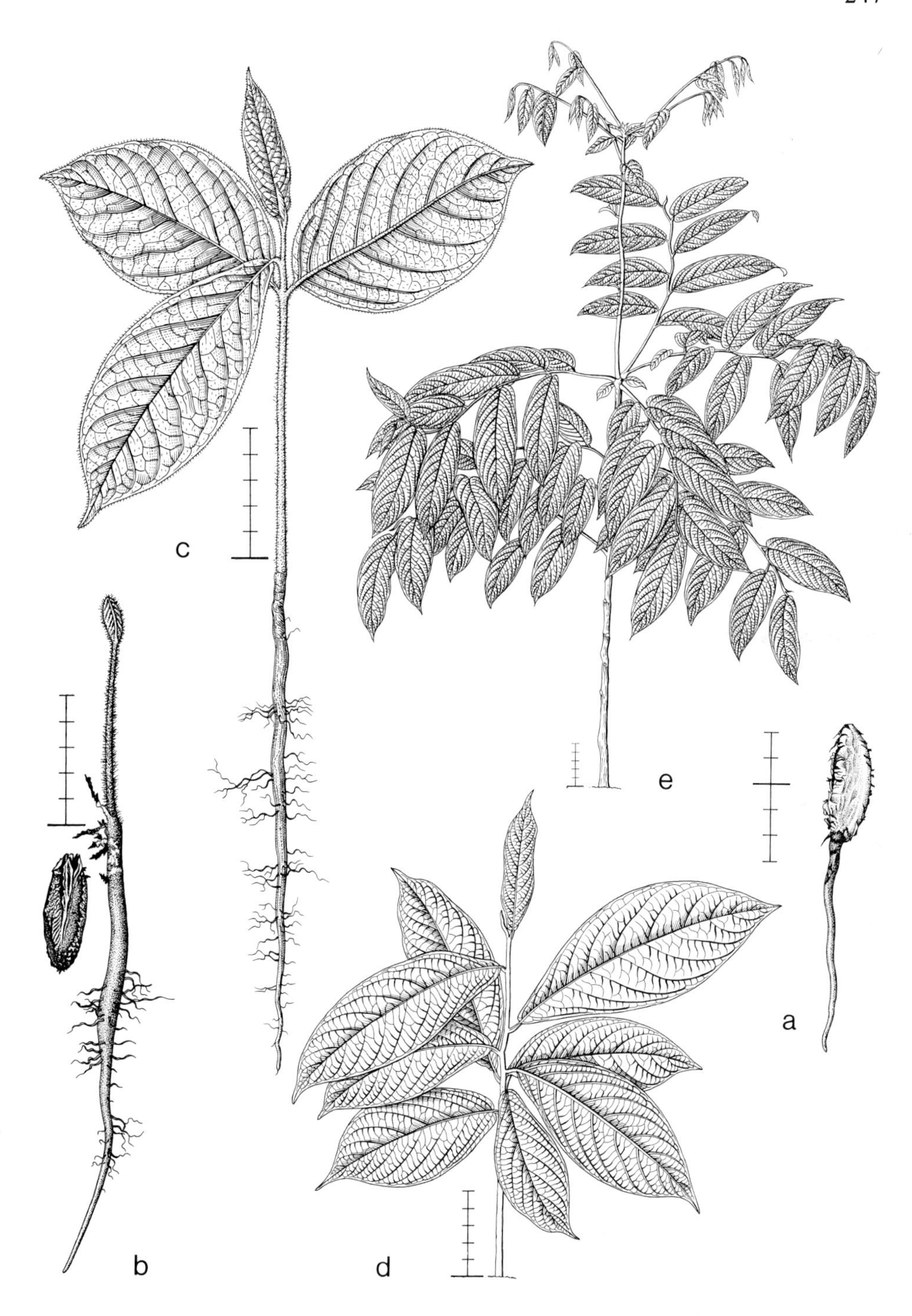

Fig. 75. *Diospyros macrophylla.* Dransfield 2535, collected 15-vi-1972, planted 20-vi-1972. a and b date?; c. 26-viii-1972; d and e date?.

acuminate; margin entire; below and on the margin with scattered, soft, hyaline, appressed hairs, on the midrib below hairy as the hypocotyl; nerves pinnate, above slightly sunken, below very prominent, lateral nerves free ending.
Next leaves spirally arranged. Petiole in 10th leaf to 10 by 3 mm. Blade oblong, the lower ones more obovate-oblong, 5th one to 12.5 by 4.3 cm, 10th one to 25 by 8.8 cm; base slightly narrowed, its tip ± obtuse; top acuminate to cuspidate; margin entire; below sparsely hairy with scattered, simple, hyaline, appressed hairs, rather densely so on the midnerve, further as the first leaves.

Specimens: Dransfield 2535 and De Vogel 1371 from W. Java, alluvial flat, primary forest on clayey coral sand, low altitude.
Growth details: In nursery germination good, ± simultaneous.
Remarks: See *D. curraniopsis*.

Diospyros sp. – Fig. 76.

Development: The taproot and hypocotyl emerge from one pole of the seed. The taproot establishes and the enclosed paracotyledons are carried above the soil by the hypocotyl becoming erect. After a resting period the endosperm is thrown off by the spreading paracotyledons. Then follows a second temporary rest.
Seedling epigeal, phanerocotylar. Macaranga type.
Taproot long, slender, fibrous, black, with few, short, unbranched, black sideroots.
Hypocotyl strongly elongating, terete, to 4.5 cm long, reddish to yellowish green turning brownish, rather densely covered with minute, simple, reddish hyaline hairs.
Paracotyledons 2, opposite, foliaceous, long persistent, in the 35th leaf stage not yet shed; simple, without stipules, shortly petiolate, herbaceous, green, almost glabrous. Petiole in cross-section elliptic, somewhat flattened, to 2 by 1.5 mm, hairy as the hypocotyl. Blade ovate, to 2.5 by 1.6 cm; base rounded, abruptly narrowed into the petiole; top acute to slightly rounded; margin entire; nerves palmate, main nerves 3–5, branching off from the thickened basal portion of the central nerve, above slightly sunken, below raised, the midnerve especially so, lateral nerves at the tip curved to the next one, forming an undulating, closed system.
Internodes: First one terete, elongating to 1.5(–3) cm, green turning brownish, densely hairy as the hypocotyl; next internodes in general not exceeding 4 mm, those of the branches towards the tip smaller. Stem slender, straight, halting growth after the development of 2–5 internodes, after which branches develop.
Leaves all spirally arranged, simple, involute when young, without stipules, petiolate, herbaceous, green, the first 1–2 in general much reduced. Petiole terete, above slightly flattened, in the third leaf to 2 by 0.3 mm, in the third leaf on the first branch to 1 by 0.6 mm, green tinged red, hairy as the hypocotyl. Blade ovate-oblong, in third one to 9 by 4 mm, those on the lateral branches larger, third one to 17 by 6 mm, 13th one to 21 by 7 mm; base obtuse, in the lowest ones usually acute; top acute, its tip rounded; margin entire; when young below hairy as the hypocotyl, soon glabrescent, but less so on the midnerve; nerves pinnate, main nerve above sunken, below slightly raised, lateral nerves above not sunken, below slightly raised, at the tip curved, forming an irregular undulating, inconspicuous marginal nerve.

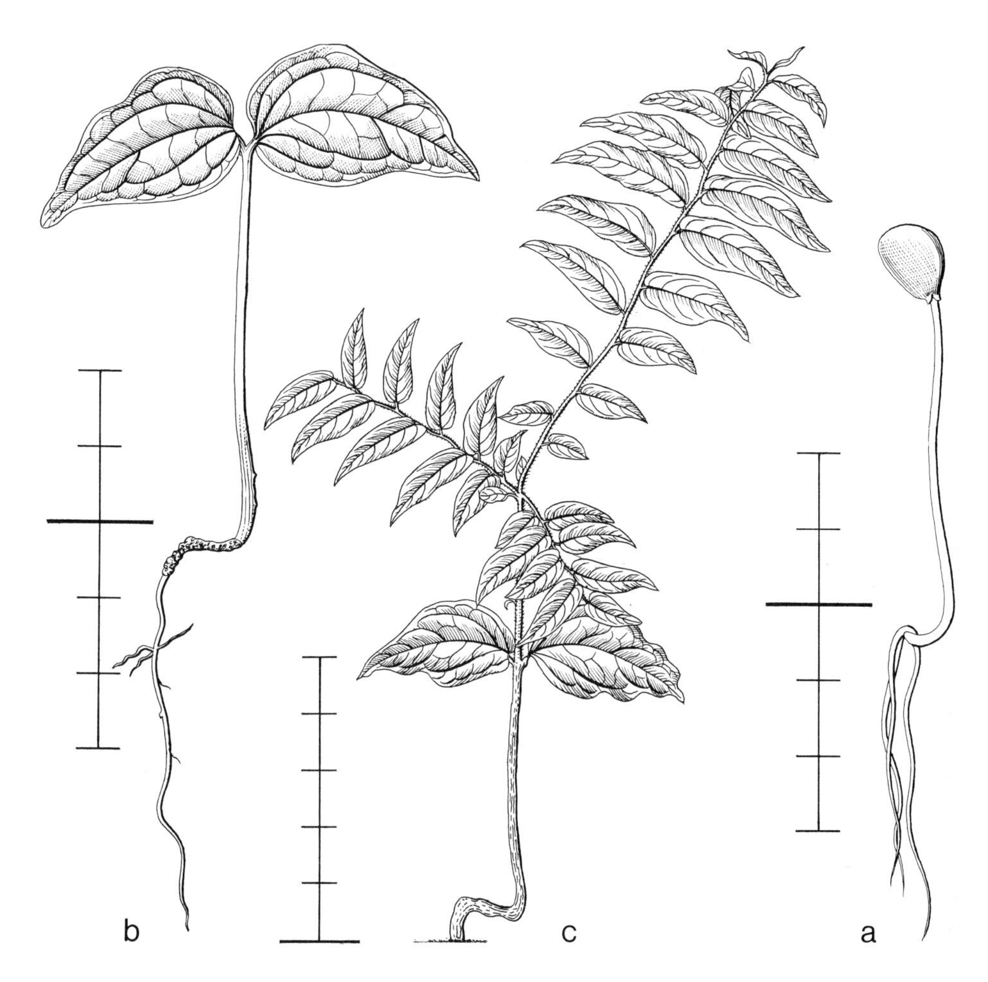

Fig. 76. *Diospyros sp.* De V. 2556, collected 10-x-1973, planted 5-xi-1973. a. 28-xi-1973; b. 19-xii-1973; c. 9-ix-1974.

Specimens: 2556 from N. Celebes, primary hill forest, low altitude.
Growth details: In nursery germination good, ± simultaneous.
Remarks: See *D. curraniopsis*.

ELAEOCARPACEAE

ELAEOCARPUS

Fl. Java 1: 396; Lubbock, On Seedlings 1 (1892) 284 (*E. oblongus*, *E. reticulatus*).
Genus in the Indian Ocean islands, India and Ceylon, throughout S.E. and E. Asia, all over Malesia, Australia, in the Pacific as far as the Societies, including the Bonins, Hawaii, and New Zealand.

Elaeocarpus petiolatus (Jack) Wall. – Fig. 77.

Development: The taproot and hypocotyl emerge from the one pole of the endocarp, which splits longitudinally. The hypocotyl becomes erect, carrying the enclosed paracotyledons above the soil. A resting stage occurs with the paracotyledons enclosed in the endocarp, after which testa and endocarp are shed by expanding of the paracotyledons. Then a second temporary rest takes place.
Seedling epigeal, phanerocotylar. Macaranga type.
Taproot long, fibrous, brownish, with rather few, few-branched sideroots as the taproot.
Hypocotyl strongly elongating, slender, terete, its upper part ± quadrangular with 4 shallow, longitudinal grooves at its top, to 5.5 cm long, reddish, turning brownish green, glabrous.
Paracotyledons 2, opposite, foliaceous, between the 10th–20th leaf stage dropped; equal, without stipules, sessile, herbaceous, green, glabrous. Blade ovate-lanceolate, to 19 by 4 mm; top obtuse; margin entire, slightly reflexed at the base; 3-nerved, midrib above slightly prominent, below slightly sunken, the two lateral nerves rather inconspicuous, ± parallel.
Internodes: First one hardly elongating, in cross-section elliptic, to 2 mm long, reddish turning brownish green, glabrous; next internodes as the first one, gradually more elongated, the 10th one to 2 cm. Stem not very slender, ± straight.
Leaves all spirally arranged, the first ± subopposite, simple, involute when young, stipulate, petiolate, herbaceous, orange-red when young, turning green above and pale green below, densely hairy with soft, appressed long hyaline hairs, quickly glabrescent. Stipules inconspicuous, triangular, minute, in lower leaves to 1 by 1 mm, often one margin descending down the stem as a minute ridge, rather soon caducous, those beyond the 20th leaf turned to the axil, succulent, to 2 by 2 mm, covered with brownish, hard resin. Petiole almost absent in the lower leaves, in the higher leaves gradually larger, in the 12th leaf to 3 by 0.2 cm, in cross-section semi-orbicular, channelled above, in higher leaves just flat, distinctly thickened at base and top. Blade in the lowest leaves ± lanceolate, to 2.4 by 0.7 cm, in higher leaves oblong, in

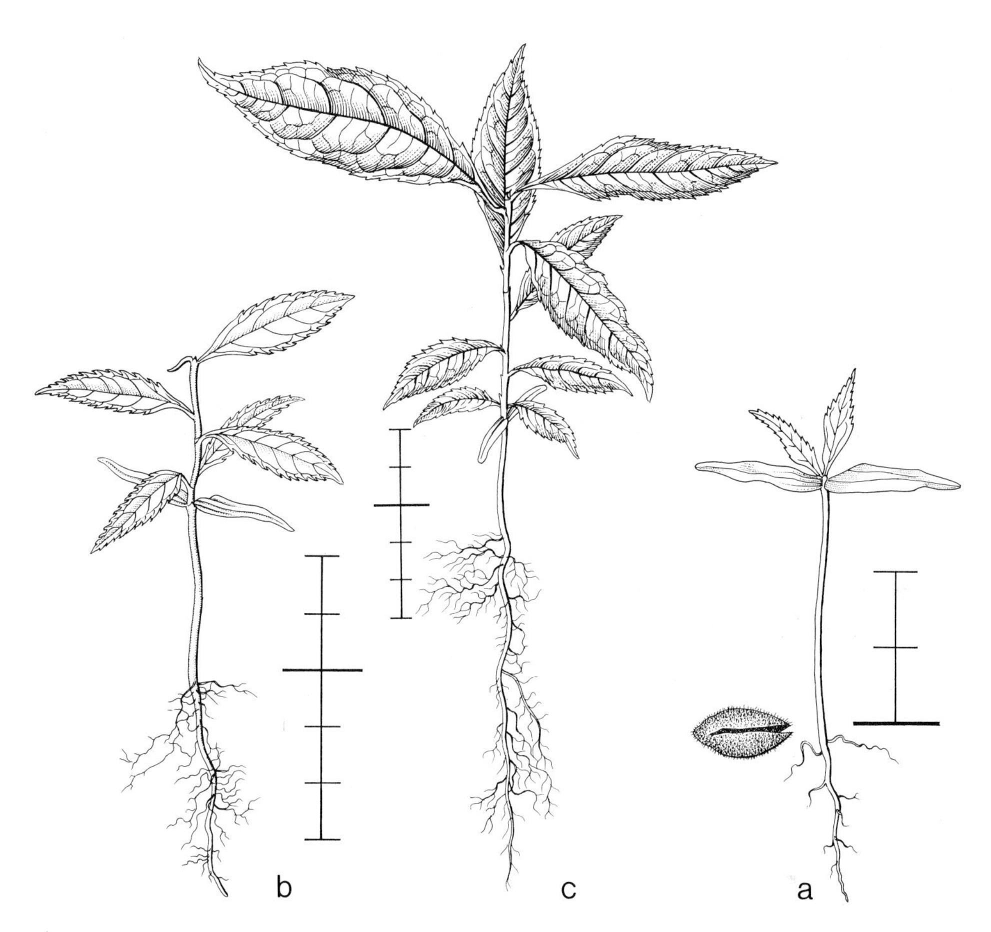

Fig. 77. *Elaeocarpus petiolatus.* De V. 1348, collected 20-iii-1972, planted 23-iii-1972. a, b, and c date?.

12th leaf to 20 by 7 cm, in 20th leaf to 30 by 13 cm, often irregularly formed; base narrowed; top obtuse, margin serrate, teeth broadly triangular, with a ± rounded shoulder, its tip subulate, creamy-white, turned to the margin, those in the higher leaves rather inconspicuous; nerves pinnate, midrib above raised, beneath prominent, lateral nerves above inconspicuous to sunken, veins ending free in the teeth.

Specimens: 1348 from S. Sumatra, primary forest on deep clay, low altitude.
Growth details: In nursery germination fair, ± simultaneous.
Remarks: The four described species, and *E. grandiflorus* also grown for the project, are similar in germination pattern and general morphology, they differ in details only.

Elaeocarpus teysmannii K. & V. – Fig. 78.

Development: The endocarp splits longitudinally, hypocotyl and taproot emerge from the slit. After establishment of the root, the hypocotyl becomes erect, pulling the paracotyledons free. A short resting stage occurs with expanded paracotyledons.
Seedling epigeal, phanerocotylar. Macaranga type.
Taproot rather short, sturdy, hard fleshy, creamy-white, turning brown, with many long, slender, much-branched, creamy-white sideroots.
Hypocotyl strongly enlarging, terete, clearly separated from the root, up to 9 cm long, green, glabrous, at the top laterally flattened, with a longitudinal groove from between the cotyledons.
Paracotyledons 2, opposite, foliaceous, when shed not known, between 6th and 13th leaf stage; simple, without stipules, shortly petiolate, herbaceous, green, paler below, glabrous. Petiole distinctly flattened, to 3 by 4 mm. Blade ovate, to 5 by 4 cm; base obtuse; top obtuse; margin entire; 3-nerved, main nerves above raised, yellow, hardly raised below, lateral nerves distinctly curved at the tip to the next one, forming an intricate pattern.
Internodes: First one elongating to 1.3 cm, terete, green tinged reddish, densely hairy with short, simple, hyaline hairs; next internodes as the first one, second one to 1.5 cm, higher ones smaller, 10th one to 1 cm long. Stem slender, straight.
Leaves all spirally arranged, simple, spreading, stipulate, shortly petiolate, herbaceous, green. Stipules 2 at the base of each petiole, caducous, minute, triangular, to 0.5 by 0.2 mm. Petiole in cross-section semi-orbicular, in the first leaf to 2 by 1.5 mm, 10th one to 7 by 1 mm, hairy as the first internode. Blade ± oblong, later ones more obovate-oblong, in the first leaf to 7 by 2.8 cm, 5th one to 7 by 2 cm, 10th one to 8.5 by 3 cm; base acute, its tip sometimes obtuse; top acuminate; margin dentate, teeth minutely apiculate with a red tip; above on the midnerve, below on the nerves, and less so on the blade hairy as the hypocotyl; nerves pinnate, midrib above slightly prominent, below more so, lateral nerves above sunken, below prominent, free ending in the teeth.

Specimens: 2655 from N. Celebes, primary forest on deep clay, medium altitude.
Growth details: In nursery germination poor.
Remarks: See *E. petiolatus*.

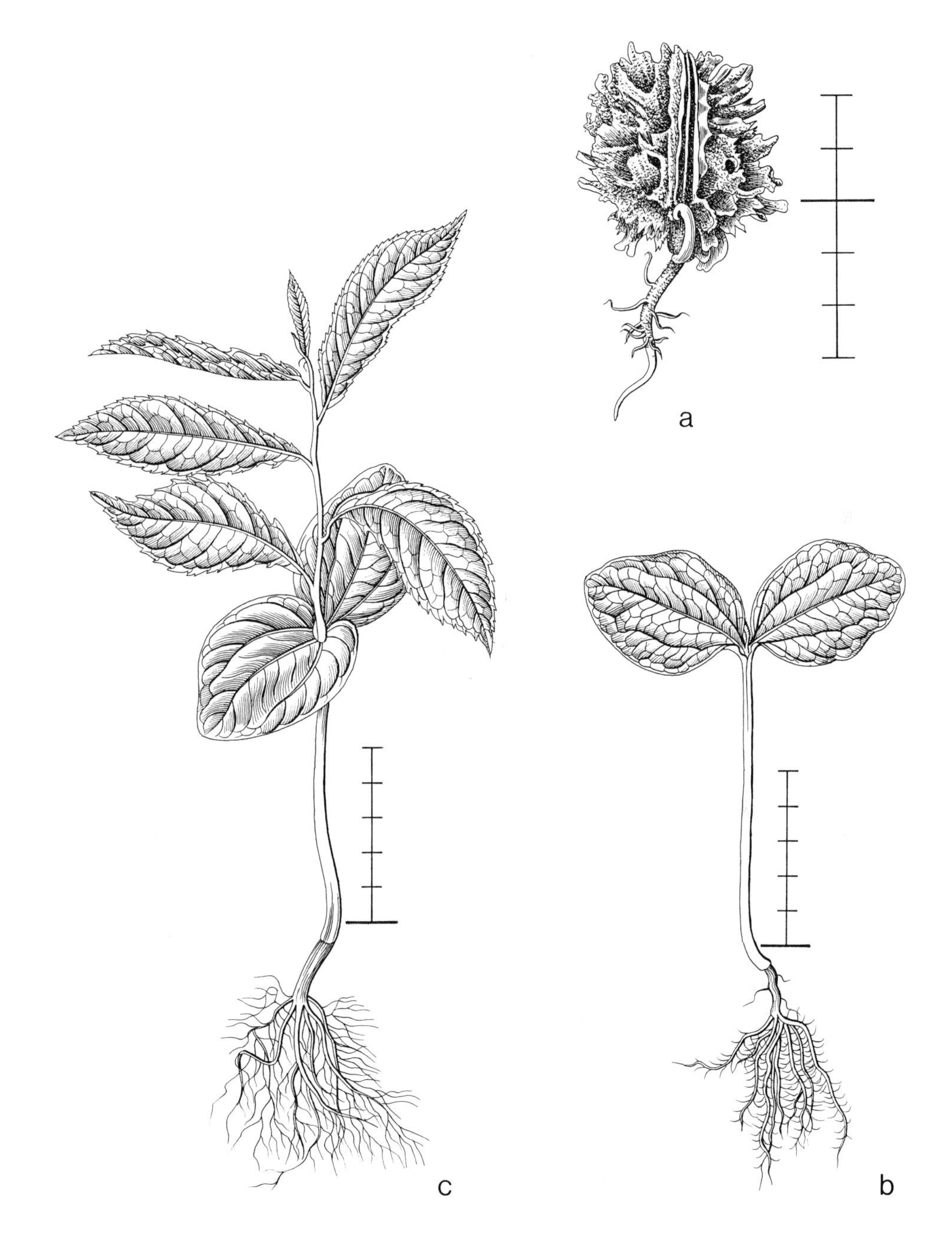

Fig. 78. *Elaeocarpus teysmannii.* De V. 2655, collected 1-xi-1973, planted 5-xi-1973. a. 19-ii-1974; b. 23-iii-1974; c. 20-v-1974.

SLOANEA

Fl. Java 1: 400; Duke, Ann. Miss. Bot. Gard. 52 (1965) 342 (*S. berteriana*).
Genus in Madagascar, India and Ceylon, throughout S.E. and tropical E. Asia, all over Malesia, E. Australia, in the Pacific as far as the New Hebrides, Central and tropical S. America.

Sloanea javanica (Miq.) Szysz. – Fig. 79.

Development: The taproot and hypocotyl emerge from one pole of the seed. The enclosed cotyledons are carried up, and after a short resting period free themselves partly from the testa when spreading. A second resting stage occurs with the cotyledons and the first pair of leaves developed.
Seedling epigeal, phanerocotylar. Sloanea type and subtype.
Taproot rather long, rather slender, creamy-white with yellow root-tip, sideroots as the taproot, shortly branched.
Hypocotyl strongly elongating, terete, to 9 cm long, green, densely hairy with minute, simple, hyaline hairs.
Cotyledons 2, opposite, succulent, dropped in the 2nd–3rd leaf stage; spreading more or less, sessile, without stipules, hard fleshy, pale green, glabrous. Blade elliptic, to 2 by 1.5 cm; base slightly auriculate, auricles minute, ± acute, 1 by 1mm; top broadly rounded; margin entire; below slightly concave, nerves not visible.
Internodes: First one elongating to 4 cm, often much shorter, terete, green, hairy as the hypocotyl; next internodes as the first one, irregular in length, 2nd to 20th one 1.5–3 cm long. Stem straight, slender.
First two leaves opposite, simple, stipulate, shortly petiolate, herbaceous, green. Stipules as those of the higher leaves. Petiole laterally slightly appressed, to 5 by 1.5 by 3 mm, hairy as the hypocotyl. Blade ovate to elliptic, to 11 by 7 cm; base broadly auriculate, auricles 3 by 15 mm; top cuspidate, its tip minutely apiculate; margin serrate, especially towards the top; on the nerves and along the margins hairy as the hypocotyl; nerves pinnate, midrib above slightly raised, below distinctly so, lateral nerves above sunken, below prominent, each curved at the end towards the next one; lateral veins free ending in the teeth.
Next leaves spirally arranged, stipulate, petiolate. Stipules 2 at the base of each leaf, long persistent, subulate, linear, to 4 by 1 mm, in 15th leaf to 8 by 1 mm, including the up to 3 mm subulate cusp, hairy as the hypocotyl. Petiole terete, in the third leaf to 15 by 1 mm, in 8th leaf to 35 by 1.5 mm, at base and top swollen. Blade obovate-oblong, in higher leaves often more oblong, in the third leaf to 13 by 5.5 cm, often smaller, in 11th leaf to 19 by 9 cm; base obtuse, in the lower leaves where passing into the margin with a distinct angle; top cuspidate; margin entire; indument and nervation as in the first leaves.

Specimens: 2212 and 2365 from S.E. Borneo, primary hill forest on deep clay, low altitude.
Growth details: In nursery germination poor.
Remarks: Two species described, similar in germination pattern and general

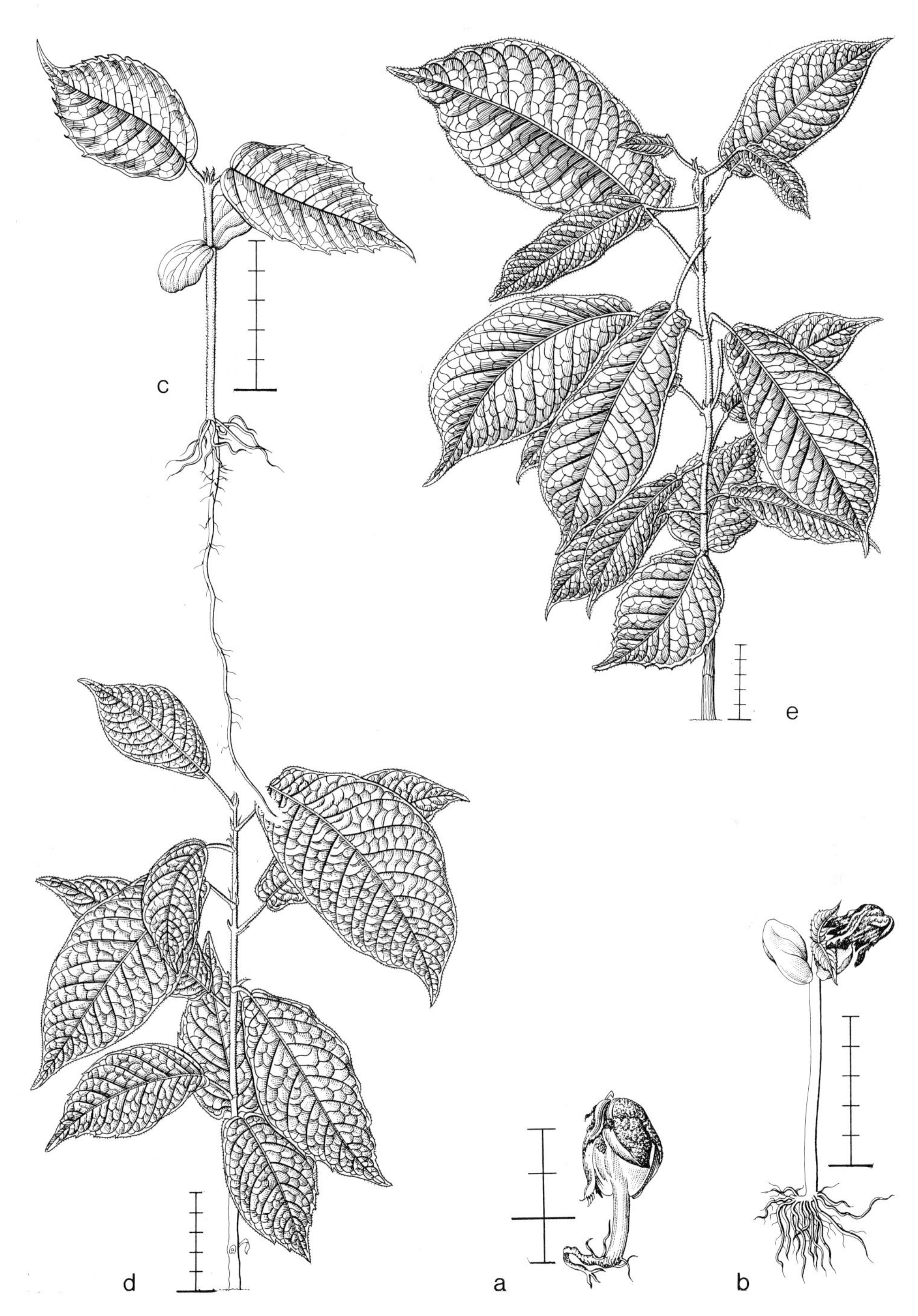

Fig. 79. *Sloanea javanica.* De V. 2212, collected 7-iii-1973, planted 22-iii-1973 (b, d, and e); De V. 2365, collected 16-iii-1973, planted 22-iii-1973 (a and c). a, b, and c. 17-iv-1973; d. 5-xi-1973; e. 4-iii-1974.

morphology. *S. berteriana* differs in details only.

EUPHORBIACEAE

APORUSA

Tree Fl. Malaya 2: 58.
Genus in India and Ceylon, throughout S.E. and E. Asia, all over Malesia, doubtfull in N. Australia, in the Pacific in Bismarck, the Solomons, W. Carolines, and Fiji.

Aporusa prainiana King ex Gage – Fig. 80.

Development: The taproot and hypocotyl emerge from one pole of the seed. After establishment of the root the enclosed paracotyledons are carried up by the hypocotyl which becomes erect. After a short resting stage with the paracotyledons enclosed in the testa, these are freed from the seedcoat by spreading. A second resting stage occurs with the paracotyledons expanded.
Seedling epigeal, phanerocotylar. Macaranga type.
Taproot long, slender, flexuous, brownish green, with many rather long, slender, branched, brownish cream sideroots.
Hypocotyl strongly enlarging, in cross-section quadrangular, to 6.5 cm long, at the angles with a distinct, sharp longitudinal ridge descending from the lower side of the cotyledonary petioles, green, turning brownish, glabrous.
Paracotyledons 2, opposite, foliaceous, in the 9th–12th leaf stage dropped; simple, without stipules, petiolate, subcoriaceous, green. Petiole in cross-section quadrangular, to 4 by 1 mm, above slightly hairy with short, stiff, brownish hairs. Blade elliptic, almost orbicular, to 2.6 by 2.4 cm; base truncate, abruptly narrowed into the petiole; top irregularly broadly rounded to truncate and emarginate; margin entire; at the base slightly hairy above on the nerves; nerves palmate, main nerves 3, above and below raised, the lateral ones at the top broadly curved to the central one, lateral nerves ending in anastomoses.
Internodes: First one terete, elongating to 1.2 cm long, brownish by dense indument of minute, simple, brown hairs; next internodes as the first one, irregular in length, the lowest ones 0.5–1 cm long, 10th one to 1 cm long. Stem straight, slender.
Leaves all spirally arranged, simple, stipulate, shortly petiolate, herbaceous, green. Stipules 2 at the base of each leaf, ± erect along the stem, sessile, narrowly triangular, to 3 by 1 mm; top acute; margin entire; ciliate as the first internode; nerves not visible. Petiole in cross-section semi-orbicular, in the lower leaves to 2 by 1 mm, in 10th one to 3 by 1 mm, tinged violet, hairy as first internode. Blade oblong, in 4th one to 5 by 1.7 cm but in general not exceeding 4 cm, gradually becoming larger, 10th one to 6 by 2.2 cm; base narrowed into the petiole; top cuspidate; margin serrate; below on the nerves and along the margin slightly hairy as the first internode; nerves pinnate, midrib above prominent, below more so, lateral nerves above slightly prominent, below prominent, forming an undulating marginal nerve, teeth each

Fig. 80. *Aporusa prainiana.* De V. 2215, collected 7-iii-1973, planted 22-iii-1973. a and b. 1-v-1973; c. 5-xi-1973; d. 4-iii-1974.

provided with a vein that is swollen at its tip.

Specimens: 2215 from S.E. Borneo, primary hill forest on deep clay, low altitude.
Growth details: In nursery germination good, ± simultaneous.

BLUMEODENDRON

Tree Fl. Malaya 2: 68.
Genus in the Andamans, Malesia, and Bismarck.

Blumeodendron tokbrai (Bl.) Kurz – Fig. 81.

Development: The hypocotyl and taproot emerge from the concave margin of the seed, the hypocotyl which becomes erect carrying the enclosed cotyledons above the soil. After a resting period, the elongating shoot pushes the enclosed cotyledons off. Subsequent growth is in flushes of 2–3 leaves. In 8 months up to 45 cm high with 4 developed leaves.
Seedling epigeal, cryptocotylar. Blumeodendron type.
Taproot long, terete, sturdy, yellowish grey, with few, slender, hardly branched, yellowish grey sideroots.
Hypocotyl strongly elongating, ± terete, fusiform, tapering to the top, to 8 cm long, dark green turning greyish, glabrous, with close-set split tubercles, those at the base roundish, the ones towards the top more elongated, to 6 mm long.
Cotyledons 2, on top of the hypocotyl, pushed off by the developing shoot, without stipules, shortly petiolate, petioles not exceeding the testa. Whole body broadly cordate, to 2.5 by 3 by 1.3 cm, the brownish black, hard testa which splits at the base persistent. Blade slightly emerging from the testa at the base, with 2 ± deltoid, fleshy, yellowish auricles to 4 by 5 mm.
Internodes: First one strongly elongating, ± elliptic in cross-section, to 9 cm long, green, sparsely covered with minute, capitate glands; next internodes as the first one, in the lowest flushes the lowest one to 10 cm long, the highest ones much shorter, usually not exceeding 1 cm, in the 5th flush the lowest one to 25 cm long, the upper ones in each flush a few cm long. Stem slender, straight.
First 2 leaves ± opposite, simple, without stipules, petiolate, subcoriaceous, green, sparsely covered with glands as on the first internode. Petiole terete, above slightly channelled, to 2.5 by 0.3 cm, slightly thickened at base and top. Blade ovate, to 11 by 6.5 cm; base obtuse to slightly retuse; top acuminate; margin entire; nerves pinnate, above slightly raised, below much so, especially the midnerve, lateral nerves ending in anastomoses.
Next leaves spirally arranged, developing in flushes of 2–4, of which in general 2 subopposite, as the first leaves. Petiole terete, distinctly swollen at base and top, in second flush to 6 cm long, in 5th flush to 9 cm long. Blade in the first flush to 19 by 9.5 cm, in 5th flush to 26 by 15 cm; base acute.

Specimens: 703 and 2086 from S.E. Borneo, alluvial flat, primary forest on deep

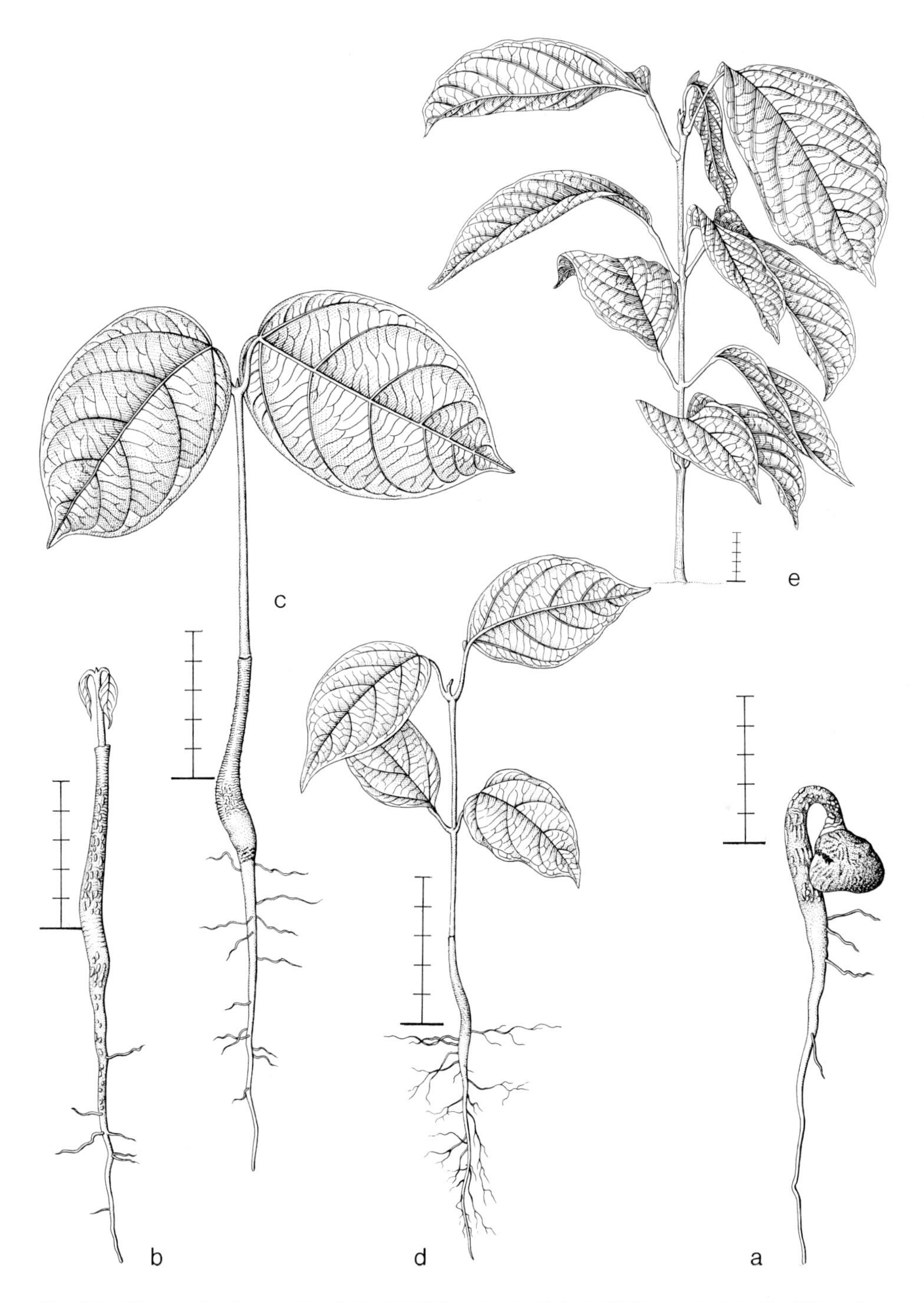

Fig. 81. *Blumeodendron tokbrai.* De V. 703, collected 10-xi-1971, planted 7-xii-1971. a, b, c, d, and e date?.

clay, low and medium altitude.
Growth details: In nursery germination fair to poor.

CROTON

Tree Fl. Malaya 2: 84; Duke, Ann. Miss. Bot. Gard. 52 (1965) 339 (*C. lucida*, *C. poecilanthus*); Mensbruge, Germination (1966) 191 (*C. aubrevillei*, *C. zambesicus*).
Genus pantropic and in the subtropics; Africa, the Near East, S.E. and E. Asia, all over Malesia, N. and E. Australia, in the Pacific as far as Tonga Arch., including the Marianas, the Carolines, and New Caledonia, also in tropical N., Central, and S. America, including the Galapagos and Revilla Gigedos.

Croton argyratus Bl. – Fig. 82.

Development: The taproot and hypocotyl emerge from one pole of the seed. The hypocotyl which becomes erect carries the enclosed paracotyledons above the soil. After a short resting stage the paracotyledons are withdrawn from the testa by spreading. A second resting stage occurs with the paracotyledons spread.
Seedling epigeal, phanerocotylar. Macaranga type.
Taproot long, slender, fibrous, creamy-white to white, with many very long, long-branched, cream-coloured to white sideroots.
Hypocotyl strongly elongating, terete, to 6.5 cm long, rather densely covered with short, simple, brownish hyaline hairs.
Paracotyledons 2, opposite, foliaceous, not known when dropped, but before the 13th leaf stage; simple, without stipules, petiolate, herbaceous, green. Petiole in cross-section semi-orbicular, to 10 by 2 mm, above channelled, hairy as the hypocotyl. Blade ovate, to 4.5 by 3.3 cm; base obtuse, abruptly narrowed into the petiole; top rounded; margin entire; above and below on the nerves and along the margin sparsely hairy as the hypocotyl; nerves palmate, main nerves 5, above slightly raised, below more so, lateral nerves ending in anastomoses.
Internodes: First one elongating, terete, to 1 cm long, green, hairy as the hypocotyl with simple and stellate hairs; next internodes as the first one, 5th one to 1.5 cm long, the 20th one to 3 cm long. Stem straight, not very slender.
Leaves all spirally arranged, simple, stipulate, petiolate, herbaceous, above green, below silvery white. Stipules 2 at the base of each petiole, rather long persistent, subulate, in 3rd leaf to 1 by 0.2 mm, in 35th leaf to 12 by 1 mm, hairy as the first internode. Petiole terete, in lowest leaves to 22 by 1.5 mm, in 20th one to 28 by 0.4 cm, above slightly channelled, hairy as the first internode. Blade elliptic, in third leaf to 7 by 3.5 cm, in 10th leaf to 11.5 by 6 cm, 20th one to 22 by 13 cm; base auriculate, auricles broadly rounded, in 10th leaf to 2 by 4 mm, in 20th one to 8 by 10 mm; top acuminate to cuspidate; margin with irregular, distant, vague teeth; above with few scattered hairs, soon glabrescent, below densely covered and silvery with hyaline, stellate hairs; nerves pinnate, in the higher leaves with 4 footnerves, above not prominent, the midrib slightly sunken, below nerves prominent, the midrib more so, lateral nerves ending in anastomoses.

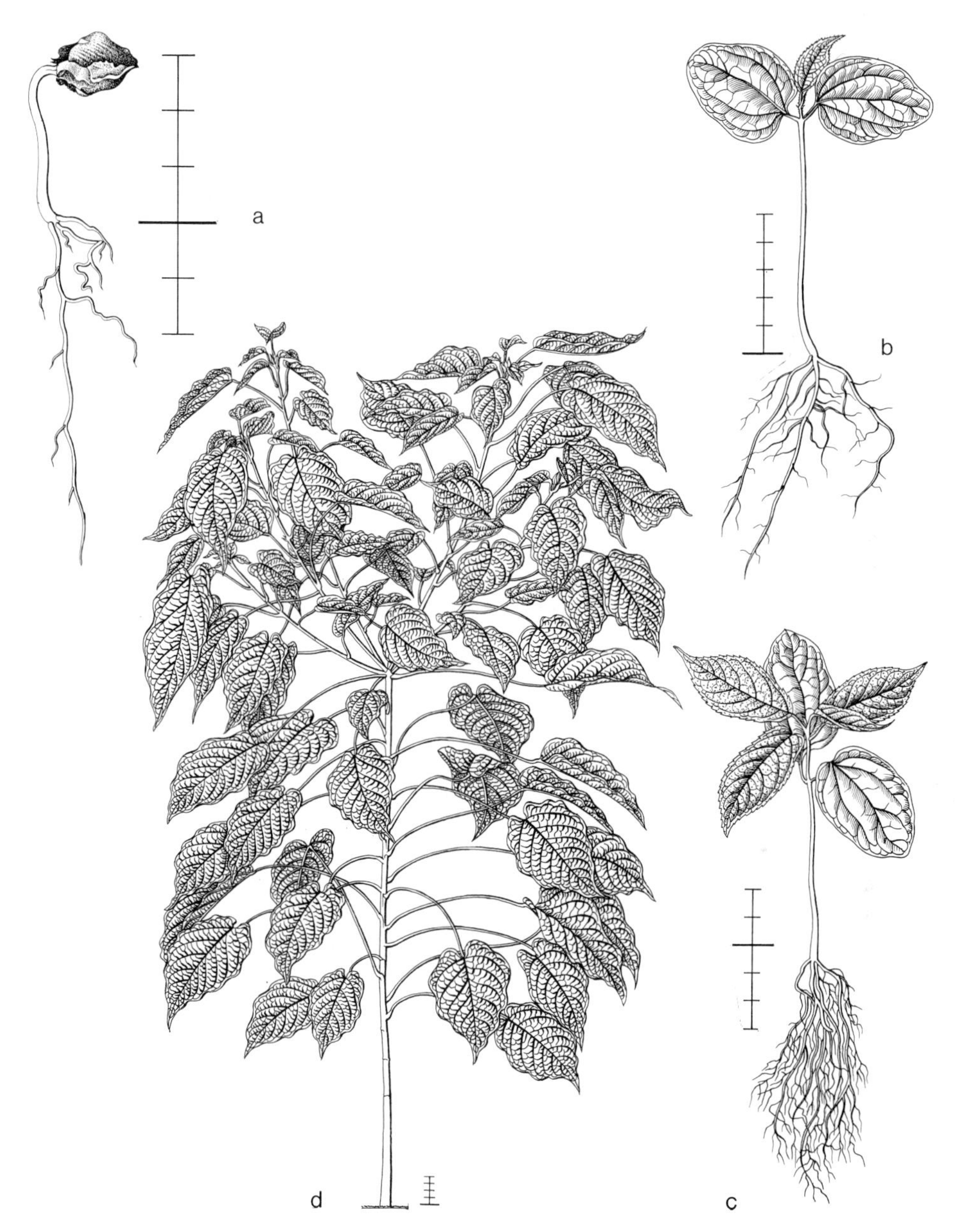

Fig. 82. *Croton argyratus.* De V. 2270, collected 10-iii-1973, planted 22-iii-1973. a date?; b. 2-v-1973; c. 21-vii-1973; d. 21-vii-1974.

Specimens: 2270 from S.E. Borneo, primary hill forest on deep clay, low altitude.
Growth details: In nursery germination fair, ± simultaneous.
Remarks: Five species described, and *C. tiglium* also grown for the project, all similar in germination pattern and general morphology. They differ in details only.

DRYPETES

Tree Fl. Malaya 2: 87; Duke, Ann. Miss. Bot. Gard. 52 (1965) 339 (*D. glauca*); Mensbruge, Germination (1966) 191 (*D. aylmeri*, *D. klainei*, *D. sp.*).
Genus pantropic; in tropical Africa, India, and Ceylon, S.E. and E. Asia, Formosa, all over Malesia, N. and E. Australia, in the Pacific as far as Fiji, including the Bonins, the Marianas, the Carolines, and New Caledonia, tropical N., Central, and S. America.

Drypetes kikir Airy Shaw – Fig. 83.

Development: The hypocotyl and taproot emerge from the seed. The hypocotyl becomes erect by which the paracotyledons are withdrawn from the testa. A short resting stage occurs with expanded paracotyledons.
Seedling epigeal, phanerocotylar. Macaranga type.
Taproot long, slender, fibrous, cream-coloured, with rather few, slender, branched, cream-coloured sideroots.
Hypocotyl strongly enlarging, terete, slender, to 5 cm long, green, with minute, brownish hyaline, simple hairs.
Paracotyledons 2, opposite, foliaceous, in the 8th–10th leaf stage dropped; simple, without stipules, shortly petiolate, subcoriaceous, green. Petiole flattened, to 1 by 1 mm, hairy as the hypocotyl. Blade ± orbicular, to 2.8 by 2.8 cm; base obtuse to slightly retuse; top rounded; margin entire; slightly hairy along the nerves as on the hypocotyl; above shining; 3-nerved, nerves above and below slightly prominent, ending in anastomoses.
Internodes: First one terete, to 2 cm long, hairy as the hypocotyl; next internodes as first one, to 1.5 cm long, often smaller. Stem slender, straight, lateral branches slightly zig-zag.
Leaves all spirally arranged, simple, without stipules, petiolate, sometimes the first one abortive, subcoriaceous, green. Petiole in cross-section semi-orbicular, to 4 by 0.7 mm, hairy as the hypocotyl. Blade elliptic to oblong, first one to 5 by 2.5 cm, 4th one to 8.5 by 3.2 cm; base obtuse to acute, top caudate; margin entire; when young hairy as the hypocotyl, glabrescent; above shining; nerves pinnate, midrib above slightly raised, below more so, lateral nerves not raised, rather inconspicuous, forming an inconspicuous marginal nerve.

Specimens: 2138 from S.E. Borneo, alluvial flat, primary forest on deep clay, low altitude.
Growth details: In nursery germination poor.
Remarks: Four species described, and *D. sp.* (De Vogel 2463) also grown for the

Fig. 83. *Drypetes kikir*. De V. 2138, collected 10-xi-1972, planted 17-xi-1972. a date?; b. 21-viii-1973; c. 9-vii-1974.

project, all similar in germination pattern and general morphology. They differ in details only.

ELATERIOSPERMUM

Tree Fl. Malaya 2: 91.
Genus in S.E. Asia and W. Malesia.

Elateriospermum tapos Bl. – Fig. 84.

Development: The taproot and hypocotyl emerge from one pole of the seed; the cotyledons swell, cracking the testa, and are withdrawn by the hypocotyl becoming erect. Resting stages after development of each flush of leaves, the endbud and the young developing leaves turning bright red with the start of each new flush.
Seedling epigeal, phanerocotylar. Sloanea type and subtype.
Taproot long, rather sturdy, pale brownish, with rather few, little branched, pale brownish sideroots.
Hypocotyl strongly elongating, terete, to 8 cm long, with two inconspicuous longitudinal ridges, green when young, with many elongated, slightly elevated, pale brownish lenticels.
Cotyledons 2, opposite, distinctly succulent, at the end of the first resting stage in 4th–6th leaf stage dropped; when old distinctly shrivelled and spreading, sessile, with broad swollen foot clasping the stem, without stipules, hard fleshy, reddish or yellowish green turning dark green, glabrous. Blade ± half ellipsoid, to 3.5 by 1.5 by 1.0 cm; base with small triangular ears; top obtuse to slightly retuse; margin entire; below distinctly concave, above slightly convex to flattened, without visible nerves.
Internodes: First one strongly elongating, terete, to 17 cm long, bright red turning green, glabrous, with scattered longitudinal, whitish lenticels; next internodes of first flush hardly enlarging, a few mm long; in next 2–3 flushes the lowest internode distinctly elongating, 2.5–9 cm long, the higher ones much smaller, to a few cm long; the 4th flush often at the base with 1–2 short internodes with abortive leaves, their axillary buds developing into 12 cm long lateral shoots, at the tip provided with two leaves; later flushes up to the 8th one not developing new lateral shoots.
First 4–6 leaves developed in the first flush, decussate, simple, stipulate, petiolate, shining reddish when young, turning green, glabrous. Stipules sessile, dropping soon, ovate to triangular, to 2 by 1.5 mm; top acute, little incised, ending in a free, long, subulate cusp; margin slightly dentate; nerve one, below swollen, ending in the cusp. Petiole terete, above channelled, to 25 by 1.5 mm, at base and top swollen. Blade ovate-oblong, the lower ones to 16 by 6 cm, the higher ones slightly smaller; base acute to slightly acuminate; top cuspidate to caudate; margin entire, slightly undulating; at the base with 2 elliptic, elevated, green glands; nerves pinnate, above slightly prominent, below much so, marginal nerve ± inconspicuous.
Next leaves spirally arranged, in 2nd and 3rd flush 2–3, in 4th flush at the base 2 abortive ones, 2–3 at the top, and 2 opposite ones at the end of the shoots, 5th to 8th flush 4–8-leaved. Leaves as those of the shoot of the first flush. Petiole in 4th flush to

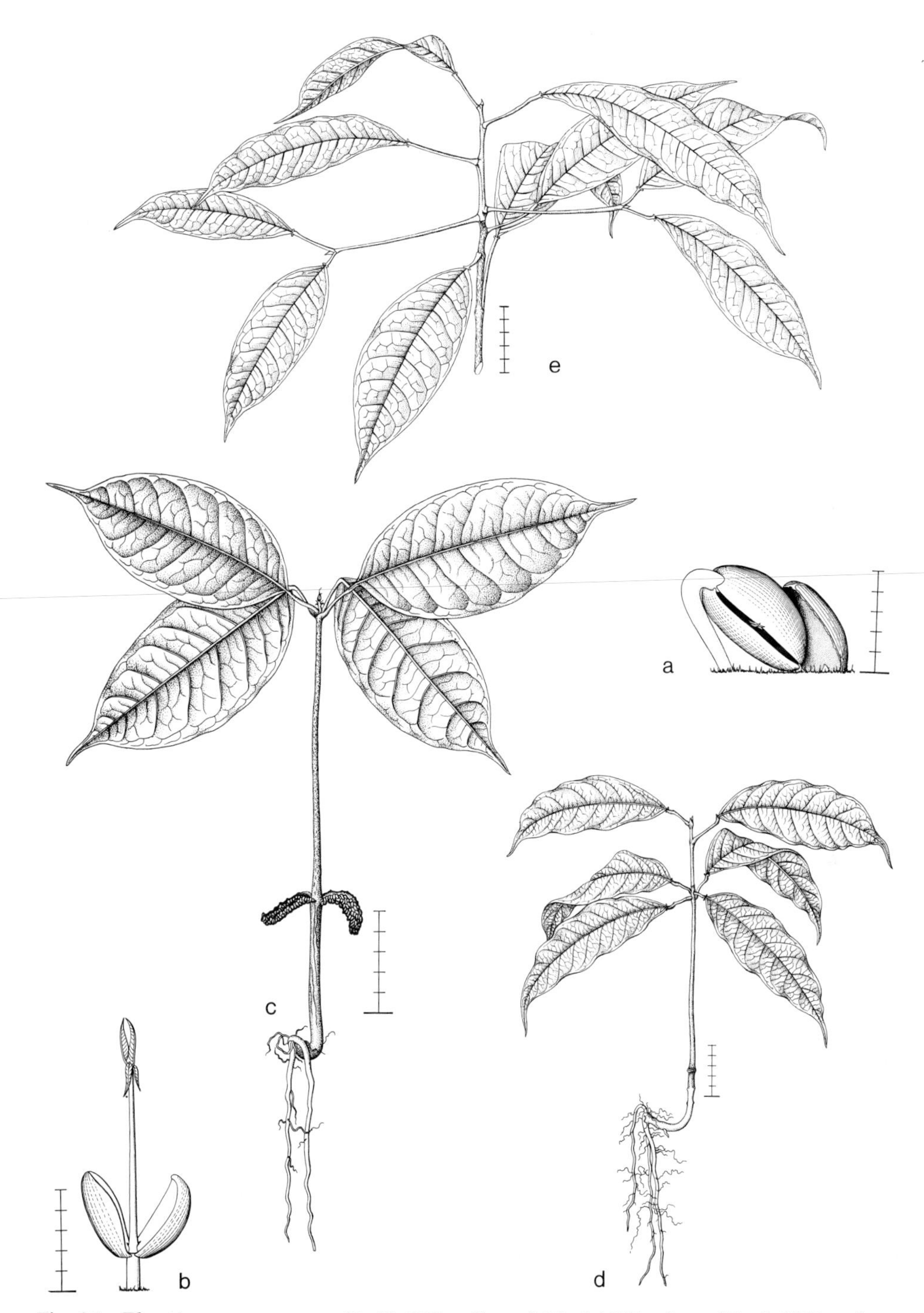

Fig. 84. ***Elateriospermum tapos.*** De V. 813, collected 15-xi-1971, planted 7-xii-1971. a, b, c, d, and e date?.

6.5 cm long, with a short, swollen base and over 1.3 cm swollen with cross-cracks, turning brown. Blade more lanceolate, in 4th flush to 21 by 6.5 cm.

Specimens: 813 from S.E. Borneo, primary hill forest on deep clay, low altitude.
Growth details: In nursery germination good, ± simultaneous.

MACARANGA

Tree Fl. Malaya 2: 105; Burger, Seedlings (1972) 103 (*M. tanarius*); Mensbruge, Germination (1966) 191 (*M. barteri, M. spinosa*).
Genus in tropical Africa and Madagascar, India and Ceylon, throughout S.E. and E. Asia up to the Ryukyus, all over Malesia, N. and E. Australia, in the Pacific as far as Tubuai and the Societies, including the Marianas, the Carolines, and New Caledonia.

Macaranga hispida M.A. – Fig. 85.

Development: The hypocotyl and taproot emerge from the seed. By spreading the paracotyledons free themselves from the testa, and are carried above the soil by the hypocotyl which becomes erect. A short resting stage occurs with expanded paracotyledons.
Seedling epigeal, phanerocotylar. Macaranga type.
Taproot rather short, slender, flexuous, cream-coloured, with many long, slender, shortly branched, cream-coloured sideroots.
Hypocotyl strongly enlarging, terete, to 3.8 cm long, pinkish, towards the top more green, densely covered with minute, simple, hyaline hairs.
Paracotyledons 2, opposite, foliaceous, in the 5th–6th leaf stage dropped; simple, without stipules, petiolate, herbaceous, green. Petiole in cross-section semi-orbicular, to 3 by 1 mm, hairy as the hypocotyl. Blade ± orbicular, to 1.2 by 1.1 cm; base truncate, abruptly narrowed into the petiole; top broadly rounded; margin entire; nerves pinnate, with footnerves, midrib above slightly prominent, below more so, lateral nerves above sunken, below raised, at the end curved to the next ones.
Internodes: First one hardly elongating, terete, to 1 mm long, green, hairy as the hypocotyl; next internodes as first one, second one to 0.5 mm long, 5th one to 1 mm long. Stem rather stunted, ± straight.
Leaves all spirally arranged, simple, stipulate, petiolate, herbaceous, green. Stipules 2 at the base of each petiole, sessile, persistent, triangular, to 1 by 0.7 mm; top acute; margin entire; hairy as the hypocotyl, nerve not visible. Petiole in cross-section terete, in first leaf to 7 by 1 mm, in 5th leaf to 13 by 1 mm, hairy as the hypocotyl. Blade ovate, in first leaf to 2.5 by 1.8 cm, in 5th one to 3 by 2.6 cm; base retuse, distinctly auriculate, auricles rounded, in 5th leaf to 3 by 11 mm; top acuminate; margin dentate, rather sparsely short hairy as the hypocotyl; nerves pinnate, midrib above slightly raised, below prominent, lateral nerves above slightly sunken, below raised, free ending in the teeth.

Specimens: 2635 from N. Celebes, primary hill forest on deep clay, medium altitude.

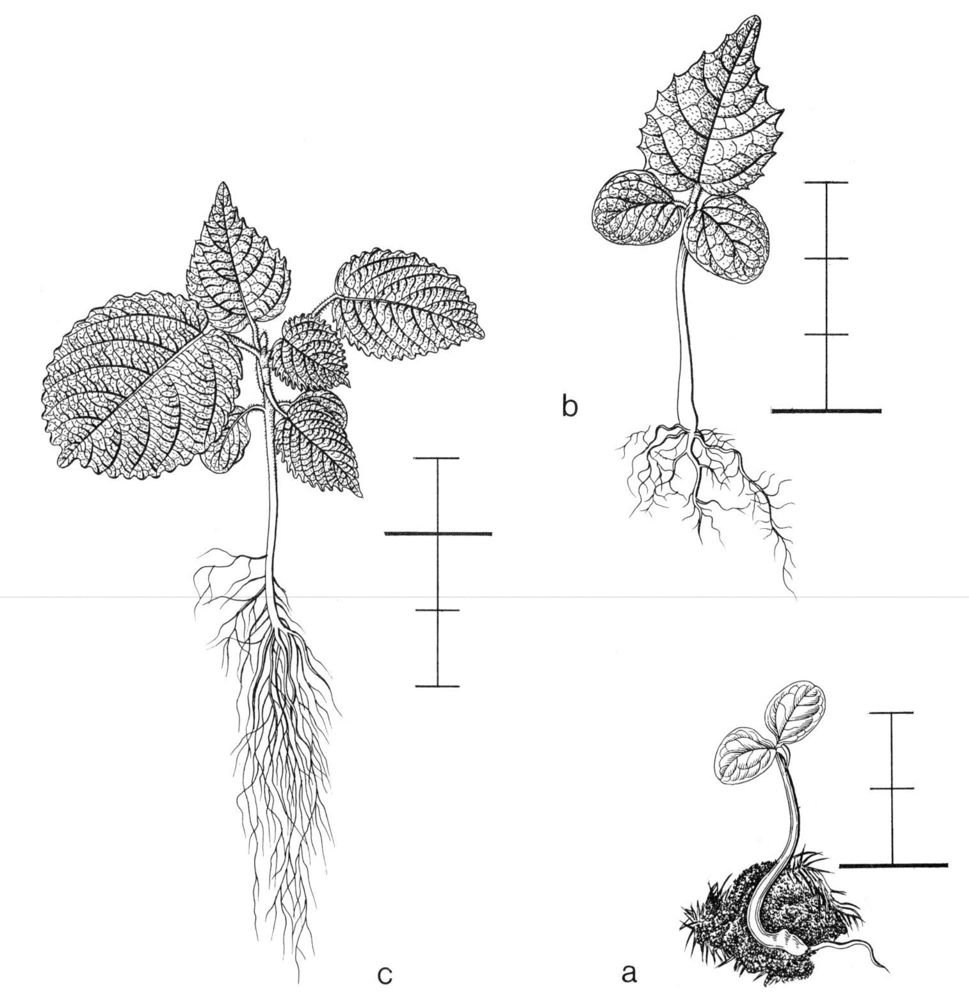

Fig. 85. *Macaranga hispida.* De V. 2635, collected 31-x-1973, planted 5-xi-1973. a. 16-i-1974; b. 20-iii-1974; c. 21-v-1974.

Growth details: In nursery germination poor.
Remarks: Four species described, all similar in germination pattern and general morphology. They differ in details only.

MALLOTUS

Tree Fl. Malaya 2: 113; Troup, Silviculture 3 (1921) 837 (*M. philippinensis*).
Genus in tropical Africa and Madagascar, India and Ceylon, throughout S.E. and E. Asia to Japan, all over Malesia, N. and E. Australia, in the Pacific as far as Samoa, including the Carolines and New Caledonia.

Mallotus sp. – Fig. 86.

Development: The hypocotyl and taproot emerge from a pole of the seed. By spreading of the paracotyledons the testa is shed, in the meantime these are carried up by the hypocotyl which becomes erect. A short resting stage occurs with the paracotyledons expanded.
Seedling epigeal, phanerocotylar. Macaranga type.
Taproot slender, flexuous, white, especially in the upper part with many long, slender, much-branched sideroots.
Hypocotyl terete, slender, strongly elongating, tapering to the top, to 4.5 cm long, green, densely hairy with minute, simple, hyaline hairs.
Paracotyledons 2, opposite, foliaceous, not exactly known when dropped, before the 8th leaf stage; without stipules, petiolate, slightly asymmetric, herbaceous, green. Petiole in cross-section semi-orbicular, to 2 by 1 mm. Blade broadly elliptic, to 27 by 19 mm; base at the narrow side acute, at the broad side truncate; top rounded; margin entire; sparsely hairy as the hypocotyl; 3-nerved, above nerves prominent, below much so, the two lateral ones descending into the petiole, midrib with lateral nerves, ending ± free.
Internodes: First one slightly enlarging, terete, to ± 3 mm long, green, hairy as the hypocotyl; next internodes as first one, the lowest ones to 5 mm, gradually becoming larger, 10th one to 1 cm long, 20th one to 3 cm long, hairy as the hypocotyl but also with hyaline, stellate hairs.
Leaves all spirally arranged, but beyond the first lateral branch (25th–35th leaf) decussate, simple, stipulate, petiolate, herbaceous, green, the ones developed in a dry period much smaller. Stipules 2 at the base of each leaf, subulate, caducous, sessile, narrowly triangular, in the lower leaves to 1 by 0.3 mm, gradually the higher ones larger, 20th one to 7 by 1.5 mm; top acute; margin entire. Petiole terete, tapering to the top, higher ones slightly thickened at the base, to 11 by 1 mm, rather densely hairy as the hypocotyl with simple and stellate hairs, often tinged red. Blade ovate to ovate-oblong, in first leaf to 5 by 2.6 cm, higher ones gradually larger, in 8th leaf 6.5 by 3.5 cm, in 20th one to 19 by 12 cm, in 30th one to 27 by 16 cm; base slightly retuse, in the higher ones where attached to the petiole with two small, slightly sunken, shining, in young leaves ant-frequented glands; top acuminate to cuspidate; margin serrate, teeth pale green, slightly hairy above with scattered stellate hairs, below more densely hairy with on the nerves stellate hairs, on the blade scattered, minute, capitate, yellow glands; nerves pinnate, above hardly raised, below much so, lateral nerves ending in anastomoses, each vein ending in a tooth.

Specimens: 1528 from N. Sumatra, alluvial flat, disturbed primary forest on deep clay, low altitude.
Growth details: In nursery germination poor, plants showing vigorous growth.
Remarks: Two species described, similar in germination pattern and general morphology. They differ in details only.

Fig. 86. *Mallotus sp.* De V. 1528, collected 21-vii-1972, planted 27-vii-1972. a, b, c, and d date?.

SAUROPUS

Tree Fl. Malaya 2: 130.
Genus in India and Ceylon, throughout S.E. and tropical E. Asia, all over Malesia and N. Australia.

Sauropus rhamnoides Bl. – Fig. 87.

Development: The hypocotyl and taproot emerge from the acute pole of the seed. The hypocotyl becomes erect by which the cotyledons enclosed in the testa are carried up above the soil. After a short resting period the cotyledons free themselves from the testa. Then follows a second temporary rest.
Seedling epigeal, phanerocotylar. Sloanea type and subtype.
Taproot long, slender, flexuous, whitish, with many long, slender, branched, whitish sideroots.
Hypocotyl strongly elongating, terete, to 6 cm long, green, with minute, red, longitudinal lines, glabrous.
Cotyledons 2, opposite, inequal, succulent, the larger one shed during development of the 3rd branch, the smaller one to 6 mm higher on the stem than the bigger one, and much earlier dropped; without stipules, petiolate, hard fleshy, dark (bluish-) green, glabrous. Petiole in cross-section flattened, to 3 by 2 mm. Lower blade irregularly broadly elliptic, in cross-section broadly triangular, to 1.4 by 1.1 cm; base rounded; top irregularly truncate; margin entire; above at the base with an elliptic depression where in the seed the smaller cotyledon rests, in the centre with a longitudinal groove. Higher blade much smaller than the lower one, ± orbicular, to 6 by 5.5 mm; base and top rounded; margin entire; below with 5 rather inconspicuous, longitudinal nerves.
Internodes: First one strongly elongating, slender, in cross-section quadrangular, to 4.5 cm long, green, glabrous; next internodes as the first one, the second one to 1 mm long, next ones of main axis to 2.5 cm long, those of the side-branches to 1.5 cm long. Stem straight, slender, soon branching.
First two leaves opposite, simple, stipulate, petiolate, herbaceous, green, glabrous. Stipules 4 at the base of the leaves, long persistent, sessile, broadly triangular, ± 1 by 1 mm; top acuminate; margin entire; membranaceous, pale green, glabrous. Petiole in cross-section semi-orbicular, to 3 by 0.8 mm, clearly jointed just above the stipules. Blade ovate-oblong, to 5.5 by 2 cm; base rounded; top slightly acuminate; margin entire; nerves pinnate, midrib above slightly prominent, below much so, lateral nerves above sunken, below raised, ending in anastomoses.
Next leaves on main stem all reduced, scale-like, somewhat smaller than the stipules with which they form a semi-whorl of 3 scales at the base of the side-branches, persistent. Lateral branches in cross-section quadrangular, each with a serial bud in the axil, slender, slightly zig-zag. Leaves of lateral branches at first smaller than the first two leaves, to 3.5 by 1.5 cm, higher ones larger.

Specimens: 1534 from N. Sumatra, alluvial flat, primary forest on deep clay, low altitude.

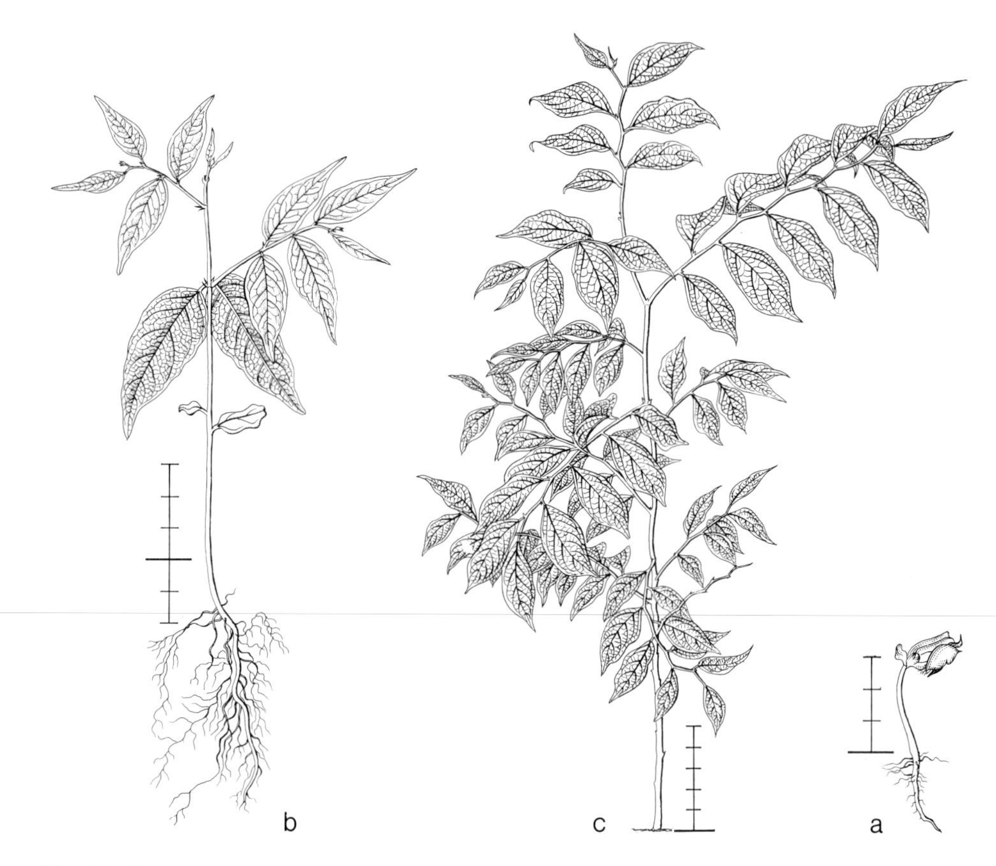

Fig. 87. *Sauropus rhamnoides.* De V. 1534, collected 21-vii-1972, planted 27-vii-1972. a and b date?; c. 13-viii-1973.

Growth details: In nursery germination good, ± simultaneous, seedlings and saplings liable to insect attack, and suffering from drought. Within two years flowering.

TREWIA

Tree Fl. Malaya 2: 134; Troup, Silviculture 3 (1921) 841 (*T. nudiflora*).
Genus in India and Ceylon, throughout S.E. Asia, and W. Malesia, including Java and the Philippines.

Trewia nudiflora L. – Fig. 88.

Development: The taproot and hypocotyl emerge from the seed; after establishment of the root the paracotyledons spread and are withdrawn from the testa, carried up by the hypocotyl which becomes erect. Sometimes the shrivelled endosperm still adheres to one of the blades. A short resting stage occurs with the paracotyledons expanded.
Seedling epigeal, phanerocotylar. Macaranga type.
Taproot quite long, fleshy, white, with many long, much-branched white sideroots, all densely covered by small white root-hairs.
Hypocotyl strongly enlarging, terete, to 11 cm long, pale green, below glabrous, in the upper part densely covered with minute, simple, hyaline, curved hairs.
Paracotyledons 2, opposite, foliaceous, in the 6th–8th leaf stage dropped; simple, without stipules, petiolate, herbaceous, green. Petiole in cross-section semi-orbicular, to 30 by 1.5 mm, above channelled, densely hairy as the hypocotyl. Blade elliptic, to 4.5 by 3 cm; base and top rounded; magin entire, slightly undulating; along the margin and below on the nerves slightly hairy as the hypocotyl; nerves palmate, main nerves 5, above and below slightly prominent.
Internodes: First one elongating to 5 mm, terete, green, hairy as the hypocotyl; next internodes as the first one, to 5 mm long, gradually the subsequent ones larger, 10th one to 1 cm long, 15th one to 1.5 cm long. Stem rather slender, slightly zig-zag when young, later straight.
Leaves all spirally arranged, conduplicate-induplicate when very young, simple, stipulate, petiolate, herbaceous, green. Stipules 2 at the base of each petiole, subulate, to 1 by 0.2 mm in the lower leaves, in 13th leaf to 1.5 by 0.5 mm, shrivelling soon but persistent. Petiole terete, in the lower leaves to 5 by 1.5 mm, in 25th one to 16 by 2.5 mm, hairy as the hypocotyl. Blade ovate, in lowest leaves to 5.5 by 4 cm, in 25th one to 16 by 13 cm; base obtuse to retuse, in higher ones broadly auriculate, auricles rounded, in 25th leaf to 1.5 by 5 cm; top acuminate; margin serrate, each tooth with a swollen tip; below on the nerves and on the margin hairy as the hypocotyl, with simple stellate hairs; nerves pinnate, with footnerves, above slightly raised, below much so, lateral nerves ending in anastomoses, veins ending free in the teeth.

Specimens: 2192 from S.E. Borneo, alluvial flat, primary forest on deep clay, low altitude.

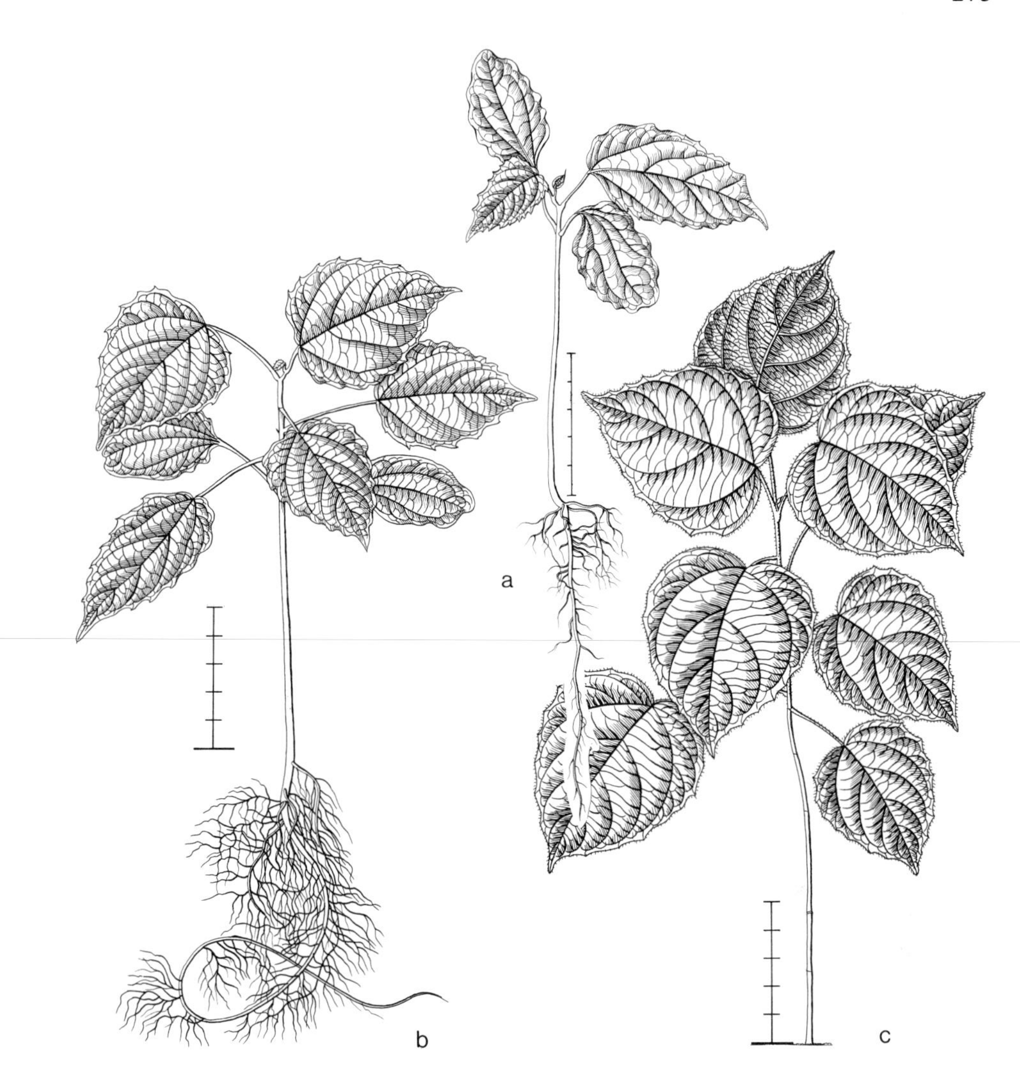

Fig. 88. *Trewia nudiflora.* De V. 2192, collected 6-iii-1973, planted 22-iii-1973. a. 28-v-1973; b. 16-vii-1973; c. 15-vii-1974.

Growth details: In nursery germination poor, ± simultaneous.

FLACOURTIACEAE

HYDNOCARPUS

Fl. Males. I, 5: 14; Troup, Silviculture 1 (1921) 13 (*H. wightiana*).
Genus in W. and S. India, Ceylon, S.E. Asia, W. Malesia including W. and C. Java, Celebes, and the Philippines.

Hydnocarpus polypetala (Sloot.) Sleum. – Fig. 89.

Development: Taproot and hypocotyl emerge from one pole of the seed. The paracotyledons remain within the testa and endosperm, and are carried up by the hypocotyl which becomes erect. After a short resting period, the testa and endosperm are thrown off by the expanding paracotyledons. A second resting stage occurs with the paracotyledons expanded.
Seedling epigeal, phanerocotylar. Macaranga type.
Taproot slender, hard fleshy, yellowish brown, with few, long, slender, unbranched, yellowish brown sideroots.
Hypocotyl strongly elongating, terete, slender fusiform, to 7.5 cm long, greenish cream, with minute, rather close-set, brownish hairs, in the upper part with 2 longitudinal grooves starting from between the cotyledons.
Paracotyledons 2, opposite, foliaceous, not known when shed, after the 6th leaf stage; without stipules, shortly petiolate, herbaceous, green. Petiole in cross-section semi-orbicular, to 3 by 2 mm, densely hairy as the hypocotyl. Blade ovate to almost cordate, to 5.5 by 4.5 cm; base exsculptate, abruptly narrowed into the petiole; top acute to blunt; margin entire; the basal part of the nerves hairy above; 3-nerved, main nerves above slightly prominent, below more so, marginal lateral nerves broadly curved, each reaching the next one.
Internodes: First one strongly elongating, in cross-section slightly ovate, to 4 cm long, green, densely hairy with minute, simple, brownish hairs; next internodes as the first one, the 3rd one to 1.5 cm long. Stem slender, straight.
Leaves all spirally arranged, simple, stipulate, shortly petiolate, subcoriaceous, green. Stipules 2 at the base of each petiole, sessile, narrowly triangular, in the lower leaves to 3.5 by 1 mm, in 5th one to 5 by 1.5 mm; top acute; margin entire; below and along the margin hairy as the first internode; nerves inconspicuous. Petiole in cross-section semi-orbicular, in the lower leaves to 5 by 1 mm, in 5th one to 10 by 2 mm. Blade ovate-oblong, the lowest one to 8.3 by 3.3 cm, the higher ones gradually larger, 5th one to 11.5 by 5 cm; base wedge-shaped; top caudate; margin entire; hairy as the first internode, but mainly on the nerves and along the margin; nerves pinnate, midrib above slightly prominent, below more so, lateral nerves above sunken, below prominent, forming a regular marginal nerve.

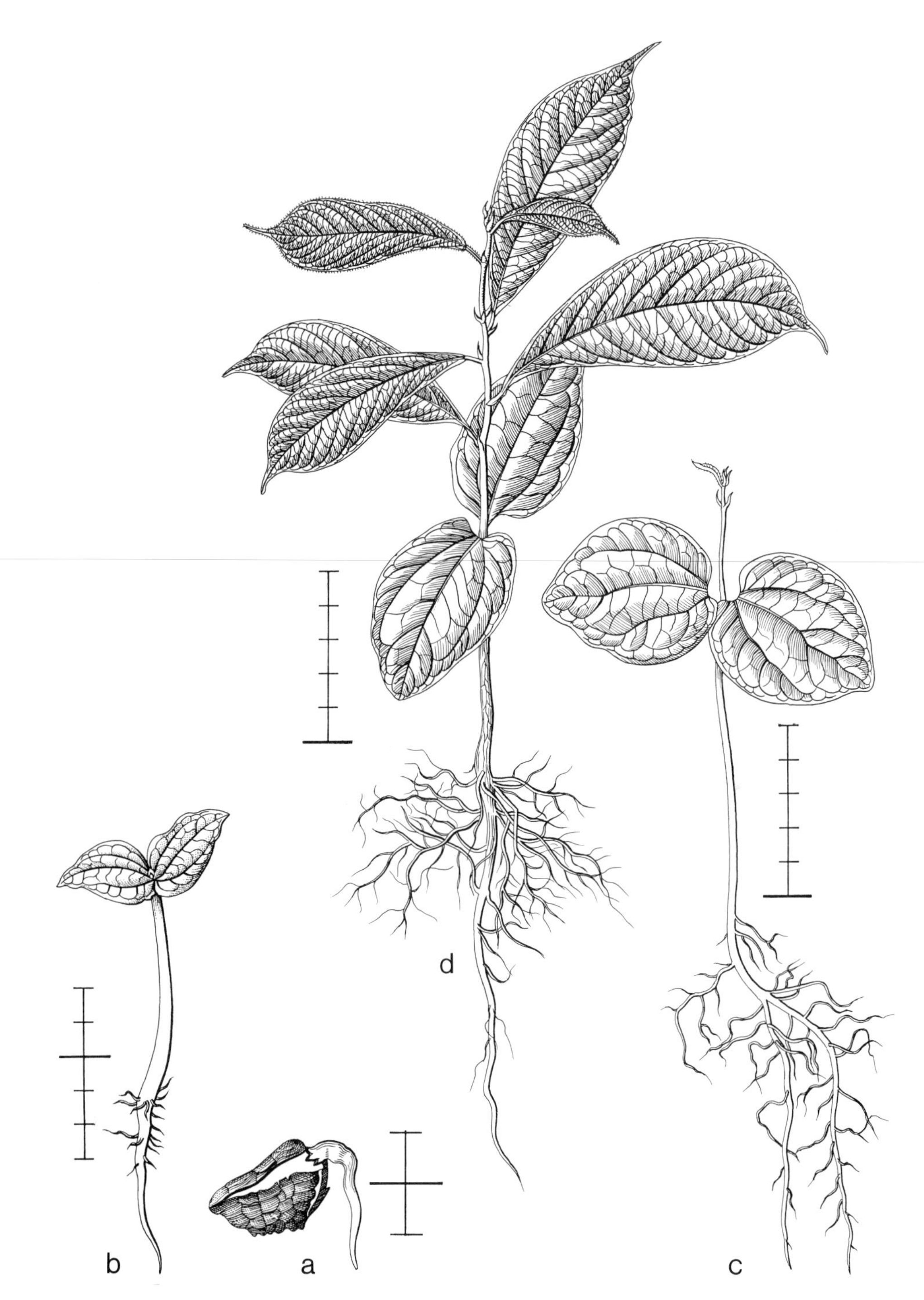

Fig. 89. *Hydnocarpus polypetala.* De V. 2242, collected 9-iii-1973, planted 22-iii-1973. a. 7-iv-1973; b and c. 10-vii-1973; d. 6-v-1974.

Specimens: 2242 from S.E. Borneo, primary hill forest on deep clay, low altitude.
Growth details: In nursery germination good, ± simultaneous, but some seeds lagging behind for a long time. Endosperm very attractive to rodents.
Remarks: Two species described, similar in germination pattern and general morphology. *H. wightiana* differs in details only.

GNETACEAE

GNETUM

Fl. Males. I, 4: 337; Bower, Quart. J. Microsc. Sci. 22 (1882) 278 (*G. gnemon*); Fröschel, Österr. Bot. Z. 6 (1911) 1 (*G. funiculare*, *G. sp.*); Hill and De Fraine, Ann. Bot. 24 (1910) 319 (*G. gnemon*, *G. moluccense*, *G. scandens*); Lehmann-Baerts, Cellule 66 (1967) 332 (*G. africanum*); Martens, Handb. Pflanzenanat. XII, 2 (1971) 168 (general, *G. africanum*); Martens and Lehmann-Baerts, Cellule 66 (1967) 343 (*G. africanum*).
Genus pantropic; W. tropical Africa, India and Ceylon, S.E. and E. Asia, throughout Malesia, in the Pacific in the Bismarcks, Solomons, St. Cruz, and Fiji, tropical S. America.

Gnetum latifolium Bl. var. *blumei* Markgr. – Fig. 90.

Development: The taproot emerges from one pole of the seed, the shoot emerging from the same pole. A resting stage occurs with 2 developed leaves (or paracotyledons).
Seedling epigeal, phanerocotylar. Not comparable to a Dicot seedling type, resembling Horsfieldia type and subtype.
Taproot long, sturdy, pale brown turning dark brown, with rather few, strongly flexuous, pale brown, pale tipped sideroots.
Hypocotyl strongly elongating, terete, to 6 cm long, slender, green, glabrous. Hypocotyledonary outgrowth enlarging, secund in the testa. The seed envelope is shed in the 2nd–4th leaf stage, ellipsoid, apiculate at both poles, to 25 mm long. Whole body ellipsoid, apiculate at both poles, the persistent, fibrous, golden outer integument is provided with irritant longitudinal fibres.
?*Paracotyledons* probably much reduced and then not observed, or represented by the first two leaves and morphologically not different from the subsequent leaves.
Internodes: First one strongly elongating, terete, to 12 cm long, slender, green, glabrous; next internodes as first one, second one to 4.5 cm long, higher ones not seen. Stem straight, slender, slightly thickened on the nodes.
Leaves all decussate, simple, without stipules, petiolate, herbaceous, green, often tinged red, glabrous. Petiole in cross-section semi-orbicular, to 10 by 1.5 mm. Blade oblong, in first two pairs to 11 by 5.3 cm, higher ones not seen; base slightly narrowed; top cuspidate; margin entire, irregular; nerves pinnate, above hardly prominent, below much so, especially the midrib, lateral nerves ending in anastomoses.

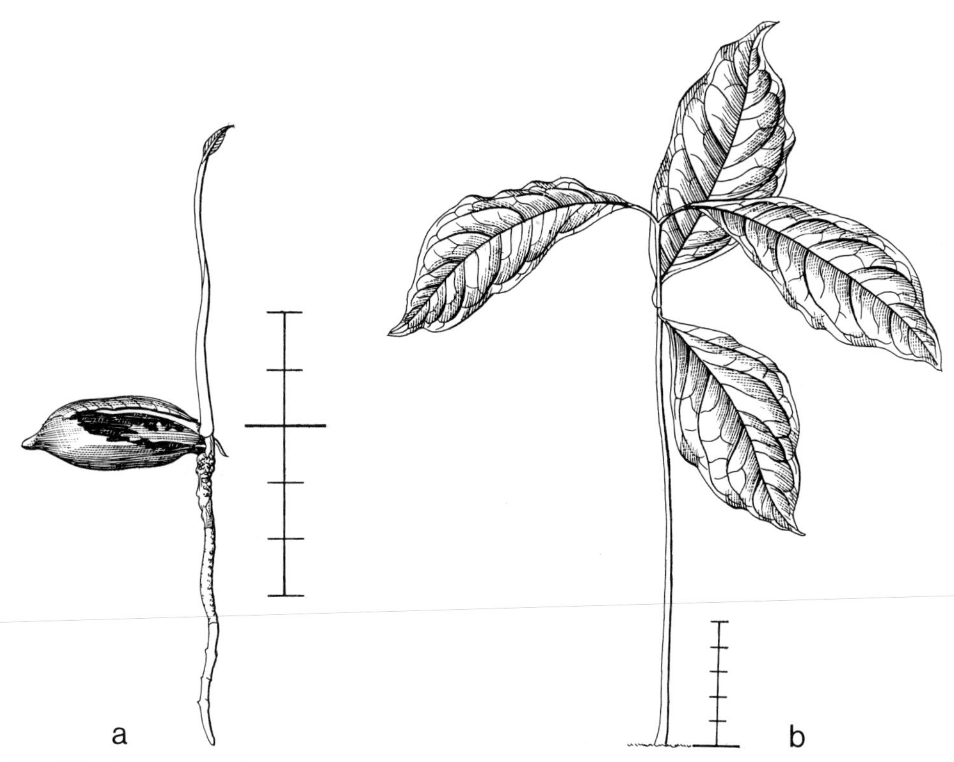

Fig. 90. *Gnetum latifolium* var. *blumei.* De V. 2542, collected 19-x-1973, planted 5-xi-1973. a. 6-iii-1974; b. 10-ix-1974.

Specimens: 2542 from N. Celebes, primary forest along small stream, low altitude.
Growth details: In nursery germination poor.
Remarks: Seedling different from normal Angiosperm seedlings in the development of a haustorium from the base of the hypocotyl. In the embryo the cotyledons are present as fleshy scales on top of the hypocotyl; they may remain reduced when extracted, or develop into foliar leaves resembling the normal leaves (*C. africanum*).

GONYSTYLACEAE

GONYSTYLUS

Fl. Males. I, 4: 349.
Genus all over Malesia except the Lesser Sunda Islands, in the Pacific in the Solomons and Fiji.

Gonystylus macrophyllus (Miq.) Airy Shaw – Fig. 91.

Development: The taproot emerges from one pole of the seed. The cotyledonary petioles elongate slightly, usually not carrying the plumule free from the testa. The hypocotyl grows in thickness, pushing aside the bases of the cotyledons, rupturing in that part the testa. The shoot is enabled to emerge through the slit. A resting stage occurs with 3–5 developed leaves.
Seedling hypogeal, cryptocotylar. Horsfieldia type and subtype.
Taproot long, slender, sturdy, brownish, with in the upper part few, slender, reddish brown, cream-tipped, few-branched sideroots.
Hypocotyl not elongating.
Cotyledons 2, secund, succulent, in the 11th–13th leaf stage shed; simple, without stipules, petiolate, hard fleshy, green where exposed, glabrous. Whole body ellipsoid, to 5.5 by 2.5 by 2.5 cm, the dark brownish grey, tough testa persistent. Petiole in cross-section flattened, to 5 by 3 mm. Blade at the base auriculate, auricles ± triangular, to 7 by 7 mm.
Internodes: First one strongly elongating, terete, to 14 cm long, dark green, rather sparsely hairy with minute, simple, appressed, hyaline hairs; next internodes as the first one, those of the first flush of 3–5 leaves to 2 cm long, the higher ones irregular in length, ultimately to 5 cm long. Stem rather slender, straight.
Leaves all spirally arranged (sometimes lower ones subopposite), simple, conduplicate-induplicate when young, without stipules, petiolate, subcoriaceous, green, sometimes intercalated with some much reduced ones. Reduced leaves subulate, in cross-section semi-orbicular, above widely channelled, hairy as first internode, dropped soon. Petiole almost terete, ultimately to 2.5 by 0.4 cm, in the lower leaves smaller, distinctly jointed at the base, hairy as first internode, glabrescent. Blade in the lower leaves elliptic, in the higher ones more oblong, the lowest ones to 15 by 8.3 cm, the 10th one to 18 by 8.8 cm, 15th one to 27.5 by 12.5 cm; base rounded, in the higher ones acute; top acuminate; margin entire; almost glabrous; nerves pinnate, midrib above sunken, below very prominent, lateral nerves very inconspicuous, not raised, in the lower leaves curved to the next ones, in the higher ones at the base ending in anastomoses; punctate with minute hyaline dots.

Specimens: 2399 from W. Java, primary hill forest on deep clay, medium altitude.
Growth details: In nursery germination good, but sometimes difficulties in establishing, the shoot dying off.
Remarks: Two species cultivated, similar in germination pattern and general morphology. *G.* cf. *consanguineus* (De Vogel 2233) differs in details only.

GOODENIACEAE

SCAEVOLA

Fl. Males. I, 5: 339, 567; ibid I, 6: 951.

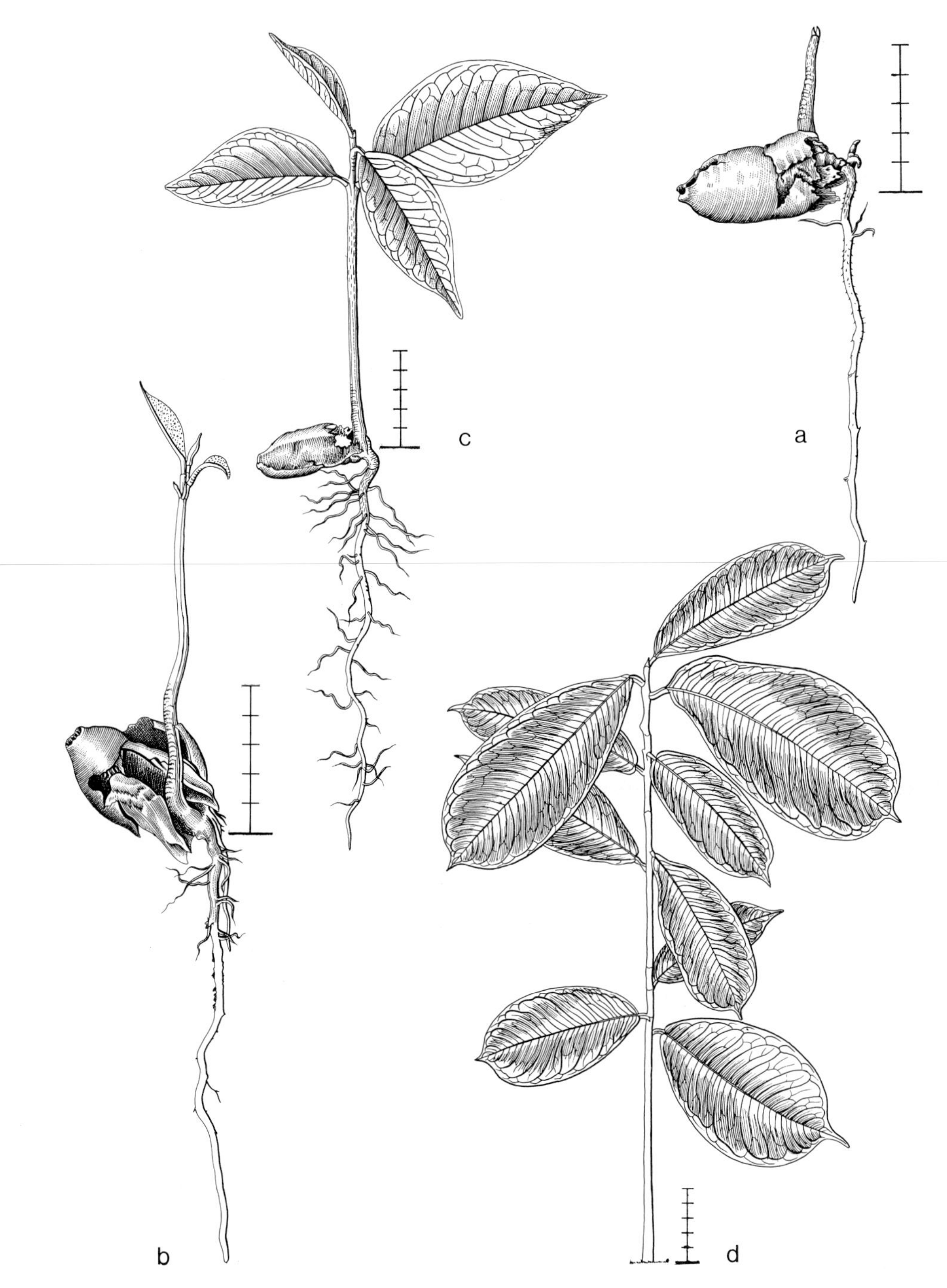

Fig. 91. *Gonystylus macrophyllus.* De V. 2399, collected and planted 20-iv-1973. a. 9-vii-1973; b and c. 10-vii-1973; d. 18-iv-1974.

Genus pantropic; Africa, the Near East, throughout the Indian Ocean, India and Ceyon, S.E. and tropical E. Asia including the Ryu Kyu Islands, throughout Malesia, N. and E. Australia, in the Pacific as far as Tuamotu Arch., including Bonin Islands, Hawaii, and New Caledonia, also in Central and tropical S. America.

Scaevola taccada (Gaertn.) Roxb. – Fig. 92.

Development: At the acute pole the fruit detaches in valves from where the hypocotyl and taproot emerge. The hypocotyl becomes erect by which the paracotyledons are pulled free. A short resting stage occurs with expanded paracotyledons only.
Seedling epigeal, phanerocotylar. Macaranga type.
Taproot long, slender, flexuous, cream-coloured, with rather few, slender, somewhat branched, cream-coloured sideroots.
Hypocotyl strongly enlarging, terete, to 3.5 cm long, green, rather densely covered with small hyaline, simple hairs.
Paracotyledons 2, opposite, foliaceous, not known when dropped; simple, without stipules, petiolate, herbaceous, green. Petiole in cross-section semi-orbicular, to 3 by 1 mm, hairy as the hypocotyl. Blade ± spathulate, to 15 by 7 mm; base narrowed into the petiole; top broadly rounded; margin entire; rather sparsely hairy as the hypocotyl; nerves pinnate, midrib above slightly sunken, below prominent, lateral nerves very inconspicuous.
Internodes: First one terete, elongating to 6 mm long, green, hairy as the hypocotyl; next internodes as the first one, only gradually increasing in length, 10th one to 7 mm long, 20th one to 8 mm long, 40th one to 1.5 cm long, a few lower ones hairy as the hypocotyl, the higher ones entirely glabrous. Stem straight, rather slender, thickened on the nodes.
Leaves all spirally arranged, simple, revolute when young, without stipules, shortly petiolate, herbaceous, green, the lower ones hairy as the hypocotyl, higher ones entirely glabrous. Petiole in cross-section semi-orbicular, in the lowest leaves to 3 by 1 mm, in the higher ones the blade descending, in the 15th one to 7 by 2 mm, in 35th one to 10 by 6 mm. Blade spathulate, first one to 2.5 by 1.5 cm, 15th one to 5.5 by 2 cm, 30th one to 15.5 by 5 cm; base narrowed into the petiole; top ± acute; margin in the lower leaves distinctly, in the higher ones more obscurely crenulate, each tooth provided with a paler thickening; nerves pinnate, midrib flat and above slightly sunken, below very prominent, lateral nerves rather inconspicuous, above slightly raised, below not so, ending in anastomoses, with one vein ending in the thickening of the teeth.

Specimens: 1573 from W. Java, open beach forest, on coral sand, low altitude.
Growth details: In nursery germination poor.

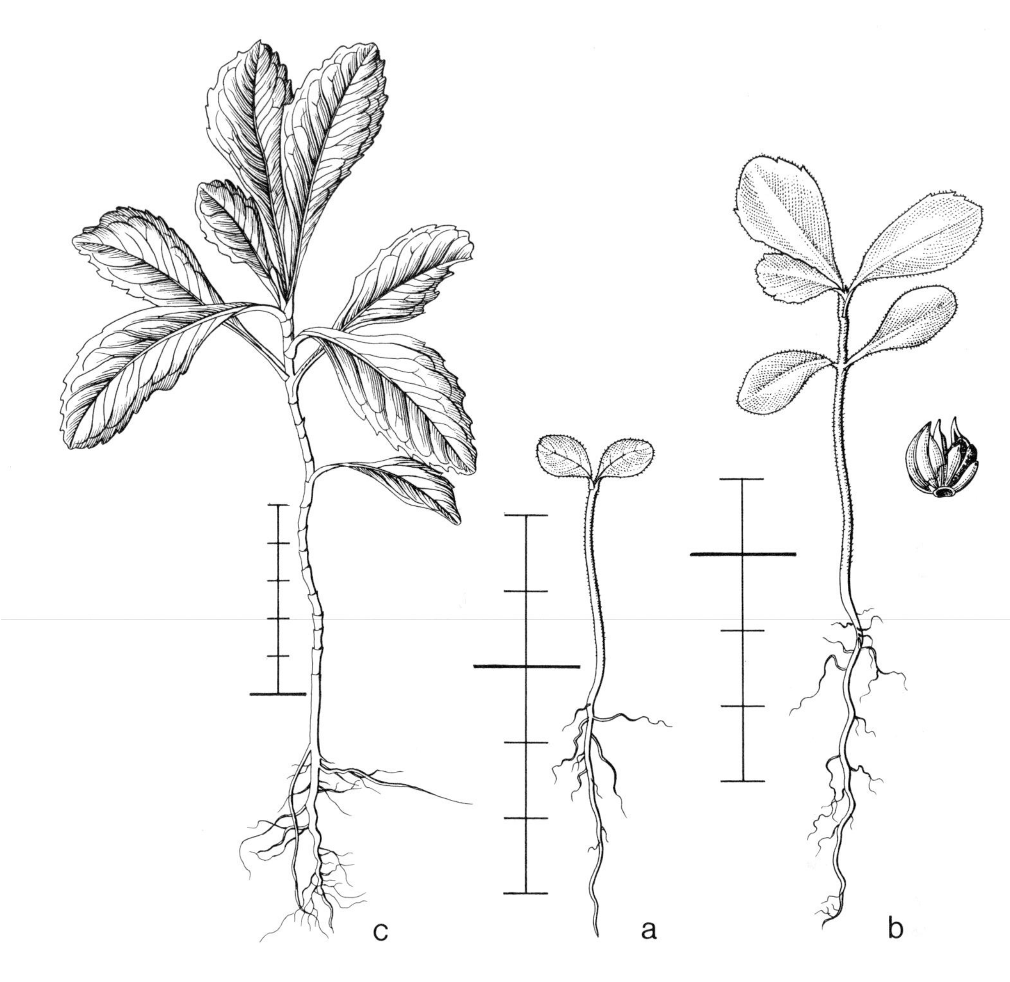

Fig. 92. *Scaevola taccada.* De V. 1573, collected 28-viii-1972, planted 4-ix-1972. a and b date?; c. 14-viii-1973.

GUTTIFERAE

MAMMEA

Fl. Java 1: 384; Brandza, Ann. Sci. Nat. Bot. sér. 9, 8 (1908) 285 (*M. americana*); Mensbruge, Germination (1966) 268 (*M. africana*); Voorhoeve, Liberian high forest trees (1965) 108 (*M. africana*).
Monograph: Kostermans, Comm. For. Res. Inst. Bog. 72 (1961) 1.
Genus pantropic; tropical W. and E. Africa and Madagascar, India and Ceylon, throughout S.E. and E. Asia, all over Malesia, in the Pacific as far E. as the Marquesas, including the Bonins, the Marianas, Central Polynesia and New Caledonia, in Central America in the W. Indies.

Mammea odorata (Raf.) Kosterm. – Fig. 93.

Development: The taproot emerges from one pole of the seed, the shoot develops from the same pole. A short resting period occurs in the 4th leaf stage.
Seedling hypogeal, cryptocotylar. Horsfieldia type and subtype.
Taproot long, sturdy, hard, brownish, with many, few-branched, slender, brownish, paler tipped sideroots.
Hypocotyl not elongating.
Cotyledons secund, covered by persistent, but eroding testa. Whole body ellipsoid in outline, flattened at one side, to 7 by 3.5 by 2 cm, base asymmetric, rounded, top blunt. With slightly yellow, very thick and sticky latex.
Internodes: First one not elongating; next internodes elongating, terete, in the leafless part to 3.5 cm long, in the leafy part irregular in length, 0.5–2 cm long, green, glabrous. Stem sturdy, straight.
Leaves all decussate, simple, growing in flushes, without stipules, shortly petiolate, coriaceous, dark green above, paler below, glabrous. First flush with 5 pairs of scales and 2 pairs of leaves. Scales much reduced, sessile, triangular, to 5 by 7 mm, the higher ones more narrowly triangular, to 10 by 5 mm; 1-nerved, nerve below prominent. Petiole of developed leaves terete, to 10 by 4 mm, above slightly flattened. Blade elliptic, to 13 by 7.3 cm; base acute to obtuse; top acute to acuminate; margin entire; nerves pinnate; midrib robust, above hardly raised, below prominent, pale green, lateral nerves minute and fine, at the end broadly curved to the next one, veins ending in anastomoses; with very little pale yellow latex. Next flushes with 2 pairs of scales and 1–2 pairs of leaves, as those of first flush.

Specimens: Bogor Botanic Garden VII. B. 101, from Kei Islands.
Growth details: In nursery germination poor.
Remarks: Two species described, similar in germination pattern and general morphology. *M. americana* differs in details only.

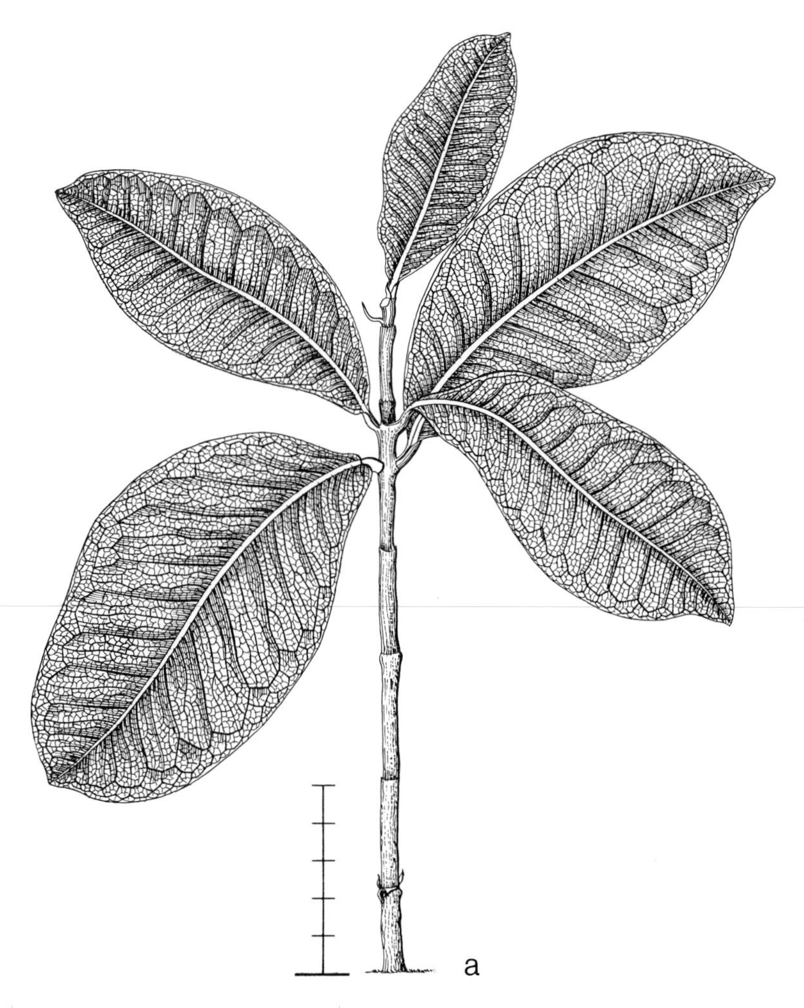

Fig. 93. *Mammea odorata.* Bogor Bot. Gard. VII. B. 101, collected and planted i-1973. a. 8-ix-1973.

HERNANDIACEAE

HERNANDIA

Tree Fl. Malaya 2: 245; Duke, Ann. Miss. Bot. Gard. 52 (1965) 332 (*H. sonora*).
Genus pantropic; W. Africa, E. Africa, including Madagascar, throughout the islands of the Indian Ocean, India and Ceylon, throughout S.E. and E. Asia, up to the Ryu Kyu Islands and Bonin, throughout Malesia, N.E. Australia, in the Pacific as far as Tuamotu Arch., including the Bonins, the Marianas, Central Polynesia, and New Caledonia.

Hernandia nymphaeifolia (Presl) Kubitzki – Fig. 94.

Development: The fruit wall splits along the suture of the valves, these spread slightly to distinctly by the development of the taproot. The cotyledonary petioles elongate, enabling the shoot to emerge. No resting stage occurs.
Seedling hypogeal, cryptocotylar. Heliciopsis type and subtype.
Taproot sturdy, fleshy, creamy-white, with many sturdy, hardly branched sideroots.
Hypocotyl not elongating.
Cotyledons 2, secund, succulent and ruminate, in the 7th–9th leaf stage shed, in general remaining enclosed. Whole body ± ellipsoid, 2.5 by 2 by 2 cm, the often splitting, brownish to blackish leathery endocarp in general persistent. Petioles elongating, in cross-section flattened, to 2.5 by 0.5 cm, greyish brown, glabrous. Exposed part of the ruminate cotyledons brownish, deeply grooved and wrinkled.
Internodes: First one somewhat elongating, ± terete, to 2.3 cm long, shining pale green, slightly warty with lenticels, with minute, scattered, simple, hyaline hairs and much smaller glandular hairs, glabrescent; next internodes as the first one, irregular in length, 1–2.5 cm long in the scaly part, in the leafy part 20th one to 2.5 cm, 30th one to 4.5 cm long. Stem rather sturdy, with pale, long lenticels, straight.
Leaves all spirally arranged, without stipules, petiolate, soft coriaceous, shining dark green above, pale green below, rather densely hairy as the first internode, glabrescent. Lower up to six ones much reduced, scale-like, gradually passing in the normal leaves. Petiole in cross-section semi-orbicular, channelled above, in the lower developed leaves to 3 cm long, in the higher ones gradually much longer, terete, the 10th one to 15 by 0.4 cm, 30th one to 19 by 0.5 cm, pale green, soft fleshy, with long pale lenticels, at base and tip reddish, the base slightly swollen. Blade in lower leaves quite small, rhomboid to ovate, 6 by 4.8 cm, passing in the gradually much bigger, peltate lower leaves, 10th one to 23 by 18 cm, 30th one to 30 by 23 cm; base truncate, in the not-peltate lower leaves slightly auriculate with broadly rounded base; top acuminate; margin entire; nerves palmate, above slightly prominent, below much so, pale yellowish green, curved at their tip towards the next nerves.

Specimens: 1384 and 1396 from W. Java, beach forest, primary forest on coral sand, low altitude.
Growth details: In nursery germination fair.
Remarks: Two species described, similar in germination pattern and general

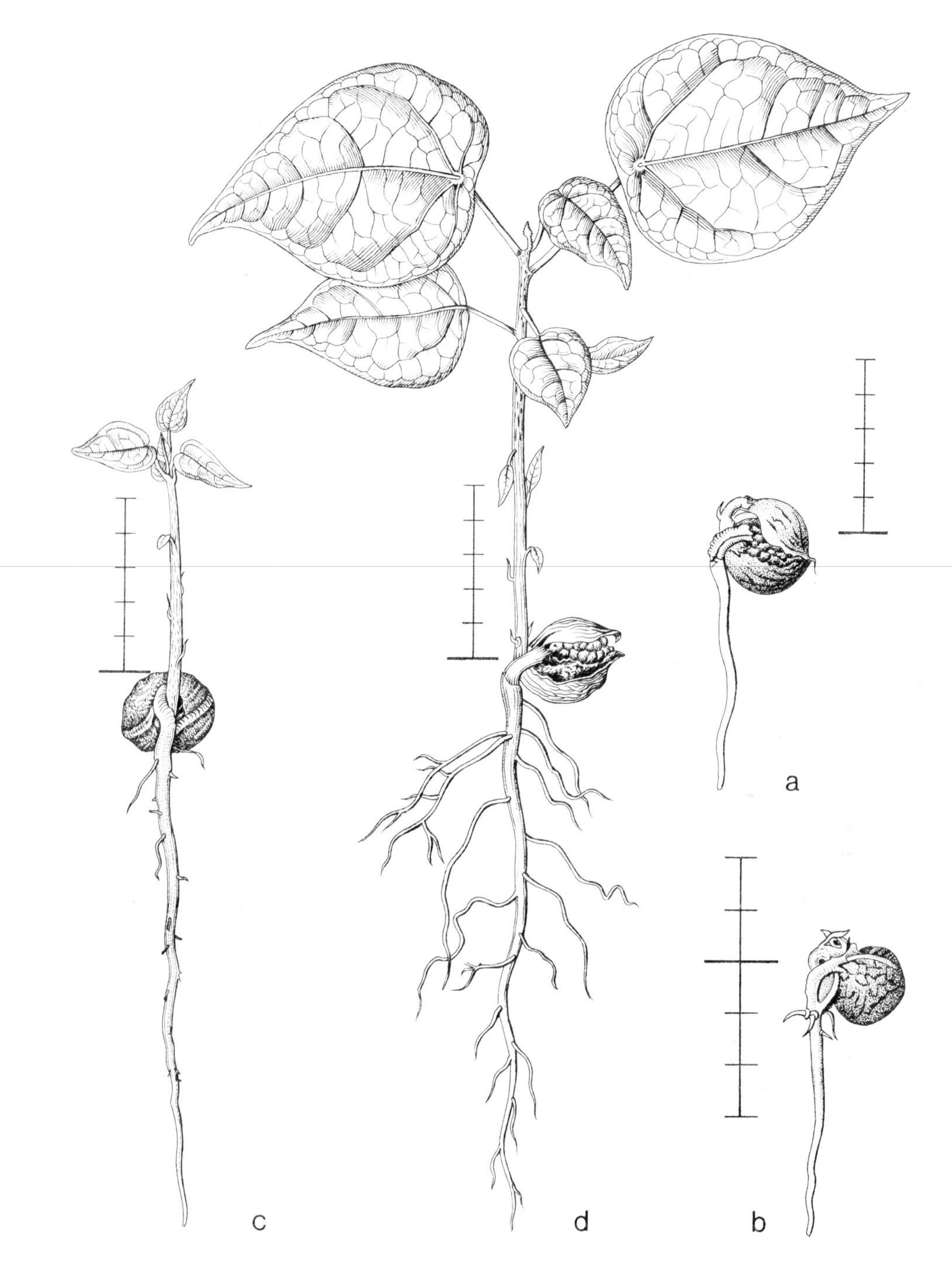

Fig. 94. *Hernandia nymphaeifolia.* De V. 1384, collected 21-vi-1972, planted 2-vii-1972. a, b, c, and d date?.

morphology. *H. sonora* differs in details only.

ICACINACEAE

GOMPHANDRA

Fl. Males. I, 7: 21.
Genus in India and Ceylon, throughout S.E. Asia, all over Malesia, N.E. Australia, in the Pacific to St. Cruz.

Gomphandra mappioides Val. – Fig. 95.

Development: The taproot and hypocotyl emerge from one pole of the fruit. The paracotyledons enclosed in the endocarp are carried above the soil by the hypocotyl which becomes erect. After a short resting stage the paracotyledons spread, and the endocarp is shed. Then follows a second temporary rest.
Seedling epigeal, phanerocotylar. Macaranga type.
Taproot rather long, sturdy, fleshy, creamy-white, with rather many, long, slender, few-branched, creamy-white sideroots.
Hypocotyl strongly enlarging, terete, to 6 cm long, green, rather sparsely hairy with short, hyaline, appressed hairs.
Paracotyledons 2, opposite, foliaceous, long persistent, when shed not known; simple, without stipules, shortly petiolate, subcoriaceous, green, glabrous. Petiole flattened, to 3 by 3 mm, with below the three main nerves descending. Blade cordate, to 4 by 3 cm; base broadly auriculate, auricles rounded, to 2 by 10 mm; top acute, its tip rounded; margin entire; nerves palmate, main nerves 3, above sunken, the lateral ones only so for the lower half, the lower half below prominent, free ending.
Internodes: First one strongly elongating, terete, to 4.5 cm long, green, rather densely hairy as the hypocotyl; next internodes as first one, 3rd one to 3 cm long. Stem slender, straight.
Leaves all spirally arranged, simple, conduplicate-induplicate when very young, without stipules, petiolate, herbaceous, green. Petiole in cross-section semi-orbicular, in the second leaf to 5 by 1.5 mm, above channelled, hairy as the hypocotyl. Blade lanceolate, in the second leaf to 7 by 2 cm, higher ones not seen; base ± acute, narrowed shortly into the petiole; top acute to slightly acuminate; margin entire; with scattered hairs as on the hypocotyl, above soon glabrescent; nerves pinnate, above sunken, midrib below prominent, lateral nerves hardly so, curved at the tip and forming a slightly undulating marginal nerve.

Specimens: 2670 from N. Celebes, primary hill forest on deep clay, medium altitude.
Growth details: In nursery germination good, ± simultaneous.

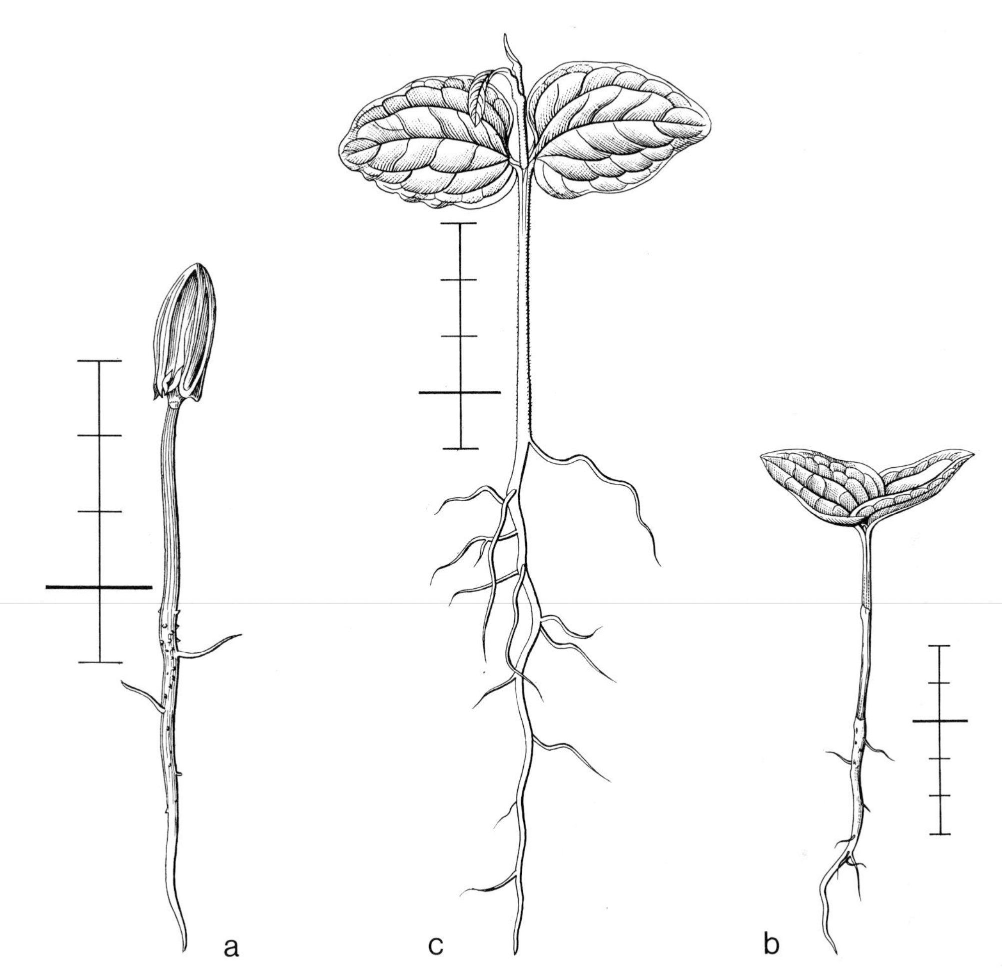

Fig. 95. *Gomphandra mappioides.* De V. 2670, collected 31-x-1973, planted 5-xi-1973. a. 16-i-1974; b. 7-ii-1974; c. 27-iii-1974.

GONOCARYUM

Fl. Males. I, 7: 14.
Genus in S.E. Asia and E. Asia as far N. as Formosa, all over Malesia except for Java and the Lesser Sunda Islands.

Gonocaryum littorale (Bl.) Sleum. – Fig. 96.

Development: The taproot emerges from one pole of the persistent endocarp. By slight elongation of the cotyledonary petioles the plumule is carried free from the endocarp, and the shoot is able to elongate. No resting stage occurs.
Seedling hypogeal, cryptocotylar. Heliciopsis type and subtype.
Taproot long, slender, sturdy, hard, creamy-white, with few, rather long, slender, few-branched, creamy-white sideroots.
Hypocotyl hardly enlarging, subterranean, terete, to 5 mm long, brownish cream, glabrous.
Cotyledons 2, secund, succulent, when shed not known, but probably long persistent; without stipules, petiolate. Whole body in cross-section ellipsoid, dorsally flattened, to 6.5 by 4.5 by 3 cm, the fibrous, hard, distinctly 3-ribbed, ochre endocarp persistent. Petiole in cross-section flattened, to 3 by 4 mm, brownish cream, glabrous.
Internodes: First one terete, slightly elongating, to 1 mm long, dull brownish green, rather sparsely hairy with minute, simple, hyaline hairs; next internodes distinctly longer, in the leafless part of the stem to 2.5 cm long, lowest ones in leafy part to 1.5 cm long, higher ones not seen. Stem slender, straight.
Leaves all spirally arranged, simple, conduplicate-induplicate when very young, without stipules, petiolate, subcoriaceous, dull brownish green when young, turning green, almost glabrous. The lower 8–9 much reduced, persistent, scale-like, sessile, triangular, to 1 by 1 mm, rather abruptly passing into the normal leaves. Petiole in cross-section semi-orbicular, in the second leaf to 8 by 1.5 mm, distinctly jointed at the base, ochre, sparsely hairy as the first internode. Blade oblong, in second leaf to 8 by 4 cm; base ± acute; top acuminate; margin entire; nerves pinnate, midrib above sunken, below prominent, lateral nerves not sunken, below slightly raised, free ending, slightly descending along the midnerve.

Specimens: Bogor Botanic Gardens III. G. 106, from New Guinea.
Growth details: In nursery germination poor.

STEMONURUS

Fl. Males. I, 7: 56; Sleumer in ibid. (1971) 58 (*S. scorpioides*).
Genus in Ceylon, throughout S.E. Asia and Malesia, except for the Lesser Sunda Islands and the Moluccas, in the Pacific as far as the Solomons and the W. Carolines.

Stemonurus malaccensis (Mart.) Sleum. – Fig. 97.

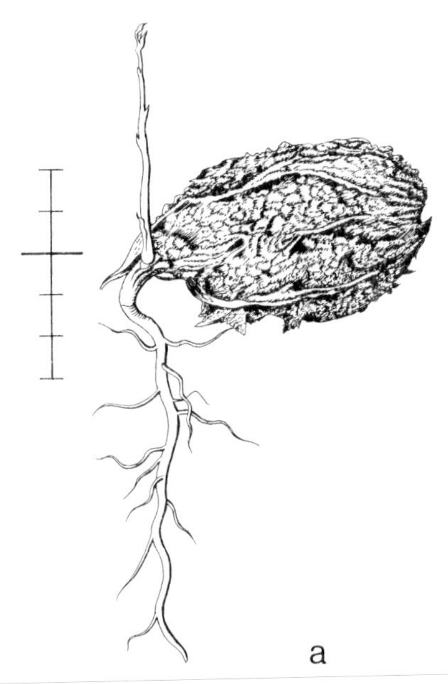

Fig. 96. ***Gonocaryum littorale***. Bogor Bot. Gard. III. G. 106, collected and planted ix-1973. a. 16-i-1974.

Development: The taproot and hypocotyl emerge from one pole of the fruit. The hypocotyl becomes erect by which the paracotyledons enclosed in testa and fruit wall are carried above the soil. During the first resting stage the much swelled endosperm breaks the fruit wall, after which the developing paracotyledons free themselves. A second long resting stage occurs with developed paracotyledons only.
Seedling epigeal, phanerocotylar. Macaranga type.
Taproot long, sturdy, fleshy, creamy-white, with in the upper part very few fleshy, creamy-white, unbranched sideroots.
Hypocotyl strongly elongating, clearly distinguishable from root and stem, terete, to 11 cm long, slightly tapering to the top, shining green, glabrous, later with grey-brown, longitudinal, raised lenticels.
Paracotyledons 2, opposite, foliaceous, in the 8th–12th leaf stage dropped; simple, shortly petiolate, without stipules, coriaceous, green, glabrous. Petiole in cross-section semi-orbicular, to 5 by 3 mm, above convex. Blade ovate, to 5.5 by 5 cm; base obtuse, abruptly narrowed in the petiole; top acute; margin entire; above shining dark green, below dull pale green, glabrous; nerves pinnate, midrib below raised, lateral nerves inconspicuous.
Internodes: First one strongly elongating, to 4.5 cm long, terete, shining green, glabrous; next internodes as the first one, irregular in length, second one to 3.5 cm long, 8th one to 6.5 cm long, but often shorter. Stem rather slender, straight.
Leaves all spirally arranged, conduplicate-induplicate when young, simple, without stipules, petiolate, coriaceous, above shining dark green, below dull pale green, glabrous. Petiole in cross-section semi-orbicular, in higher ones more terete, in first

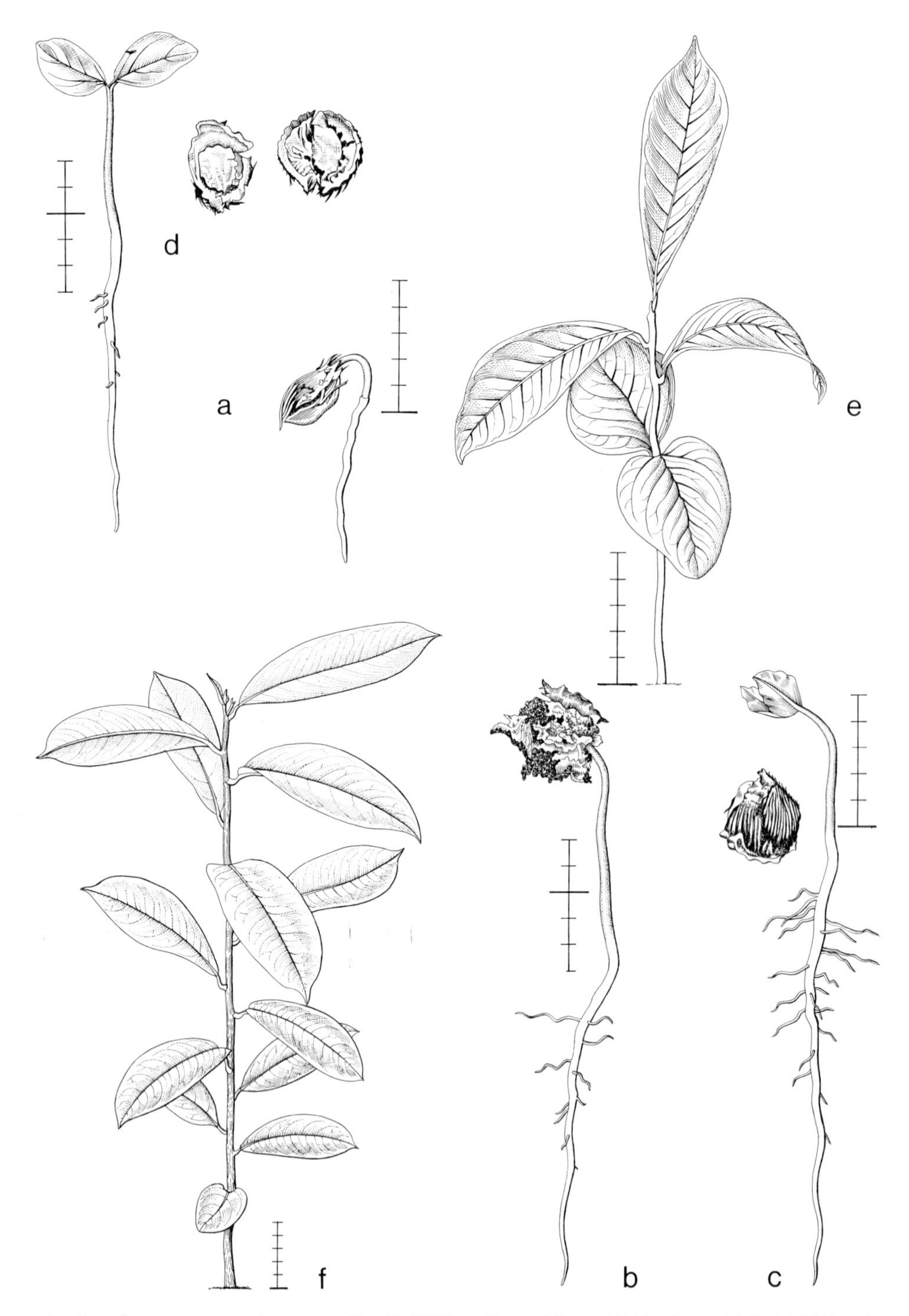

Fig. 97. *Stemonurus malaccensis.* De V. 1812, collected 23-x-1972, planted 17-xi-1972. a, b, c, d, and e date?; f. 14-viii-1973.

leaf to 15 by 2.5 mm, in 5th one to 10 by 3.5 mm, higher ones gradually larger, above channelled. Blade oblong to obovate-oblong, lowest ones to 10 by 4 cm, gradually becoming larger, 5th one to 13 by 5 cm, 10th one to 19 by 6.5 cm; base acute; top acuminate; margin entire; nerves pinnate, midrib below much raised, lateral nerves inconspicuous, forming an inconspicuous marginal nerve.

Specimens: 1812 from S.E. Borneo, hill ridge, primary forest on limestone rock, medium altitude.
Growth details: In nursery germination fair, ± simultaneous.
Remarks: Two species described, similar in germination pattern and general morphology. *S. scorpioides* differs in details only.

JUGLANDACEAE

ENGELHARDIA

Fl. Males. I, 6: 143.
Genus in N. India, throughout S.E. Asia, S. China, and Malesia to the Louisiades, also in Central and tropical N. America.

Engelhardia spicata Lech. ex Bl. – Fig. 98.

Development: The taproot and hypocotyl emerge from between the wings of the fruit. The hypocotyl becomes erect by which the paracotyledons are withdrawn from fruit wall and testa. A resting stage occurs with developed paracotyledons.
Seedling epigeal, phanerocotylar. Macaranga type.
Taproot rather short, rather slender, white, with few short, relatively sturdy, white sideroots.
Hypocotyl elongating, terete, to 1.2 cm long, at the collet narrowed into the root, rather densely hairy with relatively long, simple, hyaline hairs.
Paracotyledons 2, opposite, foliaceous, when dropped not known; deeply lobed, without stipules, petiolate, herbaceous, green. Petiole in cross-section semi-orbicular, to 1 by 0.5 mm, hairy as the hypocotyl. Blade wider than long, elliptic in outline, to 10 by 7 mm, palmatifid; base narrowed into the petiole, ciliate along the margin; lobes 4, in groups of 2 by the deeper central incision, elliptic, the lateral ones smaller, to 4 by 2 mm, the central ones to 5 by 2.5 mm; inconspicuously 2-nerved, nerves branching, not very prominent.
Internodes: First one terete, to 5 mm long, green, rather sparsely hairy with coarse, patent, hyaline, simple hairs; next internodes as the first one, 2nd one to 3 cm long, hairy as the first one. Stem ± straight, slender.
Leaves all spirally arranged, simple, without stipules, petiolate, herbaceous, green. Petiole terete, to 2 by 0.5 mm, hairy as the first internode. Blade ovate, to 12 by 7 mm; base rounded to obtuse; top acute; margin coarse-serrate, in the lower leaves with 3 pairs of apiculate teeth, those in higher leaves not seen; above on the blade

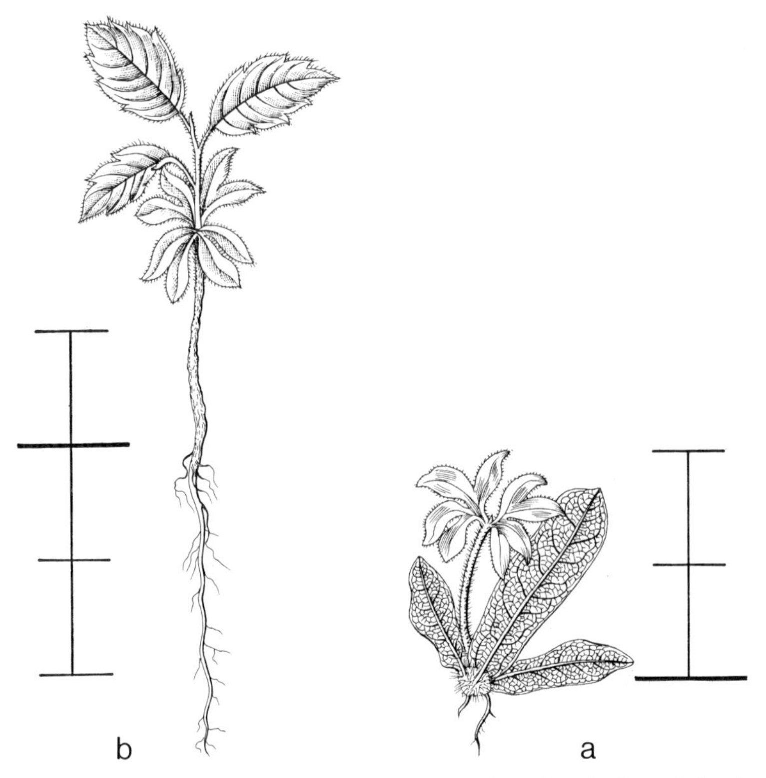

Fig. 98. *Engelhardia spicata.* Sukasdi 161, collected and planted date?. a. 26-ix-1973; b. 17-ix-1973.

sparsely, below on the midrib and along the margin more densely hairy as the first internode, and below on the blade with scattered, globular, minute, hyaline glands; nerves pinnate, midrib above prominent, below more so, lateral nerves above and below slightly raised, ending free in between and in the top of the teeth.

Specimens: Sukasdi 161 from Central Java, high altitude.
Growth details: In nursery germination poor; all specimens died when few leaves had developed.

LAURACEAE

BEILSCHMIEDIA

Fl. Java 1: 130; Mensbruge, Germination (1966) 121 (*B. mannii*, *B. sp.*).
Genus pantropic; tropical Africa, India, and Ceylon, S.E. and tropical E. Asia, including Formosa, all over Malesia, E. Australia, in the Pacific in Bismarck,

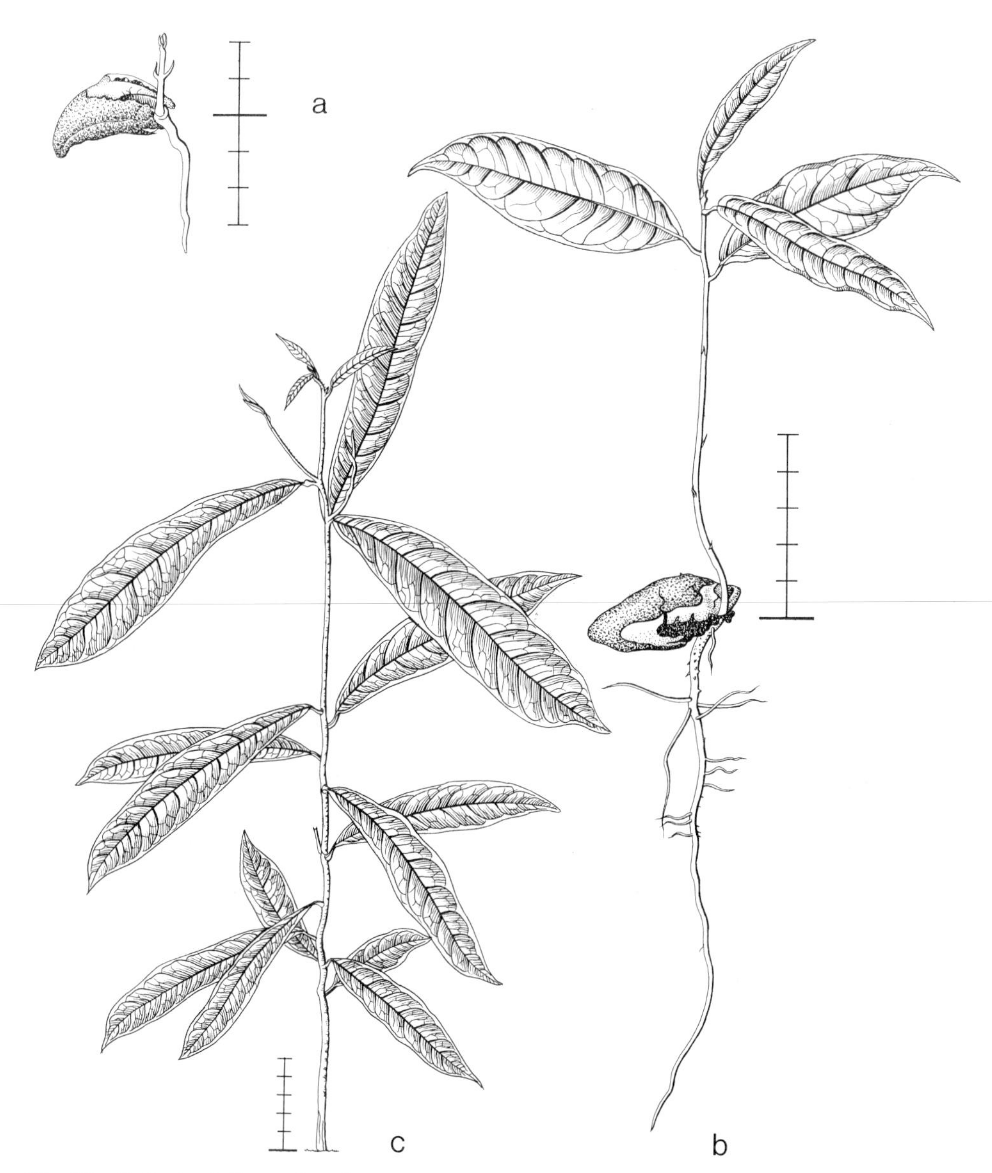

Fig. 99. ***Beilschmiedia roxburghiana.*** **Bogor Bot. Gard. IX. D. 128a, collected and planted** i-1973. a and b. 27-i-1973; c. 29-viii-1973.

Solomons, and New Zealand, doubtful in St. Cruz, New Hebrides, and New Caledonia, tropical S. America.

Beilschmiedia roxburghiana Nees – Fig. 99.

Development: The fruit splits at one pole and the taproot emerges. By slight elonga-

tion of the cotyledonary petioles the plumule is pushed free from the envelopments, after which the shoot is able to elongate. No resting stage occurs.
Seedling hypogeal, cryptocotylar. Heliciopsis type and subtype.
Taproot long, slender, fibrous, orange-brown; sideroots many, long, slender, with few long branches, its tips creamy-white.
Hypocotyl hardly enlarging, terete, to 5 mm long, glabrous.
Cotyledons 2, secund, succulent, when shed not known, before the 15th leaf stage; simple, without stipules, petiolate, hard fleshy. Whole body ellipsoid, to 4 by 1.7 by 1.7 cm, dark brownish, the fruit wall and testa persistent. Petioles slightly elongating, in cross-section flattened, to 3 by 2.5 mm.
Internodes: First one slightly elongating, terete, to 1 cm long, green, hairy with minute, simple, appressed, hyaline hairs, later with warty, pale brown lenticels; next internodes as the first one, irregular in length, to 4 cm long, below in the leafy part to 2 cm long, higher up irregular in length, sometimes to 8 cm long. Stem straight, slender.
Leaves all spirally arranged, simple, without stipules, petiolate, subcoriaceous, reddish when young, turning green. The lower 6–8 scale-like, sessile, lanceolate, to 4 by 0.8 mm, dropping soon. Petiole in cross-section semi-orbicular, in the lower leaves to 8 by 1.5 mm, in 20th one to 15 by 3 mm, hairy as the hypocotyl. Blade oblong to lanceolate, 2nd one to 11 by 4 cm, 7th one to 14 by 4.3 cm, 20th one to 20 by 6.3 cm; base acute to slightly narrowed; top acuminate; margin entire; above shining, almost glabrous; nerves pinnate, above slightly raised, below much so, lateral nerves free ending, especially those in the higher leaves.

Specimens: Bogor Botanic Gardens IX. D. 128a, from Bangka.
Growth details: In nursery germination good, ± simultaneous.
Remarks: The three species described are similar in germination pattern and general morphology. They differ in details only.

CINNAMOMUM

Fl. Java 1: 120; Burger, Seedlings (1972) 133 (*C. iners*, *C. parthenoxylon*).
Genus in India and Ceylon, throughout S.E. and E. Asia up to S. Japan, all over Malesia, N. Australia, in the Pacific as far as Samoa, including the Bonins, E. Carolines, and St. Cruz, tropical S. America.

Cinnamomum sintoc Bl. – Fig. 100.

Development: The fruit wall splits longitudinally, and by secondary growth of the cotyledonary node the bases of the cotyledons are pushed aside. The taproot and shoot emerge from the slit at the blunt pole of the fruit. No resting stage occurs.
Seedling hypogeal, cryptocotylar. Heliciopsis type and subtype.
Taproot long, slender, flexuous, pale brown, with few, rather long, unbranched sideroots.
Hypocotyl not enlarging.

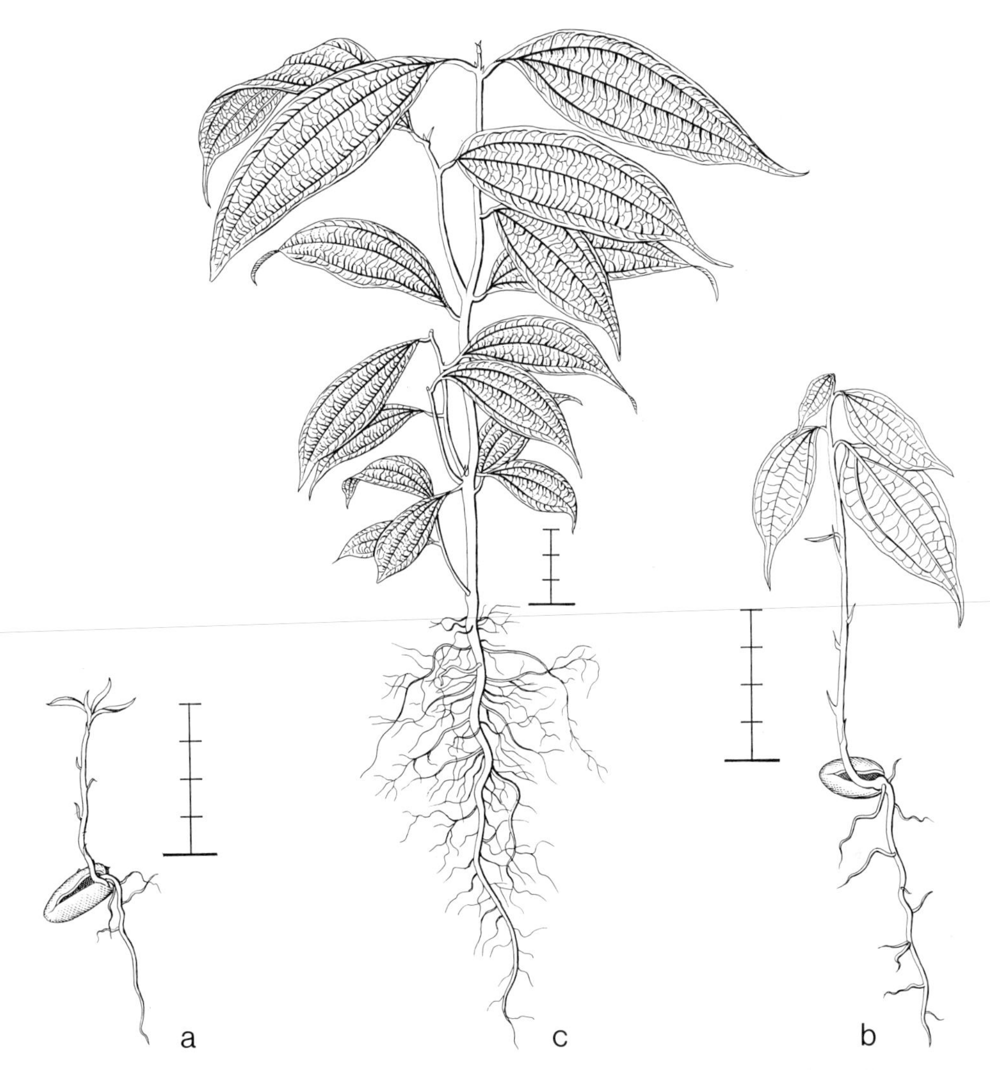

Fig. 100. *Cinnamomum sintoc.* De V. 1470, collected 16-viii-1972, planted 27-viii-1972. a and b date?; c. 9-viii-1973.

Cotyledons 2, secund, succulent, in the 5th–6th leaf stage shed; ± peltate, sessile, without stipules. Whole body ellipsoid, to 1.6 by 0.6 by 1 cm, the greyish brown fruit wall, which splits irregularly along the margin of the cotyledons, persistent.
Internodes: First one terete, elongating to 1.3 cm, green, with rather scattered, minute, simple, hyaline hairs; next internodes as the first one, irregular in length, to ± 2 cm long. Stem slender, straight.
Leaves all spirally arranged, simple, spreading, without stipules, petiolate, sub-coriaceous, green. The lower 5–7 much reduced, scale-like, soon dropped, linear, to 5 by 0.5 mm. Petiole in cross-section semi-orbicular, first one to 5 by 1 mm, 11th one

to 8 by 1.5 mm, green to reddish green, densely hairy as the 1st internode. Blade ovate-oblong, higher ones lanceolate, first one to 6 by 2.5 cm,11th one to 12 by 3.5 cm; base slightly narrowed, in later ones acute; top cuspidate; margin entire; below rather densely hairy with greyish, minute hairs, above almost glabrous except for the nerves; nerves pinnate, main lateral nerves 2, branching off some distance from the base, above all sunken, but in the lower leaves somewhat prominent, below very prominent, with many, rather inconspicuous cross-veins, from the base of the blade with 2 inconspicuous footnerves.

Specimens: 1470 from N. Sumatra, primary hill forest on deep clay, high altitude.
Growth details: In nursery germination ± simultaneous, rather poor.
Remarks: C. sintoc has the cotyledons secund at soil level, covered by the persistent fruit wall. In *C. iners* and *C. parthenoxylon* the cotyledons are either opposite at each side of the stem, or they are together at one side of the stem, their upper sides facing, and covered.

CRYPTOCARYA

Fl. Java 1: 131.
Genus pantropic; tropical Africa, India and Ceylon, S.E. and tropical E. Asia including Formosa, all over Malesia, N. and E. Australia, in the Pacific as far as Tonga, including Hawaii, New Caledonia and Lord Howe, Central and tropical S. America.

Cryptocarya sp. – Fig. 101.

Development: The taproot emerges from one pole of the often polyembryonic fruit. The cotyledonary petioles elongate, bringing the plumule out of the fruit, after which the shoot starts elongating. No resting stage occurs.
Seedling hypogeal, cryptocotylar. Heliciopsistype and subtype.
Taproot long, sturdy, cream-coloured, with rather few, rather sturdy, few-branched, cream-coloured, pale tipped sideroots.
Hypocotyl not enlarging.
Cotyledons 2, secund, succulent, ineffective after the 12th–15th leaf stage, but often longer persistent; simple, without stipules, petiolate. Whole body ellipsoid, to 2.3 by 1.2 by 1.2 cm, the tough, ochre, longitudinally ribbed endocarp persistent. Petioles in cross-section flattened, to 8 by 4 mm, glabrous.
Internodes: First one strongly elongating, terete, to 4 cm long, green, rather densely hairy with short, simple, hyaline hairs; next internodes as the first one, irregular in length, those in the leafless part of the stem to 3.5 cm, those below in the leafy part of the stem 1 to 1.5 cm long, 10th one to 3 cm long. Stem slender, straight, in the lower 3 internodes with much reduced leaves.
Leaves all spirally arranged, simple, without stipules, petiolate, subcoriaceous, green. First two leaves much reduced, represented by fleshy, ± triangular warts. Petiole in cross-section semi-orbicular, in lower leaves to 5 by 1 mm, in 10th leaf to 7

Fig. 101. *Cryptocarya sp.* De V. 2628, collected 27-x-1973, planted 5-ix-1973. a. 21-ii-1974; b. 20-iii-1974; c. 10-ix-1974.

by 1.5 mm, above channelled, hairy as the first internode, but more or less glabrescent. Blade ± ovate-oblong, in 3rd leaf to 7.5 by 2.7 cm, in 10th leaf to 9 by 3 cm; base slightly narrowed into the petiole; top cuspidate, its tip ± rounded; margin entire; at the base midnerve slightly hairy as the first internode; tri-nerved, above nerves sunken, below prominent, especially the midrib, lateral nerves rather inconspicuous, near the margin with a vague, marginal nerve.

Specimens: 2628 from N. Celebes, primary forest on deep clay, high altitude.
Growth details: In nursery germination good, ± simultaneous. Many fruits polyembryonic.

EUSIDEROXYLON

Fl. Java 1: 123; Meijer, Bot. Bull. Sandakan 11 (1968) 112 (*E. zwageri*).
Genus in Sumatra and Borneo.

Eusideroxylon zwageri T. & B. – Fig. 102.

Development: The taproot ruptures the woody endocarp, emerging from the apiculate pole of the fruit. The endocarp cracks irregularly, partly the hard wall is dropped. The shoot emerges from the apiculate pole. No resting stage occurs.
Seedling hypogeal, cryptocotylar. Heliciopsis type and subtype.
Taproot long, sturdy, hard, brownish orange, ochre-tipped, with rather many long, rather slender, few-branched, ochre-tipped sideroots.
Hypocotyl not enlarging.
Cotyledons 2, often partly fused, secund, hard succulent, long persistent, not yet shed in the 50th leaf stage. Whole body thick fusiform, in cross-section orbicular, to 12 by 5.5 by 5.5 cm, apiculate at the tip with an up to 7 mm long, thick cusp, rounded at the base, the thick and hard, woody, longitudinally grooved endocarp persistent, partly dropped, light brown.
Internodes: First one slightly elongating, terete, to 3 mm long, rather sparsely covered with minute, simple, appressed, brownish hairs; next internodes as the first one, those in the scaly lower part of the stem to 4 cm long, green; lowest one in the leafy part 4.5 cm long, subsequent ones 5–6 cm long, gradually increasing in length. Stem slender, straight, each leaf on the main stem provided with a lateral branch, which develops at the same time as the young leaf.
Leaves all spirally arranged, simple, conduplicate-induplicate when very young, without stipules, petiolate, herbaceous, reddish (green) when young, turning green. Lowest up to 12 ones much reduced, sessile, scale-like, appressed to the stem, soon caducous, the margins descending as a small ridge along the internode. Petiole terete, in the first developed leaf to 10 by 1.5 cm, often smaller, in higher ones not much larger. Blade oblong, in the first leaf to 11 by 4.5 cm, in higher one not much larger, the basal ones of each lateral branch smaller, the terminal ones of these branches larger, that in first lateral branch to 15 by 6.5 cm, in 8th lateral branch to 19.5 by 7.5 cm; base acute; top acuminate to cuspidate; margin entire; below on the nerves hairy

Fig. 102. *Eusideroxylon zwageri.* Bogor Bot. Gard., without number, collected and planted i-1973. a. 1-vii-1974; b and c. (cross-section through fused cotyledons) 2-vii-1974.

as the first internode; nerves pinnate, above sunken, below prominent, especially the midrib, lateral nerves curved at the tip, forming a much undulating marginal nerve.

Specimens: Bogor Botanic Gardens IX. C. 8, from Borneo.
Growth details: In nursery germination good, long delayed, over a long stretch of time.

LITSEA

Fl. Java 1: 125; Burger, Seedlings (1972) 135 (*L. glutinosa*).
Genus pantropic; Africa, S.E. and tropical E. Asia, all over Malesia, E. Australia, in the Pacific as far as Tonga, tropical Central and N. America.

Litsea castanea Hook. f. – Fig. 103.

Development: The taproot emerges from the acute pole of the fruit. By secondary growth of the cotyledonary node the bases of the cotyledons are pushed aside, rupturing the endocarp. The shoot emerges from the slit, the cotyledons remain secund. A short resting stage occurs with 3 developed leaves.
Seedling hypogeal, cryptocotylar. Heliciopsis type and subtype.
Taproot slender, fibrous, pale brownish, with rather few slender, branched sideroots.
Hypocotyl not elongating.
Cotyledons 2, secund, succulent, when shed not known, between the 8th and 12th leaf stage; simple, without stipules, peltate, the exposed part creamy yellow. Whole body ellipsoid, ± 2.5 by 1.2 by 0.6 cm, the persistent endocarp dark brown with vague ridges, at the acute pole splitting and becoming separated.
Internodes: First one slightly elongating, terete, to 5 mm long, densely covered with simple, white to brownish hairs; next internodes as the first one, irregular in length, the higher ones usually the longer, in 20th leaf stage sometimes to 11 cm long, in general not exceeding 5 cm. Stem slender, creamy-white when young, gradually turning to red, later green.
Leaves all spirally arranged. The lower ± 8 scale-like, sometimes subopposite, sessile, narrowly triangular, to 4 by 0.6 mm, herbaceous, reddish green to green, caducous, the upper ones more leaf-like. Higher ones developed, close together, simple, conduplicate-induplicate in the very young stage, gradually in the higher ones quickly becoming larger, without stipules, shortly petiolate, herbaceous, greenish red turning green. Petiole terete, above with a longitudinal groove, in the lower leaves to 3.5 by 1.5 mm, in 15th leaf to 25 by 5 mm, hairy as the first internode. Blade oblong, 3rd one 8 by 3.2 cm, 10th one to 14 by 5.5 cm, 15th one to 32 by 15 cm; base acute; top acuminate; margin entire; distantly hairy as the first internode, especially on the nerves and along the margin; nerves pinnate, above slightly prominent, below much so, lateral nerves free ending, blade next to the nerves bright green, and there punctate with minute hyaline dots.

Specimens: 1330 from S. Sumatra, primary forest on deep clay, low altitude.

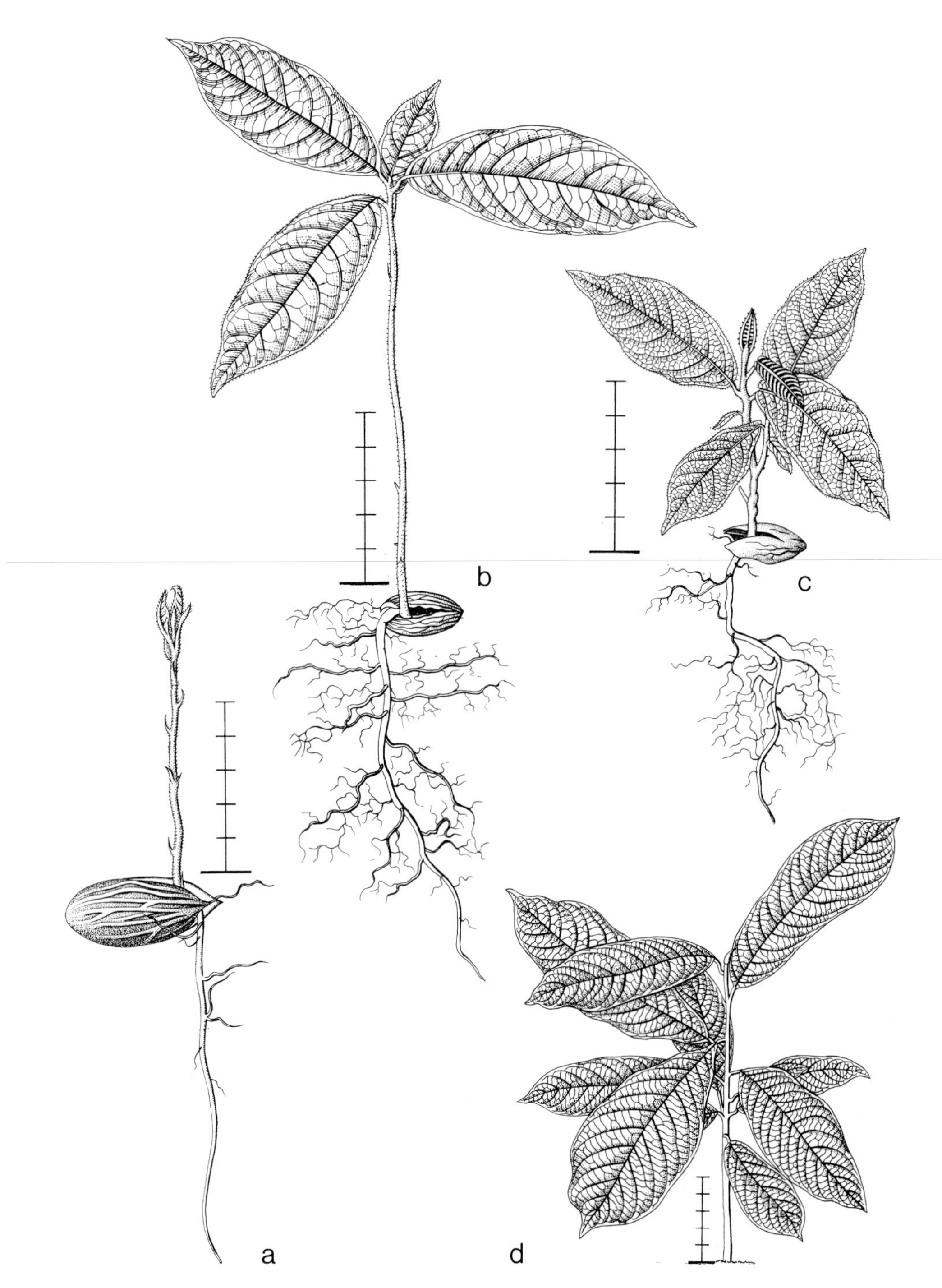

Fig. 103. *Litsea castanea*. De V. 1330, collected 17-iii-1972, planted 23-iii-1972. a, b, c, and d date?.

Growth details: In nursery germination good, ± simultaneous.
Remarks: Three species described, and two others (*L. sp.*, De Vogel 2645 and De Vogel 2647) cultivated for the project. The most common way of germination is that the fruit wall splits lengthwise at the base. The bases of the secund cotyledons are pushed aside by secondary growth of the hypocotyl (*L. castanea*, *L. glutinosa* p.p., De Vogel 2645 & 2647). *L. noronhae* and *L. glutinosa* p.p. are different in the opening of the fruit, which splits across in the centre. The opposite cotyledons are pushed aside by secondary growth of the cotyledonary node.

Litsea noronhae Bl. – Fig. 104.

Development: The fruit splits across in the centre, the cotyledons are pushed aside by secondary growth of the cotyledonary node. The taproot emerges from below, the shoot above from between the slit. No resting stage occurs.
Seedling hypogeal, cryptocotylar. Cynometra ramiflora type.
Taproot long, slender, flexuous, pale brown, with rather few, few-branched, flexuous sideroots.
Hypocotyl not enlarging.
Cotyledons 2, opposite, succulent, in the 5th–6th leaf stage shed, without stipules, peltate, almost sessile, slightly spreading, slightly unequal, hard fleshy. Blade semi-ellipsoid, to 7 by 6 by 6 mm, the bigger one to 11 mm long, above flat, below concave, the brown fruit wall persistent on the cotyledons.
Internodes: First one elongating to 10 mm long, terete, green, rather densely covered by minute, simple, hyaline hairs; next internodes as the first one, irregular in length, in the lower ones to 1.8 cm long, ultimately in the leafless part the longest one to 5 cm, those in the leafy part finally to 2–4.5 cm. Stem slender, straight, later warty by elevated, brownish lenticels, sometimes growing in flushes.
Leaves all spirally arranged, simple, without stipules, petiolate, subcoriaceous, green. First ± 6 much reduced, scale-like, sessile, linear, to 5.5 by 0.5 mm, hairy as the first internode, abruptly passing into the normal leaves; each new flush starting with some reduced leaves. Petiole terete, in 5th leaf to 5 by 1 mm, in 15th leaf to 15 by 2.5 mm, brownish hairy as the first internode. Blade obovate, in 5th one to 5.5 by 3.3 cm, in 10th one to 10 by 5.8 cm, in 15th one to 16 by 9 cm; base obtuse; top acute, in higher ones acuminate and abruptly apiculate; margin entire; below on the nerves with scattered hairs as on the first internode; nerves pinnate, above in the lower leaves sunken, in the higher ones slightly raised, below prominent, in the lower leaves lateral nerves distinctly curved at the tip, forming an irregular, much curving marginal nerve, in higher leaves much less curved, ending in anastomoses.

Specimens: 1443 from N. Sumatra, alluvial flat along river, depleted primary forest on deep clay, low altitude.
Growth details: In nursery germination poor.
Remarks: See *L. castanea*.

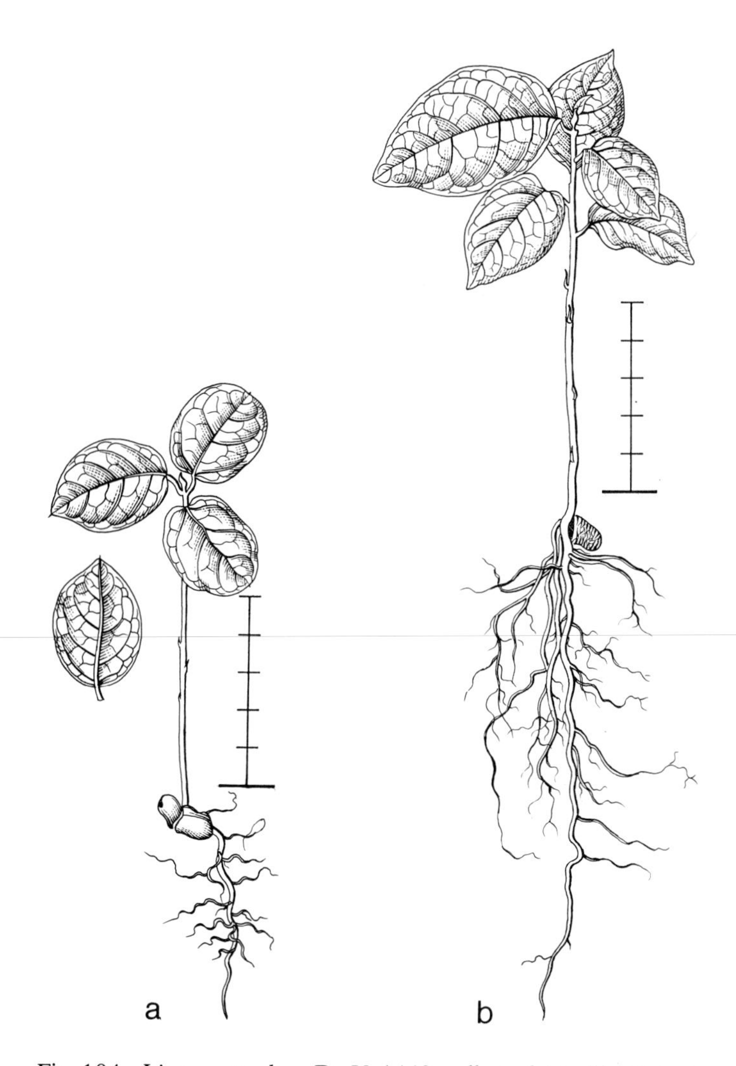

Fig. 104. *Litsea noronhae.* De V. 1443, collected 13-viii-1972, planted 27-viii-1972. a and b date?.

LEGUMINOSAE

ABAREMA (Mimos.)

Fl. Java 1: 550 (under *Pithecellobium* s.l.).
Monograph: Kostermans, Bull. Org. Sci. Res. Indon. 20 (1954) 1–122.
Genus in India and Ceylon, S.E. and E. Asia to Formosa, all over Malesia, tropical Australia, in the Pacific in New Caledonia and the W. Carolines, Central and (sub)tropical S. America.

Abarema elliptica Kosterm. – Fig. 105.

Development: The testa splits lengthwise along the margin of the cotyledons, and remains persistent. The cotyledons spread somewhat, enabling the root and shoot to emerge from one pole. No resting stage occurs.
Seedling hypogeal, semi-cryptocotylar. Endertia type, Chisocheton subtype.
Taproot long, slender, fibrous, creamy-white with a strong 'Parkia smell', with few to many long, slender, more or less branched sideroots, nodules few to many, white, to 4 mm big.
Hypocotyl slightly enlarging, terete, to 1.5 cm long, green, glabrous.
Cotyledons 2, opposite to more or less secund, succulent, shed in the 3rd leaf stage; sessile, ± spreading, without stipules, elliptic in outline, to 3 by 1.6 cm; base distinctly auriculate within the entire testa, auricles to 4 by 5 mm, broadly triangular, with an acute apex; top broadly rounded; margin entire; the convex outside covered by the persistent, smooth, dark blue testa, inside smooth, flat, green; glabrous, smelling like the roots.
Internodes: First one strongly elongating, terete, to 4.5 cm long, green, when older slightly greyish brown, slightly hairy with minute, simple, appressed brown hairs, soon glabrescent; next internodes as the first one, to 3 cm long, often smaller. Stem not very slender, slightly zig-zag, slightly thickened at the nodes below the petiole.
Leaves all spirally arranged, double compound, without stipules, petiolate, herbaceous, green, hairy as the internodes, soon glabrescent. The lowest leaf generally much reduced, scale-like. Petiole terete, in lowest one to 30 by 1.5 mm, 5th one to 6 cm long, 10th one to 9 cm long, distinctly thickened at the base over a length of up to 4 mm, at the top distinctly broadened, with terminal cusp. Terminal cusp below between the petioles 2nd order, subulate, to 0.8 by 0.3 mm. Petioles 2nd order two, in a plane parallel to the axis, terete, to 35 by 1 mm, thickened at base and top as the petiole 1st order, with terminal cusp; apicula 2nd order below between the petiolules, to 3 by 0.3 mm, above at its tip with a paler, round gland; petiolules opposite, terete, up to 4 mm long, in a panel ⊥ to that of the petioles 2nd order. Leaflets 4 in each leaf, ± ovate to ovate-oblong, in 5th leaf to 11.5 by 6 cm, in 10th one to 17 by 9.5 cm; base narrowed, more or less asymmetric, one side broader than the other and more rounded; top cuspidate; margin entire; nerves pinnate, above not raised, in higher leaves slightly raised, main nerve below prominent, sidenerves much less so, darker green, free ending.

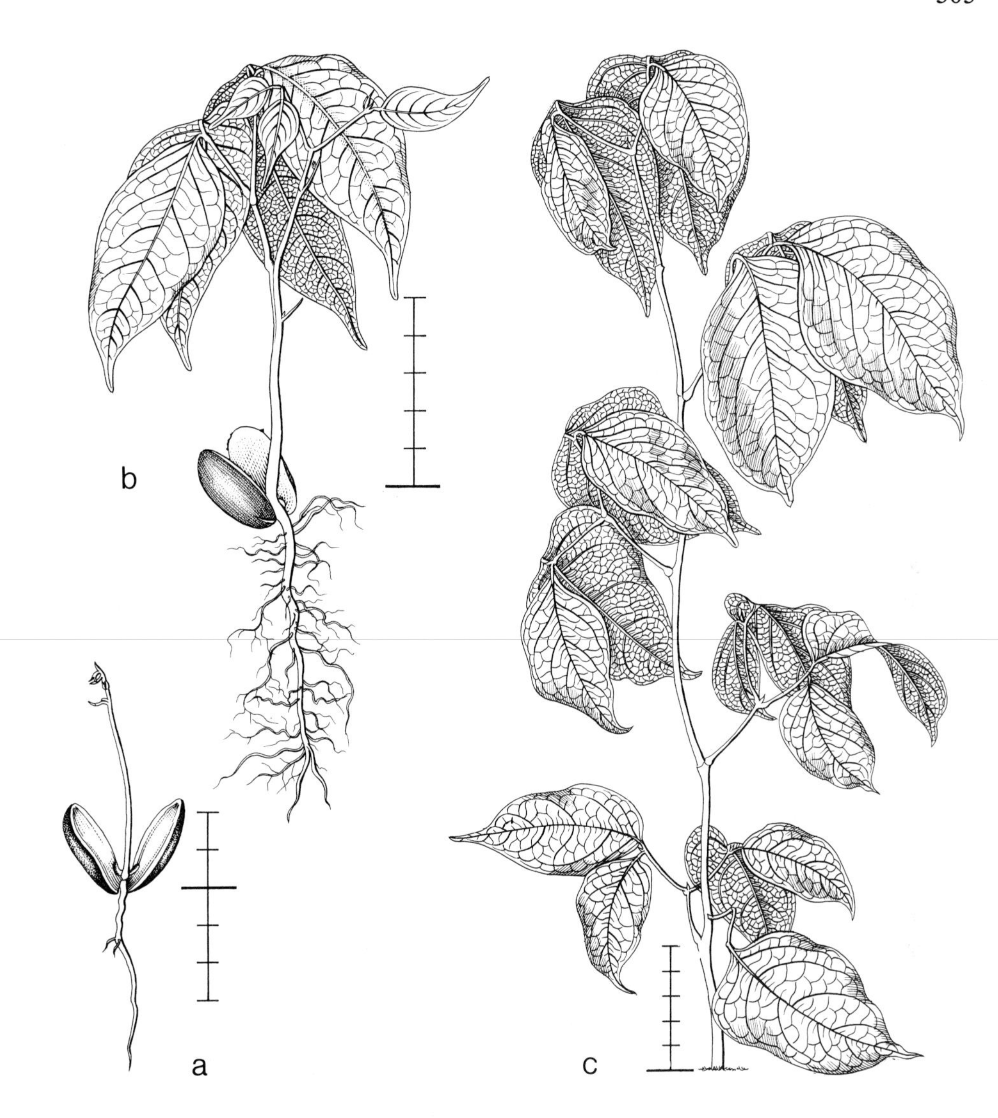

Fig. 105. *Abarema elliptica.* Bogor Bot. Gard. XIX. M. 33, collected and planted i-1973. a date?; b. 24-iii-1973; c. 3-ix-1973.

Specimens: Bogor Botanic Gardens XIX. M. 33, from Borneo.
Growth details: In nursery germination good, ± simultaneous.

AFZELIA (Caesalp.)

Fl. Java 1: 529 (*Pahudia*); Léonard, Mém. Ac. Roy. Belg., Cl. Sci., 30 (2) (1957) 106 (*A. africana*, *A. bella*, *A. bipindensis*, *A. bracteata*, *A. cuanzensis*, *A. pachyloba*, *A. peturei*); Mensbruge, Germination (1966) 145 (*A. africana*, *A. bella*); Voorhoeve, Liberian high forest trees (1965) 129 (*A. bracteata*).
Revision (*Pahudia*): De Wit, Bull. Bot. Gard. Btzg III, 17 (1941) 139.
Genus in tropical Africa, throughout the Indian Ocean, India and Ceylon, throughout S.E. Asia, and W. Malesia including W. Java and the Philippines.

Afzelia javanica (Miq.) Léonard – Fig. 106.

Development: The taproot and hypocotyl emerge from one pole of the seed. By swelling of the cotyledons the seedcoat ruptures in an irregular way, and is shed. The cotyledons are carried high above the soil by the hypocotyl which becomes erect. A short resting stage occurs with the cotyledons and the first two leaves developed.
Seedling epigeal, phanerocotylar. Sloanea type and subtype.
Taproot long, slender, fibrous, brownish cream, with many long, slender much-branched, creamy-white sideroots.
Hypocotyl strongly elongating, terete, to 14 cm long, bright red when young, turning brown, glabrous.
Cotyledons 2, opposite, succulent, in the 2nd–3rd leaf stage dropped; simple, hardly spreading, more or less appressed along the stem, without stipules, sessile, peltate, hard fleshy, outside bright red turning brownish red, inside green, glabrous. Blade half ellipsoid, to 3.8 by 1.2 by 1 cm; base rounded, its top emarginate, below the place of attachment with a broad channel; top rounded; margin entire; outside convex, inside flat, nerves not visible.
Internodes: First one strongly elongating, slender, terete, to 15 cm long, green, soon turning silvery by dying of the bark, glabrous, rather soon with many warty, lengthwise split lenticels; next internodes as the first one, second one 7 cm long, third one to 4.5 cm long, 15th one to 6 cm long. Stem slender, slightly pendulous and distinctly zig-zag at the top.
First two leaves opposite, compound, stipulate, petiolate, herbaceous, green, glabrous. Stipules interpetiolate, dropping very soon, narrowly triangular. Petiole terete, above channelled, to 4.5 by 0.1 cm, base distinctly swollen, jointed below the thickening. Rhachis as petiole, to 4 cm long, not channelled. Apical cusp articulate, dropping very soon, leaving a short, blunt, to 0.5 mm big base, subulate, terete, above channelled, to 9 by 0.5 mm. Blade paripinnate. Leaflets petiolulate, without stipellae, opposite in two (rarely 3) pairs. Petiolule terete, to 5 by 1 mm. Blade of leaflets ovate-oblong, to 10.5 by 3.5 cm; base obtuse, its tip abruptly slightly narrowed; top caudate; margin entire; nerves pinnate, midrib above sunken, below very prominent, lateral nerves below hardly prominent, irregularly curved at the end.

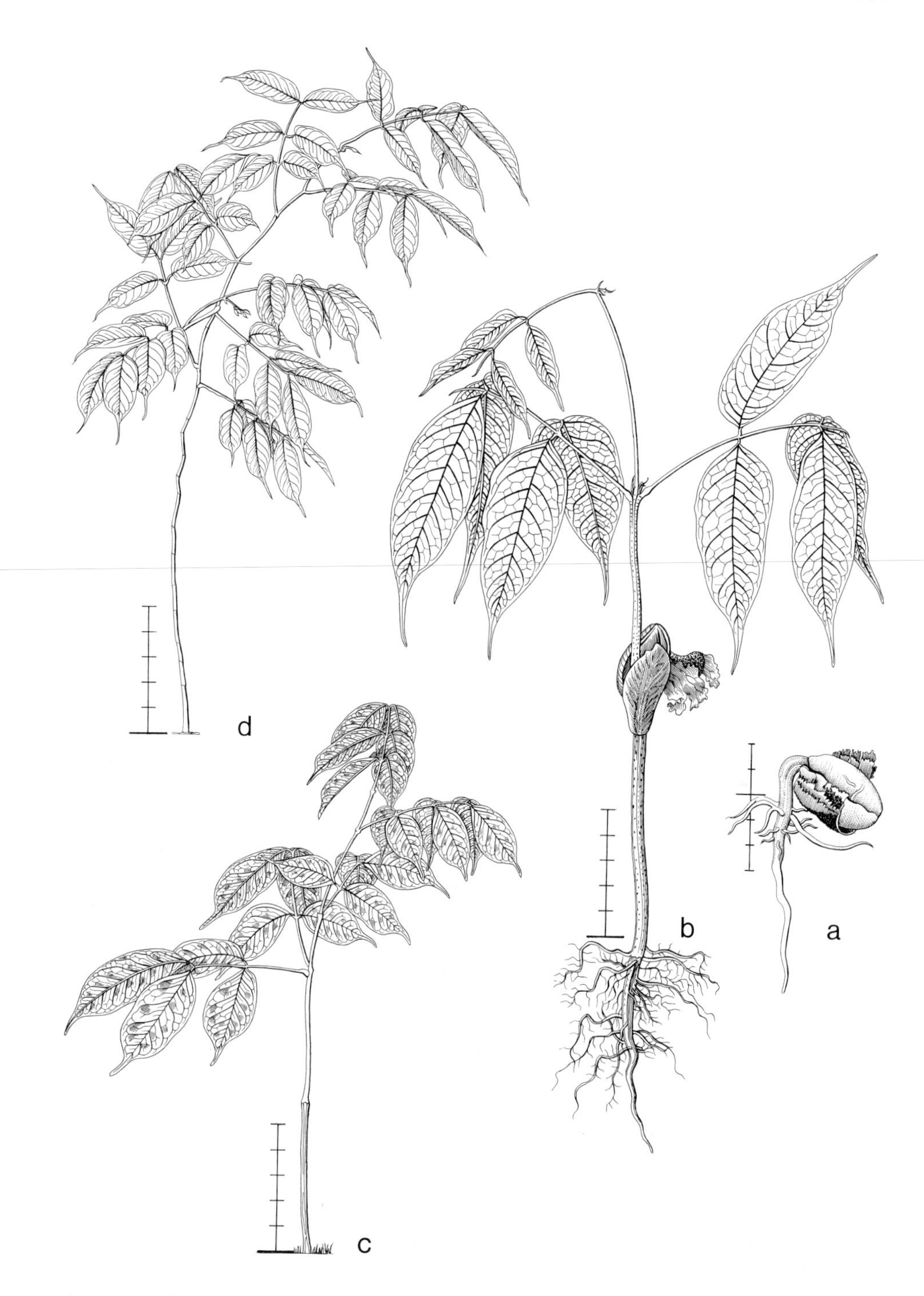

Fig. 106. *Afzelia javanica.* De V. 2272, collected 10-iii-1973, planted 22-iii-1973. a. 7-iv-1973; b. 5-v-1973; c. 18-vii-1973; d. 21-vii-1973.

Next leaves spirally arranged. Stipules 2 at the base of each leaf, dropping very soon, narrowly triangular, to 3.5 by 0.8 mm, slightly curved along and appressed to the petiole. Petiole terete, faintly channelled above, as those of first leaves. Rhachis and apical cusp as those of first leaves. Blade paripinnate, the lower 2–3 with 2 pairs of leaflets, the subsequent ones with up to 3 pairs, beyond the 15th one with 4 pairs of leaflets. Leaflets as those of first leaves, somewhat smaller.

Specimens: 2272 from S.E. Borneo, primary hill forest on deep clay, low altitude.
Growth details: In nursery germination fair, ± simultaneous.
Remarks: All species described show the same germination pattern and morphology, they differ in details only.

CAESALPINIA (Caesalp.)

Fl. Java 1: 545; Compton, J. Linn. Soc. Bot. 41 (1912) 19 (*C. coriaria*, *C. pulcherrima*, *C. sappan*, *C. sepiaria*); Hickel, Graines et Plantules (1914) 280 (*C. sepiaria*, *C. tinctoria*); Lubbock, On Seedlings 1 (1892) 456 (*C. tinctoria*).
Monograph: Hattink, Reinwardtia 9 (1974) 1.
Genus pantropic; tropical Africa, throughout the islands of the Indian Ocean, S.E. and E. Asia, all over Malesia and the Pacific, tropical America.

Caesalpinia bonduc (L.) Roxb. – Fig. 107.

Development: The cotyledons, which before germination leave a cavity in the centre of the seed, swell considerably by the intake of water, the length of the seed increasing from 2.0 up to 3.2 cm. The grey, hyaline outer testa splits off in small, rectangular scales. The taproot emerges from one pole. The cotyledonary petioles elongate, pushing the plumule out of the seed, thus enabling the shoot to emerge. No resting stage occurs.
Seedling hypogeal, cryptocotylar. Horsfieldia type and subtype.
Taproot strong, fibrous, brownish red, with many rather short, hardly branched sideroots like the taproot.
Hypocotyl slightly elongating, terete, ± 3 mm long, distinctly swollen, fleshy, reddish, glabrous.
Cotyledons 2, secund, succulent, in the 3rd–4th leaf stage shed; simple, without stipules, petiolate. The whole body ellipsoid, to 3.5 by 2.2 by 2 cm, in general the irregularly splitting inner testa persistent. Blade half ellipsoid, reddish cream, turning green where exposed, below concave, above flat, upper sides facing. Petioles flattened, their base connate, clasping the stem, the part extended from the testa to 8 by 4 mm, pale green, glabrous.
Internodes: First one strongly elongating, in cross-section rhomboid, sturdy, to 10 cm long, in the upper part often with scattered, sharp prickles on a swollen base to 0.5 cm long; next ones small, gradually becoming longer, the lowest ones to 2 cm long, the 10th one sometimes to 20 cm long, as first one. Stem rather sturdy, slender, erect and slightly zig-zag when young, later climbing.

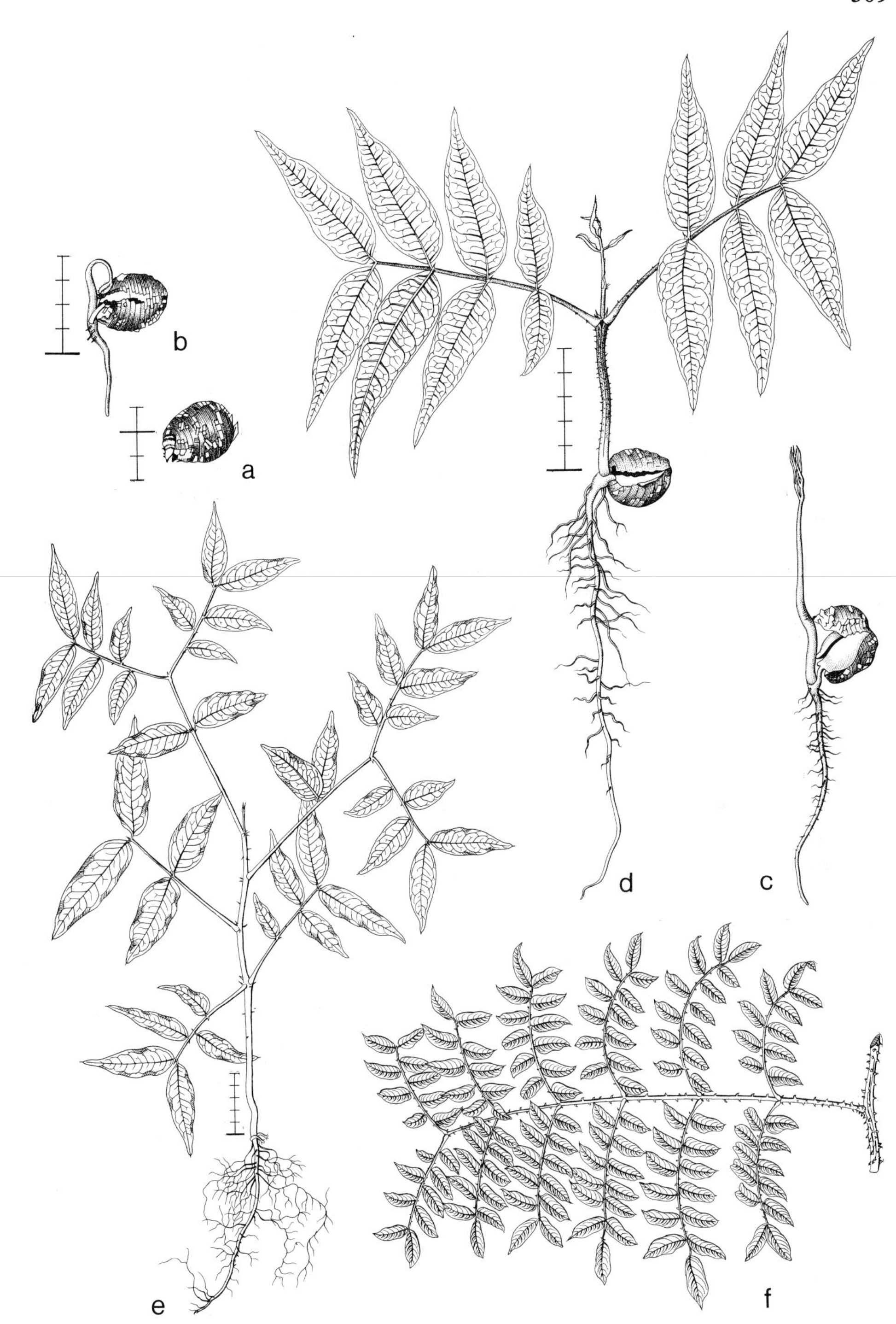

Fig. 107. *Caesalpinia bonduc*. De V. 1395, collected 23-vi-1972, planted 2-vii-1972. a, b, c, and d date?; e. 4-viii-1973; f. leaf of a sapling ± 2 years old.

First two leaves opposite, compound, paripinnate, 3–4-jugate with endcusp, stipulate, stipellate, petiolate, petiolulate, herbaceous, green. Stipules two pairs of erect, subulate cusps, to 1.5 by 0.1 mm, green. Petiole to 3.5 by 0.2 cm, ± terete, slightly thickened at the base, slightly flattened above, and there with short, simple, reddish hairs, reddish green. Rhachis as the petiole, to 7 cm long, internodes quite regular, 2–2.5 cm long. Stipellae like the stipules. Terminal cusp subulate, 4 by 0.3 mm, further as the stipules. Blade ovate-lanceolate, to 18 by 17 cm. Petiolule terete, with a constriction at the base, to 2 by 1 mm, hairy above as the petiole. Leaflets ovate-oblong to ovate-lanceolate, to 8 by 2.5 cm, but often smaller; base obtuse; top acuminate to cuspidate, abruptly apiculate, cusp 2 by 0.2 mm; margin entire; base, margin, and main nerve above hairy as the petiole, glabrescent; nerves pinnate, midrib very prominent below, lateral nerves rather inconspicuous.
Next leaves all spirally arranged, third one 2–4-jugate, next ones double compound, 2-jugate, the lower jugae in the form of simple leaflets, the upper ones compound with 3 pairs of leaflets, ultimately higher ones turning into double compound leaves with 6 pairs of jugae first order, each with 5–7 subopposite to alternate pairs of leaflets, further as the first leaves.

Specimens: 1395 from W. Java, alluvial flat, primary forest on coral sand, low altitude.
Growth details: In nursery germination good, not simultaneous, over a long stretch of time. Seedlings showing vigorous growth.
Remarks: C. coriaria and *C. pulcherrima* have the thick, fleshy, exposed cotyledons above the soil on a long hypocotyl. *C. tinctoria* is similar but the cotyledons are subfleshy. *C. bonduc*, *C. sappan*, and *C. sepiaria* have the enclosed cotyledons at soil level. All species have the first two leaves opposite.

CYNOMETRA (Caesalp.)

Fl. Java 1: 525; Mensbruge, Germination (1966) 145 (*C. ananta*, *C. megalophylla*); Voorhoeve, Liberian high forest trees (1965) 167 (*C. ananta*, *C. leonensis*).
Revision: Knaap-Van Meeuwen, Blumea 18 (1970) 1.
Genus pantropic; tropical Africa, India and Ceylon, throughout S.E. Asia, all over Malesia, N.E. Australia, in the Pacific as far as Fiji, including the Marianas, the Carolines, and New Caledonia, tropical N., Central, and S. America.

Cynometra ramiflora L. – Fig. 108.

Development: The taproot emerges from below at the margin of the fruit. The persistent, thick fruit wall ruptures along the margin of the valves. The cotyledons spread slightly, thus enabling the shoot to emerge from between them. No resting stage occurs.
Seedling hypogeal, semi-cryptocotylar. Cynometra ramiflora type.
Taproot very long, slender, rather sturdy, flexuous, dirty orange-brown, with numerous long, slender, wiry, short-branched, dirty orange-brown sideroots.

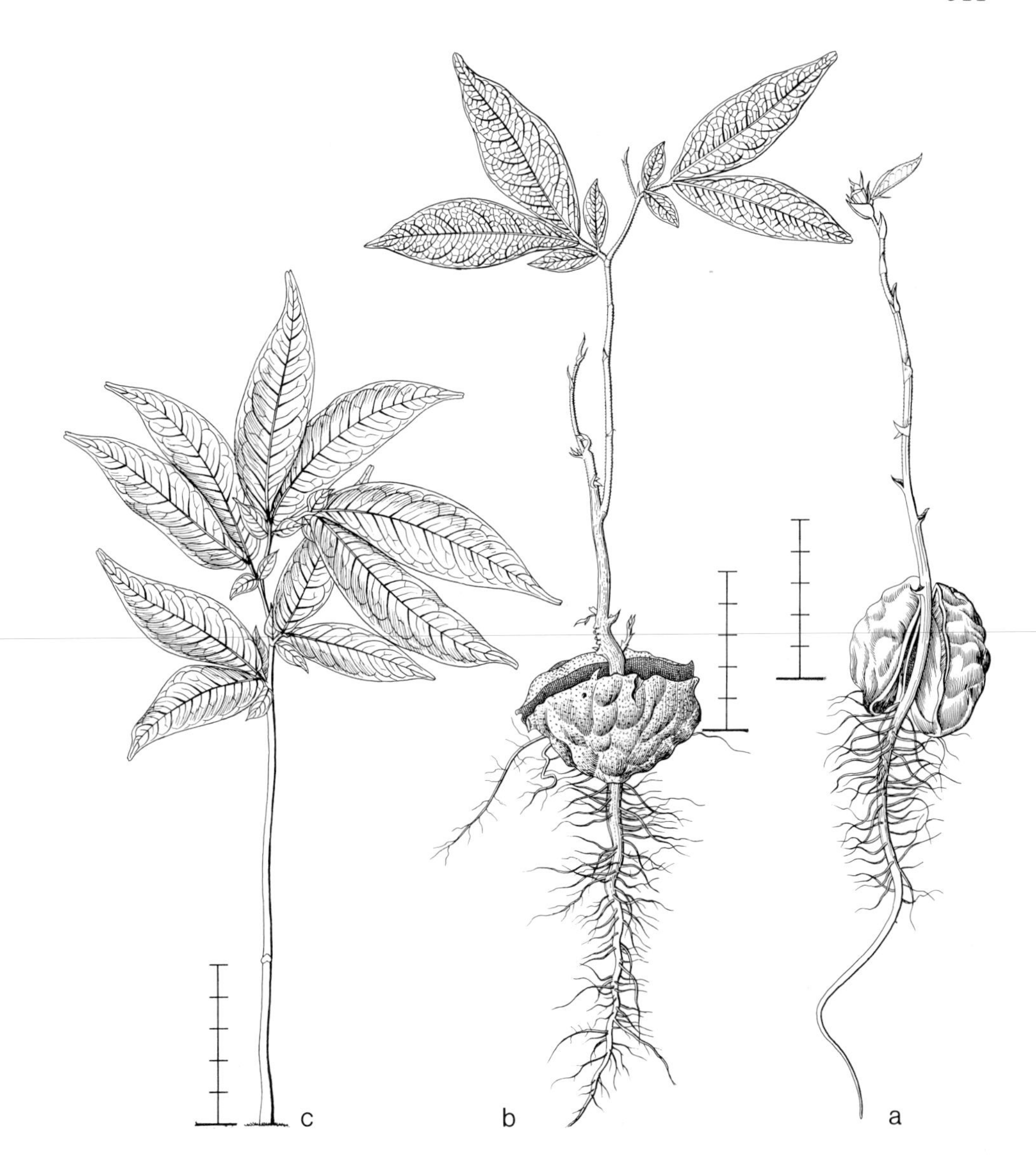

Fig. 108. *Cynometra ramiflora.* Bogor Bot. Gard., without number, collected and planted xi-1973. a. 17-xii-1973; b. 11-ii-1974; c. 11-ix-1974.

Hypocotyl not enlarging.
Cotyledons 2, opposite, succulent, not exactly known when dropped, between 2nd and 8th leaf stage; simple, sessile, slightly spreading, without stipules, hard fleshy, covered by the persistent, thick, woody, brown fruit wall. Blade broader than long, together with the fruit wall to 5.5 by 4 by 1.8 cm, below concave, inside ± flat.
Internodes: First one somewhat elongating, terete, to 1.5 cm long, green, densely covered with small, brown, simple hairs; next internodes as the first one, those in the scaly part not exceeding 2 cm long, the lower ones in the leafy part to 2.5 cm long, usually smaller. Stem rather slender, slightly zig-zag, usually dying off, then from the base with lateral branches, taking over the function of main shoot.
Leaves all spirally arranged, higher ones compound, stipulate, shortly petiolate, subcoriaceous, pale yellow with red margin when young, turning green, often the developed ones intercalated with more or less reduced ones. Lowest up to 10 much reduced, scale-like, triangular, often long persistent; each lateral shoot showing the same sequence of reduced scaly lower leaves and higher developed ones. Stipules 2 at the base of each developed leaf, shrivelling and dropping soon, linear, in 7th leaf to 6 by 1 mm, papyraceous, 1-nerved, nerve slightly prominent. Petiole terete, in first leaf to 3 by 1 mm, in higher ones not much larger. Rhachis as the petiole, in lower ± 5 leaves not exceeding 1 cm, slightly swollen at base and top. Blade bijugate without terminal cusp, ± obtriangular in outline, in first one to ± 7 by 7 cm. Leaflets shortly petiolulate, conduplicate-induplicate when very young. Petiolules terete, in first leaf to 1 by 0.5 mm. Basal leaflets much reduced in size, oblong, in the first leaf to 2.2 by 0.9 cm. Higher leaflets fully developed, lanceolate, in first leaf to 7 by 2.1 cm, in 7th one to 10.2 by 3.3 cm; base distinctly asymmetric; top acuminate; margin entire; glabrous; nerves pinnate, midrib above and below prominent, lateral nerves hardly so, curved at their tip and forming an inconspicuous marginal nerve.

Specimens: Bogor Botanic Gardens without number, locality not known.
Growth details: In nursery germination poor, ± simultaneous. In almost all specimens the main shoot died off, and was replaced by one or more lateral shoots from the base taking over the function of main shoot.
Remarks: C. ananta and *C. leonensis* have the free fleshy cotyledons on a long epigeal hypocotyl, the first two leaves are opposite, with several pairs of leaflets. In *C. megalophylla* and *C. ramiflora* the seedling is hypogeal, the opposite cotyledons at the outer side covered by the thick fruit wall. *C. megalophylla* has the first two leaves opposite, with 3 pairs of opposite leaflets. *C. ramiflora* has all leaves spirally arranged, the lowest up to 10 much reduced and scale-like.

DESMODIUM s.l. (Papil.)

Fl. Java 1: 602; A. P. de Candolle, Mém. Légum. (1825) pl. 13, f. 63 (*D. sp.*); Compton, J. Linn. Soc. Bot. 41 (1912) 46 (*D. canadense*, *D. gangeticum*, *D. sp.*); Lubbock, On Seedlings 1 (1892) 436 (*D. canadense*); Ohashi, Ginkgoana 1 (1973) 26 (*D. amethystinum*, *D. caudatum*, *D. elegans*, *D. heterocarpon*, *D. laxum*, *D. oldhamii*, *D. podocarpum*, *D. triflorum*).

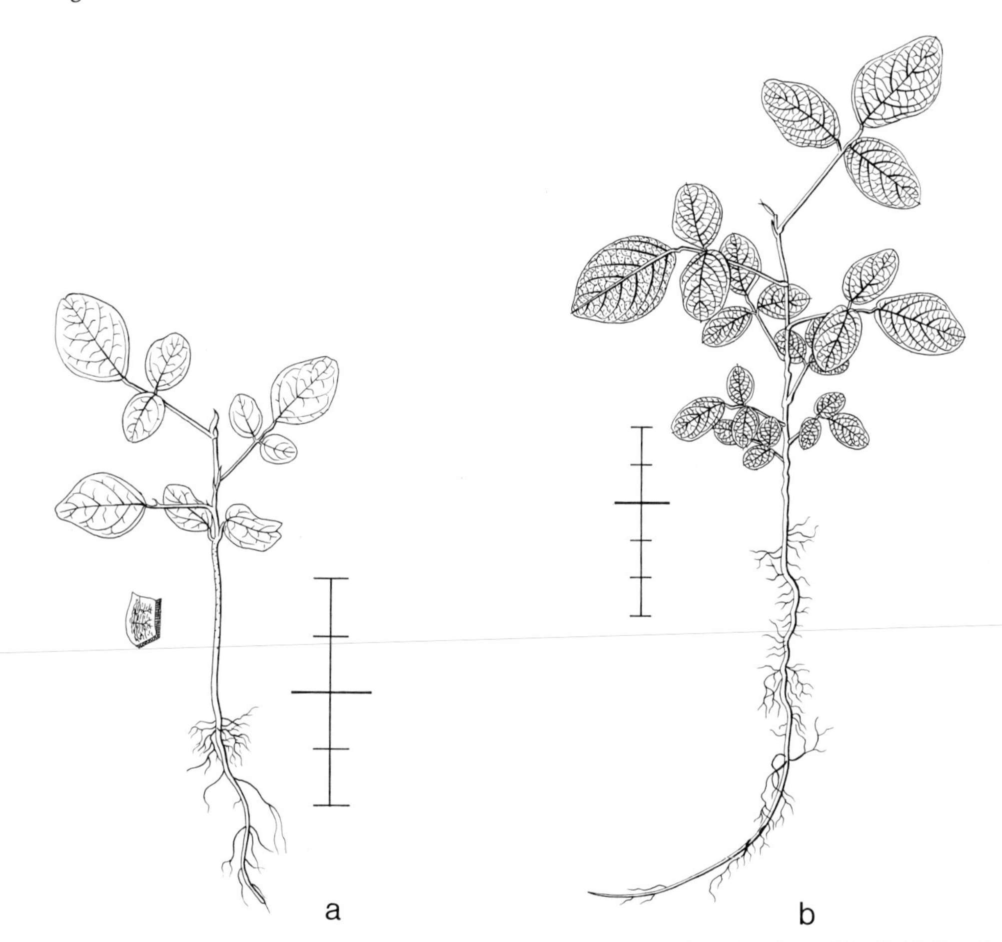

Fig. 109. *Desmodium umbellatum.* De V. 1381, collected 21-vi-1972, planted 2-vii-1972. a and b date?.

Revision: Ohashi, Ginkgoana 1 (1973) 1.
Genus pantropic; in tropical Africa, India, Ceylon, throughout S.E. and E. Asia including Korea and Japan, all over Malesia, N. and E. Australia, in the Pacific as far as Tonga, including New Caledonia, the Marianas, and the Carolines, also in tropical N., Central, and S. America, including Cocos and the Galapagos.

Desmodium umbellatum (L.) DC. (= *Dendrolobium* sensu Ohashi) – Fig. 109.

Development: The taproot and hypocotyl emerge from the segment of the pod. The hypocotyl becomes erect, and by spreading of the paracotyledons the latter are withdrawn from fruit wall and testa. A short resting stage occurs with the paracotyledons exposed.
Seedling epigeal, phanerocotylar. Macaranga type.
Taproot long, slender, flexuous, brownish, with rather few, little branched, short sideroots.

Hypocotyl strongly elongating, terete, to 3 cm long, green, slightly hairy with short, appressed, simple, hyaline hairs.
Paracotyledons 2, opposite, foliaceous, when shed not known, before the 10th leaf stage; simple, without stipules, petiolate, herbaceous, green, glabrous. Petiole in cross-section semi-orbicular, to 0.3 cm long. Blade ± ovate, often irregular in shape, to 1.2 by 0.7 cm; base and top obtuse; margin entire; nerves pinnate, lateral nerves ending in anastomoses.
Internodes: First one terete, to 0.6 cm long, hairy as the hypocotyl; next internodes as the first one, gradually becoming larger, 15th one to 2.3 cm long, 30th one to 8.5 cm long. Stem rather straight, slightly drooping.
Leaves all spirally arranged. First 1–3 leaves simple or with abortive lateral leaflet(s), stipulate, petiolate, herbaceous, green. Stipules 2 at the base of each leaf, narrowly triangular, to 3 by 1 mm, dropping rather soon. Petiole in cross-section semi-orbicular, to 1.2 cm long, hairy as the hypocotyl. Blade ± ovate to elliptic, to 1.6 by 1.2 cm; base rounded; top obtuse, slightly retuse, distinctly apiculate; margin entire; below with scattered, minute, simple, hyaline, appressed hairs; nerves pinnate, midrib above slightly sunken, below very prominent, lateral nerves not raised, below slightly prominent, ending free. Next leaves trifoliolate, gradually becoming larger. Stipules as those of lower leaves, gradually increasing in length, in the higher leaves fused at the base, in 20th leaf to 7 mm long, dropped rather soon after full development of the leaf. Petiole as those of first leaves, gradually increasing in length, in 15th leaf to 4 by 0.1 cm, in 30th leaf to 5.7 cm long, at the base distinctly thickened, hairy as the hypocotyl. Rhachis as the petiole, in 15th leaf to 11 by 1 mm, in 30th leaf to 1.8 cm long. Petiolule minute, terete, in 15th leaf to 3 mm, in 30th leaf to 4 mm, at the base distinctly jointed, hairy as the petiole. Blade ± triangular in outline. Blade of lateral leaflets smaller than the top one, slightly asymmetric, ± ovate, in 15th leaf to 4.5 by 3 cm, in 30th leaf to 7.5 by 4.5 cm; base obtuse; top obtuse, slightly apiculate. Top leaflet symmetric, as lateral leaflets, in 15th leaf to 6.5 by 4 cm, in 30th leaf to 10.5 by 6.5 cm, further as the first leaves.

Specimens: 1381 from W. Java, beach forest, primary forest on coral sand, low altitude.
Growth details: In nursery germination poor.
Remarks: According to Ohashi, in subgenus *Podocarpium* (*Desmodium* s.s.) the seedlings are hypogeal, with relatively fleshy cotyledons, the first two leaves simple and opposite, subsequent leaves spirally arranged, and trifoliolate (*D. laxum*, *D. oldhamii*, *D. podocarpum*). All other subgenera of *Desmodium* s.s. show a distinct epigeal hypocotyl, the cotyledons are somewhat fleshy, the first two leaves simple and opposite. Subsequent leaves are usually spirally arranged and trifoliolate (*D. amethystinum*, *D. caudatum*), but sometimes the second pair also opposite and simple (*D. heterocarpon*, *D. triflorum*) or opposite and trifoliolate (*D. elegans*). *D. umbellatum* (= *Dendrolobium* sensu Ohashi) differs in the leaf-like and thin paracotyledons, and all leaves are spirally arranged.

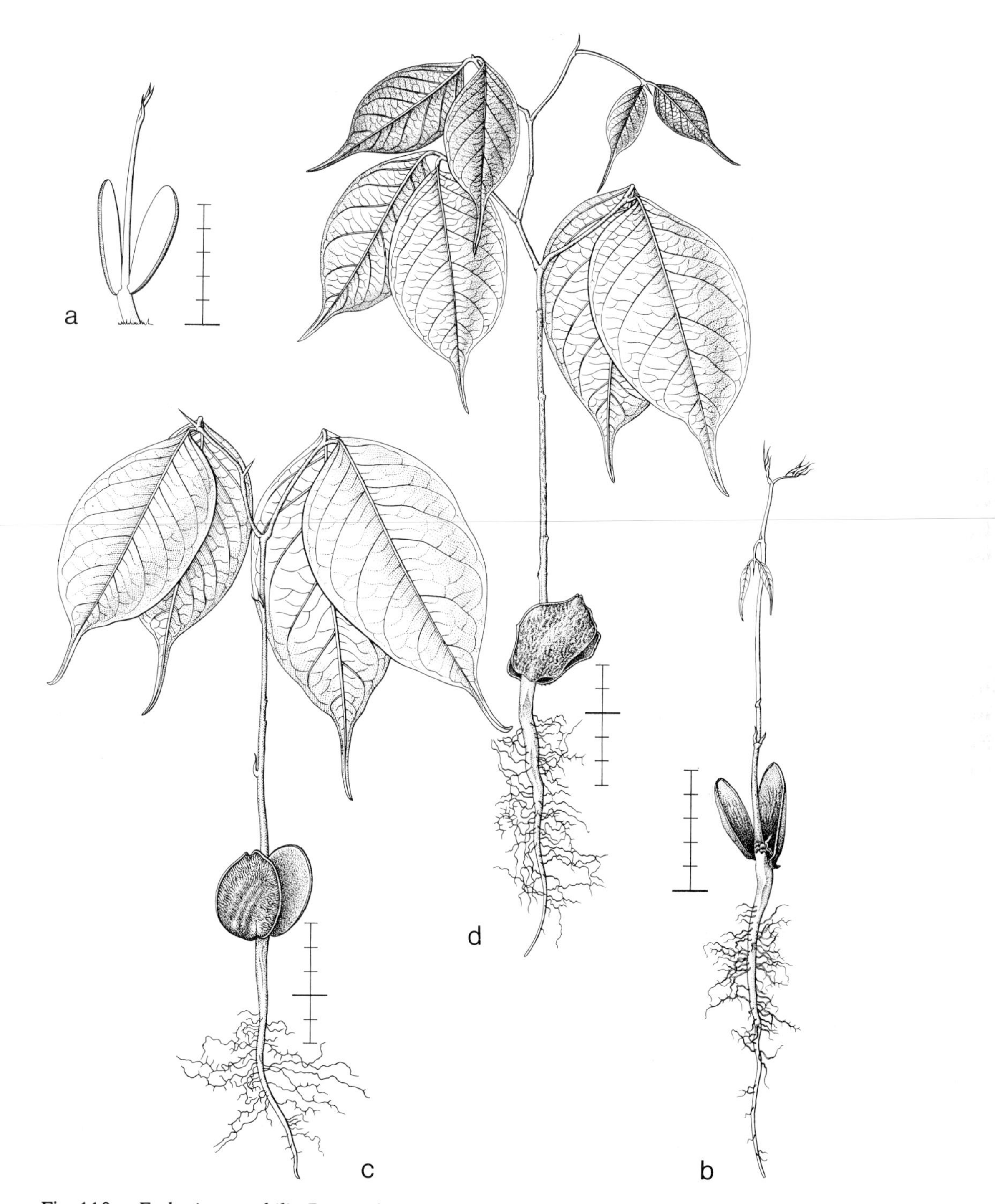

Fig. 110a. *Endertia spectabilis*. De V. 1018, collected 23-ix-1971, planted 7-xii-1971. a, b, c, and d date?.

ENDERTIA (Caesalp.)

Revision: De Wit, Bull. Bot. Gard. Btzg III, 17 (1947) 323.
Genus endemic in Borneo.

Endertia spectabilis De Wit – Fig. 110 a & b.

Development: The seed splits regularly along the margin, the taproot and hypocotyl emerge from one pole. The cotyledons are raised above soil level by the hypocotyl which becomes erect, and spread slightly, the testa persistent on the lower surface. A resting stage occurs with 2(–3) developed leaves.
Seedling epigeal, semi-phanerocotylar. Endertia type and subtype.
Taproot long, sturdy, slender, dark brown, especially in the upper part with rather long, slender, brown, rather much-branched sideroots.
Hypocotyl terete, but slightly angular by four longitudinal ridges, to 3.5 by 0.7 cm, cream-coloured, turning brown, glabrous.
Cotyledons 2, opposite, succulent, in the 2nd(–3rd) leaf stage dropped; simple, without stipules, sessile, at the outside covered by the persistent, brownish testa, slightly spreading, hard fleshy, inside light green to dark reddish, glabrous. Blade ± oblique ovate, to 4 by 3.3 cm, above flat, below slightly concave.
Internodes: First one strongly elongating, terete, to 6 cm long, but usually much smaller, pale cream-coloured to red when young, turning dark purplish red, slightly hairy with minute, simple, hyaline hairs, glabrescent; next internodes as the first one, the lower ones to 5 cm long, the higher ones to 8 cm long. Stem slender, distinctly zig-zag, later drooping.
Leaves all spirally arranged, lower ones scale-like, developed ones all compound, stipulate, petiolate, subcoriaceous, drooping and red when young, turning green. Reduced leaves 3–6, resembling the stipules of the higher leaves, long persistent, clasping the stem, sessile, slightly succulent, triangular, to 5 by 2.5 cm; top split or entire, rarely split to the base; margin entire; creamy-white to dark red, the upper ones often at the base with a linear, much reduced, soon abortive leaf-like structure. Stipules resembling the cataphylls. Petiole terete, to 4.5 by 1.5 mm, at the base distinctly swollen for ± 7 mm of its length, slightly hairy as the first internode. Rhachis in higher leaves as the petiole. Terminal cusp often soon withering, subulate, to 10 by 0.8 mm, its tip acute, above channelled. Blade paripinnate, in general up to the 4th one 1-jugate, (4–)5–7(–14)th ones 2-jugate, occasionally from 7th, usually from 8th to 10th one onwards 3-jugate. Leaflets opposite, petiolulate. Petiolules terete, to 7 by 1.5 mm, at the base distinctly jointed. Blade of leaflets elliptic, slightly asymmetric, ± oblong, to 17 by 8 cm, in the lower leaves much smaller; base at the adaxial side acute, with a depression, at the abaxial side more rounded; top caudate; margin entire; slightly hairy as the first internode, glabrescent; nerves pinnate, midrib above slightly prominent, below much so, lateral nerves below slightly prominent, free ending.

Specimens: 1018 from S.E. Borneo, alluvial flat, primary forest on deep clay, low altitude.

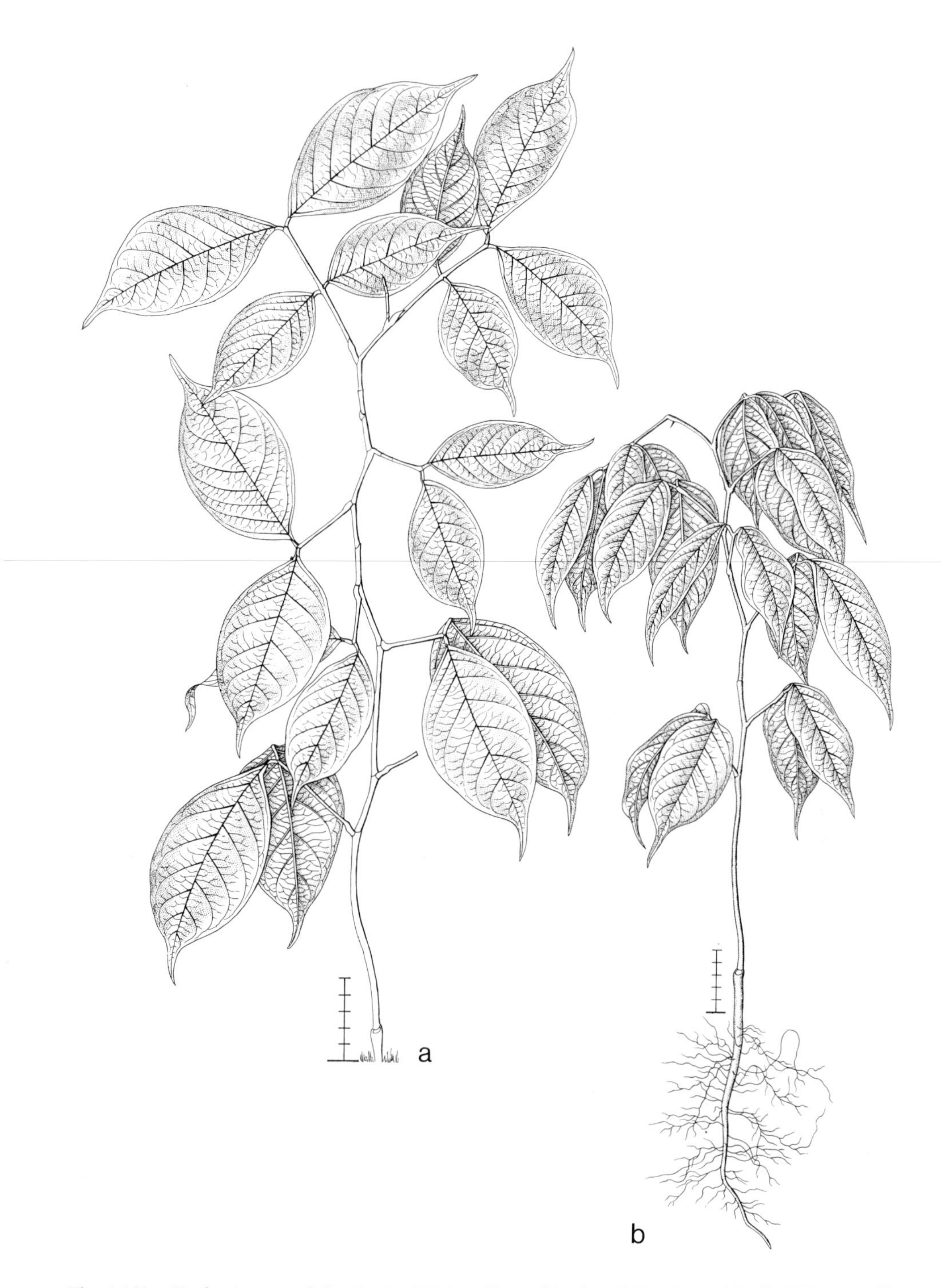

Fig. 110b. *Endertia spectabilis.* De V. 1018, collected 23-ix-1971, planted 7-xii-1971. a and b date?.

Growth details: In nursery germination good, ± simultaneous.
Remarks: Plants growing vigorous, in 8 months up to 85 cm high, with up to 16 leaves.

KOOMPASSIA (Caesalp.)

Tree Fl. Malaya 1: 202; Meijer, Bot. Bull. Sandakan 11 (1968) 112 (*K. malaccensis*).
Revision: De Wit, Bull. Bot. Gard. Btzg III, 17 (1947) 309.
Genus in Malesia: Malay Peninsula, Sumatra, Borneo, New Guinea.

Koompassia excelsa (Becc.) Taub. – Fig. 111.

Development: The taproot and hypocotyl emerge laterally from the fruit. After establishment the hypocotyl becomes erect and the cotyledons spread, thus shedding the fruit wall. Sometimes this does not succeed, and they remain secund and covered by the dry, winged fruit wall, the shoot emerging laterally. A resting stage occurs with the cotyledons and two developed leaves.
Seedling epigeal, phanerocotylar. Sloanea type and subtype.
Taproot long, slender, fibrous, creamy-brown coloured, with many quite long, few- and shortly branched, creamy-white sideroots in 4 distinct longitudinal rows.
Hypocotyl strongly elongating, terete, to 7.5 cm long, green, soon turning brown, glabrous.
Cotyledons 2, opposite, somewhat succulent, in the 2nd–3rd leaf stage dropped; simple, sessile, stipulate, widely spreading, hard fleshy, green, glabrous. Stipules 2, interpetiolarly connate, distinctly curved, sessile, narrowly triangular, to 3 by 1.5 mm; base with a minute, subulate cusp to 0.5 mm long on each side; top acute; margin entire; below in the centre at the base densely hairy with small, simple, brown hairs; nerve one, sunken above, quite distinct below. Blade oblong, to 3.5 by 1.2 cm; base auriculate, auricles turned down, those of respective leaves at both sides of the stem appressed against each other; top rounded; margin entire; nerve one, depressed above, below inconspicuous.
Internodes: First one strongly elongating, terete, slender, to 7 cm long, green, rather densely covered with minute, simple, dark brown hairs; next ones as the first one, lower ones to 3 cm long, but generally not exceeding 2 cm, 20th one to 4 cm long. Stem slender, straight.
First two leaves opposite, compound, stipulate, petiolate, herbaceous, green. Stipules long persistent, sessile, either two at the base of each leaf, linear, to 3 by 0.8 mm, or interpetiolarly connate, triangular, to 3 by 1.5 mm; top acute; margin entire; nerve one; hairy as the first internode. Petiole to 1.5 cm long, ± terete, channelled above, distinctly thickened at the base, hairy as the first internode. Rhachis as the petiole, to 5 cm long. Blade imparipinnate, ± elliptic in outline, to 9 by 6 cm. Leaflets 8–11, opposite except for the terminal one, sometimes not opposite, conduplicate induplicate when young, shortly petiolulate. Petiolule terete, to 1 by 0.5 mm, hairy as the first internode. Blade of leaflet ovate-lanceolate, to 3.5 by 1.1 cm, base acute to obtuse; top cuspidate, apiculate with a 1 mm long cusp; margin entire; below along

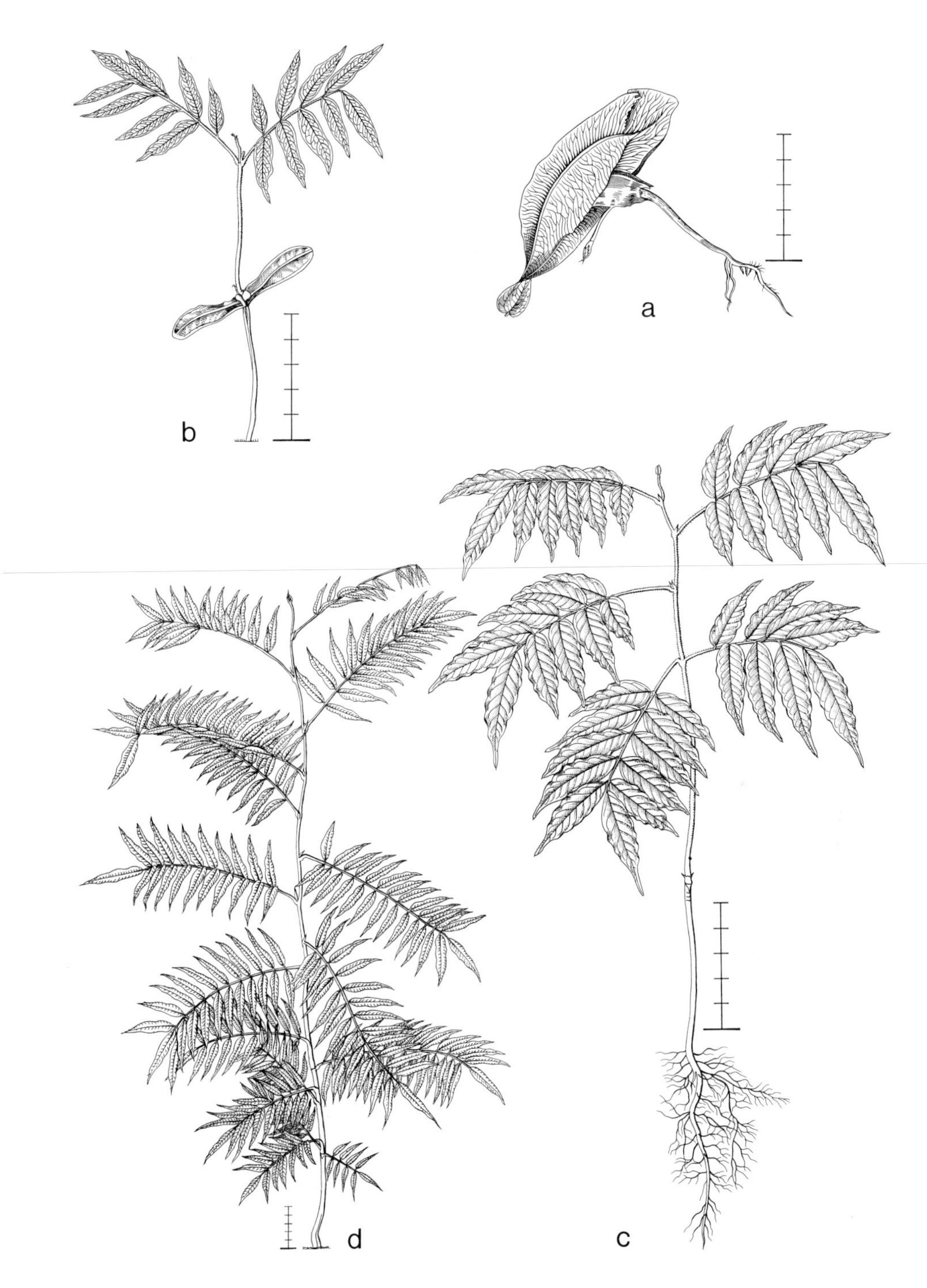

Fig. 111. *Koompassia excelsa.* De V. 2257, collected 9-iii-1973, planted 22-iii-1973. a and b. 2-v-1973; c. 14-vii-1973; d. 16-vii-1974.

the midrib and along the margin hairy like the first internode; nerves pinnate, midrib sunken above, prominent below, lateral nerves hardly so, rather inconspicuous, joined to an inconspicuous marginal nerve.
Next leaves spirally arranged. Stipules long persistent, 2 at the base of each leaf, ovate-lanceolate, to 5 by 1.5 mm, clasping the swollen base of the petiole. Petiole, rhachis, blade, petiolule, and leaflets as those of first leaf, becoming slightly larger in the following leaves. Lateral leaflets subopposite, in the fourth leaf up to 4 pairs, in 15th leaf 8 pairs, in 25th one up to 12 pairs. Blade of leaflets lanceolate, in 25th leaf to 5.5 by 1.5 cm, further as those of first leaves.

Specimens: 2257 from S.E. Borneo, low hills, primary forest on deep clay, low altitude.
Growth details: In nursery germination fair, ± simultaneous.
Remarks: Two species described, similar in germination pattern and general morphology. *K. malaccensis* differs in details only.

ORMOSIA (Papil.)

Tree Fl. Malaya 1: 299; Duke, Ann. Miss. Bot. Gard. 52 (1965) 336 (*O. krugii*); ibid. 56 (1969) 149, 150 (*O. sp.*).
Genus in S.E. and tropical E. Asia, all over Malesia, in the Pacific in Bismarck and the Solomons, Central and tropical S. America.

Ormosia sp. – Fig. 112.

Development: The taproot emerges from one pole of the seed, and establishes. The seedcoat splits and is retardedly shed by the swelling cotyledons which become entirely exposed and turn green. The first internode and two opposite leaves develop, after which a long resting stage occurs. Development slow.
Seedling ± hypogeal, phanerocotylar. Sloanea type, Palaquium subtype.
Taproot slender, fibrous, brown, with rather few, little branched, brown sideroots.
Hypocotyl not enlarging.
Cotyledons 2, ± turned lateral, succulent, in the 3rd–4th leaf stage shed; simple, without stipules, sessile, only slightly spreading, hard fleshy, pale green, glabrous. Blade half ellipsoid, to 1.6 by 1.4 by 0.4 cm, outside concave, inside flat.
Internodes: First one strongly elongating, terete, to 8 cm long, towards the base slightly thickened, green, slightly hairy with minute, stiff, appressed, hyaline hairs, glabrescent; next ones as first one, irregular in length, to 1.5 cm long. Stem rather slender, slightly zig-zag.
First two leaves opposite, compound, without stipules, petiolate, subcoriaceous, green. Petiole terete, to 1.8 by 0.1 cm, abruptly thickened at the base, hairy as the first internode. Rhachis as the petiole, to 7 by 1 mm. Blade imparipinnate, triangular in outline, trifoliolate, terminated by the top leaflet. Leaflets petiolulate, without stipellae, the lowest two opposite. Petiolules swollen, terete, to 1.5 by 1 mm, hairy as the first internode. Blade of leaflets ovate-oblong, the terminal one to 5.5 by 2 cm,

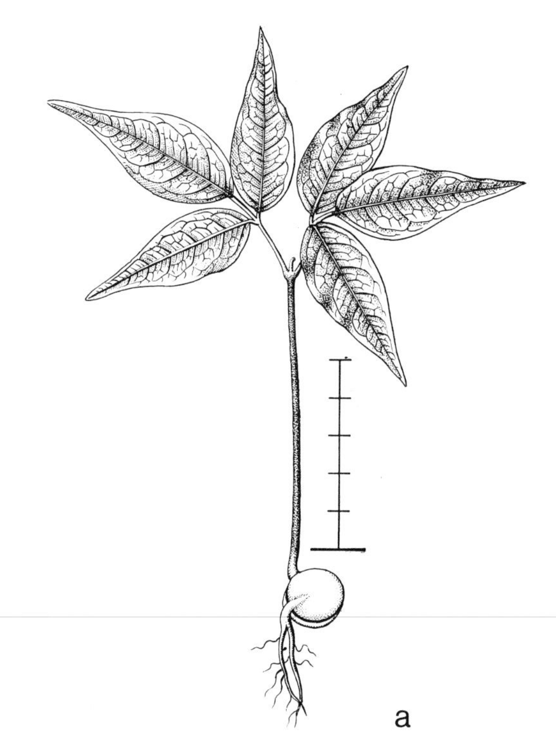

Fig. 112. *Ormosia sp.* Dransfield 2389, collected 30-i-1972, planted 5-ii-1972. a date?.

the lateral ones smaller; base obtuse; top cuspidate; margin entire; almost glabrous; nerves pinnate, midrib above slightly sunken, below prominent, lateral nerves inconspicuous.
Next leaves all spirally arranged, as the first two leaves, 5th one with petiole to 27 mm long, rhachis to 12 mm long, terminal leaflet to 7.5 by 2.5 cm, lateral ones to 6 by 2.5 cm.

Specimens: Dransfield 2389 from Malay Peninsula, medium altitude.
Growth details: In nursery germination good, ± simultaneous.
Remarks: O. krugii is probably of the Endertia type/subtype, but Duke pictures an abnormal specimen in which the axillary buds of the cotyledons have developed. All leaves on the main shoot are spirally arranged. *O. sp* (Duke 1969) has fleshy, non-photosynthetic cotyledons with the testa often adhering to their lower surface, a long first internode with two simple opposite first leaves. *O. sp.* (Dransfield 2389) has cotyledons that shed the testa and become exposed at soil level. The first internode elongates considerably, and bears two opposite, trifoliolate first leaves.

PARENTEROLOBIUM (Mimos.)

Monograph: Kostermans, Org. Sc. Res. Ind. 20 (1954) 1.
Genus endemic in Borneo.

Parenterolobium rosulatum (Kosterm.) Kosterm. – Fig. 113.

Development: The testa splits along the margin, the taproot and hypocotyl emerge from the slit. After establishment of the root the hypocotyl becomes erect and the cotyledons are withdrawn from the testa. A short resting stage occurs with 2 developed leaves.
Seedling epigeal, phanerocotylar. Sloanea type and subtype.
Taproot quite short, stunted, whitish, with many long, few-branched, white sideroots.
Hypocotyl elongating, slightly angular, to 4 cm long, pale green, in its upper part densely hairy with minute, simple, hyaline hairs.
Cotyledons 2, opposite, succulent, in the 2nd leaf stage dropped, simple, without stipules, ± sessile, fleshy, pale creamy-green; blade ovate, to 2.7 by 1.7 cm; base auriculate, auricles laterally rounded, with a rounded to acute tip, to 1 by 1 cm, usually one distinctly smaller; top rounded, with more or less irregular outgrowths; margin entire; outside concave, often with a longitudinal groove, inside flat.
Internodes: First one elongating, in cross-section ± quadrangular, to 1.3 cm long, pale green, hairy as the hypocotyl; next internodes as the first one, not seen when fully developed.
First two leaves opposite, compound, without stipules, petiolate, herbaceous, green. Petiole terete, to 5 by 1.5 mm, hairy as the hypocotyl. Blade 2-foliolate. Terminal cusp subulate, to 1 by 0.3 mm. Leaflets petiolulate. Petiolule terete, to 2 by 1.5 mm, at the base distinctly articulate. Blade of leaflets ovate-oblong, to 4.5 by 2 cm, slightly asymmetric; base at one side acute, at the other side more rounded; top cuspidate to caudate; margin irregularly undulating in the plane of the leaf, entire; on the nerves hairy as the hypocotyl; nerves pinnate, above slightly raised, below very prominent, lateral nerves ending in anastomoses.

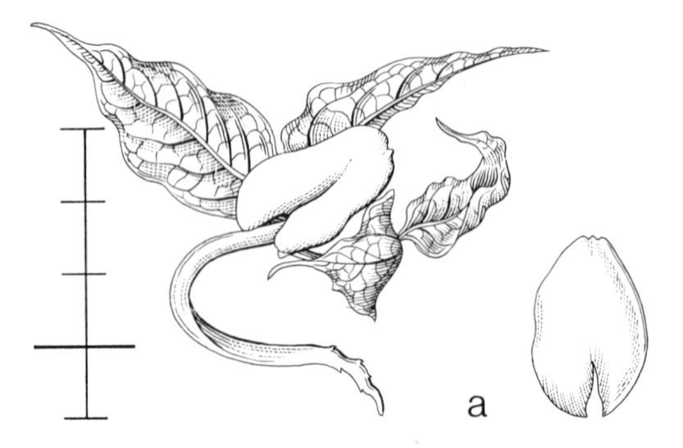

Fig. 113. *Parenterolobium rosulatum.* De V. 2305, collected 21-iii-1973, planted 22-iii-1973. a date?.

Next leaves spirally arranged, stipulate. Stipules 2 at the base of the petiole, subulate, to 1 by 0.2 mm, hairy as first internode. Developed leaves not seen.

Specimens: 2305 from S.E. Borneo, primary hill forest on deep clay, low altitude.
Growth details: In nursery germination poor.

PARKIA (Mimos.)

Fl. Java 1: 564; Burger, Seedlings (1972) 193 (*P. roxburghii*); Duke, Ann. Miss. Bot. Gard. 52 (1965) 335 (*P. biglandulosa*); Mensbruge, Germination (1966) 132 (*P. bicolor*); Voorhoeve, Liberian high forest trees (1965) 232 (*P. bicolor*).
Genus pantropic; Central Africa, S.E. Asia including E. India, W. Malesia including Celebes and the Philippines, W. New Guinea, Palau Islands, Ponape, and Fiji Islands, tropical S. America.

Parkia speciosa Hassk. – Fig. 114.

Development: The hypocotyl and taproot emerge from one pole of the seed. The hypocotyl becomes erect and the cotyledons spread, by which the testa splits and usually drops, enabling the plumule to emerge from between them. A resting stage occurs with one developed leaf.
Seedling epigeal, phanerocotylar, with strong 'Parkia smell'. Sloanea type and subtype.
Taproot long, sturdy, fleshy, cream-coloured, with many scattered, long, unbranched, cream-coloured sideroots.
Hypocotyl strongly enlarging, terete, to 5 cm long, green to brownish green, with close-set, minute, simple, hyaline hairs.
Cotyledons 2, opposite, succulent, in the first leaf stage dropped; sessile, not spreading, simple, ? without stipules, thick fleshy, green, glabrous. Blade elliptic, to 25 by 0.8 cm; base irregularly rounded; top rounded; margin entire; nerves not visible.
Internodes: First one strongly enlarging, terete, to 5 cm long, slightly tapering to the top, green to brownish green, hairy as the hypocotyl; next internodes as the first one, in the lower ones to 2–2.5 cm long, very slowly the higher ones increasing in length. Stem zig-zag, slender.
Leaves all spirally arranged, double compound, paripinnate with terminal cusp, spreading, stipulate, petiolate, herbaceous, green. Stipules two at the base of each leaf, minute, narrowly triangular, to 2 mm long. Petiole terete, in the first leaf to 0.5 cm, in second leaf to 2.5 cm, in 15th leaf to 6.5 cm long, when old slightly woody, in higher leaves at about $\beta\delta$ from the base above with an ovate, warty, shining gland, green to brownish green, hairy as the hypocotyl. Rhachis first order terete, in first leaf to 2 cm, in second leaf 6 cm, in 15th one to 30 cm, as the petiole. Pinnae first order sessile, opposite, sometimes alternate, lanceolate, in first leaf 4 pairs, in second one 4–5 pairs, in 15th leaf to 17 pairs, those in lowest leaves to 4 by 1.1 cm, in 15th one to 9.5 by 1.7 cm, without terminal leaflet. Rhachis 2nd order as rhachis first order, in lower leaves to 4 cm long, in 15th leaf to 8.5 cm long, those in the first leaf

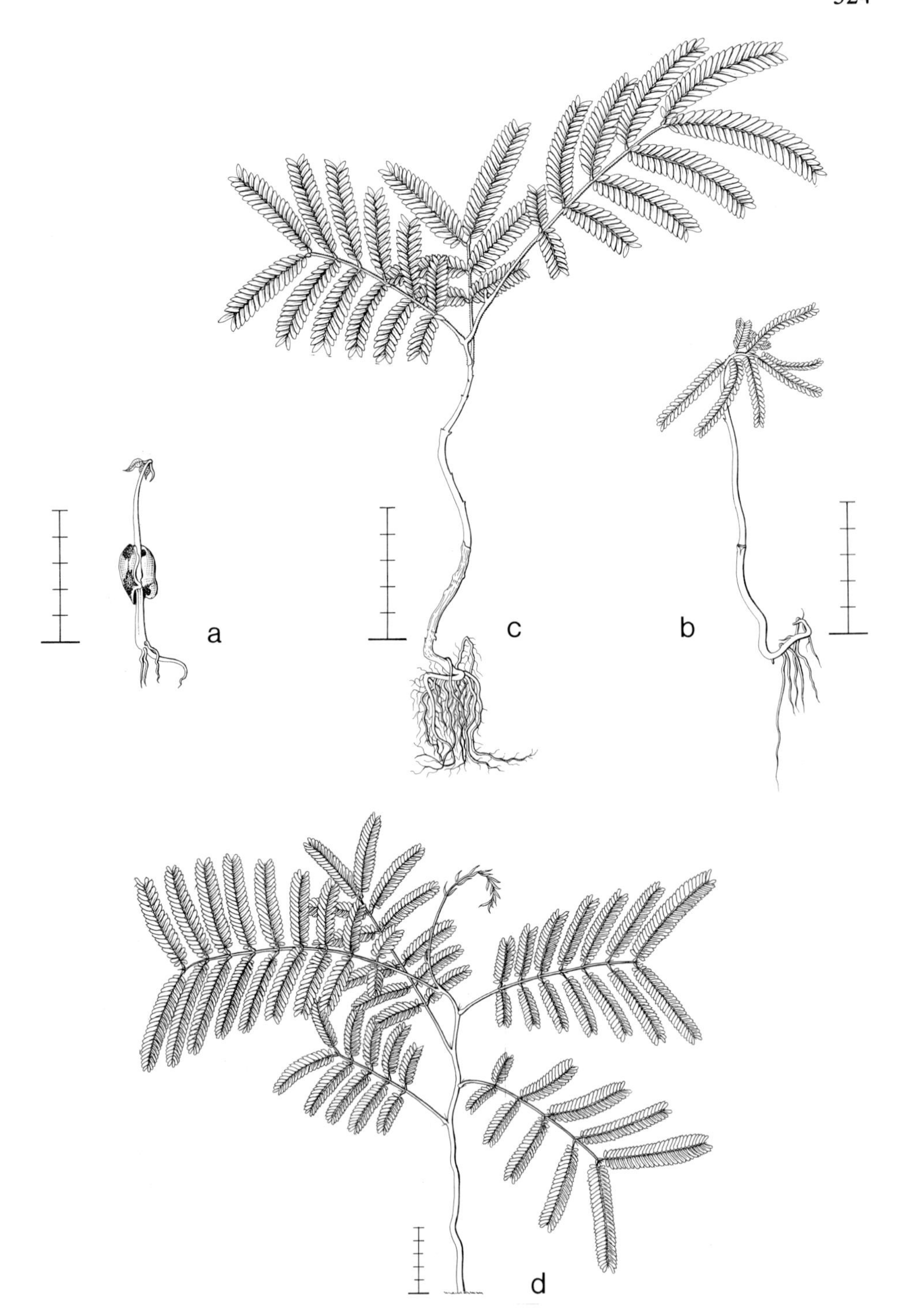

Fig. 114. *Parkia speciosa.* De V. 2088, collected 6-ix-1972, planted 17-ix-1972. a and b date?; c. 29-iii-1973; d. 18-viii-1973.

rather close together, in higher ones more distant and at the base swollen. Terminal cusp warty. Leaflets sessile, oblong, to 6 by 2 mm, in first leaf to 20 pairs/pinna, in second one to 18 pairs, in 15th leaf to 32 pairs/pinna, in each leaf the lowest pinnae with less pairs; base oblique, with an abaxial tooth; in first leaf top distinctly apiculate, in 2nd and higher leaves more rounded; margin entire; above dark green, below paler so, in 2nd leaf top slightly hairy; 1-nerved, with additional abaxial nerve, midrib above raised, below less so, lateral nerves rather inconspicuous.

Specimens: 2088 from S.E. Borneo, alluvial flat, primary forest on deep clay, low altitude.
Growth details: In nursery germination poor.
Remarks: Four species described, similar in germination pattern and general morphology. They differ in details only.

PHANERA (Caesalp.)

Fl. Java 1: 533.
Revision: De Wit, Reinwardtia 3 (1956) 381.
Genus in Africa, India and Ceylon, throughout S.E. and tropical E. Asia, all over Malesia, N.E. Australia.

Phanera semibifida (Roxb.) Benth. – Fig. 115.

Development: The taproot and hypocotyl emerge from one pole of the seed. After establishment of the root, the cotyledons are withdrawn from the testa by spreading, and carried above the soil by the hypocotyl which becomes erect. A short resting stage occurs with the cotyledons and two opposite leaves.
Seedling epigeal, phanerocotylar. Sloanea type and subtype.
Taproot long, slender, fibrous, brownish yellow, sideroots rather many, quite short, unbranched, later much-branched, yellow.
Hypocotyl strongly enlarging, terete, to 3 cm long, green, turning reddish brown, glabrous.
Cotyledons 2, opposite, slightly turned to one side, in the 2nd–3rd leaf stage dropped; slightly succulent, sessile, without stipules, hard fleshy, green, glabrous. Blade asymmetric, above concave, below convex, to 1.8 by 1.4 cm; base unequal, auriculate, the larger auricle to 2 by 1.5 mm, the smaller one 1 by 1 mm; top broadly rounded; margin entire; nerves pinnate, inconspicuous.
Internodes: First one strongly elongating, terete, to 3 cm long, green, with rather close-set, more or less appressed, simple, reddish hairs; next internodes as the first one, to 1 cm long, the higher ones gradually larger, 20th one to 4 cm long. Stem slender, erect, slightly zig-zag, later the higher part drooping.
First two leaves opposite, simple, conduplicate-induplicate when young, stipulate, petiolate, herbaceous, green. Stipules 2 between the petioles, interpetiolary connate, sessile, ± ovate, to 3 by 2.5 mm; top more or less distinctly bifid, its tips rounded; margin entire; hairy as the first internode. Petiole terete, to 2 by 0.1 cm, base and top

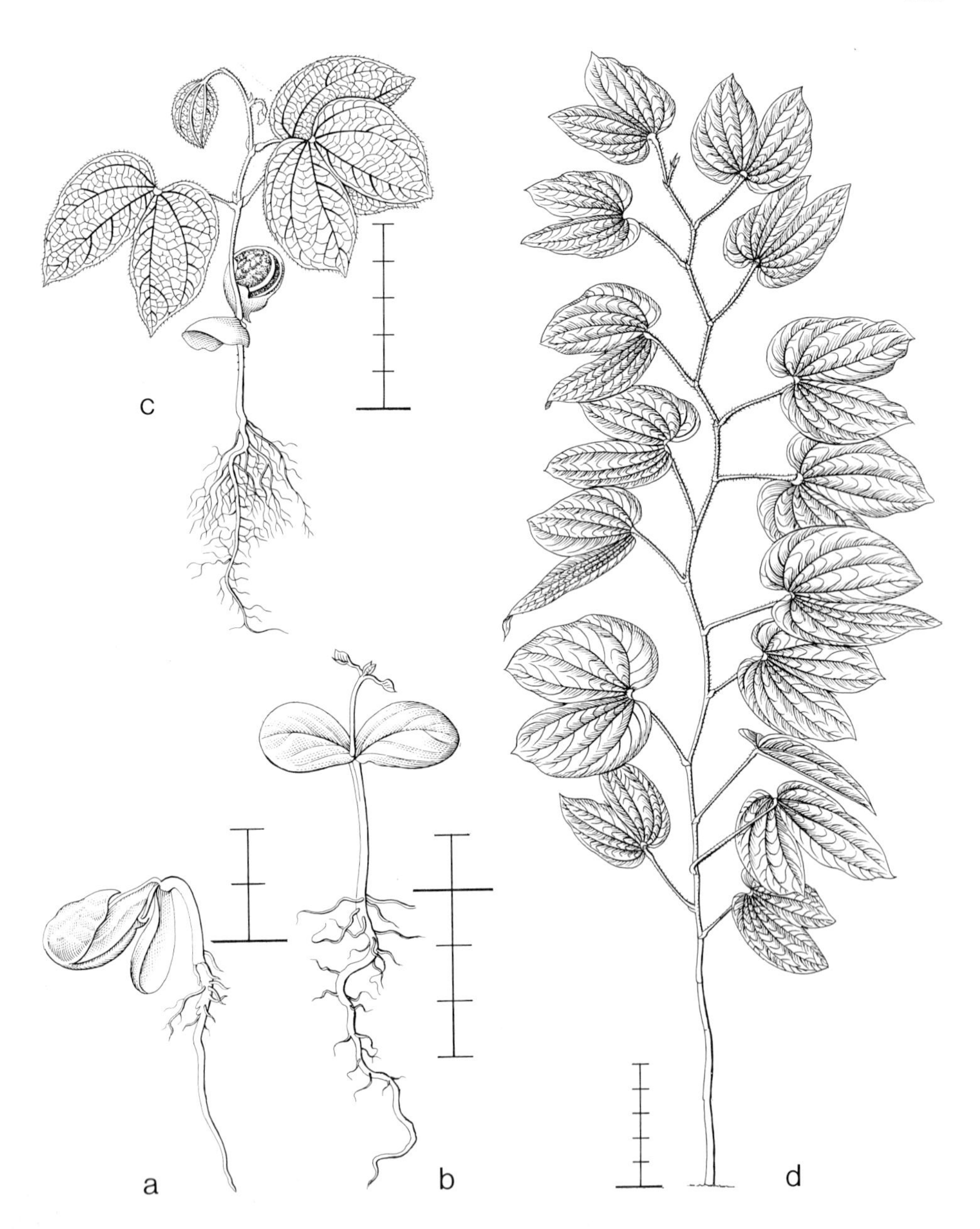

Fig. 115. *Phanera semibifida.* De V. 2190, collected 6-iii-1973, planted 22-iii-1973. a and b. 5-iv-1973; c. 27-v-1973; d. 16-vii-1974.

distinctly thickened, hairy as the first internode. Blade ± obrhomboid in outline, to 5 by 5 cm; base auriculate, auricles broadly rounded, to 0.5 by 1.5 cm; top deeply bifid, in the incision with a subulate soft cusp to 2 by 0.2 mm, each lobe 2.5 by 2.5 cm, its tip (acute to) acuminate; margin entire, above glabrous, below when young white by dense cover of white, hyaline, simple hairs which soon turn red, when older glabrescent; nerves palmate, midrib not branched, above sunken, other nerves above not prominent, all nerves below prominent, lateral nerves ending free in the margin. *Next leaves* as first ones, spirally arranged. Stipules 2 at the base of each leaf, sessile, ± ovate, in 20th leaf to 5 by 4 mm; top rounded; margin entire; further as first ones. Petiole terete, in 5th one to 25 by 0.5 mm, in 20th one to 17 by 1.5 mm, base and top distinctly swollen, hairy as the first internode. Blade ± obrhomboid in outline, in 5th leaf to 4.5 by 4.2 cm, in 20th one to 11.5 by 12 cm; auricles in 5th leaf to 2 by 10 mm, in 20th one to 1.2 by 3.5 cm; terminal cusp in the incision in 20th leaf to 4 by 0.3 mm; lobes in 5th leaf 2.3 by 2 cm, in 20th leaf to 6 by 6 cm, further as the first leaves.

Specimens: 2190 from S.E. Borneo, alluvial flat, primary forest on deep clay, low altitude.
Growth details: In nursery germination good, ± simultaneous.

SARACA (Caesalp.)

Fl. Java 1: 577; Compton, J. Linn. Soc. Bot. 41 (1912) 11 (*S. cauliflora*).
Revision: Zuijderhoudt, Blumea 15 (1967) 413.
Genus in India and Ceylon, throughout S.E. Asia, in Malesia eastwards to Celebes.

Saraca declinata (Jack) Miq. – Fig. 116.

Development: The taproot and hypocotyl emerge from one pole of the seed. The persistent testa splits along the cotyledonary margins, and the cotyledons spread. No resting stage occurs.
Seedling epigeal, semi-cryptocotylar. Endertia type and subtype.
Taproot long, sturdy, hard, orange-brown, with many scattered, long, branched orange-brown sideroots.
Hypocotyl slightly enlarging, terete, to 1.3 cm long, glabrous.
Cotyledons 2, opposite, succulent, in the 3rd leaf stage dropped; simple, ± equal, without stipules, sessile, hard fleshy, the exposed upper side green, below covered by the persistent brown testa, glabrous. Blade peltate, lanceolate, to 6 by 1.8 by 0.3 cm, distinctly asymmetric; base and top rounded; margin entire; nerves not visible.
Internodes: First one slightly elongating, terete, to 2 mm long, green, often reddish, glabrous; next internodes as the first one, irregular in length, to 2 cm long. Stem rather slender, zig-zag in the leafy part.
Leaves all spirally arranged, compound, stipulate, petiolate, subcoriaceous, above dark green, below paler, glabrous. The first ± 5 scale-like, sessile, triangular, to 3 by 3 mm, top rounded. Next leaves fully developed, paripinnate, in outline 14 by 12 cm. Stipules intrapetiolar, resembling the scales, sessile, triangular, to 5 by 4 mm; top

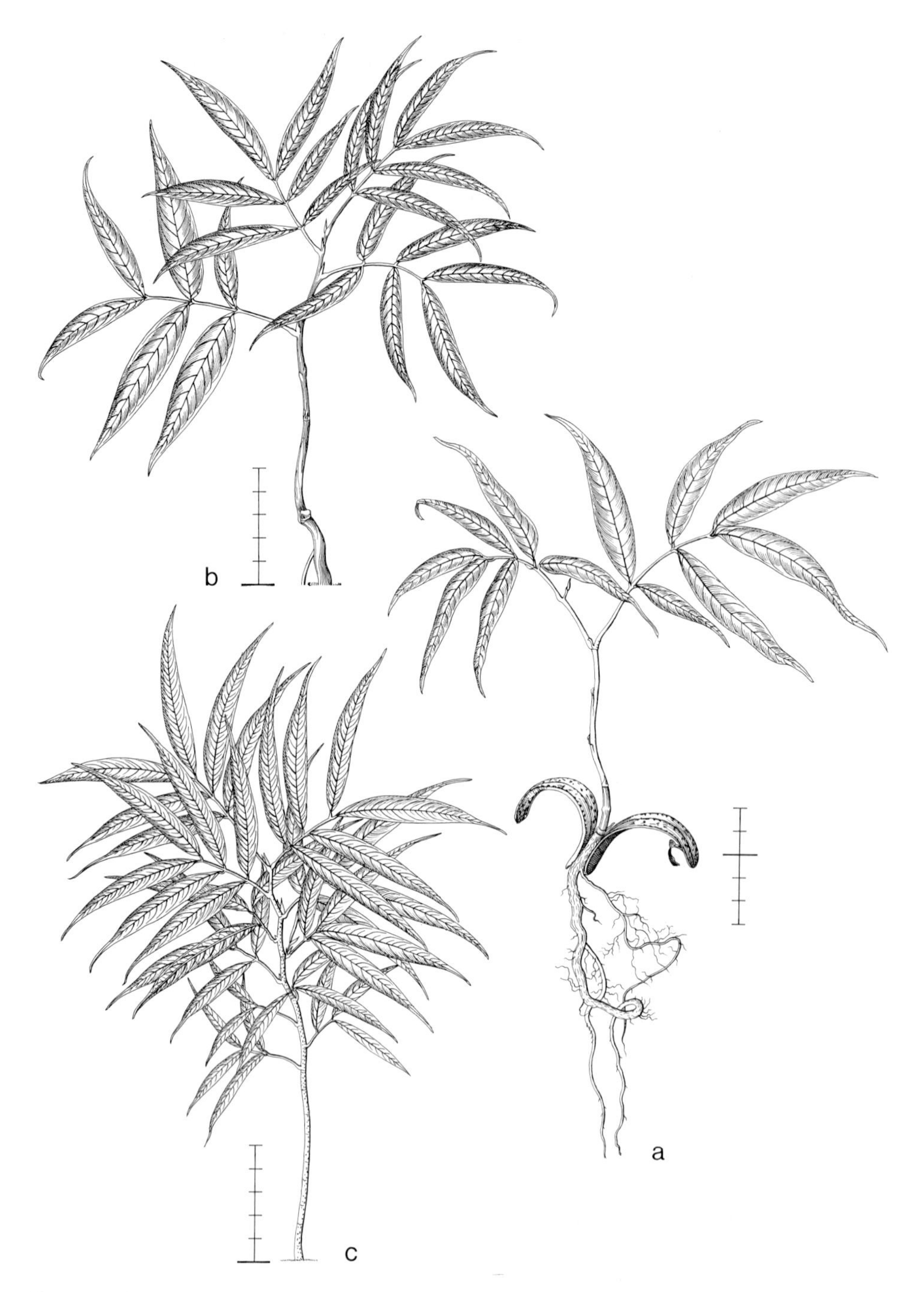

Fig. 116. *Saraca declinata.* De V. 2335, collected 14-iii-1973, planted 22-iii-1973. a. 14-iv-1973; b. 24-vii-1973; c. 5-viii-1974.

rounded to acute; margin entire. Petiole in cross-section semi-orbicular, to 30 by 1.5 mm, above channelled, at the base distinctly swollen. Rhachis as the petiole, to 45 by 1.5 mm. Leaflets in 3 opposite pairs, simple, without stipellae, shortly petiolate. Petiolule in cross-section ± semi-orbicular, to 4 by 1.5 mm, above channelled, the blade descending on the upper part. Blade of leaflets linear, in first leaf to 9.5 by 1.5 cm, in 15th leaf to 16 by 2 cm, in the lower leaves the lower leaflets smaller; base acute, slightly asymmetric; top caudate, its tip acute to rounded; margin entire; nerves pinnate, main nerve above prominent, below less so, lateral nerves hardly raised, ending free.

Specimens: 2335 from S.E. Borneo, alluvial flat, germinating seeds on sand banks in small streams, low altitude.
Growth details: In nursery germination good.
Remarks: S. cauliflora has a one-sided collar at the base of the hypocotyl. It is not stated in the short seedling description whether the testa is persistent, and the germination pattern is not described.

SOPHORA (Papil.)

Fl. Java 1: 577; A. P. de Candolle, Mém. Légum. 2 (1825) 75, 83 (*S. japonica*); Compton, J. Linn. Soc. Bot. 4 (1912) 22 (*S. chrysophylla, S. tetraptera*); Csapody, Keimlingsbestimmungsbuch der Dikotyledonen (1968) 101 (*S. japonica*); Hickel, Graines et Plantules (1914) 277 (*S. japonica, S. moorcroftiana, S. tetraptera*); Lubbock, On Seedlings 1 (1892) 452 (*S. secundiflora*).
Genus pantropic; tropical Africa, throughout the islands of the Indian Ocean, India and Ceylon, throughout S.E. and E. Asia, all over Malesia, N. and E. Australia, in the Pacific as far as Easter Isl., including the Marianas, Hawaii, New Zealand, and the Chathams, tropical Central and S. America including Revilla Gigedos and Juan Fernandez, in the Atlantic in Gough Island.

Sophora tomentosa L. – Fig. 117.

Development: The taproot emerges from the side of the seed. The cotyledonary petioles enlarge slightly, and by secondary growth of the cotyledonary node the bases of the cotyledons are pushed aside. During this process the testa is often irregularly ruptured. The shoot emerges from the same pole as the taproot. No resting stage occurs.
Seedling hypogeal, cryptocotylar. Horsfieldia type and subtype.
Taproot long, slender, flexuous, creamy-white, with few long, hardly branched, cream-coloured sideroots, rather soon with developing nodules.
Hypocotyl not enlarging.
Cotyledons 2, secund, succulent, in the 6th–8th leaf stage shed, ± sessile, without stipules, partly remaining hidden in the irregularly splitting, leathery, brownish seedcoat. Whole body ± ellipsoid, to 10 by 8 by 8 mm, creamy-white, green where exposed.

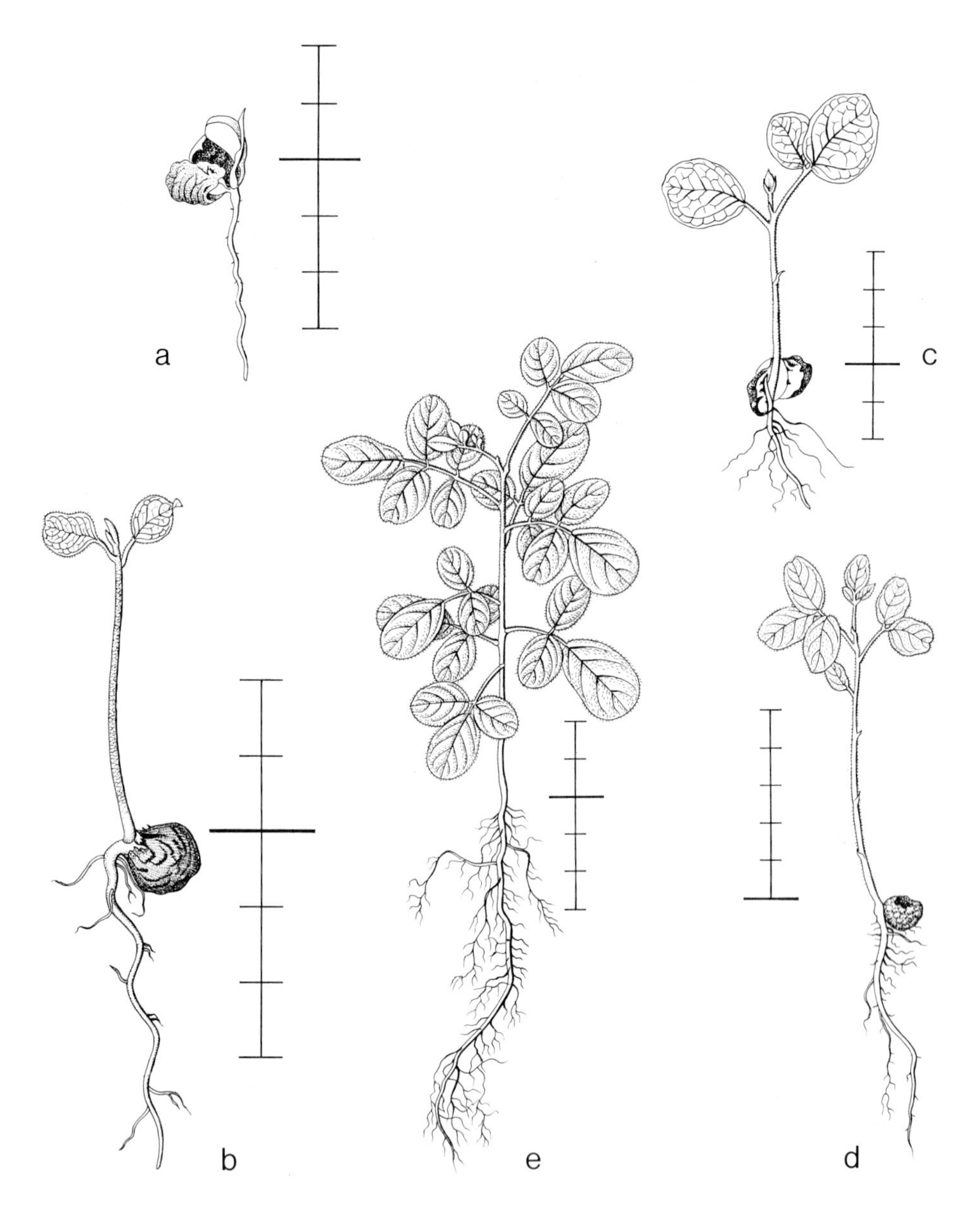

Fig. 117. *Sophora tomentosa.* De V. 1380, collected 21-vi-1972, planted 2-vii-1972. a, b, c, d, and e date?

Internodes: First one elongating to 2 cm long, terete, green, rather densely hairy with curled, simple, hyaline hairs; next internodes as the first one, irregular in length, to 15 mm long, but in general not exceeding 8 mm. Stem slender, rather zig-zag.
Leaves all spirally arranged, the lower ones unifoliolate, the higher ones compound, imparipinnate, without stipules, petiolate, subcoriaceous, green, paler below, hairy as the first internode. First 1–2 ones scale-like, ± triangular, sessile, to 1 by 0.7 mm, somewhat succulent. Next 1–3 leaves unifoliolate. Petiole in cross-section semi-orbicular, above channelled, to 8 by 0.8 mm, at the base distinctly thickened. Blade obovate, to 1.5 by 1.2 cm; base and top obtuse; margin entire, on both sides rather densely hairy as the first internode, slowly glabrescent; nerves pinnate, midrib above slightly sunken, below slightly prominent, lateral nerves less so, ending in anastomoses. Subsequent leaves compound, the lower 3–5 and occasionally some higher ones 3-foliolate (sometimes 2-foliolate), from ± 25th one onwards 7-foliolate. Petiole to 1.5 by 0.8 mm, as those of the unifoliolate leaf. Leaflets in opposite pairs, minutely petiolulate. Petiolule as the petiole, at the base distinctly jointed, ± terete, to 2 by 1 mm. Rhachis as the petiole, in 5-foliolate leaves to 12 mm long, in 7-foliolate leaves to 3 cm long. Blade of lateral leaflets elliptic to obovate, in 5-foliolate leaves to 18 by 18 mm, the lowest pair smaller, as the leaflet of the unifoliolate leaf. Terminal leaflet as lateral leaflet, in 5-foliolate leaves to 26 by 16 mm.

Specimens: 1380 from W. Java, beach forest, primary forest on coral sand, low altitude.
Growth details: In nursery germination good, but plantlets after 2 years not well developed, with most of the lower leaves dropped.
Remarks: In *S. chrysophylla*, *S. japonica*, and *S. tetraptera* the opposite, free cotyledons are borne above the soil on an epigeal hypocotyl. The first two leaves are compound and opposite (according to Hickel the leaves are all spirally arranged in *S. japonica*). *S. secundiflora*, *S. tetraptera*, and *S. tomentosa* have the cotyledons secund at soil level, clasped by the irregularly splitting testa; all leaves are spirally arranged, several lower ones unifoliolate, sometimes the lowest much reduced to abortive.

SPATHOLOBUS (Papil.)

Fl. Java 1: 630.
Genus throughout S.E. Asia, all over W. Malesia and in Celebes.

Spatholobus platypterus Merr. – Fig. 118.

Development: The winged fruit splits along the free margin, and the taproot and shoot emerge from the slit. The cotyledonary petioles elongate, carrying the plumule out of the envelopments. The shoot is then able to develop. A short resting stage occurs with two developed leaves.
Seedling hypogeal, cryptocotylar. Heliciopsis type and subtype.
Taproot long, sturdy, creamy-white, with many branched, creamy-white sideroots.

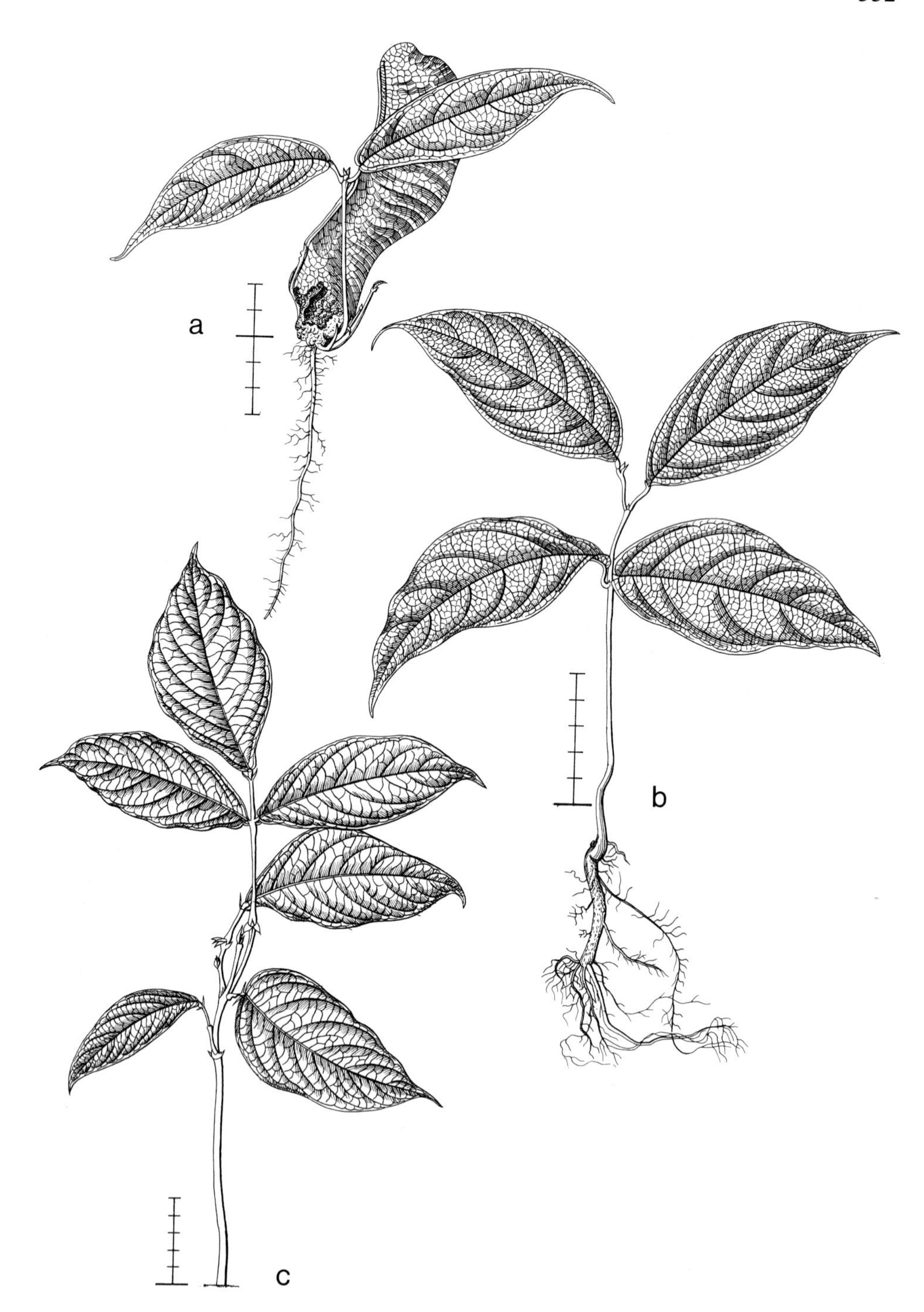

Fig. 118. *Spatholobus platypterus.* De V. 2393, collected 16-iii-1973, planted 22-iii-1973. a. 24-iv-1973; b. 26-vii-1973; c. 18-iv-1974.

Hypocotyl not enlarging.
Cotyledons 2, secund, succulent, in the 2nd–3rd leaf stage shed; sessile, without stipules, covered by the winged fruit wall. The part covering the cotyledons to 3.5 by 2 by 1.2 cm, dark brown. Wing flat, curved, at the back thickened, to 10 by 3.5 cm, dry coriaceous, brown.
Internodes: First one strongly elongating, terete, to 20 cm long, green, rather densely hairy with minute, simple, hyaline hairs; next internodes as the first one, to 1.7 cm long. Stem slender, ± straight, distinctly thickened on the nodes.
First two leaves opposite, simple, stipulate, shortly petiolate, herbaceous, green. Stipules 2, interpetiolarly connate, dropping soon, triangular, to 6 by 3 mm; top slightly narrowed; hairy as the first internode, inconspicuously 7-nerved. Petiole terete, to 10 by 2 mm, hairy as the first internode. Blade ovate-oblong, to 13 by 5 cm; base rounded; top cuspidate; margin entire; hairy when young, soon glabrescent but below on the main nerve remaining slightly hairy as on the first internode; nerves pinnate, above slightly raised, below prominent, lateral nerves free ending.
Next leaves spirally arranged, stipulate, stipellate, as the first ones. The first up to 10–12 simple. Stipules 2 at the base of each petiole, dropping soon, sessile, elliptic to triangular, in 10th leaf to 5 by 2.3 mm; top acute to acuminate; margin entire; hairy as the first internode; inconspicuously 7-nerved, in higher ones nerves increasing in number. Petiole terete, in the lower ones above channelled, in higher ones there flattened, at base and top distinctly thickened, hairy as the first internode. Stipellae 2 at the base of the apical petiolar thickening, subulate, in 10th leaf to 3 by 0.5 mm, hairy as the first internode. Blade as that of the first leaf, in 10th leaf to 14 by 6 cm, in the higher ones gradually larger, ultimately to 33 by 16 cm. Subsequent leaves compound, spirally arranged, as the first ones. Stipules 2 at the base of each leaf, sessile, narrowly triangular, in lowest ones to 8 by 3 mm. Petiole in lower compound leaves in cross-section semi-orbicular, above flattened, to 7.5 cm long, at the base distinctly swollen, as those of the simple leaves. Rhachis as the petiole, in the lower compound leaves to 3.5 cm long. Stipellae one at the base of each petiolule of the lateral leaves, and two at the base of that of the terminal one, linear, to 5 by 0.3 mm. Petiolule terete, swollen, to 4 by 3 mm. Blade of lateral leaflets oblong, to 18 by 7.5 cm, that of the terminal one oblong, to 19 by 9.5 cm; further as the blades of simple leaves.

Specimens: 2393 from S.E. Borneo, primary forest on deep clay, medium altitude.
Growth details: In the forest abundant regeneration. The shoot is liable to die, in that case a shoot develops from one of the axillary buds of the cotyledons. In the nursery the seedlings were for no apparent reason hampered in the development, and often died.

WHITFORDIODENDRON (Papil.)

Fl. Java 1: 615 (*Padbruggea*, *Millettia* p.p.); Mensbruge, Germination (1966) 176 (*Millettia lane-poolei*, *M. rhodantha*, *M. zechiana*).
Genus (incl. *Padbruggea*, *Adinobotrys*, *Millettia* p.p.) in S.E. Asia, and W. Malesia,

possibly including the Philippines.

Whitfordiodendron myrianthum (Dunn) Dunn – Fig. 119.

Development: The root emerges from near one pole of the seed which splits along the margin of the cotyledons. These spread slightly, enabling the shoot to emerge from between them. No resting stage occurs.
Seedling hypogeal, cryptocotylar. Endertia type, Chisocheton subtype.
Taproot slender, flexuous, creamy-white, with many long, slender, flexuous, much branched, creamy-white sideroots.
Hypocotyl not enlarging.
Cotyledons 2, opposite to more or less secund, succulent, long persistent, when shed not known; more or less spreading, sessile, without stipules, covered by the thin, pale brown, slightly disintegrating testa. Blade ± semi-ellipsoid, to 6 by 5 by 2.5 cm, below distinctly concave, irregularly sculptured by grooves, above smooth, flat, the flat sides facing, hard fleshy, the exposed parts turning green, glabrous, nerves not visible.
Internodes: First one terete, elongating, to 3 cm, dark green, minutely grooved and warty, with scattered, simple, hyaline, appressed hairs; next internodes as the first one, irregular in length, finally to 12 cm long. Stem in the lower part usually scaly, trailing over the soil, climbing when support is found, often dying, then from the base developing lateral, climbing shoots.
Leaves all spirally arranged, unifoliolate, later ones compound, stipulate, shortly petiolate, herbaceous, green. Lower 5–20 much reduced, resembling the stipules, scale-like, without stipules, narrowly triangular, to 5 by 2 mm; top often more or less deeply split; hairy as the first internode. Next leaves unifoliolate, the number not known. Stipules 2, situated below the articulation of the petiole of developed leaves, narrowly triangular, to 3 by 0.5 mm, hairy as the first internode. Petiole terete, in the lower leaves to 10 by 2 mm, without thickened base and top, in the higher leaves to 3 cm long by 1 mm thick, at base and top distinctly thickened to 2 mm, jointed below the basal thickening and below the apical thickening, slightly patently hairy as the first internode. Blade ovate-oblong, to 12 by 5 cm, rarely to 20 by 6.5 cm; base obtuse; top caudate; margin entire, especially below on the nerves slightly hairy as the petiole; nerves pinnate, midrib above slightly raised, below much prominent, lateral nerves above slightly sunken except for their base, below prominent, ending in anastomoses. Subsequent leaves trifoliolate, occasionally trifoliolate ones also intercalated between the simple lower ones, finally turning in 5-foliolate leaves. Petiole as those of the simple leaves, ultimately to 17 by 2 mm, thickened at the base only. Rhachis to 2 cm long, in 5-foliolate ones to 12 cm long, as the petiole, articulate at the top, there thickened. Petiolule terete, to 5 by 2 mm. Lateral leaflets to 11 by 5 cm. Terminal leaflet to 15 by 7 cm, further as in unifoliolate leaves.

Specimens: 2186 from S.E. Borneo, alluvial flat, primary forest on deep clay, low altitude.
Growth details: In nursery germination fair. Shoot often dying off, then replaced by a lateral shoot from the base.

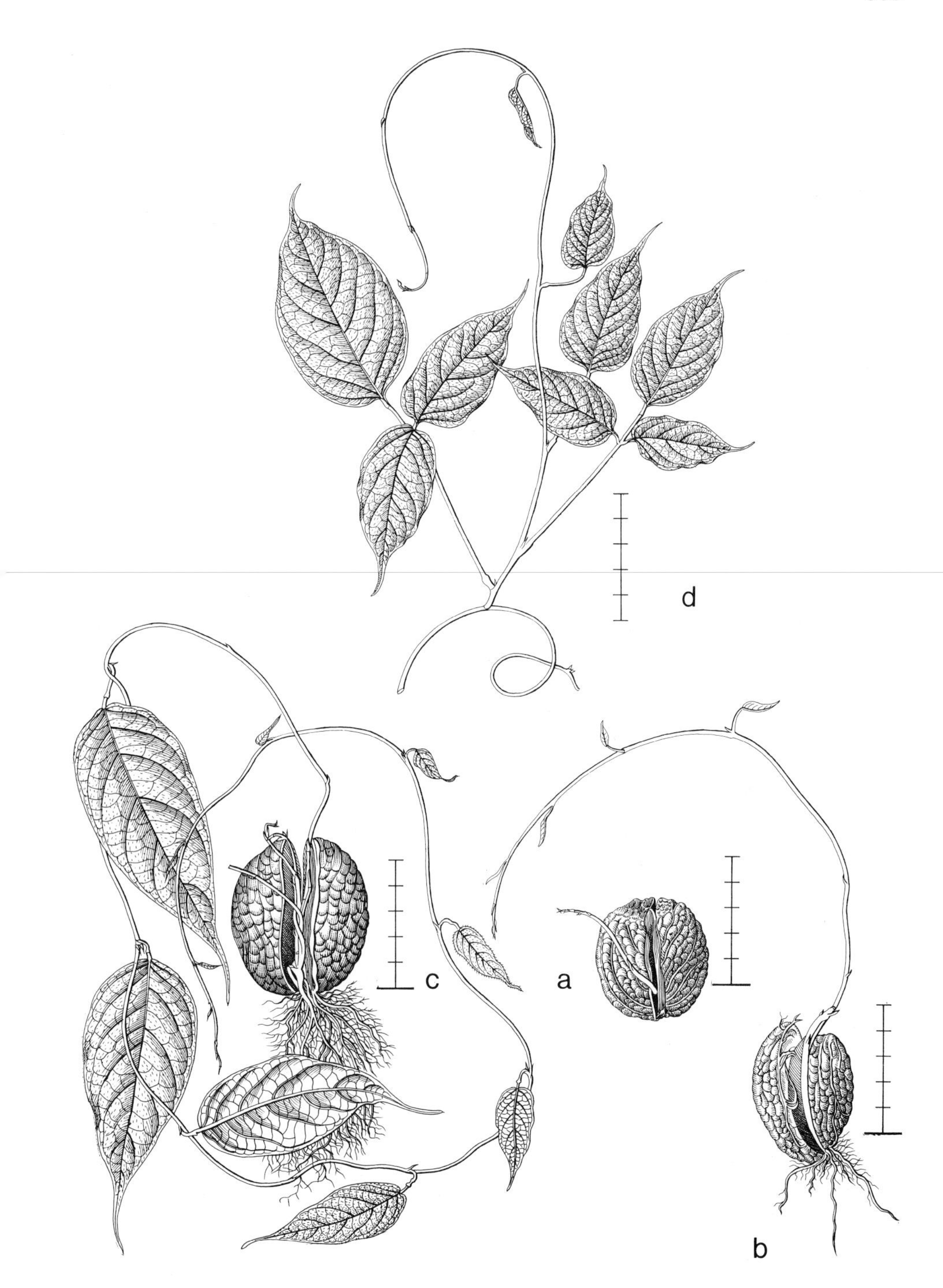

Fig. 119. *Whitfordiodendron myrianthum.* De V. 2186, collected 6-iii-1973, planted 22-iii-1973. a and b. 5-iv-1973; c. 26-iii-1974; d date?.

Remarks: Millettia lane-poolei and *M. rhodantha* have the cotyledons free, above the soil on a long erect hypocotyl, and the compound first leaves are opposite. *M. zechiana* is similar, but the opposite first leaves are simple. *Whitfordiodendron myrianthum* (syn. *Millettia nieuwenhuisii*) is entirely different in the secund cotyledons at soil level, and the climbing stem which is flexible from the beginning. According to Mr. Geesink (pers. comm.) the genus *Millettia* is heterogeneous, and some of the species should be transferred to *Whitfordiodendron*. Moreover, *Padbruggea* is probably congeneric with *Whitfordiodendron*. As most of these combinations have not yet been made, I prefer to maintain here the name *Whitfordiodendron*.

LOGANIACEAE

STRYCHNOS

Fl. Males. I, 6: 343; Cremers, Candollea 28 (1973) 269 (*S. congolana*); Leeuwenberg, Acta Bot. Neerl. 14 (1965) 227 (*S. ternata*); Troup, Silviculture 2 (1921) 674 (*S. nux-blanda*).
Genus pantropic; Africa, India and Ceylon, throughout S.E. and tropical E. Asia, all over Malesia, N. and tropical E. Australia, in the Pacific in Bismarck, the Solomons, Santa Cruz and Fiji, Central and tropical S. America.

Strychnos sp. – Fig. 120.

Development: The taproot and hypocotyl emerge from one pole of the seed. They carry the enclosed paracotyledons above the ground. The nodding upper part of the hypocotyl becomes erect, bringing the enclosed paracotyledons upright, after which the testa and endosperm are shed. A short resting stage occurs with enclosed, drooping paracotyledons, and a second one with exposed paracotyledons.
Seedling epigeal, phanerocotylar. Macaranga type.
Taproot long, flexuous, slender, creamy-white, with few short, shortly branched, creamy-white sideroots.
Hypocotyl strongly enlarging, terete, to 6 cm long, green, densely hairy with short, simple, brownish hyaline hairs.
Paracotyledons 2, opposite, foliaceous, in the 30th–50th leaf stage dropped; simple, without stipules, almost sessile, herbaceous, green. Petiole in cross-section semi-orbicular, to 1 by 1 mm, slightly hairy as the hypocotyl. Blade ovate, to 3.5 by 2.3 cm; base auriculate; auricles broadly rounded, to 2 by 4 mm; top acute, its tip obtuse; margin entire; at the base slightly hairy along the margin; nerves palmate, main nerves 3, above distinctly sunken, below prominent, lateral nerves rather inconspicuous.
Internodes: First one terete, to 2.5 cm long, green, hairy as the hypocotyl; next internodes as the first one, rather regular in length, to 1.7 cm long in the lower ones, in 30th leaf stage to 2.5 cm long. Stem slender, straight but slightly pendulous, branching soon, branches turning ± horizontal.

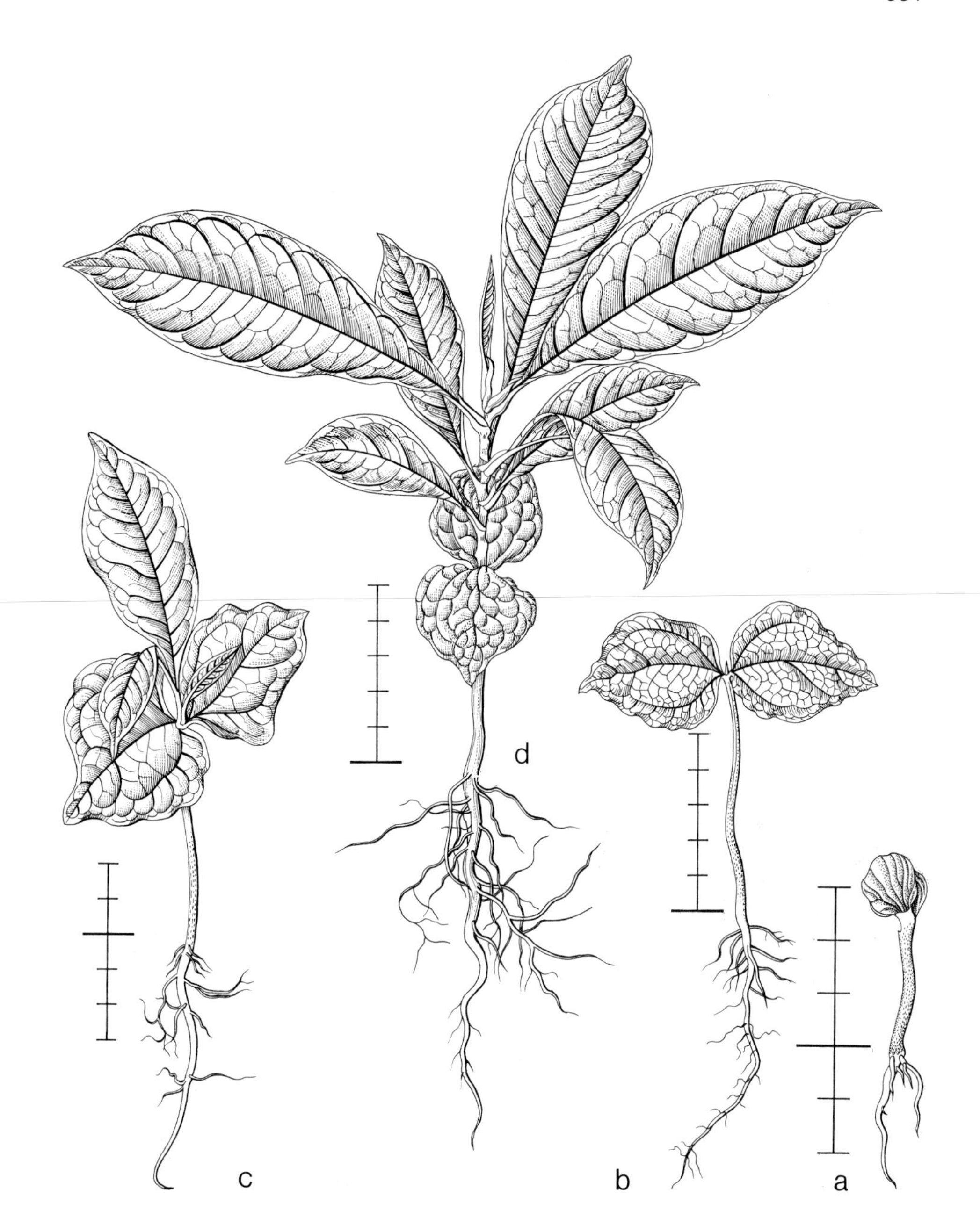

Fig. 121. *Talauma sp.* De V. 2197, collected 6-iii-1973, planted 22-iii-1973. a and b. 27-iii-1973; c. 17-vii-1973; d. 21-ii-1974.

length, 7th one to 1.5 cm long. Stem relatively sturdy, straight.
Leaves all spirally arranged, conduplicate-induplicate when very young, simple, without stipules, petiolate, herbaceous, green, glabrous. Petiole forming a sheath, enclosing the endbud, in first leaf to 7 by 1.5 mm, in 10th leaf to 20 by 3 mm, tapering to the top, in cross-section semi-orbicular with two membranaceous margins which turn brown and are shed soon after the development of the endbud. Blade oblong, later turning obovate-lanceolate, in the second leaf to 7.5 by 3.3 cm, in 8th leaf to 19 by 6 cm; base distinctly narrowed; top acuminate; margin entire; nerves pinnate, above slightly sunken, midrib below distinctly prominent, lateral nerves less so but darker green coloured, at their tip turned up and curved to the next one.

Specimens: 2197 from S.E. Borneo, alluvial flat, primary forest on deep clay, low altitude.
Growth details: In nursery germination poor.

MALPIGHIACEAE

ASPIDOPTERYS

Fl. Males. I, 5: 126.
Genus in India, throughout S.E. and the S. part of E. Asia, W. Malesia, including Celebes and the Philippines.

Aspidopterys elliptica (Bl.) Juss. – Fig. 122.

Development: The hypocotyl and taproot pierce the fruit wall and establish. The hypocotyl becomes erect, pulling the paracotyledons free from the fruit wall. A short resting stage then occurs.
Seedling epigeal, phanerocotylar. Macaranga type.
Taproot rather long and slender, rather soft, creamy-white, sideroots rather few, long, slender, creamy-white, hardly branched.
Hypocotyl strongly elongating, terete, to 4 cm long, green, densely covered with small, two-branched, hyaline hairs.
Paracotyledons 2, opposite, foliaceous, in the 4th leaf stage dropped; slightly succulent, simple, sessile, without stipules, soft fleshy, green, glabrous. Blade linear, to 20 by 3 mm; base slightly narrowed; top rounded; margin entire; 1-nerved, nerve above sunken, below slightly prominent.
Internodes: First one strongly elongating, terete, to 1 cm long, green, hairy as the hypocotyl; next internodes gradually becoming larger, second one to 1.5 cm, 5th one to 2.8 cm, 10th one to 14 cm long. Stem slender, straight and erect for the basal 8–10 internodes, beyond that climbing, with leaves reduced to scales.
Leaves all decussate, simple, without stipules, petiolate, herbaceous, above shining green, below pale green. Petiole in cross-section semi-orbicular, in first pair to 2 by 0.8 mm, in 10th pair to 15 by 1.2 mm. Blade ovate, in first pair to 3.2 by 1.6 cm, in

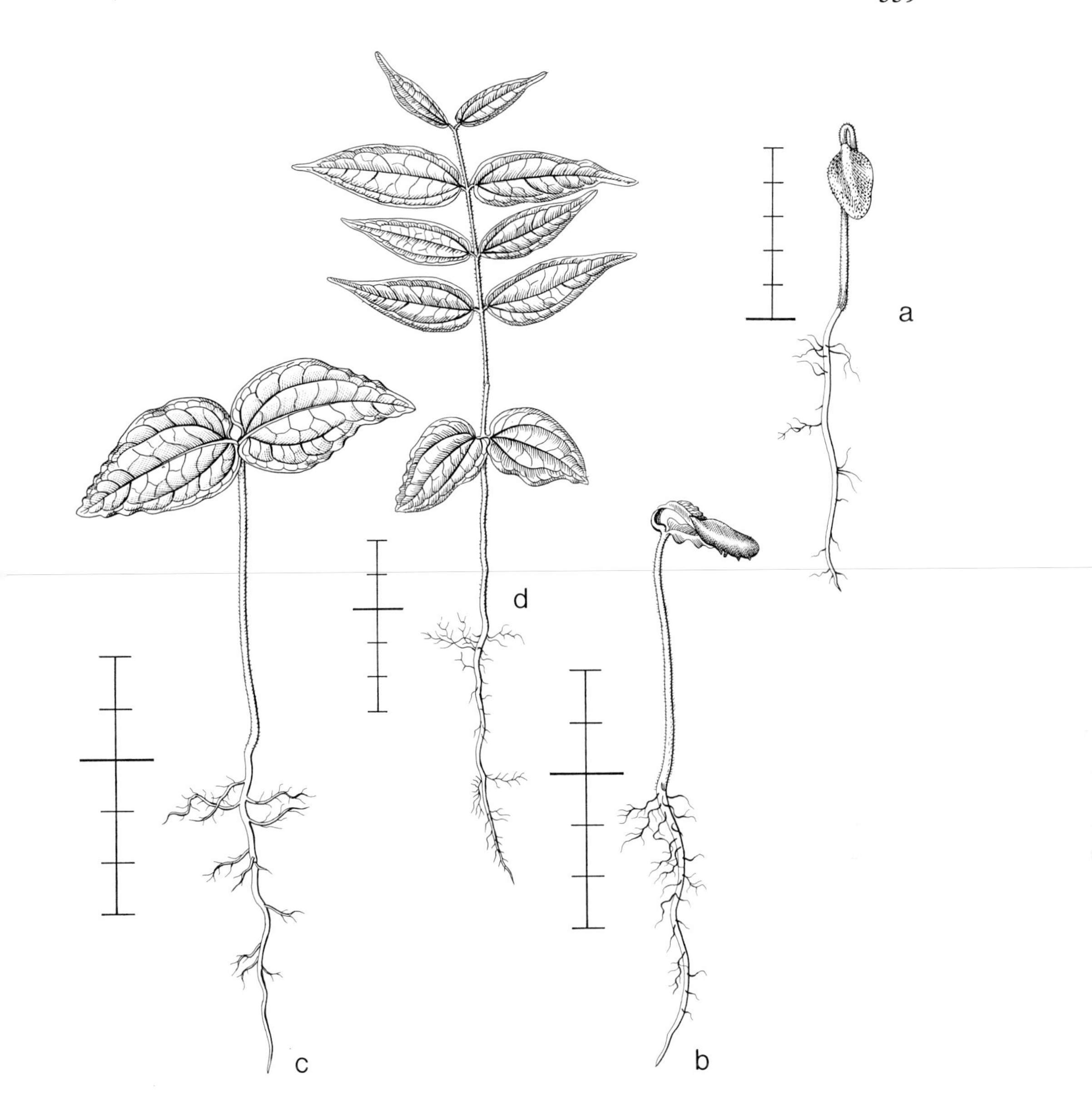

Fig. 120. *Strychnos sp.* De V. 2363, collected 16-iii-1973, planted 22-iii-1973. a, b, and c. 9-vii-1973; d. 13-xi-1973.

Leaves arranged in opposite pairs, distichous, simple, without stipules, shortly petiolate, herbaceous, green, occasionally a pair much reduced and present in the form of scales. Reduced leaves sessile, appressed to the stem, narrowly triangular, to 2.5 by 1 mm; top acute; margin entire; on the midrib densely hairy; 1-nerved. Petiole in cross-section semi-orbicular, to 2.5 by 1 mm, in higher leaves not much larger, in 30th one to 4 by 1.5 mm, hairy as the hypocotyl. Blade ovate-oblong, in the lowest leaves to 5.5 by 2.2 cm but in general much smaller, in higher leaves not much gaining in size, in 30th leaf to 8 by 2.8 cm; base rounded to obtuse; top cuspidate to caudate; margin entire; above with a few scattered minute hairs, below slightly more hairy with small, simple, brownish hairs, especially so on the nerves; nerves palmate, curvinerved, main nerves above sunken, below distinctly prominent, especially the central one, the lateral ones not so pronounced.

Specimens: 2363 from S.E. Borneo, primary hill forest on deep clay, medium altitude.
Growth details: In nursery germination fair, ± simultaneous.
Remarks: The four described species are similar in germination pattern and general morphology, they differ in details only.

MAGNOLIACEAE

TALAUMA

Tree Fl. Malaya 2: 291.
Genus in S.E. Asia and all over Malesia.

Talauma sp. – Fig. 121.

Development: The taproot and hypocotyl emerge from the margin of the seed. The hypocotyl becomes erect by which the enclosed paracotyledons are carried above the soil. After a short resting period the paracotyledons free themselves by expanding. Then follows a second temporary rest.
Seedling epigeal, phanerocotylar. Macaranga type.
Taproot long, slender, fibrous, with rather few, long, slender, little branched, creamy-white sideroots.
Hypocotyl strongly elongating, terete, to 8 cm long, with 2 longitudinal, rounded ridges starting from between the paracotyledons, green, warty with paler lenticels.
Paracotyledons 2, opposite, foliaceous, in the 8th–10th leaf stage dropped; simple, without stipules, almost sessile, herbaceous, green, glabrous. Blade rhomboid, to 5 by 4.5 cm; base obtuse; top slightly acuminate, its tip obtuse; margin entire; nerves palmate, main nerves 3, above slightly sunken, below slightly prominent, sidenerves ending in anastomoses.
Internodes: First one slightly elongating, in cross-section ± elliptic, to 2 mm long, green, glabrous; next internodes as first one, gradually becoming longer, irregular in

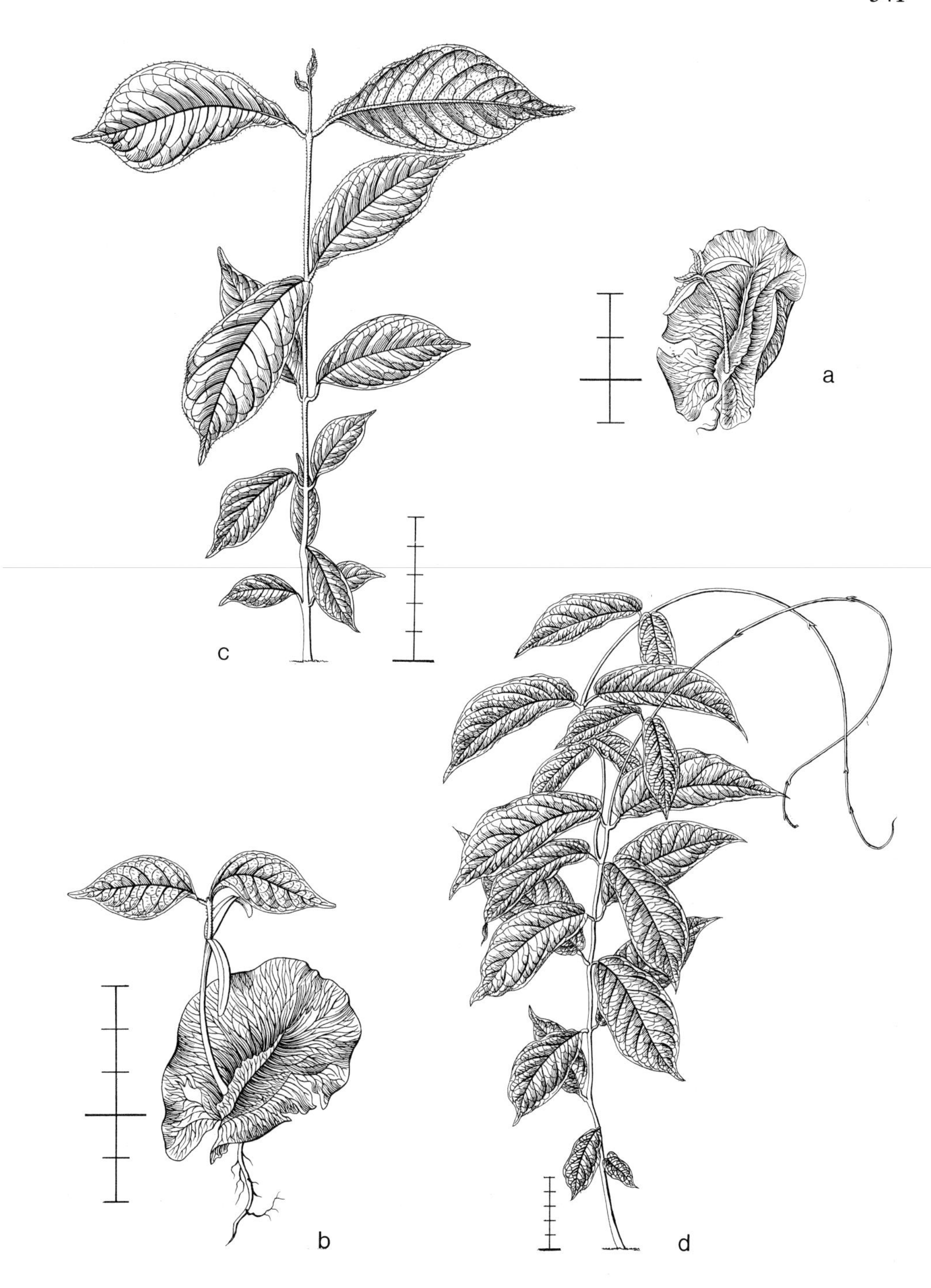

Fig. 122. *Aspidopterys elliptica.* De V. 2265, collected 10-iii-1973, planted 22-iii-1973. a and b. 18-vii-1973; c. 6-v-1974; d. 21-vii-1974.

10th pair to 10.5 by 4.8 cm; base narrowed into the petiole; top acuminate, in higher ones cuspidate; margin entire; above and below slightly hairy as the hypocotyl, below on the margin and nerves more so; nerves pinnate, midrib above slightly prominent, below much so, lateral nerves above sunken, below prominent, curved at the end and forming an undulating marginal nerve. Beyond ± the 10th pair of leaves 1–2 pairs of somewhat reduced leaves, subsequent ones reduced to small scales.

Specimens: 2265 from S.E. Borneo, primary hill forest on deep clay, low altitude.
Growth details: In nursery germination poor, ± simultaneous.

MELASTOMATACEAE

MEMECYLON

Fl. Java 1: 371; Mensbruge, Germination (1966) 294 (*M. cinnamonoides*).
Genus in tropical and subtropical Africa, throughout the islands of the Indian Ocean, India and Ceylon, throughout S.E. Asia, all over Malesia, N.E. Australia, in the Pacific in Bismarck, St. Cruz, Fiji, and Tonga.

Memecylon edule Roxb. – Fig. 123.

Development: The taproot and hypocotyl emerge from the globular seed. The hypocotyl becomes erect by which the much folded covered paracotyledons are carried above the soil. Usually during this process the paracotyledons start unfolding, by which the testa is shed. A short resting stage occurs with the paracotyledons unfolded, followed by short ones after the development of each pair of leaves.
Seedling epigeal, phanerocotylar. Macaranga type.
Taproot long, slender, flexuous, brownish cream, with especially in the upper parts rather long, much-branched creamy-white sideroots.
Hypocotyl strongly elongating, terete, to 7 cm long, with 2 distinct, sharp, wing-like longitudinal ridges ending between the paracotyledons in some minute, subulate, elongated cusps, green turning brown, glabrous.
Paracotyledons 2, opposite, foliaceous, in the 20th–30th leaf stage shed, simple, almost sessile, without stipules, subcoriaceous, above green, below paler so, glabrous. Blade almost rhomboid, to 3.1 by 2.7 cm; base truncate, abruptly narrowed where attached; top broadly truncate to retuse; margin entire, irregularly undulating; 5-nerved, the central 3 the bigger, nerves above sunken, below prominent, sidenerves ending in anastomoses.
Internodes: First one strongly elongating, in cross-section quadrangular, to 4.5 cm long, on the angles with a distinct, sharp, wing-like, slightly greenish violet longitudinal ridge, which turns interpetiolar at the top, and ends into some subulate, minute cusps, green, glabrous; next internodes as the first one, gradually increasing in length, 5th one to 3 cm long, 10th one to 5 cm long. Stem slender, straight.
Leaves all decussate, simple, without stipules, shortly petiolate, subcoriaceous,

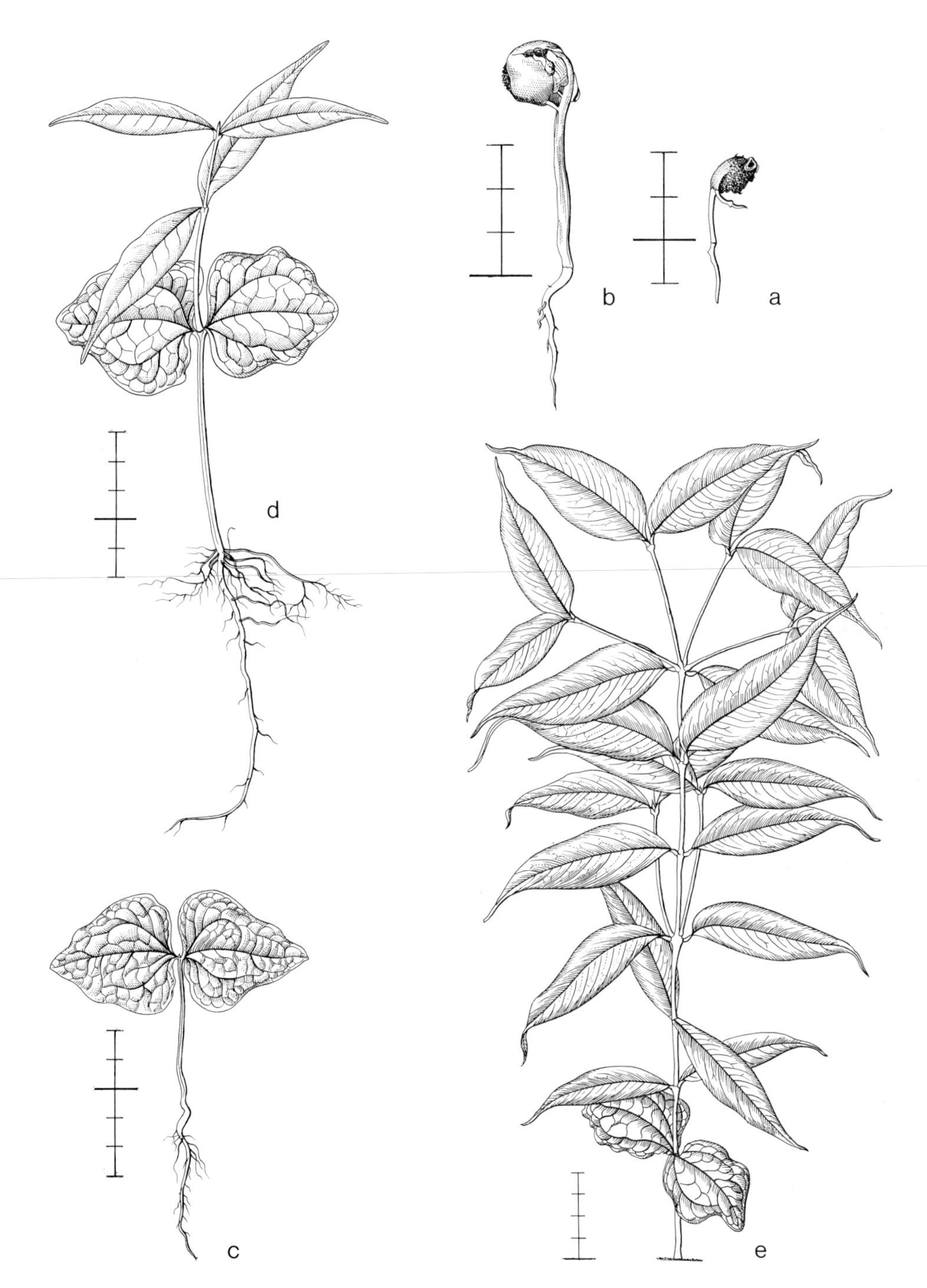

Fig. 123. *Memecylon edule*. De V. 2357, collected 16-iii-1973, planted 22-iii-1973. a. 7-vi-1973, polyembryonic fruit with a second hypocotyl emerging; b and c. 17-iv-1973; d. 14-vii-1973; e. 8-iv-1974.

rather dark greenish violet when young, turning green, glabrous. Petiole in cross-section semi-orbicular, in the lowest leaves to 1.5 by 0.7 mm, in the 5th–10th pair to 5 by 2 mm, distinctly jointed at the base. Blade ovate-lanceolate to lanceolate, the lowest ones to 7.5 by 2.2 cm, 5th pair to 12.5 by 3.5 cm, in higher ones hardly gaining in length; base acute; top cuspidate, its tip obtuse to acute; margin entire; above shining, below dull and paler; nerves pinnate, midrib above sunken, below prominent, lateral nerves below very inconspicuous, not raised, forming an inconspicuous marginal nerve.

Specimens: 2357 from S.E. Borneo, primary forest on deep clay, and 2398 from W. Java, primary forest on deep clay, medium altitude.
Growth details: In nursery germination good, ± simultaneous.
Remarks: M. cinnamonoides is according to De la Mensbruge (l.c.) hypogeal. For this investigation two species studied: *Memecylon* cf. *costatum* (two coll.) and *M. edule* (two different varieties); all epigeal, differing in details only.

MELIACEAE

AGLAIA

Fl. Java 2: 126; Burger, Seedlings (1972) 231 (*A. eusideroxylon*, *A. odoratissima*).
Monograph: Pennington and Styles, Blumea 22 (1975) 419.
Genus in India and Ceylon, throughout S.E. and E. Asia, all over Malesia, N. Australia, in the Pacific as far as Tonga Arch., including the Marianas, the Carolines, and New Caledonia.

Aglaia dookoo Griff. – Fig. 124.

Development: The seed splits across in the centre, the cotyledons are pushed aside when the hypocotylar part of the stem grows in thickness. The taproot emerges from below, the shoot from above between the slightly spreading cotyledons. No resting stage occurs.
Seedling hypogeal, semi-cryptocotylar. Endertia type, Chisocheton subtype.
Taproot long, slender, rather flexuous, brownish cream, with few slender, hardly branched, brownish cream sideroots.
Hypocotyl not elongating.
Cotyledons 2, opposite, succulent, in the 2nd–3rd leaf stage shed; simple, sessile, peltate, without stipules, semi-ellipsoid, to 1.3 by 1.5 by 5 mm; base flat; top rounded; margin entire; hard fleshy, greenish where exposed, covered by the slowly disintegrating, brownish testa.
Internodes: First one terete, elongating, to 1.8 cm long, greenish, rather densely covered with minute, simple, hyaline hairs, with many pale brown, slightly warty, small, round lenticels; next internodes as the first one, irregular in size, in the scaly part to 4 cm long, the first in the leafy part to 1 cm long, 4th one to 4 cm long. Stem

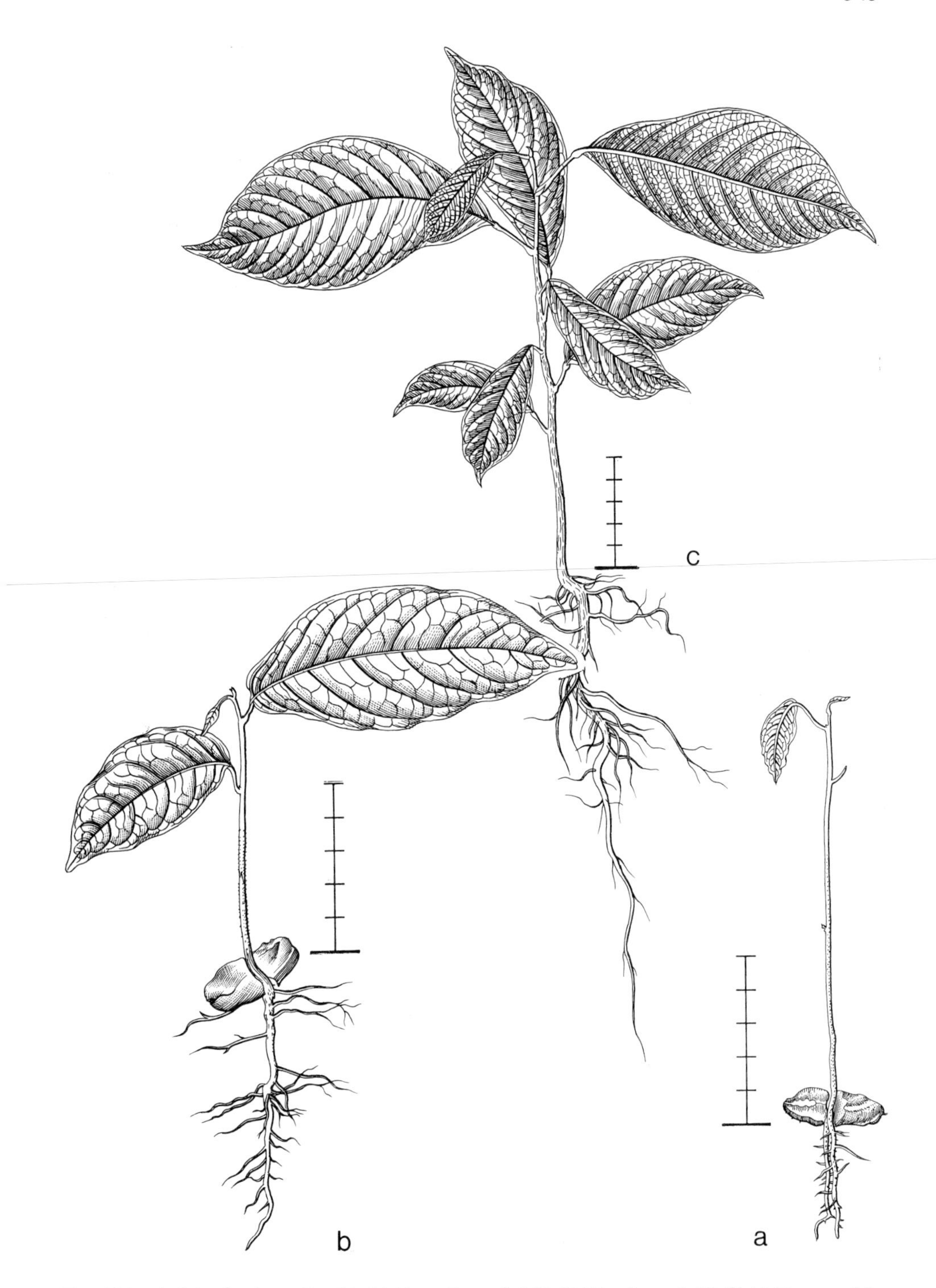

Fig. 124. *Aglaia dookoo*. De V. 2240, collected 9-iii-1973, planted 22-iii-1973. a and b. 10-vii-1973; c. 6-v-1974.

slender, ± straight.
Leaves all spirally arranged, without stipules, petiolate, subcoriaceous, green. First 3–4 leaves abortive, dropping soon. Subsequent leaves developed, simple. Petiole in cross-section semi-orbicular, in lowest one to 1.5 cm long, in 5th one to 3 cm long; base and top distinctly swollen; hairy as the first internode. Blade ovate-oblong, in first developed leaf to 9.5 by 4.5 cm, but often smaller, in 4th one to 15.5 by 7 cm; base ± rounded in the lowest leaf, in higher ones acute; top acuminate to cuspidate, its tip rounded; margin entire; below slightly hairy as the first internode, glabrescent; nerves pinnate, above sunken, below prominent, especially the midrib, lateral nerves ending in anastomoses. Next leaves compound, not seen.

Specimens: 2240 from S.E. Borneo, primary hill forest on deep clay, low altitude.
Growth details: In nursery germination poor.
Remarks: All species described show the same germination pattern. In *A. dookoo* all leaves are spirally arranged. The other species have the first two leaves opposite to subopposite. Reduction of one of these may occur. Further they differ in details only.

Aglaia tomentosa T. & B. – Fig. 125.

Development: The seed splits across in the centre, the cotyledons are pushed slightly aside when the hypocotylar part grows in thickness. First the taproot emerges from below, later the shoot emerges from above between the spreading cotyledons. No resting stage occurs.
Seedling hypogeal, semi-cryptocotylar. Endertia type, Chisocheton subtype.
Taproot long, slender, fibrous, brownish, with few, strong, few-branched sideroots as the taproot.
Hypocotyl not elongating.
Cotyledons 2, opposite, succulent, in the 8th–12th leaf stage shed, but much sooner ineffective; simple, without stipules, peltate with very short petiole, hard fleshy, dark brownish, with slowly disintegrating, persistent, cracking dark brown testa, glabrous. Blade semi-ellipsoid, to 6 by 10 by 8 mm, inside above flattened, outside irregularly concave.
Internodes: First one strongly enlarging, terete, to 7 cm long, green, densely covered with brownish, stellate hairs; next internodes 1–4.5 cm long, especially the higher ones the larger, as the first one. Stem rather slender, straight.
First two leaves opposite, simple, without stipules, petiolate, herbaceous, green. Petiole terete, to 6 by 1 mm, hairy as the first internode. Blade simple, ovate-oblong, to 6 by 2.5 cm; base retuse; top acuminate; margin entire, often irregular; pale shining green when young, dull green when old, on both sides with scattered hairs as on the first internode; nerves pinnate, above sunken, below much raised, lateral nerves ending in anastomoses.
Next leaves spirally arranged. The first 3–5(–9) simple. Petiole to 2 cm long. Blade to 13 by 4.5 cm, in the lower ones smaller. Petiole to 2 cm long. Subsequent leaves compound, gradually the higher ones larger, lowest up to six ones with 1–2 lateral leaflets, higher ones with 2–3 pairs of leaflets, further as the first leaves. Petiole terete, in higher ones gradually larger, to 4 by 0.2 cm, as the petiole of the first leaf.

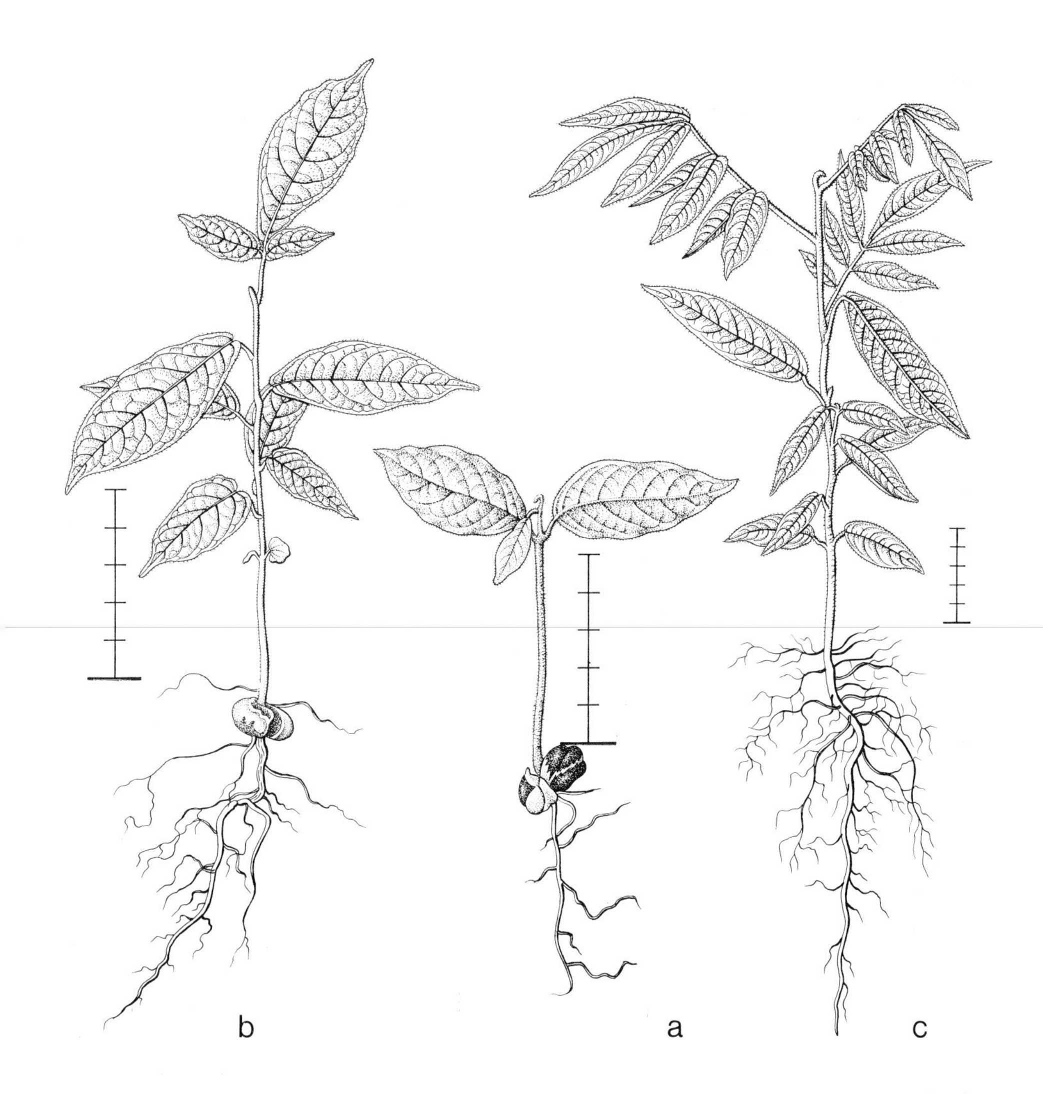

Fig. 125. *Aglaia tomentosa.* De V. 804, collected 15-xi-1971, planted 7-xii-1971. a, b, and c date?.

Rhachis as the petiole, in higher ones gradually larger, in 3rd one to 5.5 cm long. Lateral leaflets sessile, in lower leaves elliptic, to 4.2 by 2.1 cm, in higher ones oblong, to 5.5 by 1.8 cm; base in higher leaflets distinctly asymmetric, adaxial side with a rounded auricle, abaxial side obtuse; top acuminate; margin entire. Terminal leaflet oblong, in lowest leaf to 11 by 3.5 cm, in 4th one to 8.5 by 3 cm; base obtuse; top acuminate; margin entire.

Specimens: 804 from S.E. Borneo, alluvial flat, primary forest on deep clay, low altitude.
Growth details: In nursery germination poor, very long delayed. Growth slow, 8 months after planting to 12 cm high, with 7 leaves.
Remarks: See *A. dookoo*.

CHISOCHETON

Fl. Java 2: 124.
Monograph: Pennington and Styles, Blumea 22 (1975) 419.
Genus in S.E. and E. Asia, all over Malesia, N.E. Australia, Bismarck and the Solomons.

Chisocheton pentandrus (Bl.) Merr. – Fig. 126.

Development: The seed splits across in the centre. The cotyledons spread slightly when the hypocotylar part of the stem grows in thickness. From this slit the root develops from below, the shoot from above. No resting stage occurs.
Seedling hypogeal, semi-cryptocotylar. Endertia type, Chisocheton subtype.
Taproot long, slender, flexuous, orange-brown, with rather few, long, few-branched, orange-brown sideroots.
Hypocotyl not elongating.
Cotyledons 2, opposite, succulent, in the 3rd–4th leaf stage shed; simple, peltate, ± sessile, without stipules, hard fleshy, grey-brown where exposed, testa persistent. Blade semi-ellipsoid, to 7 by 12 by 8 mm, below flattened, above concave, testa grey-brown, tough, on the side turned to the soil with a broad, semi-elliptic, rough scar 5 by 9 mm, the opposite side with a semi-orbicular depression 2.5 mm wide.
Internodes: First one strongly elongating, terete, to 1.8 cm long, dull purplish yellow turning reddish purple, densely covered with minute, simple, yellowish hairs, rather soon glabrescent; next internodes as the first one, those in the leafless part of the stem to 2 cm long, first internode in the leafy part to 1.5 cm long, higher ones in general not exceeding 2 cm in length, usually much smaller. Stem slender, straight.
Leaves all spirally arranged. First 3 leaves often much reduced, succulent, scale-like, triangular, to 1 by 0.5 mm, coloured and hairy as the first internode. Next leaves fully developed, simple, conduplicate-induplicate when young, without stipules, petiolate, herbaceous, yellowish green turning green, sometimes the first one more or less irregular. Petiole in cross-section terete, in 2nd leaf to 3 by 1 mm, in 5th one to 1.5 cm long, all higher ones distinctly swollen at base and top, brownish green, hairy as the

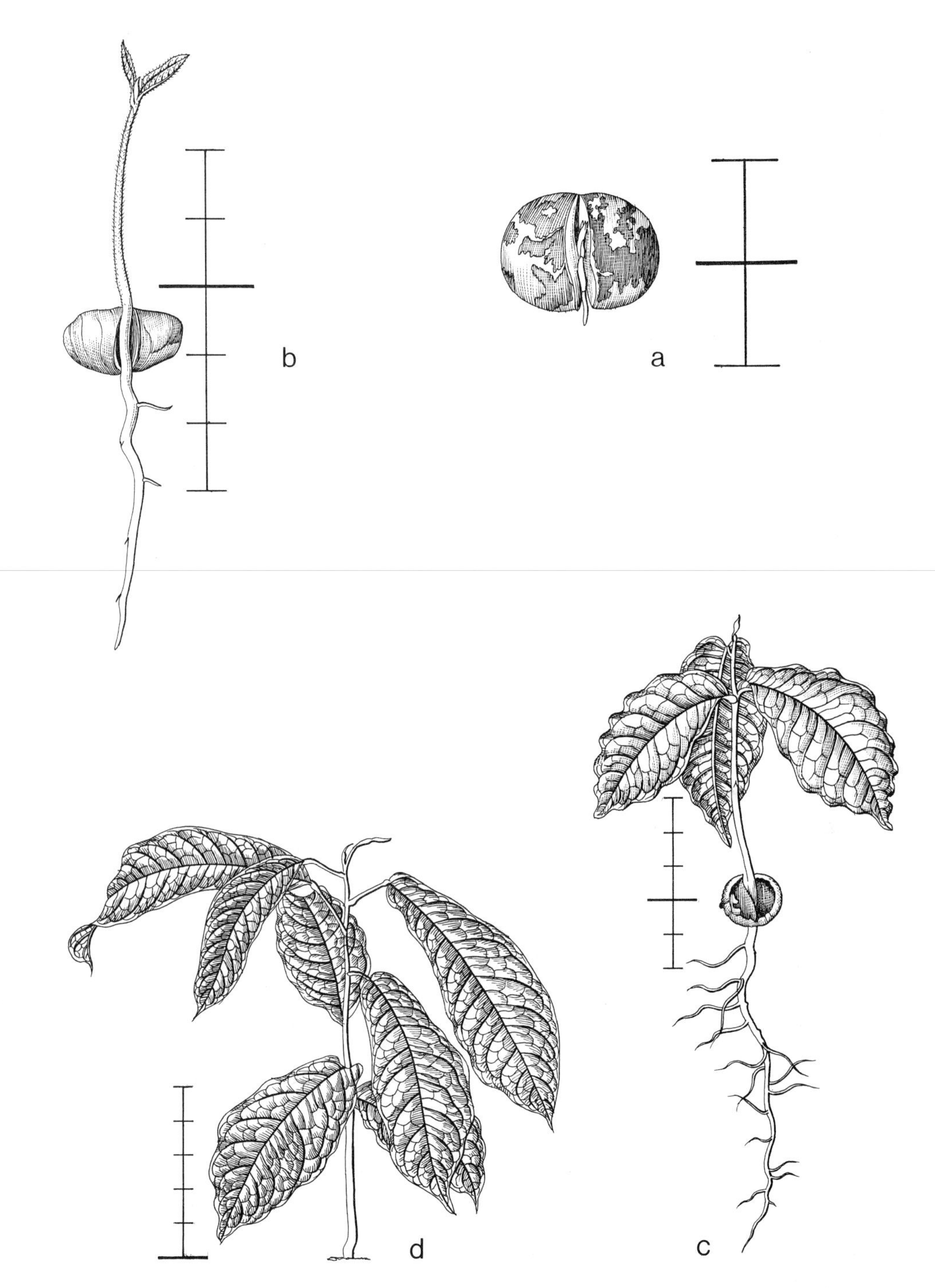

Fig. 126. *Chisocheton pentandrus*. De V. 2466, collected 7-x-1973, planted 5-xi-1973. a. 12-xii-1973; b. 9-i-1974; c. 15-ii-1974; d. 3-ix-1974.

first internode. Blade in first 3 leaves elliptic, to 6.5 by 3.2 cm, higher ones more obovate-oblong, 5th one to 9 by 3.5 cm; base wedge-shaped, its tip obtuse; top acuminate; margin entire; when young slightly hairy as the first internode, rather soon glabrescent except on the nerves; nerves pinnate, above slightly sunken, below prominent, especially the midrib, lateral nerves at the tip abruptly curved to the next one. Subsequent leaves compound, not seen.

Specimens: 2466 from N. Celebes, primary hill forest on deep clayey soil.
Growth details: In nursery germination fair, rather long delayed, ± simultaneous.
Remarks: Two species cultivated. *Ch. sp.*, Geesink 5724 is similar in germination pattern and general morphology, and differs in details only.

DYSOXYLUM

Fl. Java 2: 121; Burger, Seedlings (1972) 234 (*D. caulostachyum*, *D. densiflorum*, *D. gaudichaudianum*, *D. macrocarpum*, *D. parasiticum*); Pennington and Styles, Blumea 22 (1975) 419 (in general: cryptocotylar, less frequently phanerocotylar); Pierre, Flore Forestière Cochinchine 5 (1879–1899) pl. 352 (*D. loureiri*).
Monograph: Pennington and Styles, Blumea 22 (1975) 419.
Genus in India and Ceylon, throughout S.E. Asia, all over Malesia, N.E. and E. Australia, New Zealand, in the Pacific as far as Tonga Arch., including the E. Carolines and New Caledonia.

Dysoxylum sp. – Fig. 127.

Development: The seed splits across in the centre, the cotyledons spread slightly when the hypocotylar part of the stem grows in thickness. First the taproot emerges from below, later the shoot from above through the slit between the cotyledons. A resting stage occurs with 3–4 developed leaves.
Seedling hypogeal, semi-cryptocotylar. Endertia type, Chisocheton subtype.
Taproot long, slender, sturdy, brown, with scattered, swollen, lengthwise split warts, with rather many, not branched, lighter brown, slender sideroots.
Hypocotyl not elongating.
Cotyledons 2, opposite, succulent, when shed not known, rather long persistent even when ineffective; simple, without stipules, peltate, in the lower part with a very short petiole, hard fleshy, green to brown, the outside covered with the slowly disintegrating and flaking testa, glabrous. Blade semi-ellipsoid, to 3.5 by 3 by 2 cm, distinctly concave, below flattened, irregularly depressed, the adaxial side flattened.
Internodes: First one strongly elongating, terete, to 15 cm long, but often much shorter, green, with scattered, brownish to hyaline, simple and stellate hairs, glabrescent, with scattered, linear, pale lenticels, its basal part slightly swollen with many swollen roundish warts; next internodes as the first one, irregular in size, to 2–10 cm long, the higher ones in general the larger. Stem rather slender, rather straight.
Leaves all spirally arranged, sometimes the lower ones subopposite, without stipules, petiolate, herbaceous, green. Lower 0–2 leaves much reduced, scale-like, to 2 cm

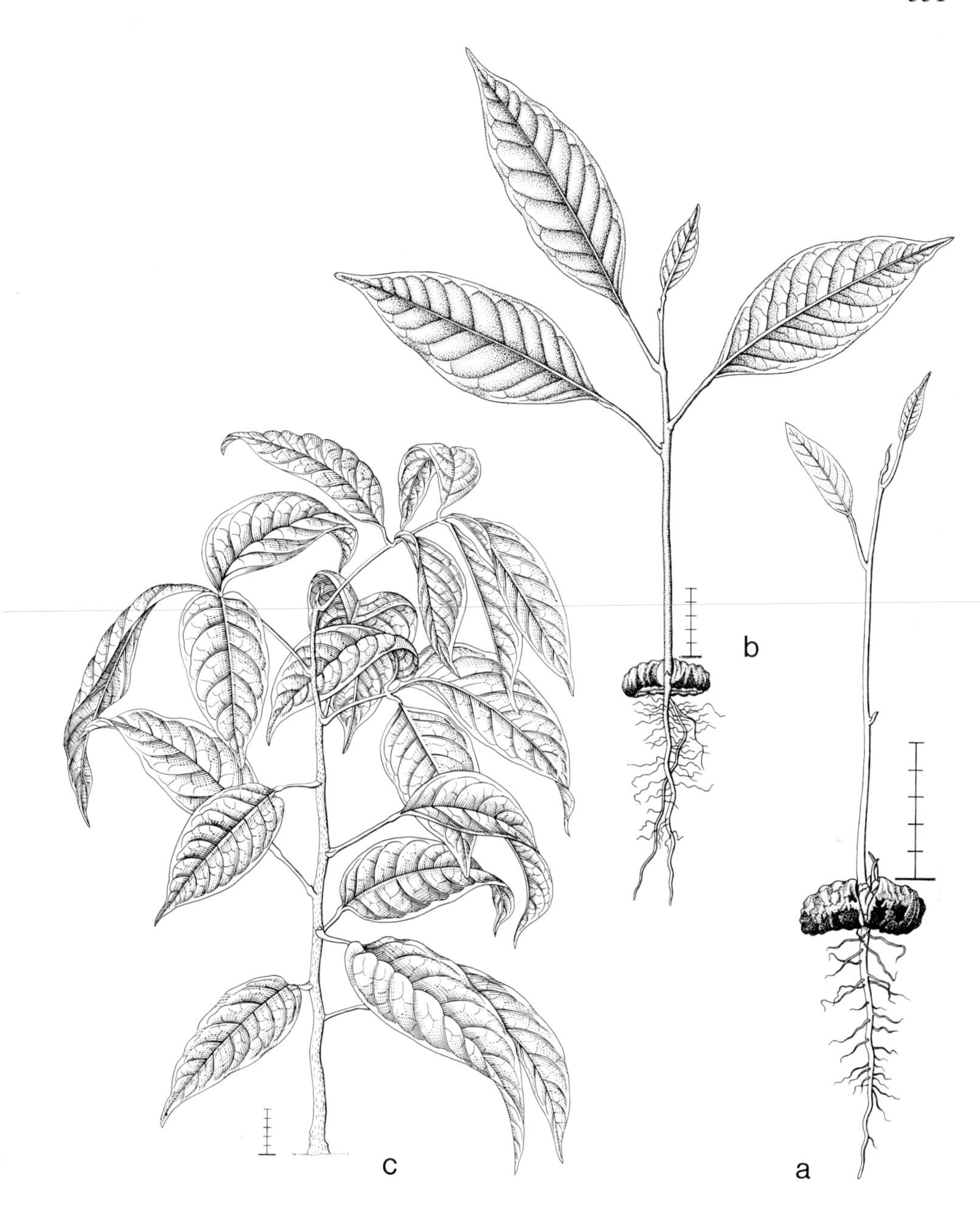

Fig. 127. *Dysoxylum sp.* De V. 1028, collected 24-xi-1971, planted 7-xii-1971. a, b, and c date?.

long, gradually passing in the normal leaves. Next 6–10 leaves simple. Petiole terete, in the lowest ones to 5.5 by 0.3 cm, in the highest ones to 10 by 0.3 cm, at base and top distinctly swollen, at the top jointed, slightly hairy as the first internode, glabrescent. Blade ovate, in the lowest one to 23 by 10 cm, in the highest ones to 35 by 15 cm; base obtuse to acute to slightly narrowed; top acuminate; margin entire; with scattered hyaline to brownish stellate hairs and minute simple hyaline ones, glabrescent; nerves pinnate, above sunken, below prominent, especially the midrib, lateral nerves ending in anastomoses. Subsequent leaves compound. Petiole in first 1-jugate leaf to 16 cm long, but usually much smaller, in higher leaves with more leaflets, leaves usually not exceeding 17 cm, above flattened. Blade imparipinnate with terminal leaflet, the first 1(–2) 1-jugate, rarely 2-jugate, next 1(–3) 2-jugate, next 1(–3) 3-jugate, in higher ones more slowly increasing in number of leaflets, gradually becoming larger. Leaflets opposite to subopposite, petiolulate. Petiolule to 1.5 by 1.5–2 mm, at the base distinctly swollen. Blade of leaflet to ovate-oblong, the lateral ones rather constant in size, to 24 by 10 cm, the terminal one gradually decreasing in size, in the lowest compound leaves to 30 by 13 cm, in 3-jugate leaves to 28 by 12 cm, but generally much shorter, in 4–jugate leaves to 22 by 9 cm, resembling the lateral leaflets in size; base acute; top acuminate; further as simple leaves.

Specimens: 1028 from S.E. Borneo, primary hill forest on deep clay, low altitude.
Growth details: In nursery germination good, ± simultaneous. In 8 months seedling to 60 cm high.
Remarks: In *Dysoxylum* the cotyledons are often relatively heavy, except in the epigeal species where they are sometimes rather small.
In *D. gaudichaudianum* and *D. parasiticum* the cotyledons are hypogeal, turned secund, and they are covered by the testa. In the first all leaves are spirally arranged; the lowest ones are much reduced, and a number of simple leaves occur before compound leaves are formed. In the second all leaves are compound, with the first two opposite, and the higher ones spirally arranged. In all other species that have been published, and also in those investigated for the project, the cotyledons are more or less peltate, and opposite on either side of the stem, with their upper sides facing. The cotyledons may be lifted from the soil like normally occurs in *D. macrocarpum*, in which the testa is shed and the blades are entirely exposed, while the first two leaves are trifoliolate and opposite, and the higher ones all compound and spirally arranged. In *D. densiflora* the cotyledons may be lifted from the soil or remain hypogeal while the testa is never shed. The first two leaves are trifoliolate and opposite, followed by some simple ones that are spirally arranged before the higher compound leaves are produced.
In most species the heavy opposite cotyledons are never lifted above the soil; the testa is not shed but may disintegrate during the life of the cotyledons, thus exposing the free surface. All leaves may be spirally arranged with the lower ones simple, and the first ones sometimes much reduced. *D. sp.* (De Vogel 1028; and also 1324, and 1908 grown for the project). In other examples with the lower leaves simple the first two leaves may be opposite (De Vogel 2087, 2091). Also examples are known with all leaves compound: here the first leaves are always trifoliolate (*D. caulostachyum*, *D. sp.* De Vogel 1228).

Dysoxylum sp. – Fig. 128.

Development: The seed splits across in the centre, the cotyledons spread slightly when the hypocotylar part of the stem grows in thickness. First the taproot emerges from below, later the shoot from above between the cotyledons. No distinct resting stage occurs.
Seedling hypogeal, semi-cryptocotylar. Endertia type, Chisocheton subtype.
Taproot fibrous, long, rather slender, yellowish white, with many rather slender, few-branched, yellowish white sideroots.
Hypocotyl not elongating.
Cotyledons 2, opposite, succulent, in the 4th leaf stage shed; simple, without stipules, peltate, with a minute petiole, hard fleshy, brownish, the outer side covered by the dark brown, persistent, disintegrating testa, pale green where exposed, glabrous. Blade semi-ellipsoid, to 1 by 1 by 1 cm, the adaxial side flat, outside concave.
Internodes: First one strongly elongating, slender, terete, to 10 cm long, dark green, covered with close-set, short, appressed, brownish hairs, at the base with protruding, pale lenticels; next internodes as the first one, rarely to 5 cm long, higher up in general not exceeding 1.5 cm. Stem slender, ± straight.
First two leaves opposite, compound, without stipules, petiolate, herbaceous, green. Petiole in cross-section semi-orbicular, above flattened, to 10 by 1 mm, hairy as the first internode. Rhachis as petiole, to 4 by 1 mm, slightly widened. Blade trifoliolate, terminated by top leaflet. Leaflets ± sessile, the lateral ones opposite to subopposite, distinctly jointed at the base, lanceolate, to 6 by 1.5 cm; base ± acuminate; top cuspidate; margin entire; above shining, below dull, glabrous; nerves pinnate, midrib above raised, below more so, lateral nerves many, ± inconspicuous, forming a regular but vague marginal nerve close to the leaf margin.
Next leaves all spirally arranged. Petiole winged, especially towards the top, tapering to the base, and there slightly swollen, in the 6th leaf to 19 by 1.8 mm, in the 25th leaf to 2.5 cm long. Rhachis as the petiole, in the 6th leaf to 10 by 1.8 mm, in 25th one to 2.5 cm long, in general distinctly winged. Blade from the 3rd to 5th leaf onwards with 2 (or 3) ± alternate (rarely opposite) lateral leaflets and a terminal leaflet, from 6th one upwards with 4 leaflets. Lateral leaflets (sub)opposite to ± alternate, often slightly larger than in earlier ones, in 3rd leaf to 5.5 by 1.6 cm, in 15th leaf lateral ones to 6.5 by 1.8 cm. Terminal leaflet in 3rd leaf to 6.5 by 2.2 cm, in 15th one to 9 by 2.5 cm, further as those of the first leaves.

Specimens: 1228 from S. Sumatra, much depleted primary forest on deep clay, low altitude.
Growth details: In nursery germination good.
Remarks: See *D. sp.*, De Vogel 1028.

SANDORICUM

Fl. Java 2: 116.
Monograph: Pennington and Styles, Blumea 22 (1975) 419.

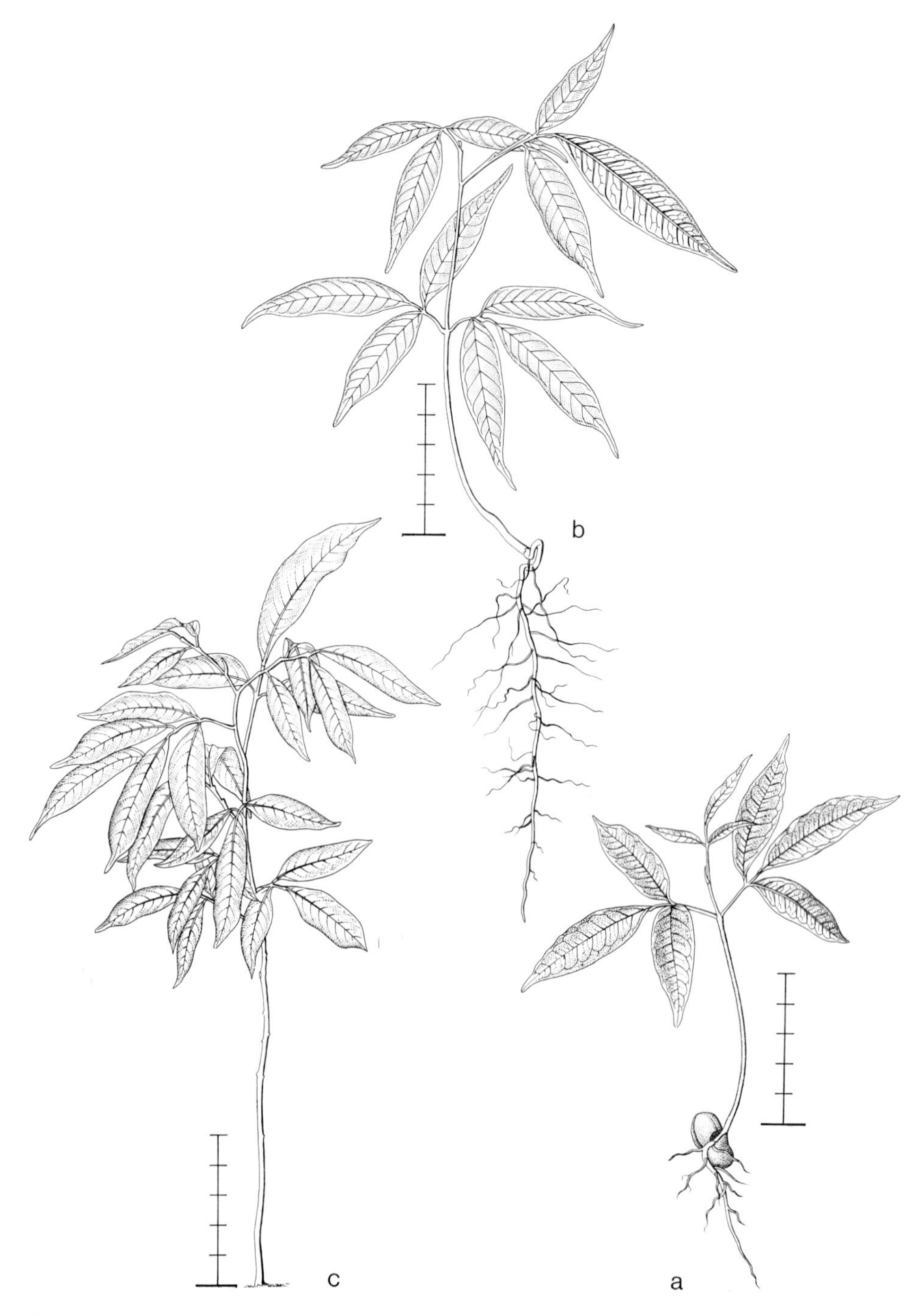

Fig. 128. *Dysoxylum sp.* De V. 1228, collected 10-iii-1972, planted 23-iii-1972. a and b date?; c. 30-vii-1973.

Genus in S.E. Asia, throughout Malesia including New Guinea.

Sandoricum koetjape (Burm. f.) Merr. – Fig. 129.

Development: The taproot and hypocotyl emerge from one pole of the seed. The cotyledons spread at an angle of 45° with the stem. In the meantime the papery outer testa splits along one margin and is thrown off, the leathery inner testa is persistent. No resting stage occurs.
Seedling epigeal, semi-cryptocotylar. Endertia type and subtype.
Taproot long, slender, flexuous, creamy-white with rather few, slender, creamy-white, hardly branched sideroots.
Hypocotyl strongly enlarging, terete, to 5 cm long, green, densely covered with minute, simple, hyaline hairs.
Cotyledons 2, opposite, succulent, in the 2nd–3rd leaf stage dropped; simple, sessile, without stipules, hard fleshy, below brown, above green, glabrous, inner testa hard leathery, persistent; outer testa papery, in general shed during development of the shoot, sometimes persistent but split, the shoot developing through the split, and then the cotyledons secund. Blade ovate, to 2.2 by 1.6 cm; base deeply split, auriculate, auricles triangular, to 5 by 5 mm; top broadly rounded; margin entire; below concave, above flat; nerves not visible.
Internodes: First one strongly elongating, terete, to 5 cm long, green, hairy as the hypocotyl; next internodes as the first one, second one to 2 cm long, next 2–3 not exceeding 1 cm, 8th one to 3 cm long. Stem slender, slightly zig-zag.
First two leaves opposite, compound, without stipules, petiolate, herbaceous, green. Petiole in cross-section semi-orbicular, to 20 by 1.5 mm, hairy as the hypocotyl. Blade trifoliolate, imparipinnate with terminal leaflet, rarely 5-foliolate, rhomboid in outline. Lateral leaflets sessile, ovate-oblong, to 3.5 by 1.2 cm; base distinctly unequal, the adaxial side beginning 2 mm from the base of the petiole; top cuspidate; margin slightly serrate with obtuse teeth; below on the main nerve hairy as the hypocotyl, slightly so below on the lateral nerves and along the margin; nerves pinnate, midrib above slightly raised, below very prominent, lateral nerves above sunken, below prominent, at the tip curved and forming an undulating marginal nerve. Terminal leaflet petiolulate. Petiolule as the petiole, to 6 by 1 mm. Blade symmetric, ovate-oblong, to 5.5 by 1.8 cm; base obtuse, further as the lateral leaflets.
Next leaves spirally arranged, compound, dull red when young, turning green. Petiole in lower ones in cross-section semi-orbicular, in higher ones more terete, in 5th one to 3.5 cm long, in 10th one to 6.5 cm long, at the base distinctly swollen, at the top slightly so. Blade trifoliolate. Lateral leaflets almost sessile, as those of the first leaves, in 5th one to 7.5 by 3.5 cm, in 10th one to 10.5 by 4.7 cm, margin ± entire. Terminal leaflet as that of the first leaves. Petiolule as the petiole, in 5th one to 1.5 cm long, in 10th one to 3 cm long. Blade of the terminal leaflet in 5th leaf to 10 by 4.7 cm, in 10th one to 11.5 by 5.8 cm; base obtuse to slightly retuse; further as the lateral ones.

Specimens: 2669 from N. Celebes, primary hill forest on deep clay, medium altitude, and Bogor Botanic Gardens, without number, locality not known.

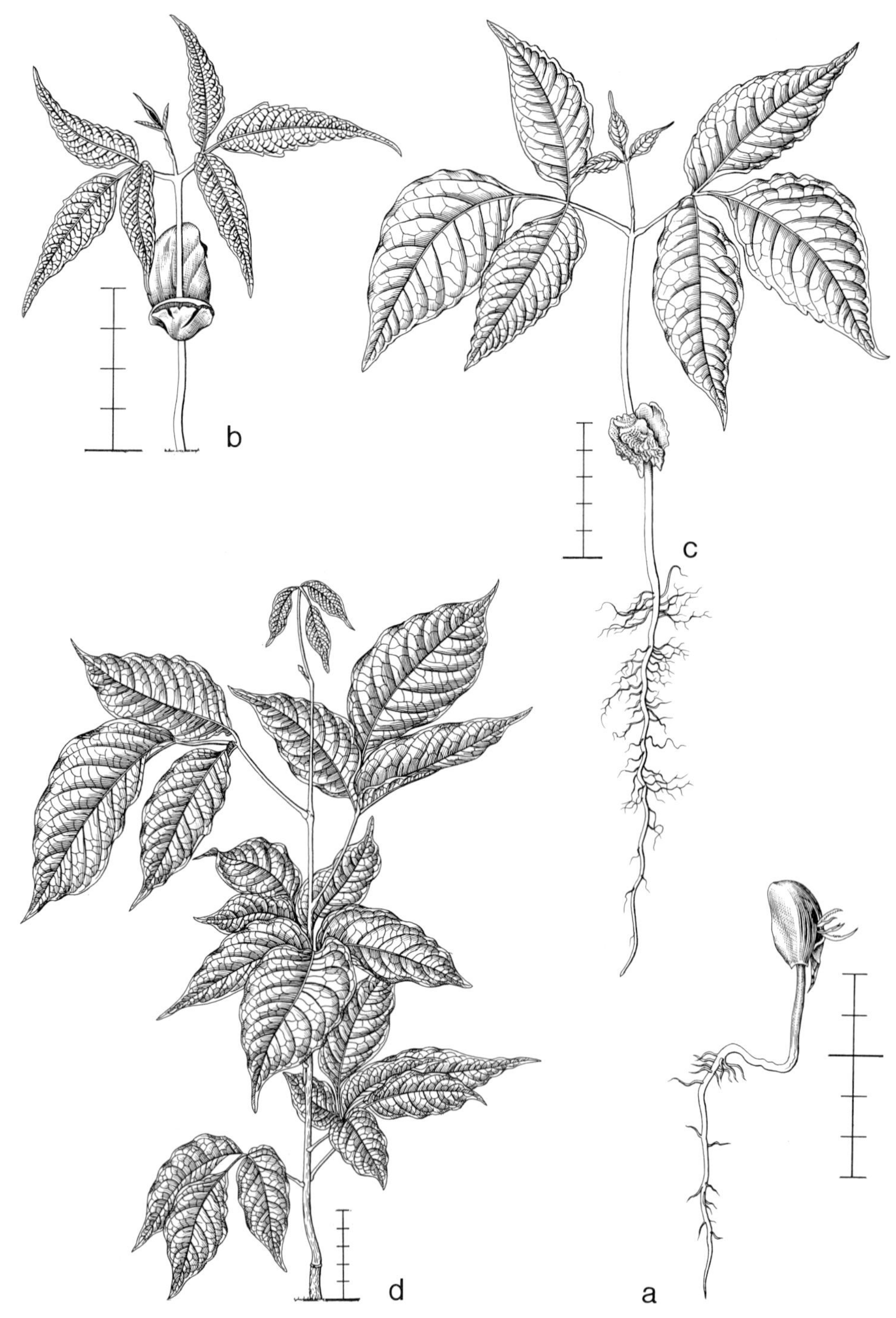

Fig. 129. *Sandoricum koetjape.* Bogor Bot. Gard., without number, collected and planted 15-i-1973 (c and d); De V. 2669, collected 31-x-1973, planted 5-xi-1973 (a and b). a and b. 29-xi-1973; c. 3-ix-1974; d date?.

Growth details: In nursery germination good, ± simultaneous. Seedlings and germinating seeds very common in the forest.
Remarks: Two species cultivated, similar in germination pattern and general morphology. *S. indicum* differs in details only.

MENISPERMACEAE

COSCINIUM

Fl. Java 1: 156.
Genus in India and Ceylon, throughout S.E. Asia, and W. Malesia.

Coscinium fenestratum (Gaertn.) Colebr. – Fig. 130.

Development: The taproot and hypocotyl emerge from the slit at the margin of the

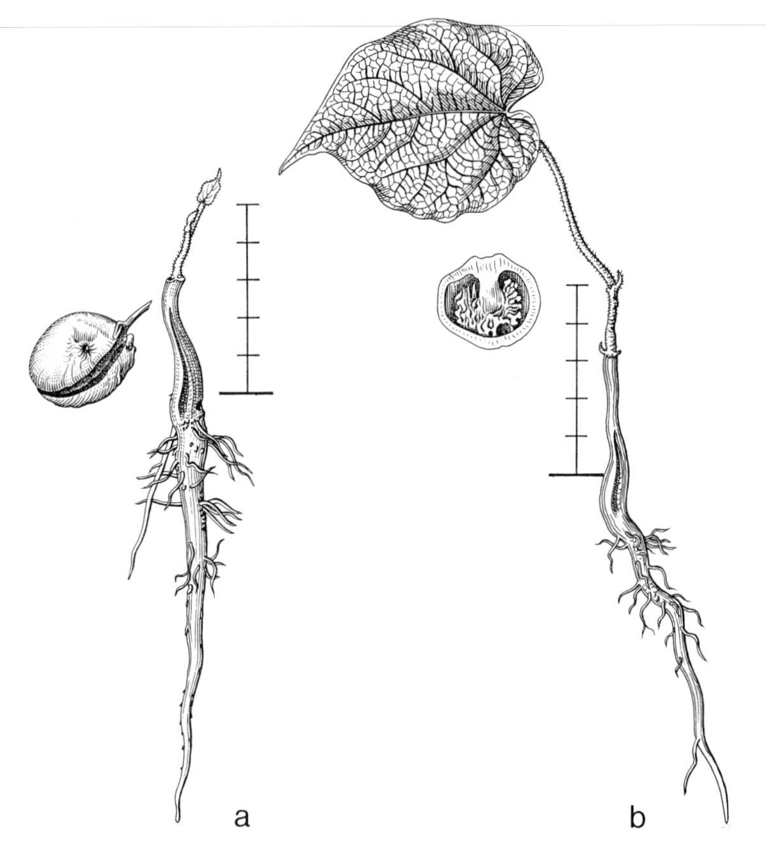

Fig. 130. *Coscinium fenestratum.* De V. 1560, collected 23-vii-1972, planted 27-vii-1972. a and b date?.

endocarp. After a rather long resting stage with the cotyledons at soil level on top of the curved hypocotyl, the cotyledons enclosed in the endocarp are shed, and the shoot is withdrawn from the envelopments.
Seedling 'epigeal', cryptocotylar. Coscinium type.
Taproot long, very sturdy, bright yellow turning brownish, with rather few, rather short, hardly branched, bright yellow sideroots.
Hypocotyl sturdy, strongly elongating, irregularly angular, tapering to the top, to 5 cm long, brownish, glabrous.
Cotyledons 2, not lifted from the soil, on top of the curved hypocotyl, dropped before the shoot has appeared, without stipules, petiolate. Petiole in cross-section semi-orbicular, to 10 by 2 mm. Whole body ± elliptic, to 3 by 3 by 2.5 cm, along the margin with a deep circumcisal groove, the woody, brownish endocarp persistent.
Internodes: First one slightly elongating, terete, to 2 cm long, densely covered with a felt-like indument of minute, white, appressed hairs; next up to 15 internodes not much elongating, to 2–2.5 cm long, as the first one; subsequent internodes strongly elongating, to 15 cm long. Stem rather slender, irregularly zig-zag and erect in the lower up to 16 internodes, higher part drooping, slender, straight, able to climb.
Leaves all spirally arranged, simple, without stipules, petiolate, subcoriaceous, green, white below by indument. Petiole terete, at the base distinctly swollen, and bent at base and top from the 3rd leaf onwards, in first one to 5.5 by 0.2 cm, in 13th one to 12 by 0.2 cm. Blade ovate, in the first leaf to 8 by 5.5 cm, in 13th one to 18 by 11 cm; base in first few leaves auriculate, auricles broadly rounded, to 7 by 15 mm, obtuse to truncate; top acuminate to cuspidate; margin entire; below with a dense felt-like indument of short, appressed, persistent, white hairs, above at the place where the petiole is attached with a tuft of small, simple, pale brown hairs; nerves palmate, above slightly sunken, below prominent, ± ending in anastomoses.

Specimens: 1560 from N. Sumatra, primary hill forest on deep clay, low altitude.
Growth details: In nursery germination rather poor.

TINOMISCIUM

Fl. Java 1: 156.
Genus in S.E. Asia and W. Malesia, also in New Guinea.

Tinomiscium phytocrinoides Kurz – Fig. 131.

Development: The endocarp splits along the suture of the valves, and the hypocotyl and taproot emerge from the slit. When established, the paracotyledons are withdrawn from the endocarp by the hypocotyl which becomes erect, and spread, often hampered by the usually adhering endosperm. A short resting stage occurs with the paracotyledons unfolded.
Seedling epigeal, phanerocotylar. Macaranga type.
Taproot rather slender, fibrous, pale brown, with rather many, branched, pale brown sideroots.

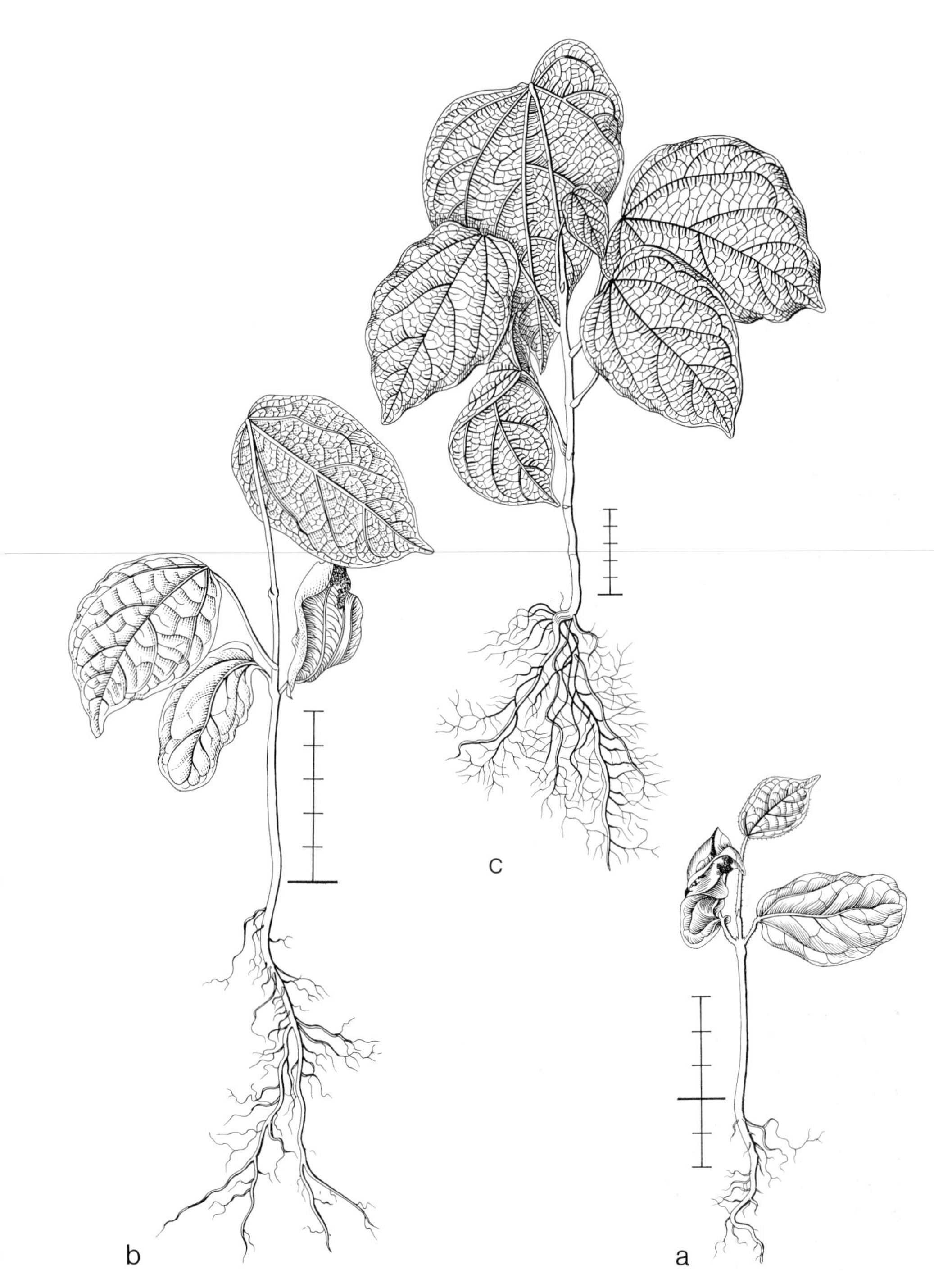

Fig. 131. *Tinomiscium phytocrinoides*. De V. 1445, collected 13-vii-1972, planted 27-vii-1972. a and b date?; c. 8-viii-1973.

Hypocotyl strongly elongating, in cross-section slightly quadrangular, to 6 cm long, green, with close-set, appressed, simple, brown hairs.
Paracotyledons 2, opposite, foliaceous, in 3rd leaf stage dropped; simple, spreading, petiole, without stipules, subcoriaceous, green, below paler. Petiole in cross-section flat elliptic, to 8 by 2 mm, hairy as the hypocotyl. Blade elliptic, to 4.8 by 2.8 cm, often partly enclosed and deformed by the usually adhering endosperm; base and top obtuse; margin entire, above on the midnerve hairy; nerves palmate, main nerves 5, midrib above and below prominent, lateral nerves much less so, descending along the main nerve in the petiole.
Internodes: First one elongating, terete, to 1 cm long, green, hairy as the hypocotyl; next internodes strongly elongating, to 2 cm long but often smaller, beyond the 10th–15th one suddenly much longer, ultimately to 20 cm long, as the first one. Stem thickened on the nodes, erect for the lower 10–15 internodes, then drooping and able to climb.
Leaves all spirally arranged, simple, without stipules, petiolate, herbaceous, later subcoriaceous, green. Petiole terete, in 3rd leaf to 8 by 2 mm, in biggest leaf to 19 cm by 3 mm, at the base distinctly thickened, higher ones above with a longitudinal groove, hairy as the hypocotyl. Blade almost peltate, ovate, higher ones cordate, 3rd one to 10 by 6.8 cm but usually much smaller, finally to 18 by 16 cm; base truncate, higher ones auriculate, auricles broad, rounded, in the biggest leaves to 7 by 2.5 cm; top acuminate, obtuse at the tip, higher ones cuspidate; margin entire; hairy above and below on the nerves and on the margin; nerves palmate, main nerves 5, above sunken, prominent below, ending in anastomoses.

Specimens: 1445 from N. Sumatra, alluvial flat, somewhat depleted primary forest on deep clay, low altitude.
Growth details: In nursery germination fair, ± simultaneous.

MONIMIACEAE

KIBARA

Fl. Java 1: 117.
Genus in the Nicobar Islands, all over Malesia including the Solomons, N. Australia.

Kibara coriacea (Bl.) Tul. – Fig. 132.

Development: The taproot and hypocotyl emerge from one pole of the seed; when established the hypocotyl becomes erect, carrying the paracotyledons enclosed by endosperm and testa above the soil. After a short resting period, testa and endosperm are shed by the spreading paracotyledons. A second resting stage occurs with the paracotyledons expanded.
Seedling epigeal, phanerocotylar. Macaranga type.
Taproot long, slender, flexuous, creamy-white, with rather few, long, slender, hardly

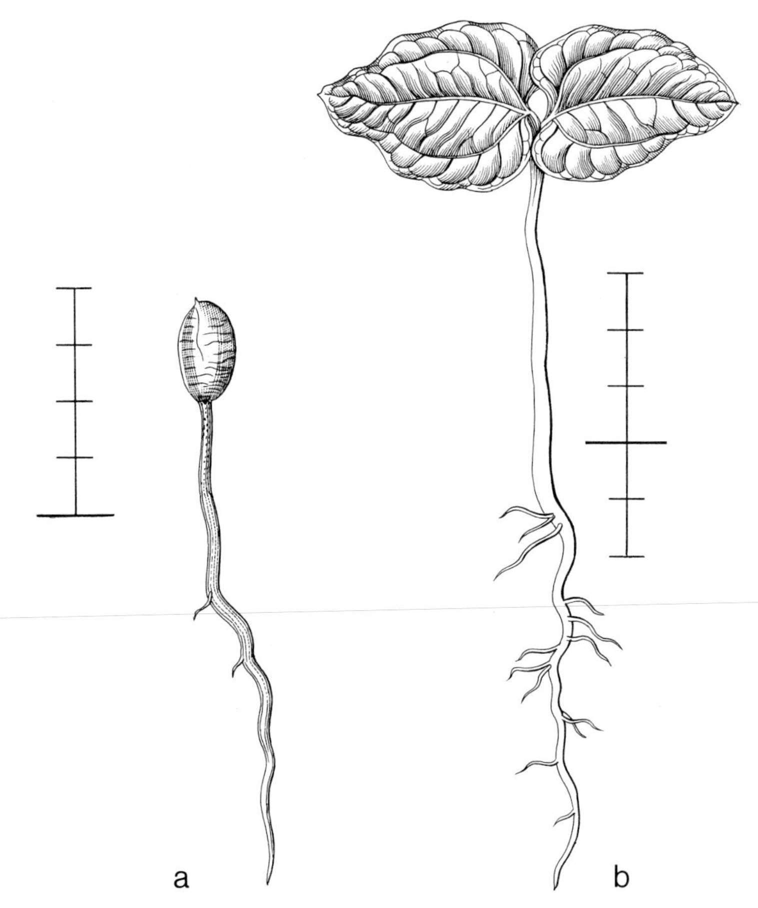

Fig. 132. ***Kibara coriacea.*** **Bogor Bot. Gard. VIII. G. 76, collected and planted i-1973.** a and b. 19-vii-1973.

branched sideroots.
Hypocotyl strongly elongating, terete, to 4.5 cm long, green, glabrous, later with many warty, brownish grey lenticels.
Paracotyledons 2, opposite, foliaceous, in the 8th–14th leaf stage dropped; simple, without stipules, petiolate, subcoriaceous, green, glabrous. Petiole in cross-section semi-orbicular, to 10 by 2 mm. Blade ovate, often irregularly curled, to 4.5 by 3 cm; base auriculate, auricles broadly rounded, to 3 by 10 mm; top rounded, at the tip with a gland-like structure; margin entire; 5-nerved, nerves above raised, below hardly so, descending into the petiole, the 3 central ones more conspicuous.
Internodes: First one strongly elongating, in cross-section elliptic, to 2.5 cm long, green, rather sparsely hairy with minute, simple, hyaline hairs; next internodes as the first one, second one to 7.5 cm long, 5th one to 8 cm long, but often much shorter. Stem straight, slender.
Leaves decussate, simple, without stipules, petiolate, herbaceous, green. Petiole in

cross-section semi-orbicular, in first leaves to 10 by 1.5 cm, in 5th pair to 15 by 2 mm, hairy as the first internode. Blade oblong, in the first leaves to 7 by 3.5 cm, in 5th pair to 18.5 by 8 cm; base acute; top acuminate; margin slightly serrate, teeth with a paler, swollen tip, slightly hairy below and along the margin as the first internode, more so on the midnerve; nerves pinnate, midrib above and below raised, lateral nerves slightly so, broadly curved to the next one, veins free ending in the teeth.

Specimens: Bogor Botanic Gardens VIII. G. 76, from Bangka.
Growth details: In nursery germination fair, ± simultaneous.

MORACEAE

ARTOCARPUS

Fl. Java 2: 18; Burger, Seedlings (1972) 259 (*A. elasticus*, *A. integra* = *A. integer*, *A. rotunda*); Duke, Ann. Miss. Bot. Gard. 52 (1965) 331 (*A. heterophyllus*); Jarrett, J. Arn. Arb. 40 (1959) 17 (general on seedlings); Troup, Silviculture 3 (1921) 876 (*A. chaplasha*, *A. hirsutus*, *A. integrifolia* = *A. integer*, *A. lakoocha*).
Monograph: Jarrett, J. Arn. Arb. 40 (1959) 1; ibid. 41 (1960) 320.
Genus in India and Ceylon, throughout S.E. and tropical E. Asia, all over Malesia, including Bismarck, the Solomons, and the Louisiades.

Artocarpus elasticus Reinw. ex Bl. – Fig. 133.

Development: The taproot emerges from the side on the margin of the seed, the cotyledonary petioles elongate slightly, bringing the plumule free from the envelopments. No resting stage occurs.
Seedling hypogeal, cryptocotylar. Horsfieldia type and subtype.
Taproot long, slender, fibrous, orange-yellowish, with many rather short, much branched, yellow sideroots.
Hypocotyl not enlarging, forming a yellow, narrow collar.
Cotyledons 2, secund, succulent, in the 3rd–5th leaf stage shed, simple without stipules, petiolate. Whole body irregular in shape, ± ellipsoid, to 1.5 by 1.2 by 1 cm, the pale, greyish testa smooth, persistent, with a brown line along its margin. Petioles hardly emerging, flattened in cross-section, to 4 mm long.
Internodes: First one terete, to 4 cm long, green, densely hairy with simple, stiff, brown hairs; next internodes as the first one, slightly irregular in length, 5th one to 3 cm long, 15th one to 5.5 cm, 20th one to 9 cm long; stem rather sturdy, straight.
Leaves all spirally arranged, simple, from the 5th–10th one onwards deeply incised, stipulate, petiolate, herbaceous, green, with abundant white latex. First 1–2 leaves in general abortive, sometimes reduced, next ones gradually increasing in size. Stipules 2 at the base of the petiole, sessile, shrivelling soon but persistent, narrowly triangular, in 4th leaf to 12 by 5 mm, in 12th one to 3 by 0.8 cm, in 20th one to 5 by 2 cm, in 30th one to 13 by 4 cm, when dropped leaving a circular scar around the stem; top

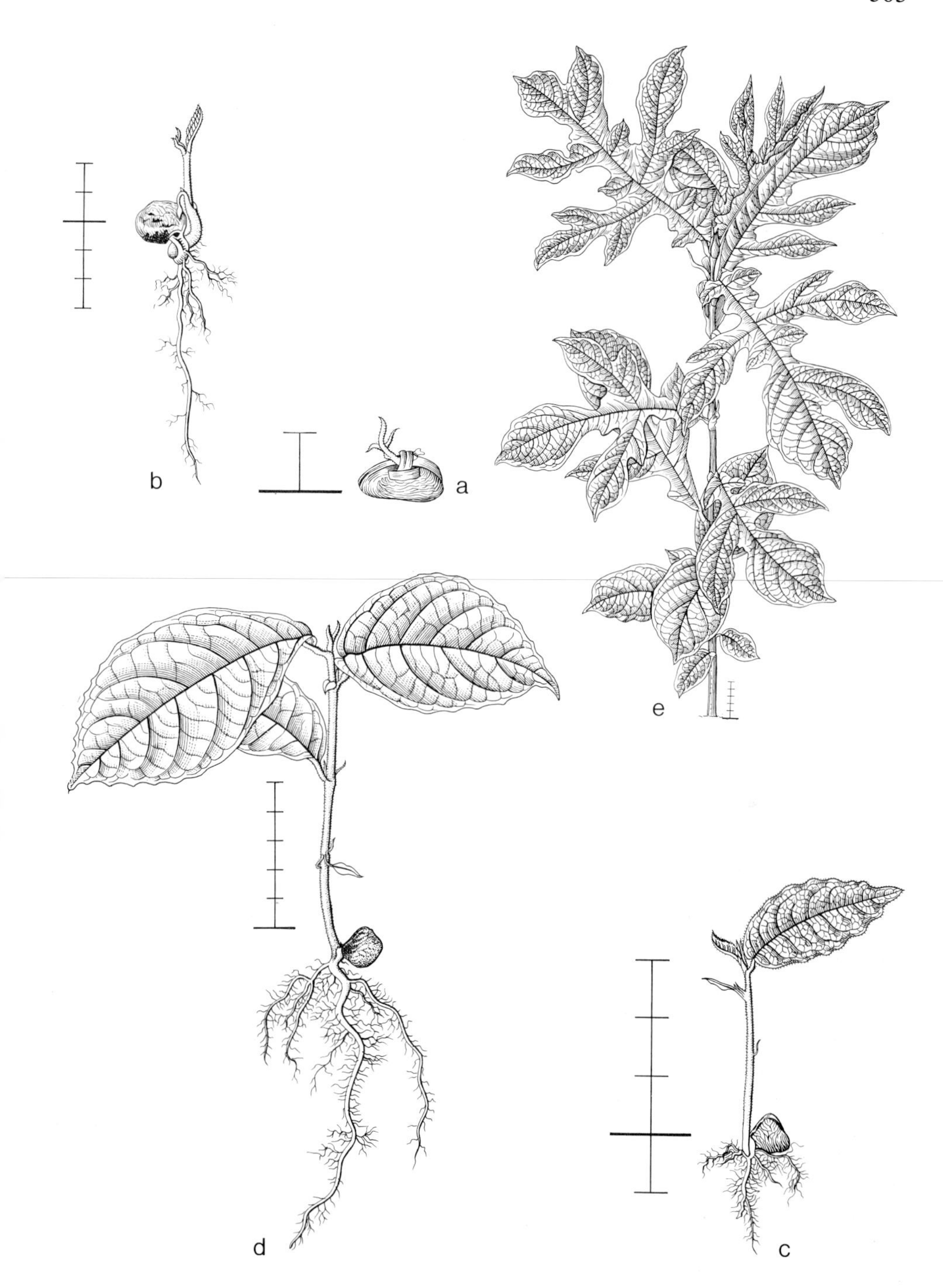

Fig. 133. *Artocarpus elasticus.* De V. 2142, collected 10-xi-1972, planted 17-xi-1972. a, b, c, and d date?; e. 22-viii-1973.

acute; margin entire; herbaceous, pale green, quite densely hairy with hairs shorter than those on the first internode; parallel-nerved, the midnerve below prominent, the lateral nerves rather inconspicuous, in higher leaves their number increasing. Petiole short, terete, in 3rd leaf to 8 by 2 mm, in 10th leaf to 2 by 0.4 cm, in 20th one to 2.5 by 0.6 cm, turned upwards, hairy as the first internode. Blade elliptic, in 3rd leaf to 11 by 6.8 cm, in 10th one 18 by 9 cm, in 20th one 32 by 26 cm, in 30th one 60 by 52 cm; base in the lowest leaves obtuse, soon acute; top acute to cuspidate; margin widely dentate towards the top; from 5th–10th one onwards lacerate; below very rough to the touch, on the nerves with hairs as on the first internode; nerves pinnate, above slightly sunken, below much prominent, lateral nerves forming an inconspicuous marginal nerve.

Specimens: 2142 from S.E. Borneo, alluvial flat, cultivated in villages, but also wild in the forest, low altitude.
Growth details: In nursery germination good, ± simultaneous, plants showing vigorous growth.
Remarks: Two types are found in the genus. Most species represent the Horsfieldia type, with the cotyledons at soil level enclosed in the testa. Some have all leaves spirally arranged, with sometimes the lowest ones reduced (*A. elasticus*, *A. lakoocha*), others have the first two leaves developed and opposite (*A. chaplasha*, *A. hirsutus*, *A. lanceifolius*, *A. rotunda*).
A. heterophyllus and *A. integer* are representatives of the Endertia type, Streblus subtype. One cotyledon is distinctly smaller than the other, and opens from the bigger one like a lid. Its lower surface is covered by the persistent testa. All leaves are spirally arranged, the lower ones are much reduced.

Artocarpus lanceifolius Roxb. ssp. *clementis* (Merr.) Jarrett – Fig. 134.

Development: The taproot emerges from one pole of the seed. The cotyledonary petioles elongate, bringing the plumule free from the envelopments. A short resting stage occurs with 2 developed leaves.
Seedling hypogeal, cryptocotylar. Heliciopsis type and subtype.
Taproot long, sturdy, orange-cream, with few long, much-branched, and many short, branched, cream-coloured sideroots.
Hypocotyl not enlarging.
Cotyledons 2, secund, succulent, in the 3rd–4th leaf stage shed; simple, without stipules, petiolate. Whole body ellipsoid, to 1.8 by 1 by 0.8 cm, the creamy-grey fruit wall persistent, below more or less flattened, at one margin with a darker longitudinal scar, one pole rounded, the other one more acute. Petiole in cross-section flattened, to 4 by 3 cm.
Internodes: First one strongly elongating, terete, to 10 cm long, green turning brown, densely covered by short, simple, hyaline to brownish, at the tip curved hairs; next internodes as the first one, in lower ones to 3.5 cm long, 10th one to 5 cm long. Stem slender, ± straight.
First two leaves opposite, simple, stipulate, shortly petiolate, subcoriaceous, green. Stipules 2, interpetiolate, sessile, shrivelling soon, triangular, to 5 by 2 mm, hairy as

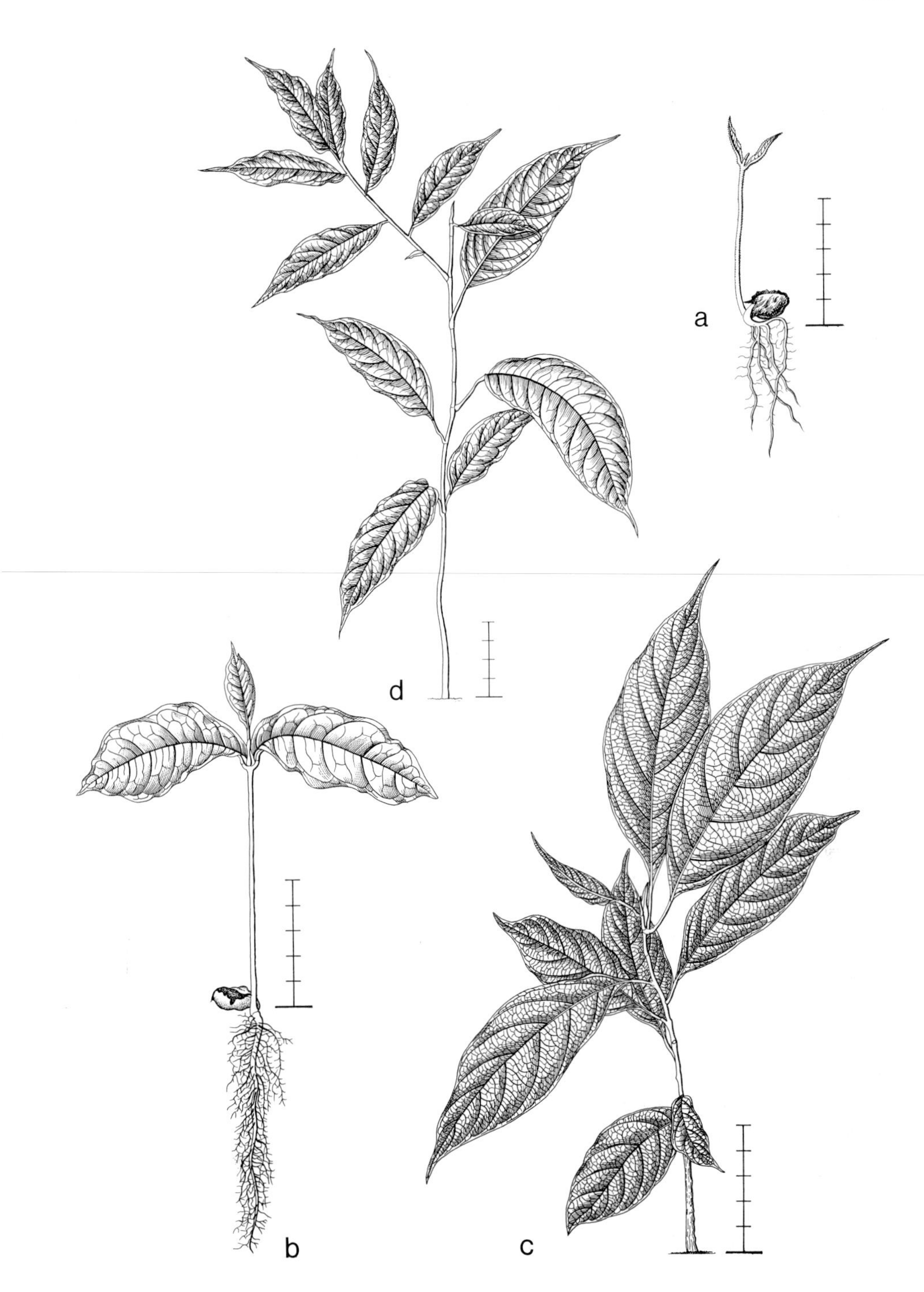

Fig. 134. *Artocarpus lanceifolius* ssp. *clementis.* De V. 2237, collected 8-iii-1973, planted 22-iii-1973. a. 5-iv-1973; b. 2-v-1973; c. 5-xi-1973; d. 16-vii-1974.

the first internode. Petiole terete, to 4 by 2 mm, brownish, hairy as the first internode. Blade slightly obovate to ovate, often rather irregular, to 8 by 4.5 cm; base acute; top acuminate, its tip rounded; margin entire; below on the nerves slightly hairy as the first internode; nerves pinnate, above slightly sunken, below prominent, lateral nerves forming an undulating marginal nerve.
Next leaves spirally arranged. The lowest up to 10 simple. Stipule 1 at the base of each petiole, covering the young leaf in the terminal bud, shed during its development, triangular, in the lower leaves to 5 by 4 mm, in 10th leaf to 18 mm long, leaving a circular scar. Petiole as that of the first leaves but distinctly swollen at the base, in 7th leaf to 4.5 cm long, above channelled. Blade oblong, in the 4th leaf to 13 by 5 cm, in 7th one to 21 by 8 cm; base distinctly narrowed into the petiole, often asymmetric; top cuspidate; further as the first leaves. Subsequent leaves compound, imparipinnate, petiolulate, as the first ones. Petiole to 5.5 cm long. Rhachis to 9.5 cm long, further as the petiole in higher leaves. Petiolules as petioles, in 12th leaf to 8 by 1 mm. Blade of leaflets ovate-oblong, in 12th leaf to 16 by 5.3 cm, further as the blade of simple leaves, often the lowest one much reduced.

Specimens: 2237 from S.E. Borneo, primary hill forest on deep clay, low altitude.
Growth details: In nursery germination good. Cotyledons very attractive to rodents.
Remarks: See *A. elasticus*.

PARARTOCARPUS

Fl. Males. (in preparation); Jarrett, J. Arn. Arb. 40 (1959) 18 (general on seedlings).
Monograph: Jarrett, J. Arn. Arb. 40 (1959) 1; ibid. 41 (1960) 320.
Genus all over Malesia, except the Lesser Sunda Islands, including Bismarck Arch., the Solomons, and the Louisiades.

Parartocarpus bracteatus (King) Becc. – Fig. 135.

Development: The taproot and hypocotyl emerge from one pole of the seed. After establishment the cotyledons spread, thus shedding the testa, and are carried above the soil by the hypocotyl which becomes erect. The testa often adheres to the bigger cotyledon. No resting stage occurs.
Seedling epigeal, phanerocotylar. Sloanea type and subtype.
Taproot sturdy, rather stunted, thick, fleshy, yellowish, densely covered with yellow hair-roots, with rather many sturdy, unbranched sideroots with hair-roots as on the taproot.
Hypocotyl strongly elongating, rather stunted and thick, terete, to 4.5 cm long, green, rather densely hairy with minute, simple, hyaline hairs and scattered longer ones, glabrescent.
Cotyledons 2, unequal, opposite, succulent, in the 2nd–3rd leaf stage dropped; simple, sessile, without stipules, petiolate, hard fleshy, green, glabrous. Smaller blade boat-shaped, to 3 by 1 by 1 cm; base narrowed; top rounded; margin entire; inside slightly convex, outside concave; nerves not visible. Larger blade boat-shaped,

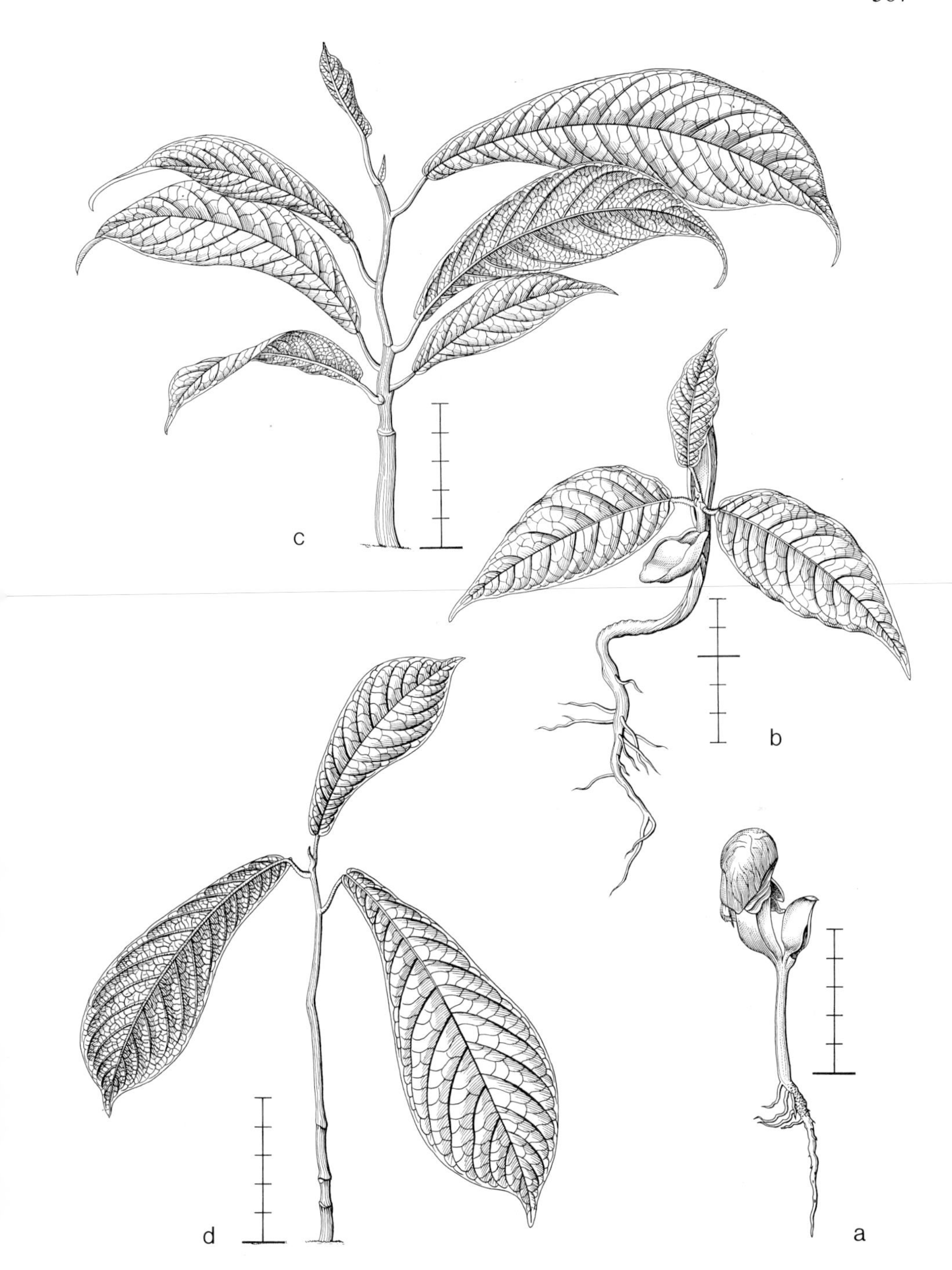

Fig. 135. *Parartocarpus bracteatus.* De V. 2287, collected 11-iii-1973, planted 22-iii-1973. a. 2-v-1973; b. 27-v-1973; c. 5-xi-1973; d. 19-viii-1974.

to 4.5 by 1 by 1 cm; base narrowed; top rounded; margin entire; further as smaller blade.
Internodes: First one elongating to 1.5 cm long, terete, green, hairy as the hypocotyl; next internodes as the first one, lower ones to 1.5 cm long, 6th one to 2.5 cm long. Stem rather stunted, ± straight.
First two leaves opposite, simple, stipulate, petiolate, subcoriaceous, green. Stipules 2 at the base of each leaf, rather soon dropped, narrowly triangular, to 5 by 1 mm, hairy as the hypocotyl. Petiole terete, to 17 by 2.5 mm, slightly tapering to the top, hairy as the hypocotyl. Blade oblong, to 11 by 4 cm; base slightly emarginate; top cuspidate to caudate; margin entire; hairy as the hypocotyl, slowly glabrescent; nerves pinnate, above sunken, below prominent, especially the midnerve, lateral nerves curved at the tip, forming an undulating, inconspicuous marginal nerve.
Next leaves spirally arranged. Stipule 1, sessile, intrapetiolar, rather soon dropped, narrowly triangular, in 7th leaf to 7 by 2.5 mm, the central side flat, erect, the lateral sides attached with an angle of 60°, clasping the developing endbud; top more or less incised; margin entire; hairy as the hypocotyl. Petiole terete, in 3rd leaf 2 cm long, hardly gaining in length. Blade more obovate-oblong, especially after the 4th one, in 3rd one to 12.5 by 4.3 cm, in 6th one to 17 by 5.5 cm; base gradually narrowed, abruptly rounded at its tip; top caudate; margin entire; above with minute, brown, glandular hairs, glabrescent, below especially on the nerves with short, simple, hyaline hairs; nervation as in the first leaves.

Specimens: 2287 from S.E. Borneo, primary hill forest on deep clay, low altitude.
Growth details: In nursery germination poor.
Remarks: Two species cultivated, similar in germination pattern and general morphology. Both have inequal cotyledons. *P. venenosus* differs in details only.

MYRISTICACEAE

HORSFIELDIA

Tree Fl. Malaya 1: 322.
Revision: Sinclair, Gard. Bull. 27 (1974) 133; ibid. 28 (1975) 1.
Genus in S.E. Asia, all over Malesia, N. Australia, in the Pacific as far as Santa Cruz, including the Carolines.

Horsfieldia wallichii (Hook. f. & Th.) Warb. – Fig. 136.

Development: The taproot emerges from a pole of the seed. The cotyledonary petioles elongate, bringing the plumule free from the envelopments, after which the shoot starts elongating. No resting stage occurs.
Seedling hypogeal, cryptocotylar. Horsfieldia type and subtype.
Taproot strong, fibrous, yellowish brown, the upper part with many protruding, lengthwise split lenticels, sideroots few, not-branched, yellowish brown.

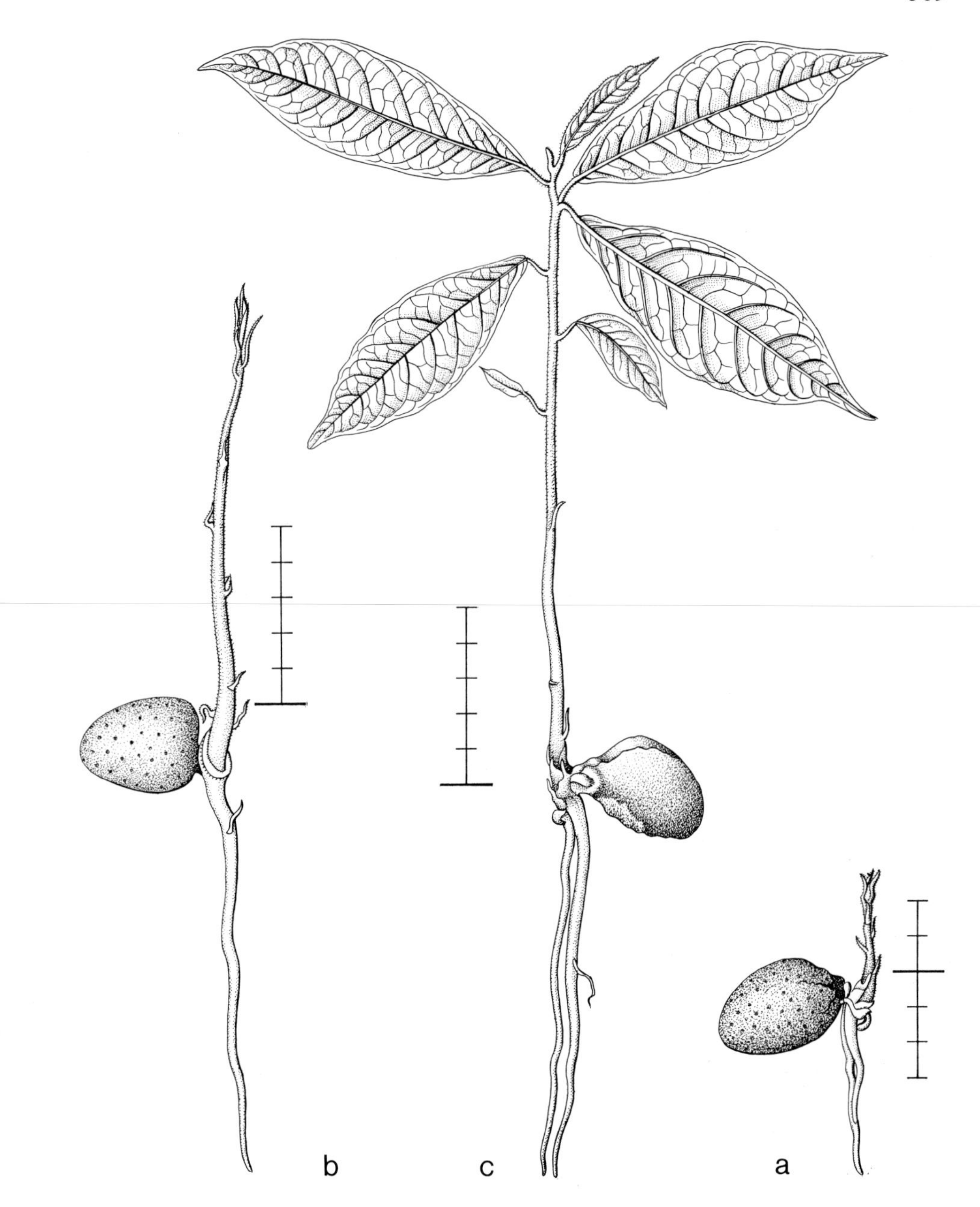

Fig. 136. *Horsfieldia wallichii.* De V. 1214, collected 9-iii-1972, planted 23-iii-1972. a, b, and c date?.

Hypocotyl a pronounced, oblique collar, the higher part on the cotyledonary side obliquely grooved, below dark brown, above smooth and light brown, densely hairy with minute, simple, light brown hairs.
Cotyledons 2, secund, succulent, when shed not known, after the 6th leaf stage; ruminate, without stipules, petiolate. Whole body ± ellipsoid, 3.3 by 2.6 by 2.2 cm, slightly flattened to the free pole, the light brownish grey testa persistent. Petiole in cross-section flattened, hairy as the hypocotyl.
Internodes: First one hardly elongating, terete, to 8 mm long, green, rather densely hairy as the hypocotyl; next internodes as the first one, first two to 4 mm long, the following ones irregular in length, sometimes to 3 cm long, those in the leafy part in general not over 2 cm long. Stem sturdy, rather slender, straight.
Leaves all spirally arranged, simple, without stipules, petiolate, herbaceous, green, the lower up to 10 much reduced, scale-like. Scales simple, sessile on a broad base, linear, to 7 by 1 mm, the lower ones crowded; top acute; hairy as the first internode, below convex, above lengthwise depressed, gradually passing into the normal leaves, shrivelling rather quick. Petioles distinctly swollen, in cross-section semi-orbicular, 3rd one to 10 by 2 mm, 25th one to 25 by 5 mm, hairy as the first internode. Blade lanceolate, higher ones more obovate-lanceolate, finally linear, in 3rd one to 12 by 3.8 cm, in 25th one to 30 by 6.5 cm; base acute to wedge-shaped; top acuminate, later cuspidate; margin entire; above shining green, below lighter so, below on the nerves hairy as the hypocotyl; nerves regularly pinnate, midrib above prominent, below much so, lateral nerves below prominent, forming an undulating marginal nerve in the higher leaves.

Specimens: 1214 from S. Sumatra, primary hill forest on deep clay, low altitude.
Growth details: In nursery germination poor.
Remarks: Two species cultivated, similar in germination pattern and general morphology. *H. macrocoma* differs in details only.

KNEMA

Tree Fl. Malaya 1: 330.
Monograph: Sinclair, Gard. Bull. 18 (1961) 110.
Genus in India, S.E. and E. Asia, throughout Malesia, including W. New Guinea.

Knema latifolia Warb. – Fig. 137.

Development: The taproot emerges from one pole of the seed. The cotyledonary petioles elongate, carrying the plumule out of the seed. The shoot then starts elongating. No resting stage occurs.
Seedling hypogeal, cryptocotylar. Horsfieldia type and subtype.
Taproot long, slender, fibrous, brownish, with very few, hardly branched, brownish sideroots.
Hypocotyl hardly elongating, not distinguishable from the root.
Cotyledons 2, secund, succulent, when shed not known, between 2nd and 7th leaf

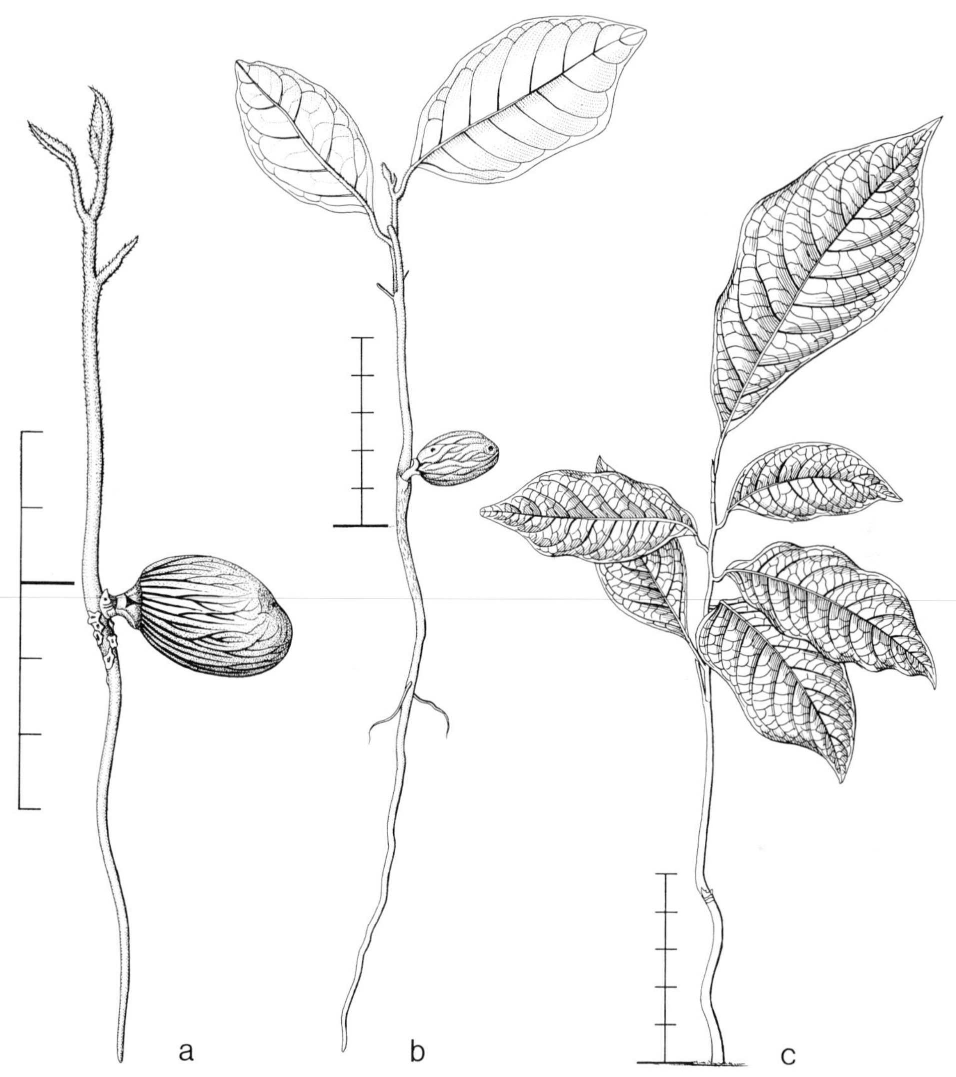

Fig. 137. *Knema latifolia.* De V. 1261, collected 11-iii-1972, planted 23-iii-1972. a and b date?, note development of a lateral shoot after damage to the growing point in b; c. 30-vii-1973.

stage; ruminate, without stipules, petiolate. Whole body ellipsoid, to 2.5 by 1.5 by 1.5 cm, the ± blackish seedcoat persistent. Petioles flattened, c. 10 by 4 mm, partly hidden in the seedcoat, warty, yellowish brown, glabrous.
Internodes: First one strongly elongating, terete, slender to 5 cm long, densely covered with tree-like, much branched hairs; next internodes as the first one, the lower ones to 1 cm long, subsequent ones usually not longer. Stem rather slender, ± straight.
Leaves all spirally arranged, simple, conduplicate-induplicate when very young,

without stipules, petiolate, subcoriaceous, green, paler and slightly bluish green below. First leaf much reduced, subulate, hairy as the first internode. Petiole in cross-section semi-orbicular, above flattened, in lower ones to 10 by 1.2 mm, in 10th one to 15 by 2.5 mm, hairy as the first internode. Blade ± oblong, later more obovate-oblong, in lower ones to 7.5 by 3.8 cm, in 10th one to 15 by 6.7 cm; base narrowed; top acute; margin entire; herbaceous, shining (yellowish) green above, dull bluish green below, above and below with scattered hairs as on the first internode, especially on the nerves and along the margin, soon glabrescent; nerves pinnate, midrib above slightly prominent, below much so, sidenerves above sunken, below prominent, curved at their tip, below in the blade free ending, towards the top of the leaf forming a much undulating marginal nerve.

Specimens: 1261 from S. Sumatra, badly disturbed primary forest on deep clay, low altitude.
Growth details: In nursery germination poor. In forest most fallen fruits attacked by boring insects.
Remarks: Two species cultivated, similar in germination pattern and general morphology. *K. cinerea* (several coll.) differs in details only.

MYRISTICA

Tree Fl. Malaya 1: 340.
Monograph: Sinclair, Gard. Bull. 23 (1968) 1–540.
Genus in India and Ceylon, S.E. Asia, Formosa, throughout Malesia, N. and N.E. Australia, the Pacific including the Carolines, extending as far as Samoa and Tonga Islands.

Myristica sp. – Fig. 138.

Development: The taproot emerges from one pole of the fruit. The cotyledonary petioles elongate, carrying the plumule out of the seed. The shoot then starts elongating. A long resting stage occurs with only the first flush of 4–6 leaves developed.
Seedling hypogeal, cryptocotylar. Horsfieldia type and subtype.
Taproot long, sturdy, brownish, lengthwise grooved, with few, slender, sturdy, unbranched, brownish sideroots.
Hypocotyl not enlarging.
Cotyledons 2, secund, succulent, in the 4th–6th leaf stage shed; ruminate, without stipules, petiolate. Whole body ellipsoid, to 4.5 by 2.5 by 2.5 cm, the dark brown testa persistent. Petiole in cross-section flattened, to 10 by 5 mm, brownish.
Internodes: First one terete, elongating to 5 mm long, dark green, densely covered with minute, simple, appressed, hyaline hairs; next internodes as the first one, the lower ones small and crowded, the higher ones irregular in size, 1–6 cm long, in the part with developed leaves to 2 cm long. Stem slender, straight, to 14 lower internodes with much reduced leaves, this part to 25 cm long.

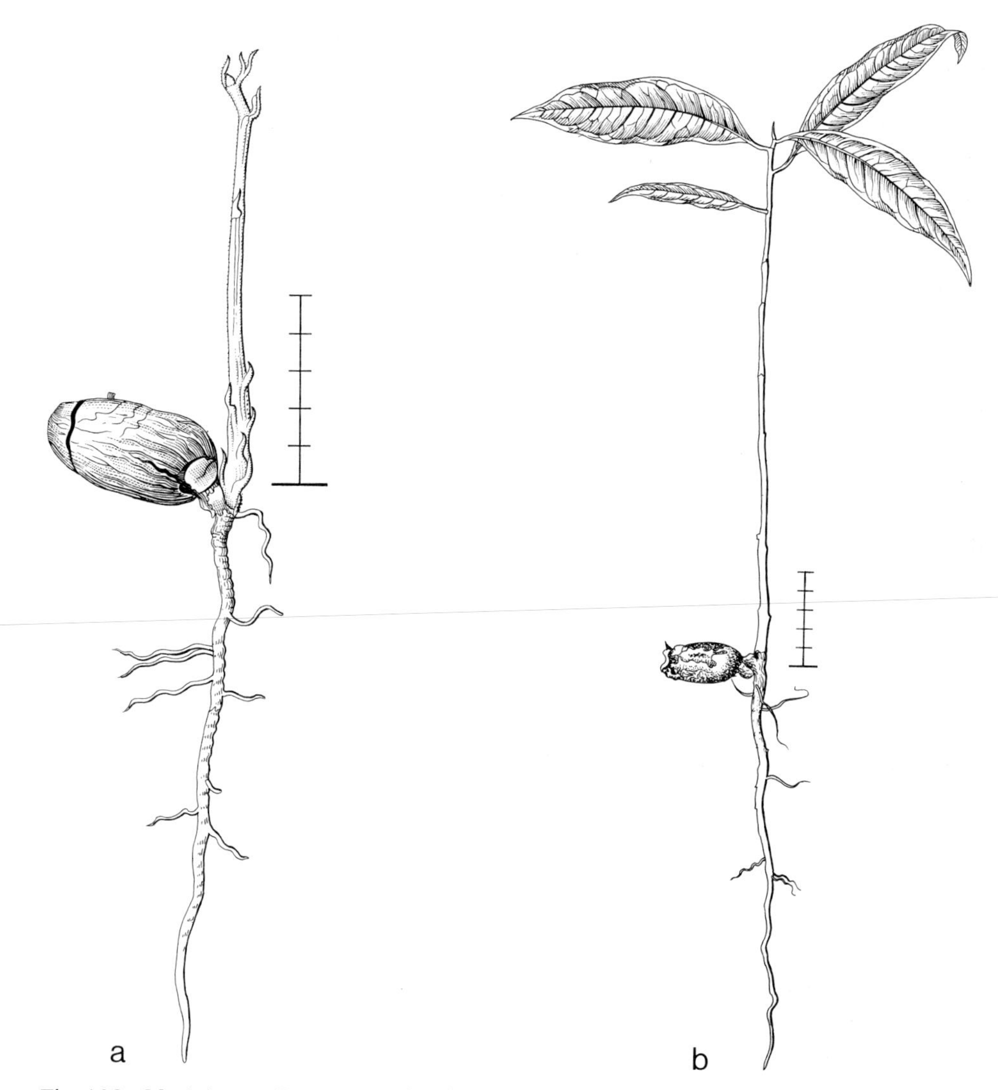

Fig. 138. *Myristica sp.* Bogor Bot. Gard. IV. G. 73, collected and planted i-1973. a. 21-vii-1973; b. 21-vii-1974.

Leaves all spirally arranged, simple, developing in flushes, without stipules, petiolate, subcoriaceous, green, much paler so below, the lower up to 14 scale-like, much reduced, soon dropping. Petiole in cross-section semi-orbicular, those in the first flush to 15 by 1.5 mm, above with descending midnerve, hairy as the first internode. Blade ovate-lanceolate, in lower developed leaves to 13 by 3.8 cm, in 10th one to 16.5 by 5.2 cm; base narrowed; top cuspidate; margin entire; below densely hairy as the first internode, especially on the nerves, above sparsely hairy with minute, branched, hyaline hairs; nerves pinnate, midrib above prominent, below much so, lateral nerves above sunken, below prominent, free ending.

Specimens: Bogor Botanic Gardens IV. G. 73, from Sumatra.
Growth details: In nursery germination fair, ± simultaneous.
Remarks: Three species cultivated, showing the same germination pattern and general morphology. *M. sp.* (Geesink 5729) and *M. fatua* (De Vogel 2568 & 2591) differ in details only.

MYRSINACEAE

ARDISIA

Fl. Java 2: 196; Burger, Seedlings (1972) 263 (*A. humilis*); Duke, Ann. Miss. Bot. Gard. 52 (1965) 346 (*A. glauciflora*); ibid. 56 (1969) 159 (*A. sp.*, two species); Lubbock, On Seedlings 2 (1892) 188 (*A. crenulata*, *A. japonica*, *A. mamillata*, *A. polycephala*).
Genus pantropic and in the subtropics; W. Africa, Madagascar, throughout the islands of the Indian Ocean, India and Ceylon, throughout S.E. and E. Asia up to Japan, all over Malesia, E. Australia, in the Pacific in Bismarcks, Solomons, Bonins, Fiji, and Tonga, tropical N., Central and S. America, including Cocos.

Ardisia macrophylla Reinw. ex Bl. – Fig. 139.

Development: The taproot and hypocotyl pierce one side of the seed. When the hypocotyl becomes erect the enclosed paracotyledons are carried above the soil. By spreading of the paracotyledons the testa is shed. A short resting stage occurs with expanded paracotyledons only. In 8 months to 12 cm high with 10 developed leaves.
Seedling epigeal, phanerocotylar. Macaranga type.
Taproot long, rather slender, flexuous, cream-coloured, with especially in the upper part rather long, slender, hardly branched, cream-coloured sideroots.
Hypocotyl strongly elongating, terete, to 3 by 0.2 cm, green, densely covered with minute, brown, simple hairs.
Paracotyledons 2, opposite, foliaceous, when shed not known; without stipules, shortly petiolate, herbaceous, green. Petiole terete, above channelled, to 1.4 by 0.8 mm, hairy as the hypocotyl. Blade ovate, often very irregular in shape, to 2-3 by 1.9 cm; base obtuse; top slightly retuse; margin entire; below, along the margin and especially on the midnerve slightly hairy as the hypocotyl; nerves pinnate, midrib above sunken, below very prominent, lateral nerves not very conspicuous, forming an inconspicuous marginal nerve.
Internodes: First one hardly elongating, terete, a few mm long, pale green, hairy as the hypocotyl; next internodes as the first one, the lower ones very crowded and short, higher ones gradually becoming larger, irregular in length, 2–8 cm long. Stem rather sturdy, straight.
Leaves all spirally arranged, simple, involute when young, without stipules, almost

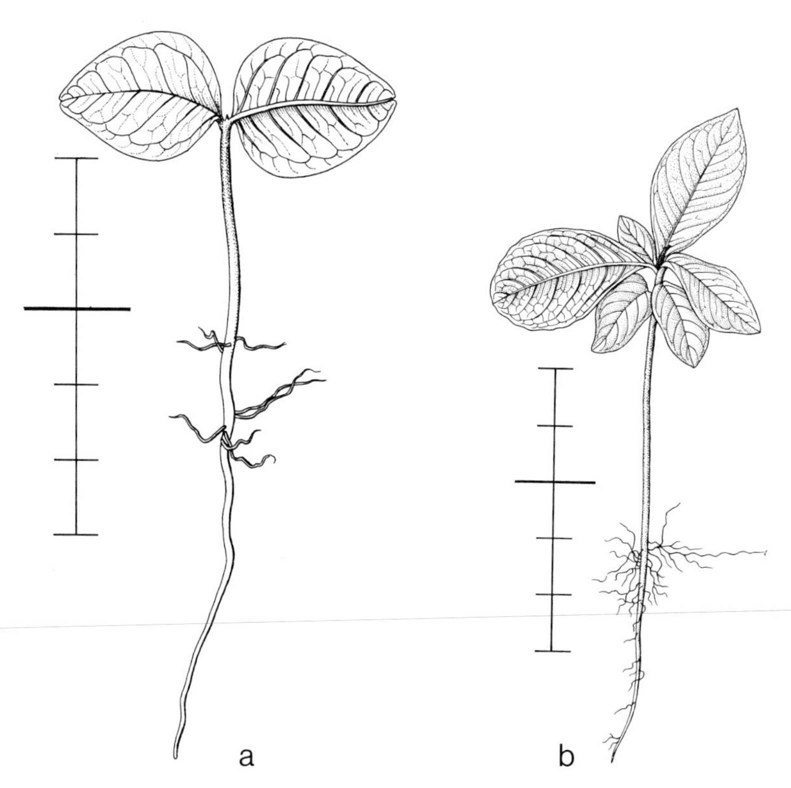

Fig. 139. *Ardisia macrophylla.* De V. 720, collected 11-ix-1971, planted 7-xii-1971. a and b date?.

sessile, the higher ones shortly petiolate, herbaceous, light green, when young below densely hairy as the hypocotyl, more or less glabrescent. Petiole in higher leaves in cross-section semi-orbicular, in 13th leaf 6 by 4 mm. Blade in lower leaves often irregularly shaped, ± ovate, to 3 by 2 cm, in the higher leaves more obovate-oblong, gradually becoming larger, in 25th leaf to 34 by 13 cm; base narrowed; in the lower leaves top roundish, in the higher ones acute to acuminate; margin entire; nerves pinnate, midrib above slightly sunken, below very prominent, lateral nerves below prominent, ending in anastomoses.

Specimens: 720 from S.E. Borneo, alluvial flat, primary forest on deep clay, low altitude.
Growth details: In nursery germination fair.
Remarks: Within the genus different germination patterns are found. *A. crenulata* has the paracotyledons secund, remaining within the envelopments, and the shoot with some leaves developed. The germination of *A. japonica* is not clear, it seems that the enclosed paracotyledons are carried on top of the hypocotyl, and are shed before the shoot starts developing. *A. glauciflora*, *A. humilis*, *A. mamillata*, *A. polycephala*, *A. macrophylla*, and *A. sp.* (2 species) have the paracotyledons finally freed from the

envelopments; they sometimes enlarge considerably, and may be morphologically very different.

MYRTACEAE

TRISTANIA

Fl. Java 1: 347; Burger, Seedlings (1972) 277 (*T. conferta*); Lubbock, On Seedlings 2 (1892) 535 (*T. conferta*).
Genus in Malesia but not in the Phillippines, N. and E. Australia, in the Pacific in New Caledonia and Fiji.

Tristania sp. – Fig. 140.

Development: The hypocotyl and taproot emerge from the free margin of the small, winged seed. After establishment of the root the paracotyledons are freed from the testa by spreading. A short resting stage occurs with the paracotyledons expanded.
Seedling epigeal, phanerocotylar. Macaranga type.
Taproot not very pronounced, with many long, slender, branched, white sideroots.
Hypocotyl elongating, in cross-section ± quadrangular, to 7 mm long, slender, cream-coloured, glabrous.
Paracotyledons 2, opposite, foliaceous, when dropped not exactly known, between 7th and 10th leaf stage; without stipules, petiolate, dark reddish turning green, herbaceous, glabrous. Petiole flattened, 0.7 by 0.5 mm. Blade obreniform, to 1.8 by 3 mm, 1-nerved; punctate with many hyaline dots.
Internodes: First one elongating to 3.5 mm long, in cross-section quadrangular, pale reddish, longitudinally winged at the angles, and there with short, simple, hyaline to brown hairs; next internodes as the first one, the higher ones in cross-section distinctly flattened to angular, lower ones to 2 mm long, the higher ones slightly increasing in length, 20th internode to 8 mm, green to cream-coloured, turning brown. Stem rather slender, straight.
First two leaves opposite, simple, sessile, without stipules, herbaceous, green, glabrous. Blade obovate, to 3.2 by 2 mm; base narrowed; top rounded, margin entire; herbaceous, green, glabrous, with many small hyaline dots; main nerve above sunken, below slightly prominent, with 2 lateral nerves.
Next leaves spirally arranged, gradually becoming shortly petiolate. Petiole in 40th leaf to 3 by 2 mm. Blade in higher leaves oblong, gradually becoming bigger, the 16th one to 15 by 5 mm, 22nd one to 24 by 8 mm; base from 40th leaf onward obtuse; top acute, slightly apiculate; margin entire; at the margin and below, especially on the midnerve, slightly hairy as the first internode; nerves pinnate, midrib above sunken, below prominent, especially in the higher ones, lateral nerves at first inconspicuous, later better visible, thin, above and below slightly raised, forming in the higher leaves a clear marginal nerve.

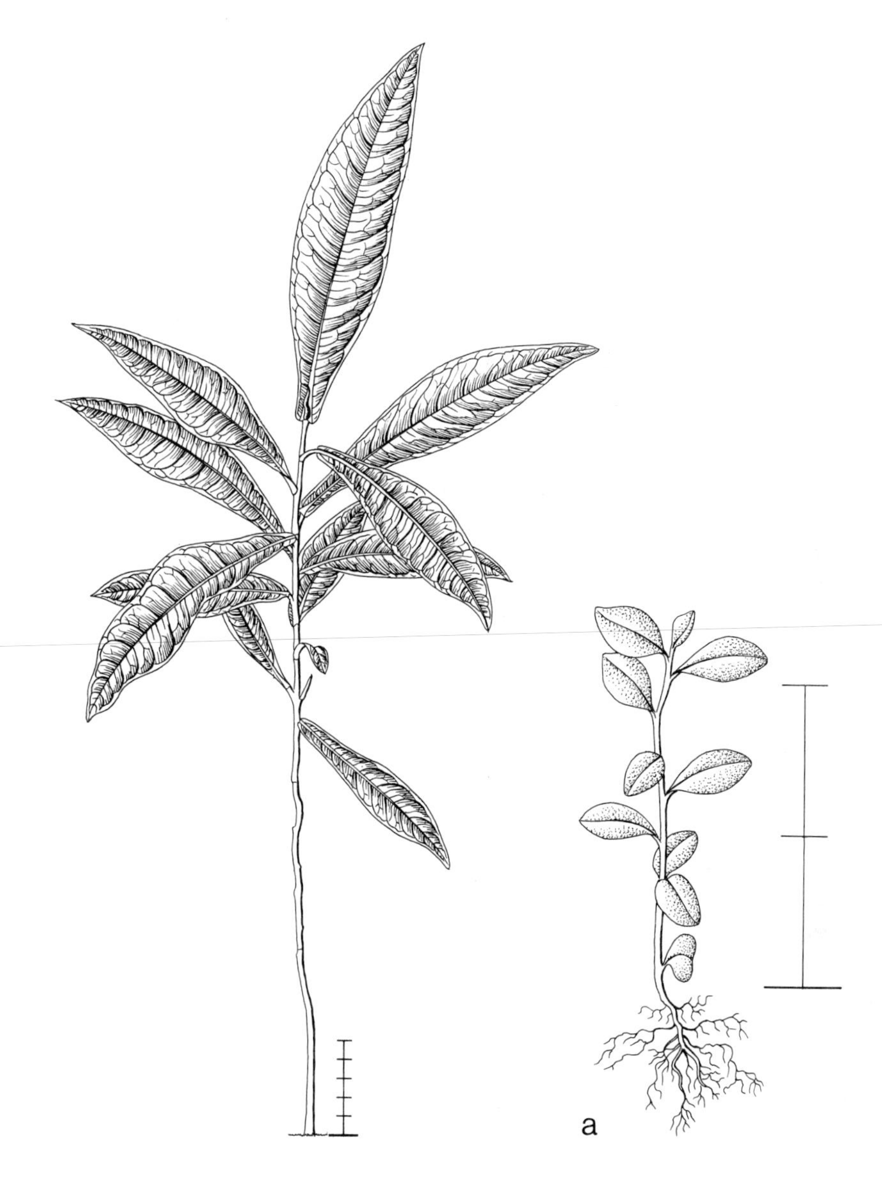

Fig. 140. *Tristania sp.* De V. 1438, collected 12-vii-1972, planted 27-vii-1972. a date?, one paracotyledon already dropped; b. 3-vii-1974.

Specimens: 1438 from N. Sumatra, alluvial flat, somewhat depleted primary forest along river, on deep clay, low altitude.
Growth details: In nursery germination poor, the small, winged seeds are easily washed away by heavy rains.
Remarks: Two species described, similar in germination pattern and general morphology. *T. conferta* differs in details only.

OLACACEAE

STROMBOSIA

Tree Fl. Malaya 2: 305; Gilbert, INEAC, Sér. Scient. 17 (1939) 1 (*S. glaucescens*); Heckel, Ann. Mus. Colon. Marseille 8 (1901) 17 (*S. javanica*, *S. lucida*, *S. membranacea*); Mensbruge, Germination (1966) 106 (*S. glaucescens*); Sleumer, Nat. Pflanzenfam. ed. 2, 16b (1935) 9 (*S. javanica*, *S. lucida*, *S. membranacea*); Voorhoeve, Liberian high forest trees (1965) 303 (*S. glaucescens*).
Genus in tropical Africa, India and Ceylon, throughout S.E. Asia, all over W. Malesia, including the Philippines.

Strombosia javanica Bl. – Fig. 141.

Development: The hypocotyl and taproot emerge from the acute pole of the seed. The hypocotyl becomes erect, carrying the cotyledons enclosed in splitting seedcoat and endosperm above the soil. After a resting period, the cotyledons which are still enclosed in the endosperm (and often partly the testa) are shed, and the shoot starts developing.
Seedling epigeal, cryptocotylar. Blumeodendron type.
Taproot long, sturdy, creamy-white, with rather few, not branched, slender, creamy-white sideroots.
Hypocotyl strongly elongating, terete, to 13 cm long, clearly marked off from the root, slightly tapering to the top, and there with two short longitudinal grooves, dark green, glabrous.
Cotyledons 2, opposite, on top of the hypocotyl, never freed, pushed off by the developing shoot. Whole body ellipsoid, to 3.5 by 2.4 by 2.4 cm, base acute, top rounded, the yellowish endosperm and the splitting, often partly shed, brownish seedcoat persistent.
Internodes: First one strongly elongating, to 3 cm long, terete, green, glabrous; next internodes as the first one, irregular in length, 10th one to 5.5 cm long, 15th one to 12 cm long, but often much smaller. Stem slender, slightly zig-zag.
Leaves all spirally arranged, simple, conduplicate-induplicate when very young, without stipules, long petiolate, herbaceous, green, glabrous. Petiole in cross-section semi-orbicular, in the first one to 1.5 by 0.1 cm, in 5th one to 2.5 by 0.15 cm, in 10th one to 6.5 by 0.15 cm, at the top thickened and there bent upwards, in higher leaves also thickened at the base. Blade in first leaves ovate, to 5.5 by 3.5 cm, in higher ones more oblong, in 5th one 12 by 5 cm, in 10th one to 18 by 7.5 cm; base acute; top cuspidate; margin entire; nerves pinnate, above sunken, below prominent, especially the midrib, lateral nerves ending in anastomoses.

Specimens: 1391 from W. Java, primary forest on clay, low altitude.
Growth details: In nursery germination good, ± simultaneous. In the forest germinating seeds abundant around the parent tree.
Remarks: In *S. javanica* the cotyledons remain within the endosperm and testa. In *S. glaucescens*, *S. lucida* and *S. membranacea* the paracotyledons become exposed.

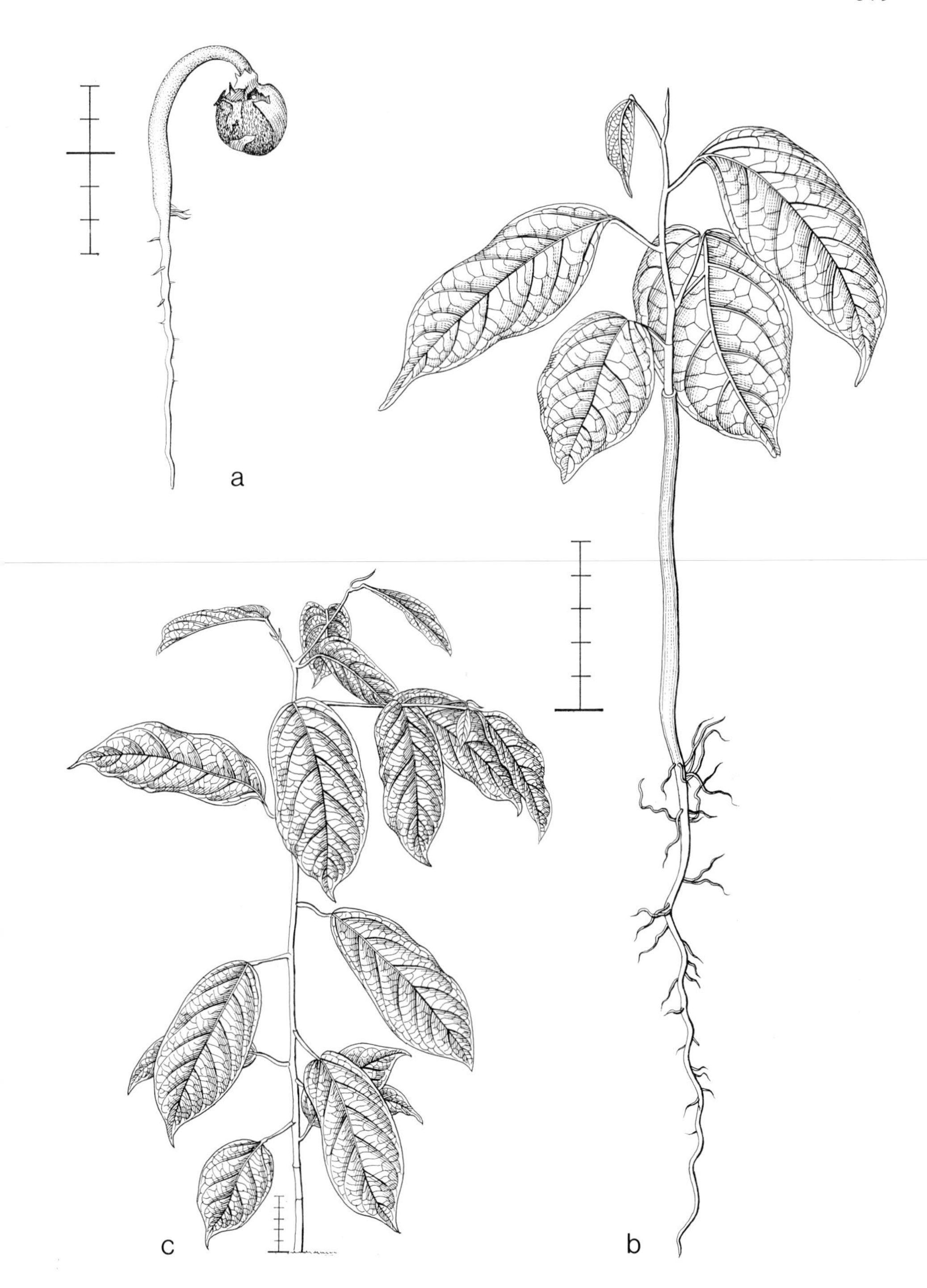

Fig. 141. *Strombosia javanica.* De V. 1391, collected 22-vi-1972, planted 2-vii-1972. a and b date?; c. 2-viii-1973.

PROTEACEAE

HELICIA

Fl. Males. I, 5: 164.
Genus in India and Ceylon, S.E. and E. Asia including S. Japan, all over Malesia, and N.E. and E. Australia.

Helicia serrata (R. Br.) Bl. – Fig. 142.

Development: The taproot emerges from the blunt pole of the fruit. The cotyledons spread slightly, at the base they are pushed aside by slight elongation of the cotyledonary petioles and because the cotyledonary node grows in thickness, thus enabling the shoot to emerge. No resting stage occurs.
Seedling hypogeal, cryptocotylar. Heliciopsis type and subtype.
Taproot rather short, sturdy, often branched, hard fleshy, pale brown, with many long, slender, branched, cream-coloured sideroots.
Hypocotyl not or little elongating.
Cotyledons 2, secund, succulent, in the 12th–20th leaf stage shed; simple, without stipules, minutely petiolate, slightly peltate, spreading at the tip and pushed aside at the base. In the forest the fruit wall in fallen fruits is always eaten by animals. Blade in outline ovate, to 3.2 by 2.2 by 1 cm; base abruptly retuse with a deep incision, auricles triangular, to 4 by 4 mm, but often smaller; top acute, but often irregular; margin entire; above ± flat, below concave, ± halfway with a cross line, the apical part more or less rough, the lower part smooth, nerves not visible.
Internodes: First one slightly elongating, terete, to 7 mm long but often smaller, green, rather densely hairy with minute, simple, hyaline hairs, rather soon glabrescent; next internodes as the first one, irregular in length, second one to 4 cm long but often smaller, 10th one usually not exceeding 2 cm. Stem slender, rather straight, often branched at the lower internodes.
Leaves all spirally arranged, simple, without stipules, petiolate, subcoriaceous, green. Petiole in cross-section semi-orbicular, in the first leaf to 3 by 1.5 mm, in the 5th leaf to 5 by 2 mm, hairy as the first internode. Blade ovate-oblong, later ones more elliptic, first one often much reduced to absent, second one to 3.5 by 1.6 cm, 5th one to 6 by 3.3 cm; base narrowed into the petiole; top acuminate, its tip obtuse and apiculate; margin widely irregularly serrate, teeth apiculate; below on the midnerve slightly hairy; nerves pinnate, midrib above and below raised, lateral nerves above slightly sunken, below slightly raised, free ending in the teeth.

Specimens: 2673 from W. Java, primary hill forest on deep clay, medium altitude.
Growth details: In nursery germination fair, ± simultaneous.

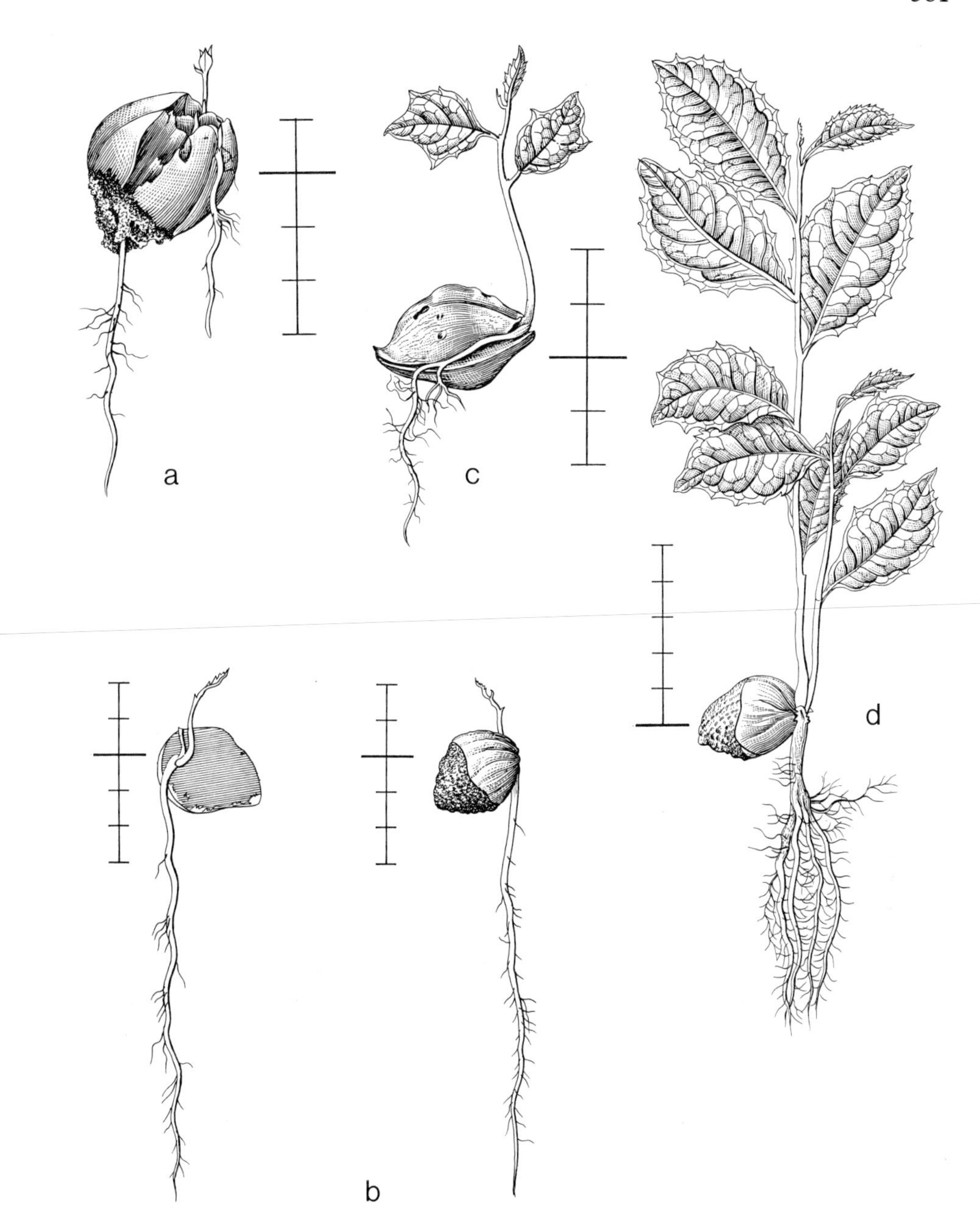

Fig. 142. *Helicia serrata.* De V. 2673, collected and planted 18-xi-1973. a and b. 7-ii-1974; c and d. 27-iii-1974.

HELICIOPSIS

Fl. Males. I, 5: 190, f. 12 on p. 165.
Genus in S.E. Asia and W. Malesia, including S. Philippines.

Heliciopsis velutina (Prain) Sleum. – Fig. 143.

Development: The taproot emerges from one pole of the fruit, the elongating cotyledonary petioles carry the plumule free from the envelopments, after which the shoot starts elongating. A resting stage occurs with 2 developed leaves.
Seedling hypogeal, cryptocotylar. Heliciopsis type and subtype.
Taproot long, slender, rather sturdy, brownish, with rather many short, slender, unbranched sideroots.
Hypocotyl slightly enlarging, terete, to 2 cm long, green with brownish flakes.
Cotyledons 2, secund, succulent, in the 4th–5th leaf stage shed; simple, without stipules, shortly petiolate, hard fleshy, green where exposed, glabrous. Whole body ellipsoid, to 4.5 by 2 by 1.5 cm, with rounded poles, the fibrous, slowly disintegrating endocarp persistent.
Internodes: First one strongly elongating, in cross-section elliptic, to 14 cm long, slightly tapering to the top, dark green, rather densely covered with simple, appressed, brownish hairs, glabrescent; next internodes as the first one, irregular in length, 1–5 cm long, but generally not exceeding 2 cm. Stem slender, rather straight.
First two leaves opposite, simple, without stipules, petiolate, subcoriaceous, green. Petiole in cross-section semi-orbicular, to 8 by 2 mm, slightly hairy as the first internode. Blade ± elliptic, but often more or less irregular in form, to 11 by 6 cm; base acute to narrowed; top acuminate; margin entire, slightly irregular; slightly hairy as the first internode, especially below on the nerves; nerves pinnate, midrib above slightly raised, below very prominent, lateral nerves above sunken, below prominent, ending in anastomoses.
Next leaves spirally arranged, slowly increasing in length. Blade more obovate-oblong, in 5th one to 14.5 by 6 cm, in 9th one to 20 by 6.5 cm; base narrowed, from the 10th one onward 3-lobed to 3-partite, further as the first leaves.

Specimens: 1074 from S.E. Borneo, primary forest on deep clay, medium altitude.
Growth details: In nursery germination fair, ± simultaneous. After damage of the main axis, sprouting from the cotyledons and lateral buds occurred.

RHAMNACEAE

VENTILAGO

Fl. Java 2: 81; Cremers, Candollea 28 (1973) 264 (*V. africana*).
Genus pantropic; tropical Africa and Madagascar, India and Ceylon, throughout S.E. and tropical E. Asia, all over Malesia, E. Australia, in the Pacific as far as Fiji,

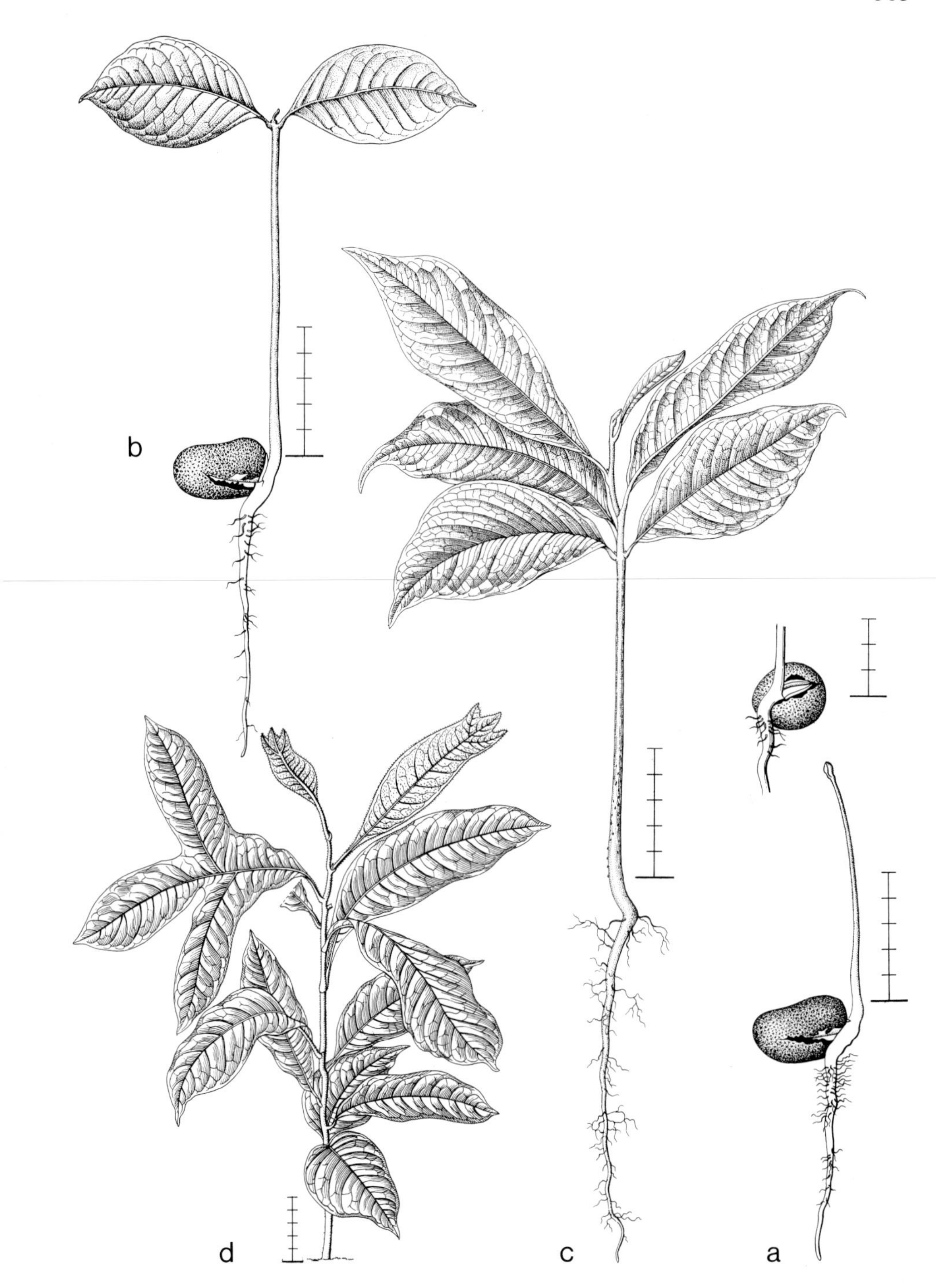

Fig. 143. *Heliciopsis velutina.* De V. 1074, collected 27-xi-1971, planted 7-xii-1971. a, b, c, and d date?.

including the W. Carolines and New Caledonia, also in tropical S. America.

Ventilago oblongifolia Bl. – Fig. 144.

Development: The taproot and hypocotyl emerge from the free pole of the fruit opposite the wing. The cotyledonary petioles elongate, pushing the plumule free from the envelopments, after which the shoot starts developing. No resting stage occurs.
Seedling hypogeal, cryptocotylar. Heliciopsis type and subtype.
Taproot long, slender, fibrous, dark brown, with rather few, slender, few-branched, dark brown sideroots.
Hypocotyl elongating to 1.3 cm long, terete, for the bigger part subterranean, the epigeal part green, rather densely covered with minute, simple, brownish hyaline hairs.
Cotyledons 2, secund, succulent, in the 4th leaf stage dropped; simple, without stipules, petiolate. Whole body globular, to 7 by 7 mm, the dark brown, winged fruit wall persistent, with a distinctly elevated collar near the free pole, the opposite pole with a lanceolate, sessile, shortly apiculate, dark brown, the 3-nerved wing to 8.5 by 2 cm. Petiole exceeding the fruit wall, in cross-section semi-orbicular, to 2 by 1 mm, above channelled, hairy as the hypocotyl. Blade at the base projecting from the fruit wall, there auriculate, auricles triangular, to 2 by 2 mm, their tip acute, yellowish green, glabrous.
Internodes: First one strongly elongating, terete, to 2.7 cm long, green, densely covered with minute, simple, brownish hairs; next internodes as the first one, to 1.5 cm long, higher on the stem to 4.5 cm long. Stem slender, zig-zag, later pendulous.
First two leaves opposite, simple, stipulate, shortly petiolate, herbaceous, pale reddish when young, turning green. Stipules 2 at the base of each petiole, narrowly triangular, to 1.5 by 0.3 mm, shrivelling soon and turning brown but persistent. Petiole ± terete, to 3 by 0.8 mm, hairy as the first internodes. Blade quite irregular in form, in general ovate, to 3 by 2 cm; base rounded to obtuse; top acute; margin entire to distinctly serrate towards the top; below at the base slightly hairy, soon glabrescent; nerves pinnate, above slightly raised, the midrib below prominent, lateral nerves slightly so, ending free.
Next leaves spirally arranged. Blade oblong, in 3rd leaf to 5.5 by 2.4 cm, in 14th leaf to 16 by 5.5 cm; base slightly exsculptate to obtuse; top acuminate to cuspidate; margin with a few, scattered, inconspicuous teeth especially near the top, further as the first ones.

Specimens: 2310 from S.E. Borneo, alluvial flat, primary forest on deep clay, low altitude.
Growth details: In nursery germination good, ± simultaneous.
Remarks: It seems that in *V. africana* the cotyledons become exposed; they are carried on an epigeal hypocotyl.

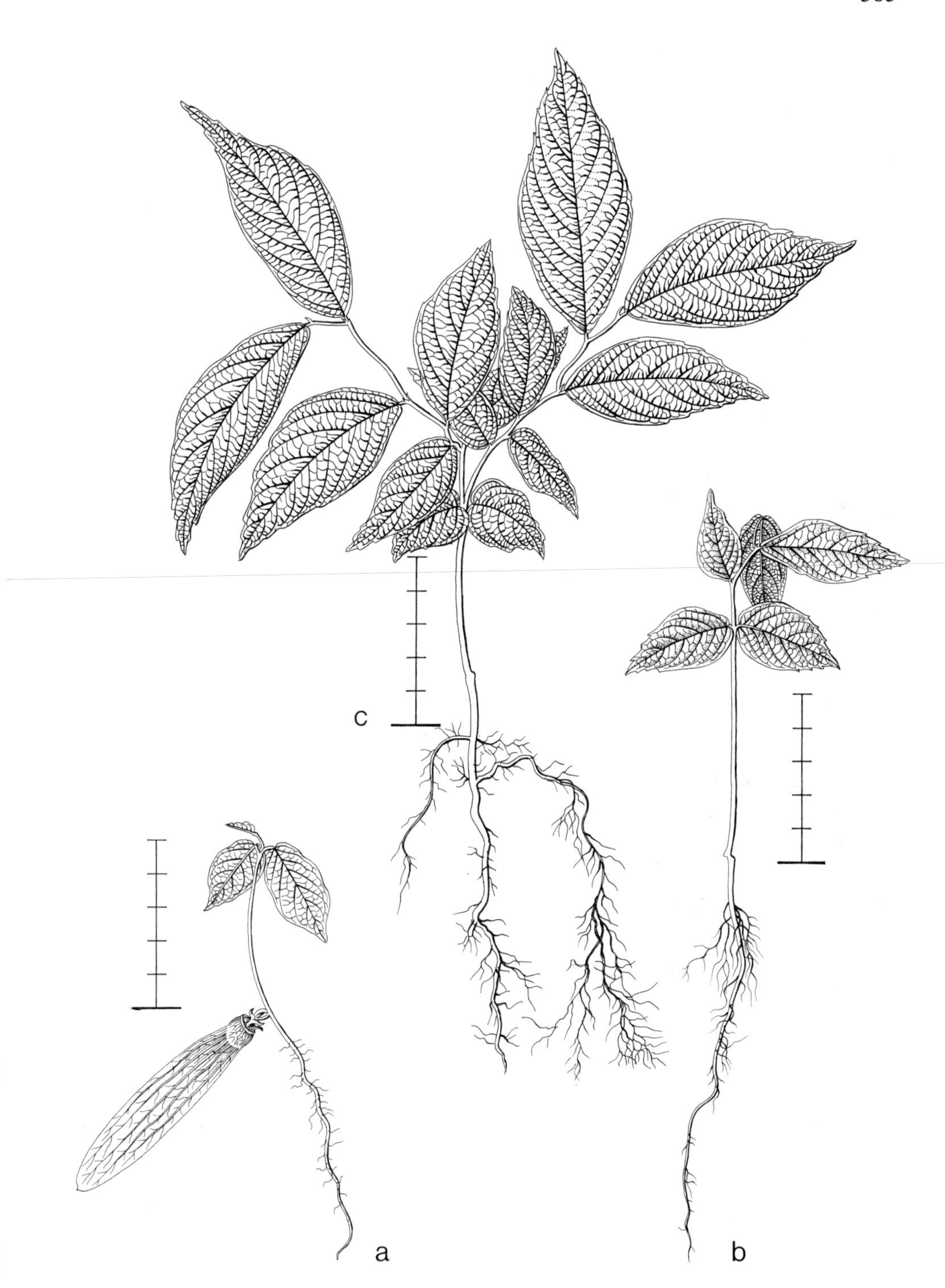

Fig. 144. *Ventilago oblongifolia.* De V. 2310, collected 13-iii-1973, planted 22-iii-1973. a. 7-iv-1973; b. 29-v-1973; c. 11-xi-1973.

ZIZIPHUS

Fl. Java 2: 81; Csapody, Keimlingsbestimmungsbuch der Dikotyledonen (1968) 135 (*Z. jujuba*); Duke, Ann. Miss. Bot. Gard. 52 (1965) 342 (*Z. jujuba*); Hickel, Graines et Plantules (1914) 300 (*Z. sativa*); Troup, Silviculture 1 (1921) 211 (*Z. jujuba*, *Z. xylopyrus*).
Genus pantropic; tropical Africa, Mediterranean, India and Ceylon, S.E. and E. Asia, all over Malesia, N. Australia, in the Pacific in Bismarck and Solomons, tropical S. and Central America.

Ziziphus angustifolius (Miq.) Hatusima – Fig. 145.

Development: The taproot and hypocotyl emerge from one pole of the seed. The swelling and slightly spreading cotyledons throw the testa off, and are withdrawn from the envelopments by the hypocotyl which becomes erect. A short resting stage occurs with the cotyledons and two leaves developed.
Seedling epigeal, phanerocotylar. Sloanea type and subtype.
Taproot long, slender, fibrous, yellowish brown, with few, hardly branched, yellow-tipped, slender sideroots.
Hypocotyl strongly elongating, terete, to 4 cm long, slender, orange-green, densely covered with minute, simple, hyaline hairs.
Cotyledons 2, opposite, succulent, in 2nd–3rd leaf stage dropped; more or less spreading, sessile, without stipules, hard fleshy, orange to greenish, glabrous. Blade in outline semi-ellipsoid, to 1.3 by 0.9 cm, base distinctly auriculate, auricles acute, 4 by 4 mm, often somewhat unequal in size; top obtuse, separated from the rest of the blade by a deep groove; margin entire; lower side concave, upper side more or less flat; nerves not visible.
Internodes: First one strongly elongating, terete, to 5 cm long, green, hairy as the hypocotyl; next internodes as the first one, 2nd one to 4 cm long, but in general much shorter, 5th one to 1.5 cm long, 10th one to 2 cm long. Stem slender, slightly zig-zag.
First two leaves opposite, simple, stipulate, shortly petiolate, herbaceous, green. Stipules 2 at the base of each leaf, subulate, to 1 by 0.2 mm, hairy as the hypocotyl. Petiole in cross-section semi-orbicular, to 4 by 1 mm, above distinctly channelled, hairy as the hypocotyl. Blade ovate, to 5 by 3 cm; base obtuse, abruptly narrowed into the petiole; top cuspidate, its tip acute; margin serrulate, with minute, yellowish hyaline glands; below on the nerves slightly hairy as the hypocotyl, glabrescent; 3-nerved, the main nerves above sunken, below very prominent, the lateral nerves forming a ⊥ angle to the main nerves, the marginal ones forming a faint marginal nerve.
Next leaves spirally arranged, involutely folded along the lateral main nerves when very young. Stipules 2 at the base of each leaf, subulate, up to the 15th leaf to 1 by 0.3 mm, hairy as the hypocotyl. Petiole in cross-section semi-orbicular, in 5th leaf to 5 by 1 mm, in 10th leaf to 7 by 1.5 mm. Blade lanceolate to oblong, in 5th leaf to 6.5 by 2 cm, in 10th one to 13 by 4.5 cm; base acute; top cuspidate; margin slightly to distinctly serrulate, especially towards the top, further as the first leaves.

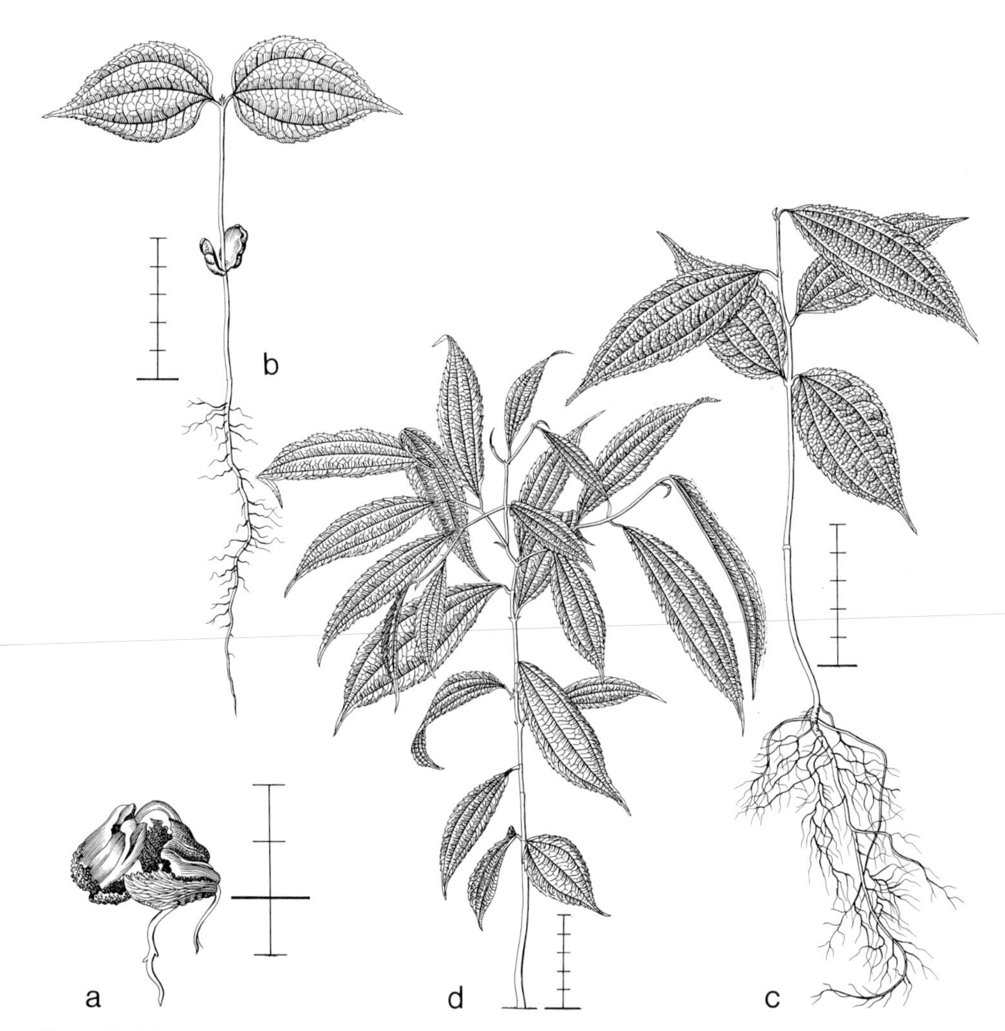

Fig. 145. *Ziziphus angustifolius.* De V. 2196, collected 6-iii-1973, planted 22-iii-1973. a and b. 27-iii-1973; c. 17-vii-1973; d. 4-iii-1974.

Specimens: 2196 from S.E. Borneo, alluvial flat, primary forest on deep clay, low altitude.
Growth details: In nursery germination fair, ± simultaneous.
Remarks: Z. angustifolius and *Z. sativa* differ from the other two described species in the more distinct hypocotyl, the succulent cotyledons, and the extraction mechanism of the seedling. *Z. jujuba* and *Z. xyloporus* have the hypocotyl relatively short, the cotyledons foliaceous with rather long petioles. In *Z. xyloporus* the first two leaves are quite small and not always opposite.

ROSACEAE

PARINARI

Tree Fl. Malaya 2: 332; Mensbruge, Germination (1966) 124 (*P. chrysophylla* = *Maranthes*, *P. congensis*, *P. excelsa*, *P. glabra* = *Maranthes*, *P. holstii*, *P. robusta* = *Maranthes*); Voorhoeve, Liberian high forest trees (1965) 314 (*P. excelsa*).
Revision: Kostermans, Reinwardtia 7 (1965) 147.
Genus pantropic; tropical Africa, India and Ceylon, throughout S.E. Asia, all over Malesia, N.E. Australia, in the Pacific in the Solomons, Fiji, Samoa and Tonga; Central and tropical S. America.

Parinari elmeri Merr. – Fig. 146.

Development: The taproot emerges from one pole of the hard, woody endocarp. By elongation of the cotyledonary petioles the plumule is pushed out of the fruit, after which the shoot starts developing. No resting stage occurs.
Seedling hypogeal, cryptocotylar. Heliciopsis type and subtype.
Taproot long, slender, flexuous, brownish cream, with rather few, quite long, shortly branched, brownish cream sideroots.
Hypocotyl not enlarging.
Cotyledons 2, secund, succulent, when shed not known, but long persistent; simple, without stipules, shortly petiolate. Whole body ellipsoid, to 5 by 3 by 2.5 cm, the dark brown, hard stone, of which the outside is finely fibrous, persistent.
Internodes: First one slightly elongating, terete, to 7 mm long, densely covered with long, more or less appressed, simple, white, woolly hairs; next internodes as the first one, the lower ones very small, gradually becoming longer, those in the scaly part to 2 cm long, the lowest ones in the leafy part to 1.5 cm long, 10th leafy internode to 1.5 cm long, the 30th one to 2 cm long. Stem erect, slender, later drooping.
Leaves spirally arranged, the higher ones alternate and turned in one plane, simple, spreading, when young enclosed by the stipules, stipulate, shortly petiolate, slightly coriaceous, the lower up to 15 scale-like. Scales sessile, appressed to the stem, without stipules, herbaceous, green, rather soon shrivelling but persistent; blade triangular, in the highest ones the largest, to 10 by 5 mm; top acuminate; margin entire; on the midnerve below hairy as the first internode, on the blade below with shorter, felt-like, appressed white indument; 1-nerved, nerve below prominent. Stipules 2 at the base of each petiole, sessile, slightly asymmetric, rather soon shrivelling but often long persistent, narrowly triangular, in lowest ones to 12 by 4 mm, in 40th one to 20 by 4 mm; top cuspidate, margin entire; below hairy as the scales, distinctly 1-nerved, nerve asymmetrically placed, below distinctly prominent, above some additional nerves visible. Petiole in cross-section semi-orbicular, to 4 by 1.5 mm, hardly gaining in length, in lower ones distinctly curved, above with a distinct band of brown, woolly hairs, from 20th leaf onwards on both sides of this band ± halfway with a black, round gland, further as in the first leaves, hairy as the first internode. Blade ovate-lanceolate, in 5th one to 10 by 3 cm, in 20th one to 13.5 by 3.6 cm, in 30th one to 14 by 4 cm; base rounded; top cuspidate, margin entire, but

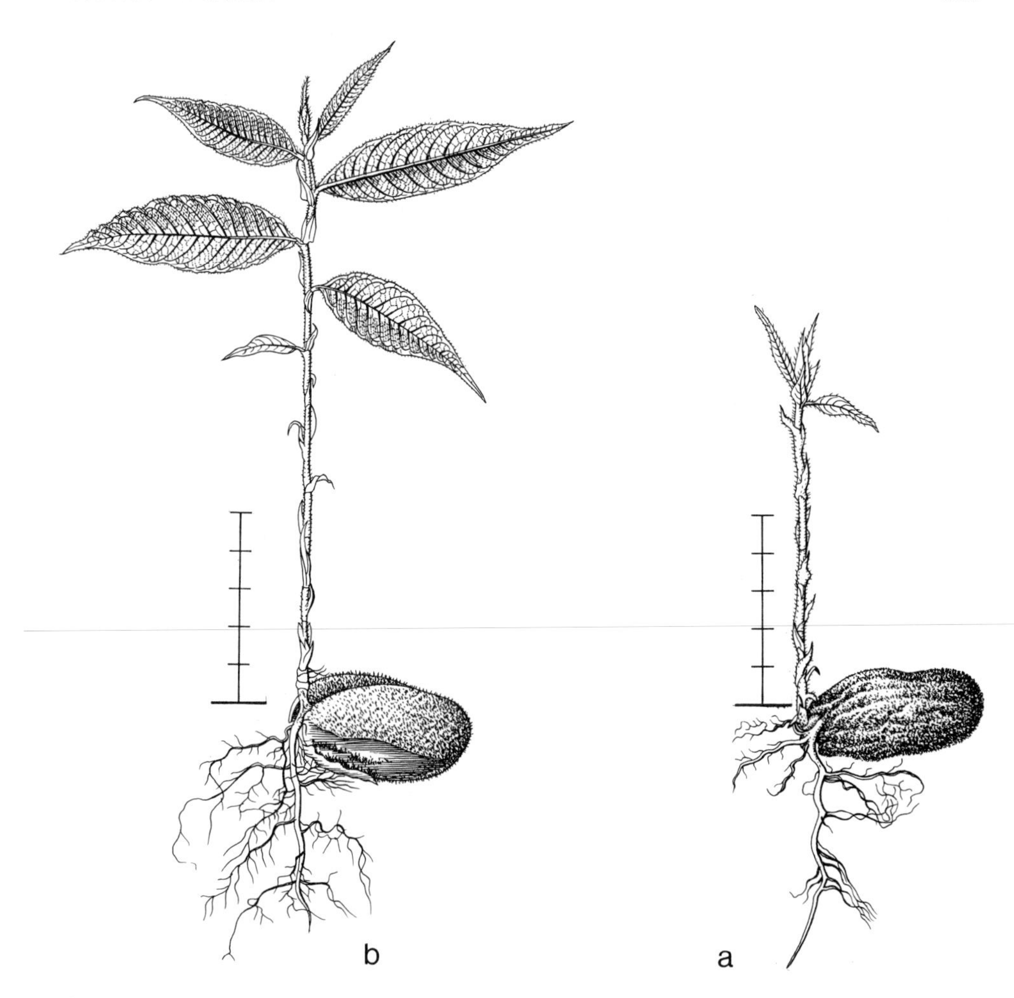

Fig. 146. *Parinari elmeri.* De V. 2355, collected 16-iii-1973, planted 22-iii-1973. a and b. 26-ix-1973.

slightly irregular, above green, below almost white by dense indument of simple, appressed, white woolly hairs, on the nerves hairy as the first internode; when young above with rather dense indument as below, soon glabrescent, above the midnerve at the base hairy as the upper side of the petiole; nerves pinnate, above hardly raised, below very prominent, especially the midrib, lateral nerves free ending.

Specimens: 2355 from S.E. Borneo, primary forest on deep clay, medium altitude.
Growth details: In nursery germination fair, long delayed.
Remarks: P. chrysophylla, *P. glabra*, and *P. robusta* have recently been transferred to the genus *Maranthes*. The foodstoring cotyledons become exposed and are borne on a long hypocotyl. The first two leaves are opposite, the subsequent ones spirally arranged (in *P. glabra* not known).

PRUNUS

Tree Fl. Malaya 2: 337; Csapody, Keimlingsbestimmungsbuch der Dikotyledonen (1968) 93, 95, 253, 255 (*P. amygdalus* = *A. communis* and *P. tenella* = *A. nana*, under *Amygdalus*; *P. armeniaca* = *A. vulgaris*, under *Armeniaca*; *P. avium*, *P. canadensis*, *P. cerasus* = *C. vulgaris*, *P. mahaleb* and *P. serrulata* under *Cerasus*; *P. laurocerasus* = *L. officinalis*, under *Laurocerasus*; *P. padus* = *Padus avium* and *P. serotina*, under *Padus*; *P. persica* = *Persica vulgaris*, under *Persica*; *P. domestica*, *P. insititia*, and *P. spinosa*, under *Prunus*); Schopmeyer, Agric. Handb. 450 (1974) 33, 658 (*P. alleghaniensis*, *P. americana*, *P. pensylvanica*, *P. virginiana*); Troup, Silviculture 2 (1921) 488 (*P. padus*).
Revision: Kalkman, Blumea 8 (1965) 1.
Genus (in broad sense) cosmopolitan; N. temperate regions, subtropical and tropical Africa, all over Asia, throughout Malesia, the Pacific, and America.

Prunus grisea (C. Muell.) Kalkm. var. *grisea* – Fig. 147.

Development: The stone splits longitudinally in the centre, beginning from the blunt pole. The taproot emerges from the blunt pole, and the shoot from between the cotyledons through the slit. In general the two valves of the stone remain connate at the acute pole. A short resting stage occurs with two expanded leaves.
Seedling hypogeal, cryptocotylar. Heliciopsis type and subtype.
Taproot rather long, slender, flexuous, pale brown, with rather few, slender, few-branched, brownish cream sideroots.
Hypocotyl not enlarging.
Cotyledons 2, opposite to secund, succulent, in the 2nd–3rd leaf stage shed; simple, slightly peltate, almost sessile, without stipules. Blades close together, flattened where touching, outside concave. Whole body heart-shaped, rounded at the base where split, at the top acute, to 1.3 by 1.2 by 0.8 cm, the hard endocarp persistent, covering the convex side of the blades.
Internodes: First one strongly elongating, terete, to 11 cm long, but usually smaller, greenish brown, densely hairy with short, simple, appressed, brownish hairs; next internodes as the first one, second one to 3.5 cm long, 5th one to 5.5 cm long. Stem slender, straight, rather soon branched.
First two leaves opposite, simple, stipulate, petiolate, herbaceous, green, but sometimes one or both more or less reduced. Stipules 2, interpetiolate between the petioles, sometimes deeply bifid, sessile, rather long persistent, narrowly triangular, to 4 by 1 mm, curved, ± succulent, below hairy as the first internode; 1-nerved, below rather inconspicuous. Petiole in cross-section semi-orbicular, to 5 by 1 mm, hairy as the first internode. Blade elliptic to oblong, to 7.5 by 2.5–4 cm; base acute; top cuspidate; margin entire; above on the midnerve and below on all nerves hairy as the first internode, below on the blade sparsely so; nerves pinnate, midrib above slightly raised, below much so, lateral nerves above slightly sunken, below raised, curved at their tip and forming an undulating marginal nerve.
Next leaves spirally arranged. Stipules 2 at the base of each leaf, sessile, patent, narrowly triangular, in 5th leaf to 5 by 2 mm, rather soon dropped, as those in the first

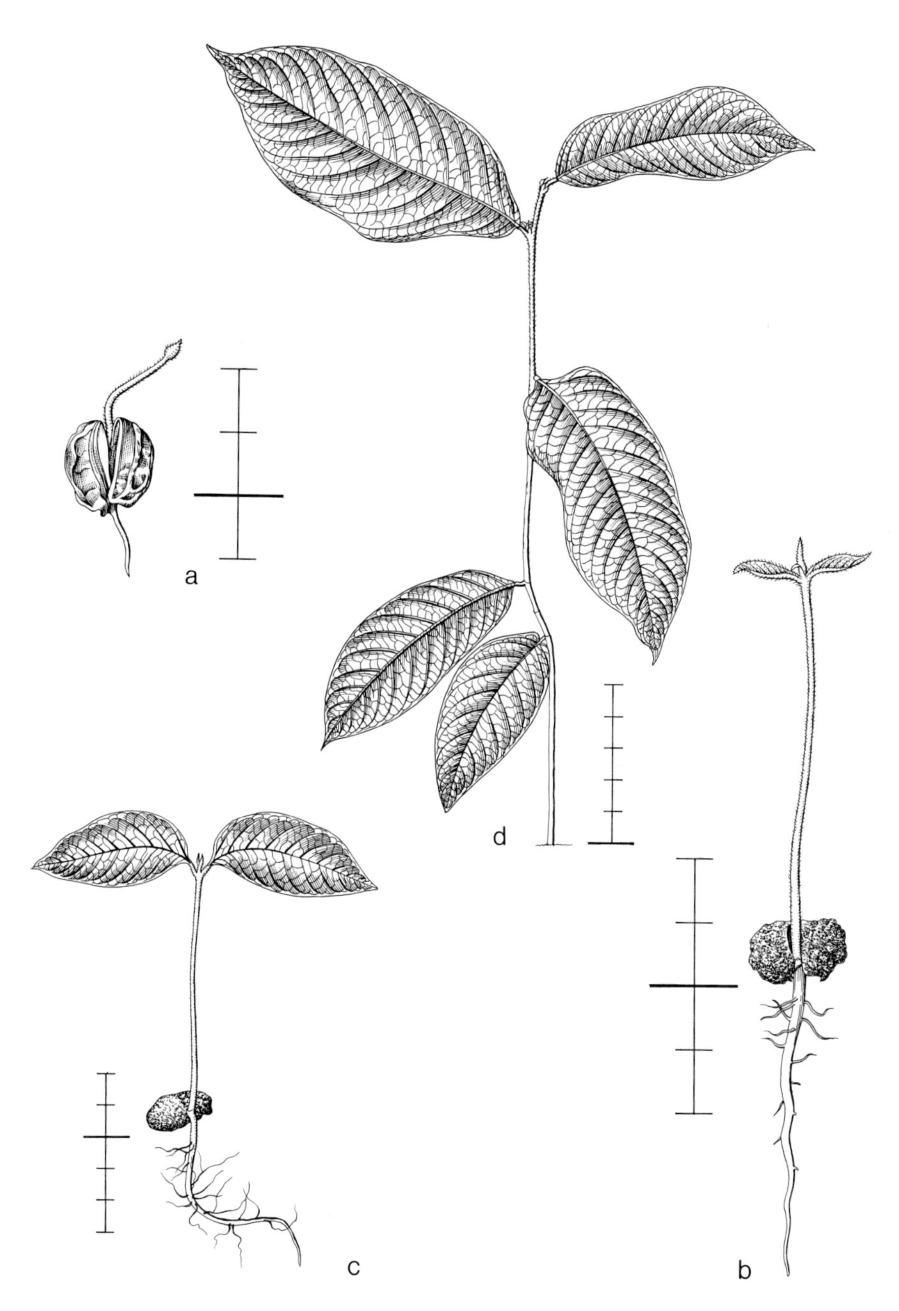

Fig. 147. *Prunus grisea* var. *grisea.* De V. 2653, collected 1-xi-1973, planted 5-xi-1973. a. 6-ii-1974; b. 16-i-1974; c. 6-ii-1974; d. 11-ix-1974.

leaves. Petiole in cross-section semi-orbicular, in higher ones more terete, in 3rd one to 6 by 1 mm, in 5th one to 8 by 2 mm, hairy as the first internode. Blade oblong, in 5th one to 11.5 by 5 cm; base obtuse, further as the first leaves.

Specimens: 2653 from N. Celebes, primary forest on deep clay, medium altitude.
Growth details: In nursery germination fair, ± simultaneous.
Remarks: Two seedling types are present. Those of the Heliciopsis type have the cotyledons at soil level, remaining in the testa and stone, the leaves are all spirally arranged or the first 2 opposite, with in some species the lower ones reduced (*P. alleghaniensis*, *P. americana*, *P. amygdalus*, *P. canadensis*, *P. persica*, *P. serotina*, and *P. tenella*).
Those of the Sloanea type with free cotyledons on a long hypocotyl may have the first 2 leaves opposite (*P. avium*, *P. cerasus*, *P. laurocerasus*, *P. mahaleb*, *P. padus*, *P. pensylvanica*, *P. serrulata*, and *P. virginica*), or the leaves are all spirally arranged (*P. domestica*, *P. insititia*, in *P. spinosa* not clear). At subgenus level the subgenera *Amygdalus*, *Armeniaca*, and *Persica* (all Heliciopsis type) are homogeneous in their seedlings as far as is known up to now. *Cerasus*, *Padus*, *Laurocerasus,* and *Prunus* have representatives of both types.

RUBIACEAE

CANTHIUM

Fl. Java 2: 319; Lubbock, On Seedlings 2 (1892) 68 (*Plectronia ventosa* = *Canthium*); Mensbruge, Germination (1966) 335 (*C. subcordatum*, *C. tekbe*).
Genus in tropical Africa, India, and Ceylon, throughout S.E. Asia, all over Malesia, E. Australia, in the Pacific as far as the Marquesas, including the Marianas, the Carolines, and New Caledonia.

Canthium dicoccum (Gaertn.) T. & B. – Fig. 148.

Development: The hypocotyl and taproot emerge from one pole of the seed. After establishment of the root the hypocotyl becomes erect, carrying the paracotyledons enclosed in the testa above the soil. After a short resting period the paracotyledons spread, thus shedding the testa. A second short resting stage occurs with the paracotyledons developed.
Seedling epigeal, phanerocotylar. Macaranga type.
Taproot slender, fibrous, creamy-white, with many, long, slender, hardly branched, creamy-white sideroots.
Hypocotyl strongly enlarging, slender, to 4.5 cm long, terete, towards the top in cross-section distinctly rhomboid, with distinct longitudinal ridges from between the cotyledons, green, slightly hairy with minute, simple, hyaline hairs.
Paracotyledons 2, opposite, foliaceous, when dropped not known, long persistent, still present in the 20th leaf stage; simple, sessile, stipulate, herbaceous, green, glabrous. Stipules interpetiolar, sessile, narrowly triangular, ± 1 mm long. Blade ±

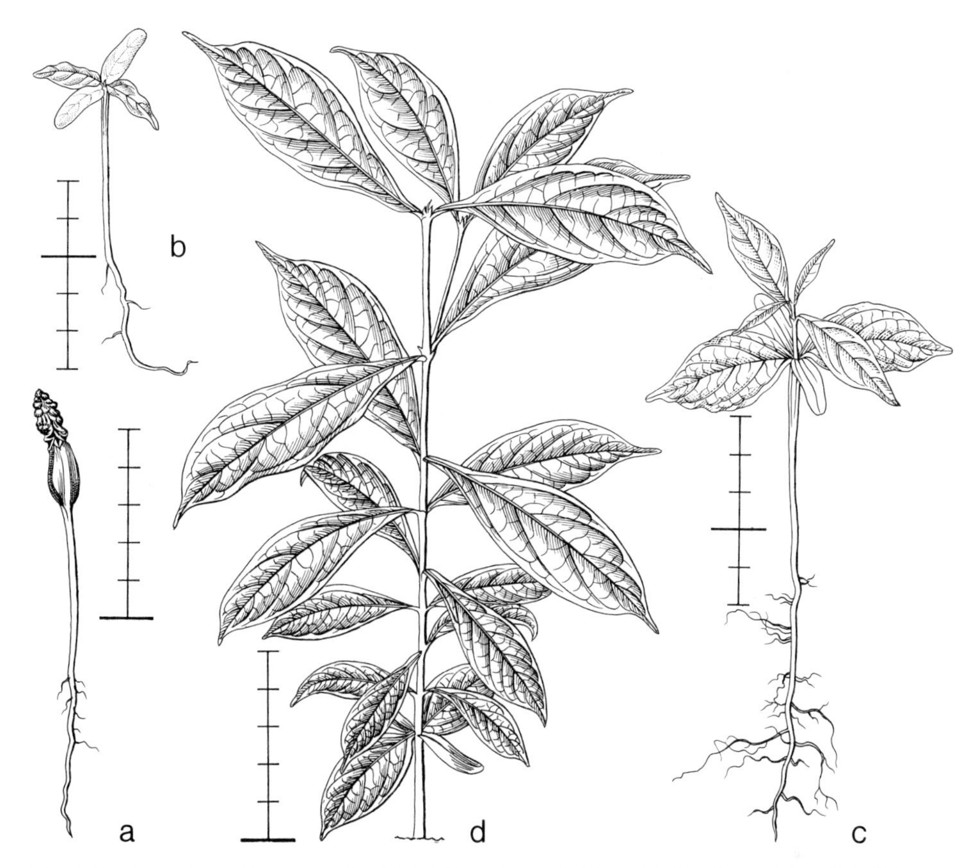

Fig. 148. *Canthium dicoccum.* De V. 1451, collected 14-vii-1972, planted 27-vii-1972. a, b, and c date?; d. 8-viii-1973.

lanceolate, to 20 by 6 mm; base and top rounded; margin entire; nerves pinnate, midrib above slightly prominent, lateral nerves rather inconspicuous.

Internodes: First one hardly elongating, to 0.5 mm long, hidden by petioles of first pair of leaves; next internodes strongly elongating, in cross-section ± quadrangular, first ones to 1.6 cm long, gradually the higher ones longer, 10th one to 4.5 cm long, green, hairy as the hypocotyl. Stem rather slender, straight.

Leaves all decussate, simple, stipulate, shortly petiolate, herbaceous, green, glabrous. Stipules interpetiolar, triangular from a broad base, in the lower leaves to 1.5 mm long, in 10th pair to 5 mm long, towards the top with an elevated, succulent ridge. Petiole in cross-section semi-orbicular, in first pair very short, in higher ones to 3 mm long. Blade in lowest leaves ovate, undulating, to 3.3 by 2 cm, in higher ones oblong to lanceolate, in 4th pair to 4 by 1.3 cm, in 10th one to 9 by 2.5 to 3.5 cm; base slightly narrowed, in higher ones wedge-shaped to narrowed; top acuminate, obtuse at the tip, in higher ones acuminate to cuspidate, its tip rounded; margin entire; nerves pinnate, midrib above and below raised, lateral nerves below raised, ending in anastomoses.

Specimens: 1451 from N. Sumatra, primary forest on deep clay, low altitude.
Growth details: In nursery germination fair, ± simultaneous.
Remarks: All species described show the same germination pattern and morphology, they differ in details only.

COFFEA

Fl. Java 2: 321; Csapody, Keimlingsbestimmungsbuch der Dikotyledonen (1968) 32 (*C. arabica*); Duke, Ann. Miss. Bot. Gard. 52 (1965) 350 (*C. arabica*); Lubbock, On Seedlings 2 (1892) 70 (*C. arabica*).
Monograph: Chevalier, Encycl. Biol. 28 (1947) 117.
Genus badly in need of revision, widely cultivated in tropics and subtropics. Native in tropical and subtropical Africa, Madagascar, India, and Ceylon, S.E. Asia, throughout Malesia including New Guinea (including *Lachnastoma*).

Coffea liberica Hiern – Fig. 149.

Development: The taproot and hypocotyl emerge from one pole of the seed. The hypocotyl becomes erect, carrying the enclosed paracotyledons above the soil. After a short resting stage the testa is shed with the spreading of the paracotyledons. A second resting stage occurs with the paracotyledons exposed.
Seedling epigeal, phanerocotylar. Macaranga type.
Taproot long, slender, flexuous, cream-coloured with rather many, slender, few-branched, cream-coloured sideroots.
Hypocotyl strongly elongating, slender, terete, to 6 cm long, green, in the upper part rather densely hairy with minute, simple, pale brownish hairs.
Paracotyledons 2, opposite, foliaceous, when dropped not known, beyond the 10th leaf stage; simple, stipulate, shortly petiolate, herbaceous, green, glabrous. Stipules interpetiolar, sessile, long persistent, scale-like, appressed to the stem, to 0.5 by 1.5 mm, top broadly rounded, green. Petiole in cross-section semi-orbicular, to 1 by 1.5 mm, above convex. Blade ± orbicular, to 3.6 by 4.3 cm; base truncate; top broadly rounded, irregular; margin entire, irregular; trinerved, nerves above slightly raised, below more so, lateral nerves at the tip curved, forming an undulating pattern.
Internodes: First one strongly elongating, in cross-section elliptic, to 2 cm long, green, on the flattened sides rather sparsely hairy with short, simple, hyaline hairs; next internodes as the first one, second one to 2 cm long, 5th one to 2.5 cm long. Stem straight, slender, thickened on the nodes.
Leaves all decussate, simple, stipulate, shortly petiolate, herbaceous, slightly red when young, turning green, glabrous. Stipules interpetiolate, and beyond the first pair also intrapetiolately connate, triangular, in the first leaves to 2 by 1.5 mm, in 5th pair tube to 3 mm long; top acute; margin entire; appressed to the stem, in top bud often with a drop of dull red, tough resin clasped between them; one-nerved, nerve prominent near the tip, ending in a small cusp. Petiole in cross-section semi-orbicular, in first leaves to 2 by 1 mm, in 5th pair to 3 by 2 mm. Blade obovate-oblong, in first leaves to 7.5 by 2.7 cm, in 5th pair to 11 by 4 cm; base slightly

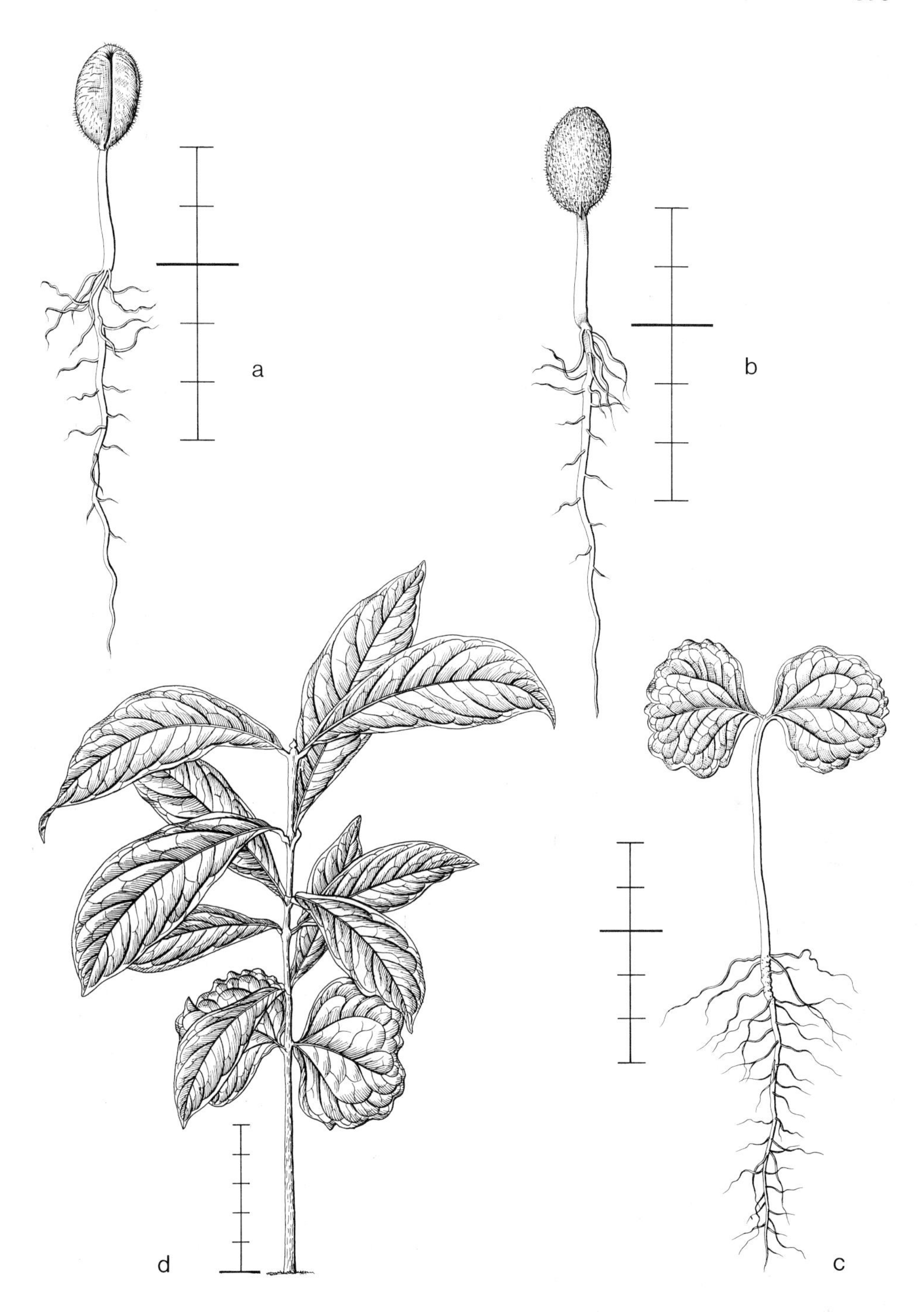

Fig. 149. *Coffea liberica.* De. V 2514, collected 12-x-1973, planted 5-xi-1973. a and b. 22-i-1974; c. 18-ii-1974; d. 5-ix-1974.

narrowed; top acuminate; margin entire; nerves pinnate, midrib above raised, below more so, lateral nerves much less so, at their tip curved to the next one.

Specimens: 2514 from N. Celebes, secondary vegetation on deep clay, also cultivated.
Growth details: In nursery germination fair, ± simultaneous.
Remarks: All species described are similar in germination pattern and general morphology; they differ in details only.

DIPLOSPORA

Fl. Java 2: 316.
Genus in tropical Africa (*Trycalysia*, sometimes combined with *Diplospora*), India and Ceylon, throughout S.E. and tropical E. Asia, throughout W. Malesia, including Java and the Philippines.

Diplospora minahasae Koord. – Fig. 150.

Development: The hypocotyl and taproot emerge from the acute pole of the seed. The hypocotyl becomes erect, carrying the paracotyledons above the soil. After a short resting stage the testa is shed by spreading of the paracotyledons. A second resting stage occurs with developed paracotyledons.
Seedling epigeal, phanerocotylar. Macaranga type.
Taproot long, very slender, flexuous, cream-coloured, with many very long, slender, flexuous, shortly branched, cream-coloured sideroots.
Hypocotyl strongly elongating, terete, to 2.5 cm long, green, glabrous.
Paracotyledons 2, opposite, foliaceous, in the 12th–16th leaf stage dropped; simple, stipulate, petiolate, herbaceous, green, glabrous. Stipules interpetiolar, sessile, scale-like, to 0.4 by 2 mm, appressed to the stem, top broadly rounded. Petiole in cross-section semi-orbicular, to 2 by 1 mm. Blade rhomboid, to 1.3 by 1.3 cm; base truncate; top broadly rounded; margin entire, slightly undulating; trinerved, main nerve above raised, below less so, lateral nerves below slightly raised rather inconspicuous.
Internodes: First one strongly elongating, in cross-section elliptic, to 1.2 cm long, below the leaves on both sides with a thickened ridge, glabrous; next internodes as the first one, second one to 1 cm long, 5th one to 1.3 cm, 8th one to 3.3 cm long. Stem rather slender, straight.
Leaves all decussate, simple, stipulate, petiolate, herbaceous, shining green above, paler below, glabrous. Stipules interpetiolar, sessile, narrowly triangular from a broad base, the lower ones to 2 by 2 mm, in 8th pair to 5 by 3 mm; top caudate; margin entire; appressed to the stem; one-nerved, nerve below raised. Petiole in cross-section semi-orbicular, in first leaves to 3 by 0.8 mm, in 8th pair to 10 by 3 mm. Blade ± oblong, in first leaves to 2.2 by 1.1 cm, gradually becoming larger, in 3rd pair to 4.5 by 2 cm, in 8th pair to 12 by 4.5 cm; base narrowed; top acuminate; margin entire; nerves pinnate, midrib above distinctly raised, below more so, lateral nerves

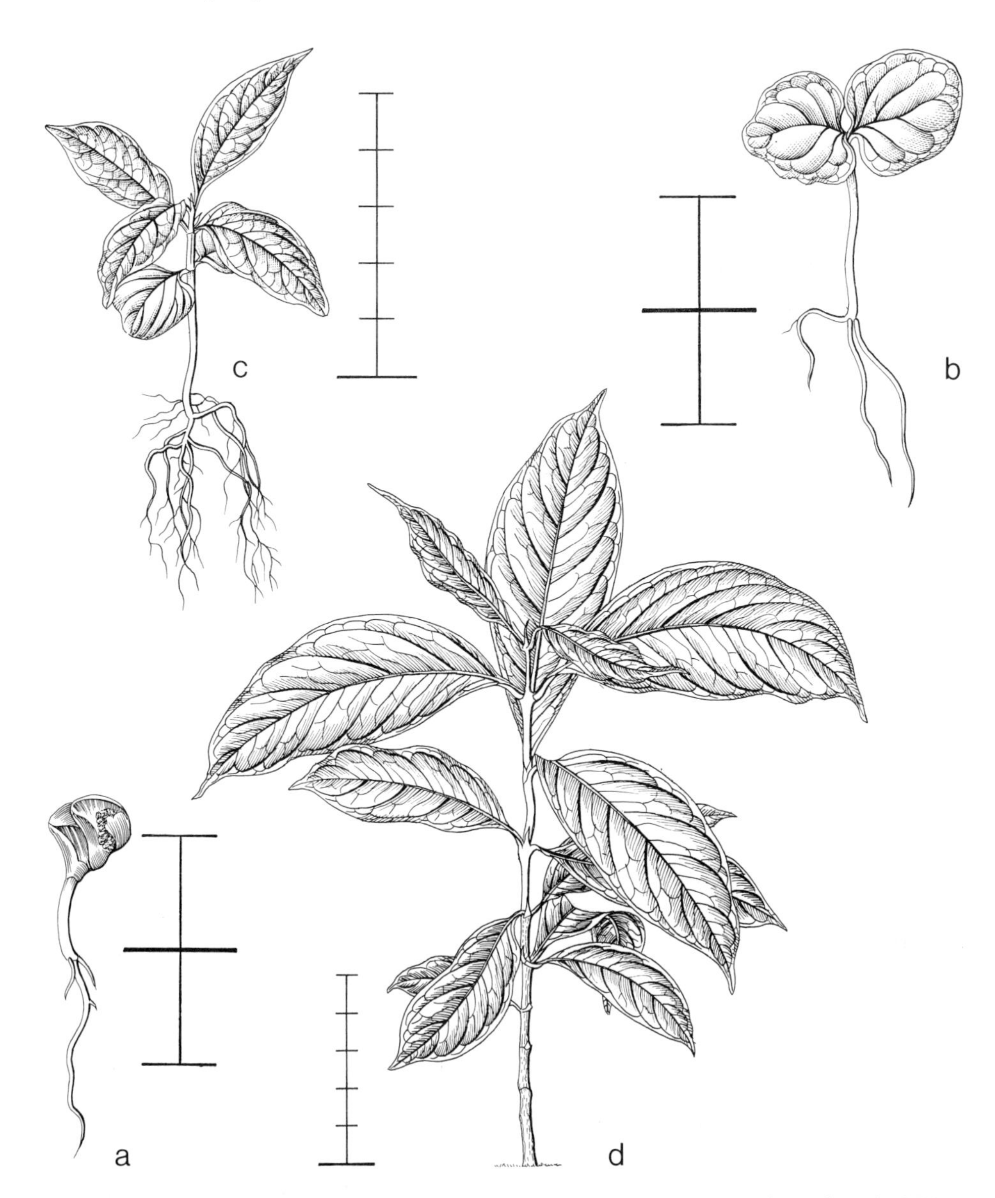

Fig. 150. *Diplospora minahasae.* De V. 2518, collected 17-x-1973, planted 5-xi-1973. a. 28-xi-1973; b. 12-xii-1973; c. 18-ii-1974; d. 5-ix-1974.

above sunken, below raised, at their tip curved but free ending.

Specimens: 2518 from N. Celebes, primary forest on deep clay, low altitude.
Growth details: In nursery germination poor.

GAERTNERA

Fl. Java 3: 655.
Revision: Van Beusekom, Blumea 15 (1967) 359.
Genus in tropical W. Africa, Madagascar and the Mascarenes, Ceylon, S.E. Asia and W. Malesia excluding the Philippines.

Gaertnera vaginans (DC.) Merr. – Fig. 151.

Development: The hypocotyl and taproot emerge from the somewhat acute pole of the seed. The for some time persistent testa is carried up when the hypocotyl becomes erect. After a resting stage, the testa is shed by spreading of the paracotyledons. Then follows a second temporary rest.
Seedling epigeal, phanerocotylar. Macaranga type.
Taproot long, fleshy, slender, whitish, with many, rather short, few-branched sideroots.
Hypocotyl strongly elongating, terete, to 4.5 cm long, herbaceous, green, glabrous.
Paracotyledons 2, opposite, foliaceous, in 6th–8th leaf stage dropped; sessile, stipulate, herbaceous, green, glabrous. Stipules interpetiolate, sessile, persistent, ± triangular, to 2.2 by 1.5 mm; top acute, with one or two cusps; margin serrate with up to 6, to 0.7 mm long subulate teeth, herbaceous, pale green, glabrous. Blade ± ovate, to 7 by 5.5 mm; base obtuse, slightly narrowed; top retuse; margin entire; nerves pinnate, midrib above slightly sunken, below somewhat prominent, lateral nerves rather inconspicuous, ending in anastomoses.
Internodes: First one elongating to 1 cm, in cross-section ± quadrangular with two longitudinal wings below the stipules, green, glabrous; next internodes as the first one, quickly the higher ones bigger, first 3 ± 1–2 cm long, 6th one to 4 cm long, 10th one to 9 cm long. Stem slender, straight.
Leaves all decussate, simple, stipulate, the higher ones shortly petiolate, herbaceous, green, glabrous. Stipules of the lower ones as those of the paracotyledons, with two long awns, those of the higher ones connate into a tube, those of the 9th pair of leaves to 13 mm long, apically between the leaves with 2 pairs of long awns, at the base with two sharp, curved ridges extending below the leaves. Petioles (in higher leaves) in cross-section semi-orbicular, in 9th leaf to 8 by 2.5 mm. Blade (ovate-)oblong to slightly lanceolate, quickly increasing in length, the lower ones to 3.7 by 1.3 cm, 9th one to 17 by 5.5 cm, 15th one to 26 by 10.5 cm; base and top acute, in the higher ones top acuminate; margin entire; above shining, below dull; nerves pinnate, midrib above prominent, below very prominent, lateral nerves in higher leaves below raised, at the tip slightly curved, ending in anastomoses.

Specimens: 1347 from S. Sumatra, primary forest on deep clay, low altitude.
Growth details: In nursery germination good, ± simultaneous.

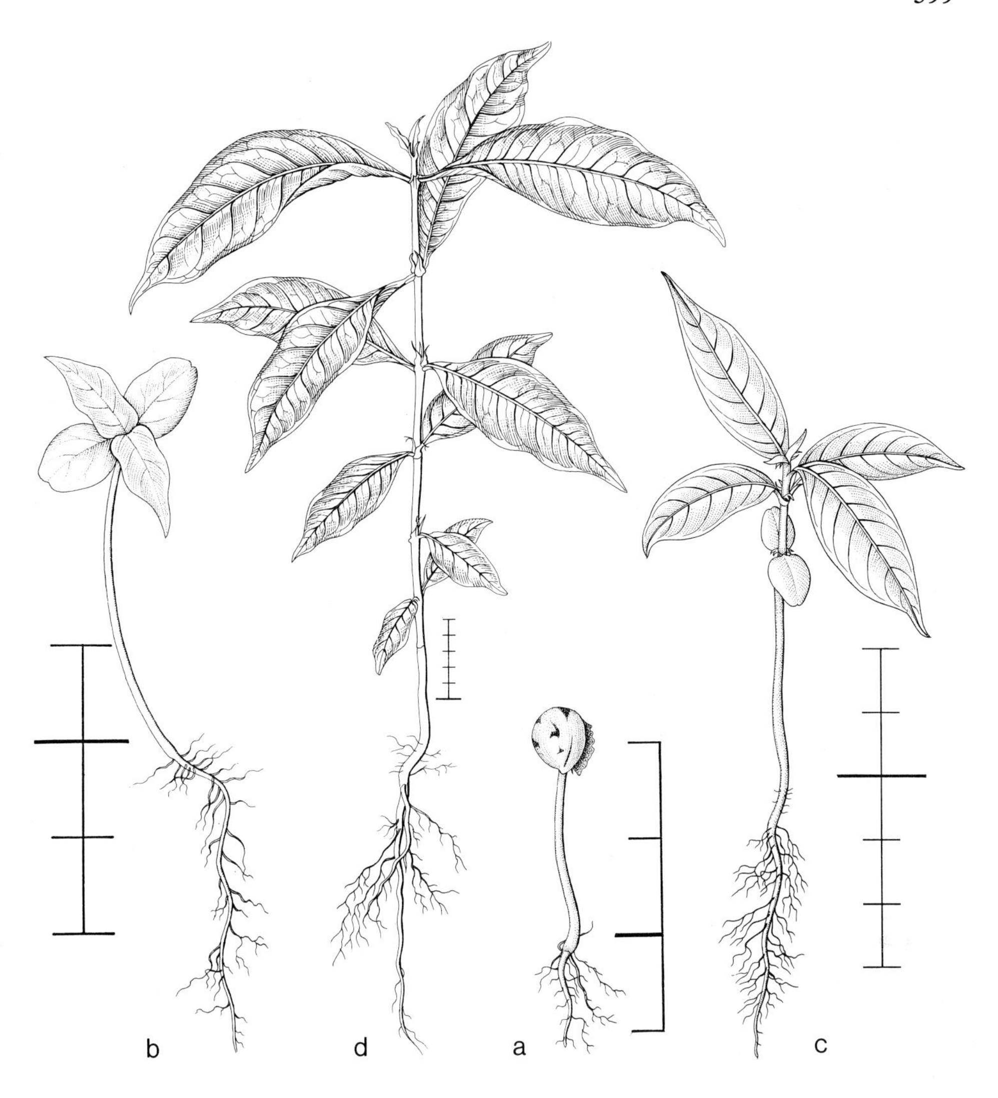

Fig. 151. *Gaertnera vaginans.* De V. 1347, collected 20-iii-1972, planted 23-iii-1972. a, b, c, and d date?.

GARDENIA

Fl. Java 2: 313; Hallé, Fl. Gabon 17 (1970) 219 (*G. imperialis*); Lubbock, On Seedlings 2 (1892) 64 (*G. thunbergia*); Troup, Silviculture 2 (1921) 628 (*G. turgida*).
Revision: Keay, Bull. Jard. Bot. Brux. 28 (1958) 15.
Genus (in broad sense) in tropical Africa, throughout the Indian Ocean, India and Ceylon, throughout S.E. and tropical E. Asia, all over Malesia, N. and E. Australia, in the Pacific as far as the Societies, including the Marianas, E. Carolines, Hawaii and New Caledonia.

Gardenia forsteniana Miq. – Fig. 152.

Development: The hypocotyl and taproot emerge from one pole of the seed. The hypocotyl becomes erect and the paracotyledons enclosed in the testa and endosperm are carried above the soil. After a resting period the paracotyledons expand, shedding testa and endosperm. Then follows a second temporary rest.
Seedling epigeal, phanerocotylar. Macaranga type.
Taproot long, slender, flexuous, creamy-white, with rather few, quite short, shortly branched, whitish sideroots.
Hypocotyl strongly enlarging, rather sturdy, terete, to 5.5 cm long, green, slightly hairy in the upper part with minute, simple, hyaline hairs.
Paracotyledons 2, opposite, foliaceous, in the 10th–14th leaf stage dropped; simple, stipulate, shortly petiolate, subcoriaceous, dark green, glabrous. Stipules interpetiolate, sessile, persistent, triangular, to 1 by 1.5 mm, appressed to the stem; top bifid; margin entire; slightly hairy as the hypocotyl; 1-nerved, nerve below prominent. Petiole flattened, to 1.5 by 3 mm. Blade rhomboid, to 1.5 by 1.5 cm; base somewhat cordate; top broadly rounded; margin somewhat irregular, wavy; 5-nerved, nerves above not raised, the midnerve at the base prominent, further slightly raised as the lateral nerves which end in anastomoses.
Internodes: First one strongly elongating, terete, to 2.5 cm long, green, rather densely covered with short, simple, hyaline hairs; next internodes as the first one, 4th one to 3.5 cm long. Stem slender, straight.
Leaves all decussate, simple, stipulate, shortly petiolate, herbaceous, green. Stipules interpetiolate, narrowly triangular, sessile on the swollen node, in the first leaves to 2.5 by 1 mm, in 4th pair to 4.5 by 1.5 mm; top acute; margin entire; appressed to the stem, hairy as the first internode; one-nerved, nerve not very conspicuous. Petiole in cross-section semi-orbicular, in the lower leaves to 2 by 1.5 mm, in the 4th pair to 5 by 1.5 mm, hairy as the first internode. Blade obovate-oblong, in first leaves to 7 by 2.7 cm, in 4th pair to 8 by 2.9 cm; base acute to slightly narrowed; top cuspidate; margin entire; slightly hairy as the first internode, especially below on the nerves; nerves pinnate, midrib above slightly prominent, below much so, lateral nerves above sunken, below prominent, curved at their tip but ± free ending.

Specimens: 2633 from N. Celebes, primary forest on deep clay, medium altitude.
Growth details: In nursery germination good, ± simultaneous, some seeds long delayed in germination.

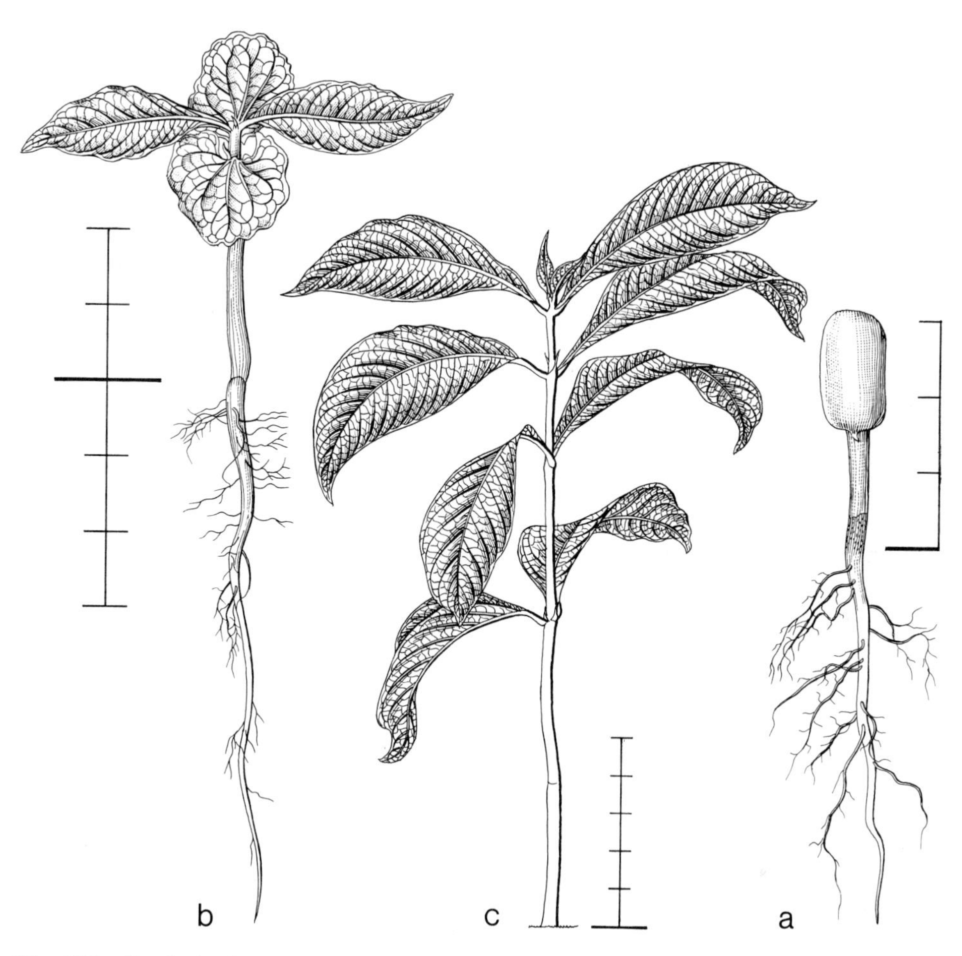

Fig. 152. *Gardenia forsteniana.* De V. 2633, collected 31-x-1973, planted 5-xi-1973. a. 16-i-1974; b. 20-iii-1974; c. 10-ix-1974.

Remarks: All species described show the same germination pattern and general morphology; they differ in details only.

Gardenia tubifera Wall. – Fig. 153.

Development: The hypocotyl and taproot emerge from one pole of the seed. The hypocotyl becomes erect, carrying the paracotyledons still enclosed in the testa above the soil. After a resting period, the testa is shed by the spreading of the paracotyledons. Then follows a second temporary rest.
Seedling epigeal, phanerocotylar. Macaranga type.
Taproot fibrous, slender, cream-coloured, with some long, slender, branched, whitish sideroots.
Hypocotyl strongly enlarging, terete, to 4 cm long, green, glabrous.
Paracotyledons 2, opposite, foliaceous, in the 5th–8th leaf stage dropped; simple, ± sessile, stipulate, herbaceous, green, glabrous. Stipules interpetiolar, sessile, broadly triangular, ± 1 by 1.2 mm; top cuspidate; margin entire; nerve one, below prominent. Blade irregularly orbicular, to 1.6 by 1.6 cm; base wedge-shaped; top ± rounded; margin entire, irregular; nerves palmate, main nerves (3–)5, above and below prominent, at the tip curved to the next one.
Internodes: First one slightly elongating, in cross-section elliptic, to 4 mm long, green, rather densely covered with minute, simple, hyaline hairs, glabrescent; next internodes each increasing in length, 5th one to 3.5 cm long, 10th one to 6 cm long, 15th one to 10 cm long. Stem slender, straight.
Leaves all decussate, simple, stipulate, almost sessile, herbaceous, hairy as the first internode. Stipules sessile, interpetiolate, the higher ones also intrapetiolarly connate and forming a tube, triangular, in the lower leaves to 2 by 1.5 mm, in the 15th pair tube to 3 cm long; margin entire; top acuminate; main nerve 1, below prominent. Petiole in cross-section semi-orbicular, in 10th pair to 10 by 2.5 mm, in 15th pair to 7 by 5 mm. Blade elliptic, later obovate-lanceolate, in first leaves 3 by 1.2 cm, in 10th pair to 22 by 5.5 cm, in 15th pair to 35 by 10.5 cm, later ones gradually becoming larger; base narrowed; top obtuse to acuminate; margin entire; above shining green, below lighter dull green, hairy as the first internode; nerves pinnate, midrib above and below prominent, lateral nerves slightly so, forming a vague, irregular marginal nerve.

Specimens: 1260 from S. Sumatra, disturbed primary hill forest on deep clay, low altitude.
Growth details: In nursery germination fair.
Remarks: See *G. forsteniana*.

GUETTARDA

Fl. Java 2: 320; Duke, Ann. Miss. Bot. Gard. 52 (1965) 350 (*G. sp.*).
Flora: N. American Flora 32, 3–4 (1934) 228.
Genus pantropic; tropical E. Africa, throughout the islands of the Indian Ocean,

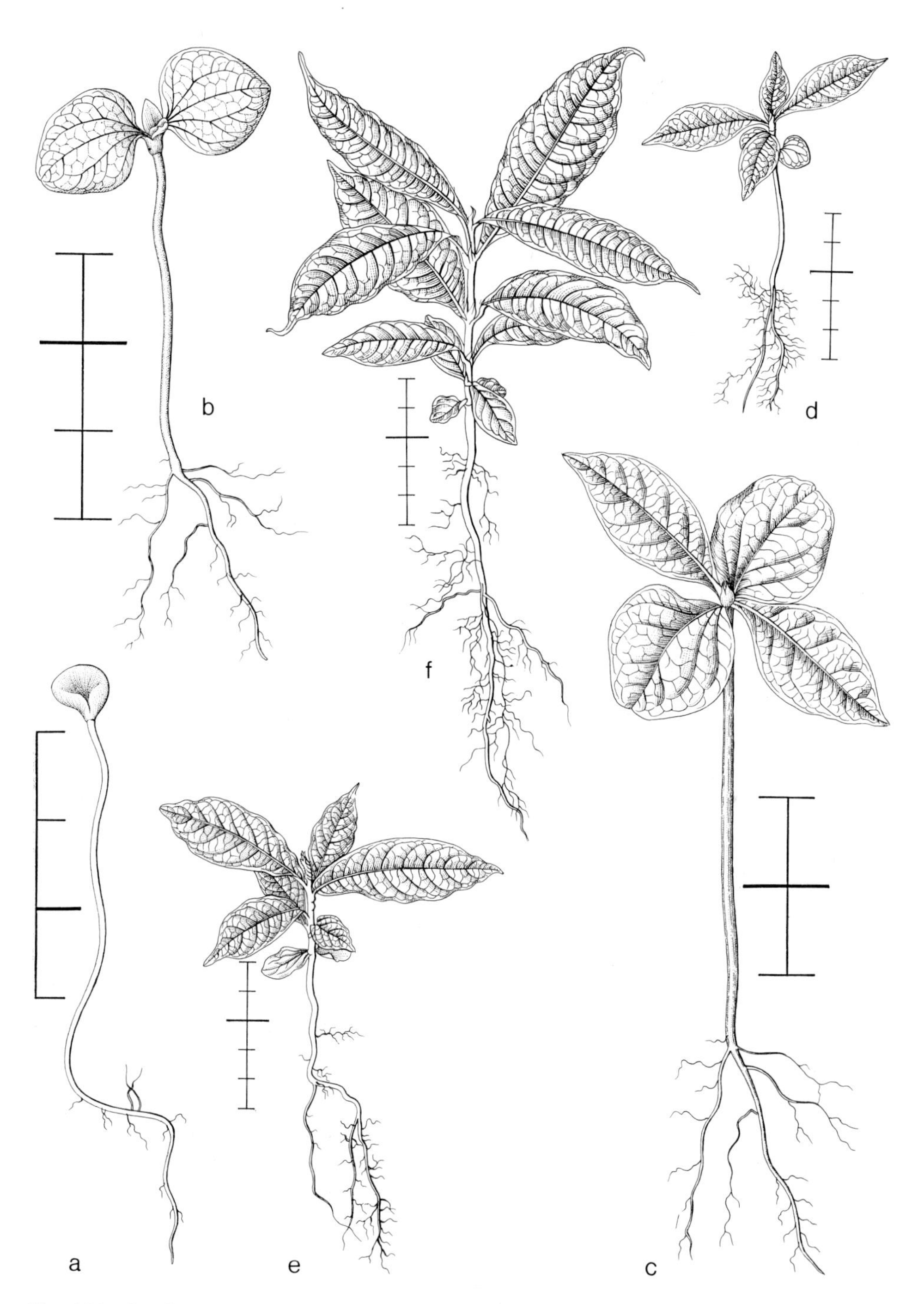

Fig. 153. *Gardenia tubifera.* De V. 1260, collected 11-iii-1972, planted 23-iii-1972. a, b, c, d, e, and f date?.

India and Ceylon, S.E. and E. Asia, all over Malesia, N.E. Australia, in the Pacific as far as the Marquesas, including the Marianas, Central Polynesia, and New Caledonia, Central and tropical S. America including Revilla Gigedos and Cocos.

Guettarda speciosa L. – Fig. 154.

Development: The taproot emerges from the side of the fibrous, woody endocarp. The outgrowth at the base of the hypocotyl hooks into a pit in the endocarp, then the hypocotyl, which becomes erect, withdraws the paracotyledons from the fruit. A short resting stage occurs with the paracotyledons expanded.
Seedling epigeal, phanerocotylar. Macaranga type.
Taproot rather feebly developed in the beginning, slender, white, with many long, slender, much-branched sideroots.
Hypocotyl strongly enlarging, terete, to 6.5 cm long, to the top slightly tapering, at the base to one side with an elliptic, fleshy outgrowth, rather densely hairy with stiff, simple, hyaline hairs.
Paracotyledons 2, opposite, dropped with the development of 4th–6th pair of leaves; simple, stipulate, shortly petiolate, herbaceous, green, hairy as the hypocotyl. Stipules intrapetiolar, narrowly triangular, sessile, to 2 by 0.8 mm. Petiole in cross-section semi-orbicular, to 0.7 by 0.7 mm. Blade ± elliptic, to 3 by 1.9 mm; base narrowed into the petiole; top rounded; margin entire; one-nerved, nerve rather inconspicuous.
Internodes: First one hardly elongating, in cross-section ± elliptic, to 1 mm long, hairy as the hypocotyl; next internodes as the first one, more elongating, especially the higher ones often irregular in length, to 1 cm long, beyond the 10th one sometimes to 2 cm long. Stem straight, slender.
Leaves all decussate, simple, stipulate, petiolate, herbaceous, green, hairy as the hypocotyl. Stipules interpetiolar, as those of the paracotyledons, in 13th pair of leaves to 2.5 by 1.5 mm. Petiole in cross-section semi-orbicular, in lowest leaves to 2 by 1 mm, in 10th pair to 5 by 1.5 mm. Blade in lowest pair to 2 by 1 cm, in higher leaves more oblong, in 3rd pair to 3 by 1.3 cm, in 10th pair to 7.5 by 2.7 cm; base obtuse to acute; top acute to slightly acuminate; margin entire; nerves pinnate, above somewhat prominent, below much so, free ending.

Specimens: 1385 from W. Java, primary beach forest on coral sand.
Growth details: In nursery germination fair.
Remarks: G. sp. (Duke 1965) is not in the possession of a fleshy basal outgrowth on the hypocotyl, and consequently will have a different extraction mechanism. Further differing in details only.

MORINDA

Fl. Java 2: 249; Lubbock, On Seedlings 2 (1892) 71 (*M. tinctoria*); Mensbruge, Germination (1966) 335 (*M. lucida*).
Genus pantropic; tropical Africa, India and Ceylon, throughout S.E. and E. Asia, all

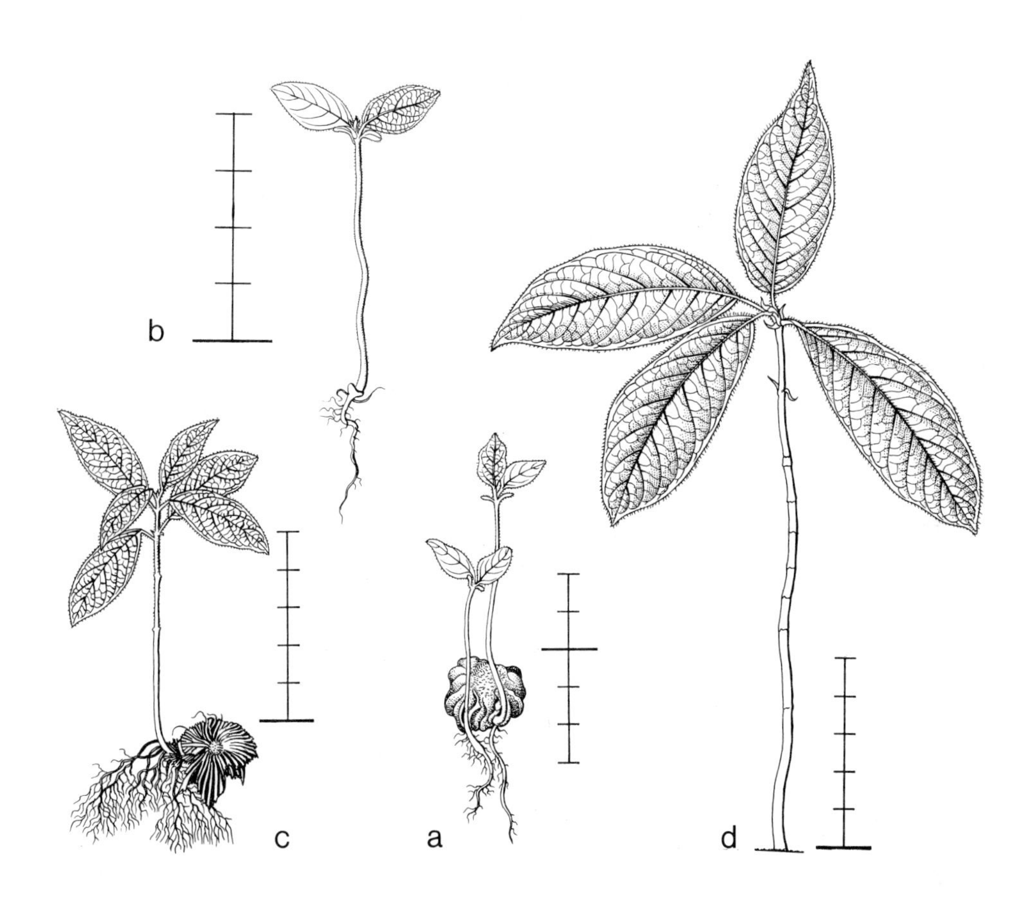

Fig. 154. *Guettarda speciosa.* De V. 1385, collected 21-vi-1972, planted 2-vii-1972. a, b, and c date?; d. 31-vii-1973.

over Malesia, N. Australia, in the Pacific as far as the Marquesas, including the Bonins, Hawaii and New Caledonia, tropical S. America.

Morinda sp. – Fig. 155.

Development: The swollen, empty chamber in the seed facilitates buoyance. The taproot and hypocotyl emerge from the flattened pole of the seed. The outgrowth at the base of the hypocotyl is clasped between the valves of the seed. The hypocotyl elongates and becomes erect, and withdraws the paracotyledons from the seed. A short resting stage occurs with the paracotyledons expanded.
Seedling epigeal, phanerocotylar. Macaranga type.
Taproot long, slender, flexuous, red-brown, with rather many long, slender, much-branched, orange to red-brown sideroots.
Hypocotyl strongly enlarging, ovate, to 4 cm long, in cross-section angularly ovate, at the base with a flattened, fleshy outgrowth, which usually is clasped by the valves of the seed, pale green, densely hairy with minute, simple, hyaline hairs.
Paracotyledons 2, opposite, foliaceous, in the 10th–12th leaf stage dropped; stipulate, petiolate, herbaceous, green, glabrous. Stipules interpetiolate between the paracotyledons, scale-like, sessile, appressed to the stem, much broader than long, to 0.5 by 1.5 mm; top broadly rounded; without nerve. Petiole in cross-section semi-orbicular, to 3 by 1 mm, above slightly convex. Blade oblong, to 2.2 by 1 cm; base narrowed; top broadly rounded; margin entire; nerves pinnate, midrib above not raised, below slightly raised, lateral nerves inconspicuous, ending free.
Internodes: First one strongly elongating, in cross-section terete, to 1 cm long, with 2 sharp longitudinal ridges ending in a cusp just below the next stipules, pale green, glabrous; next internodes as the first one, 5th one to 2.2 cm. Stem slender, straight.
Leaves all decussate, simple, stipulate, petiolate, herbaceous, green, glabrous. Stipules 2, interpetiolar, scale-like, sessile, appressed to the stem, triangular, beyond the 7th one more ovate-oblong, in the first pair of leaves to 1.5 by 1.5 mm, in 5th pair to 2.5 by 2.5 mm, in 8th pair to 7 by 3 mm; in lower ones top acute, in higher ones more rounded; 1-nerved, nerve above prominent. Petiole in cross-section semi-orbicular, in the lowest leaves to 2 by 1 mm, in 5th pair to 4 by 1.5 mm, in 8th pair to 4 by 2 mm, above convex. Blade oblong, later lanceolate, in the lowest ones to 3 by 1 cm, in 4th pair to 4.3 by 1.4 cm, in 8th pair to 11 by 2.5 cm; base slightly narrowed; top acute; margin entire; nerves pinnate, midrib above slightly raised, below much so, lateral nerves few, above slightly sunken, but in higher ones slightly raised, below raised, their tips slightly curved, forming a slightly undulating, inconspicuous marginal nerve.

Specimens: 2488 from N. Celebes, beach vegetation on coral sand.
Growth details: In nursery germination good, ± simultaneous, rather long delayed.
Remarks: M. lucida is not in possession of a fleshy basal outgrowth of the hypocotyl, and consequently must have a different extraction mechanism. Further the three described species differ in details only.

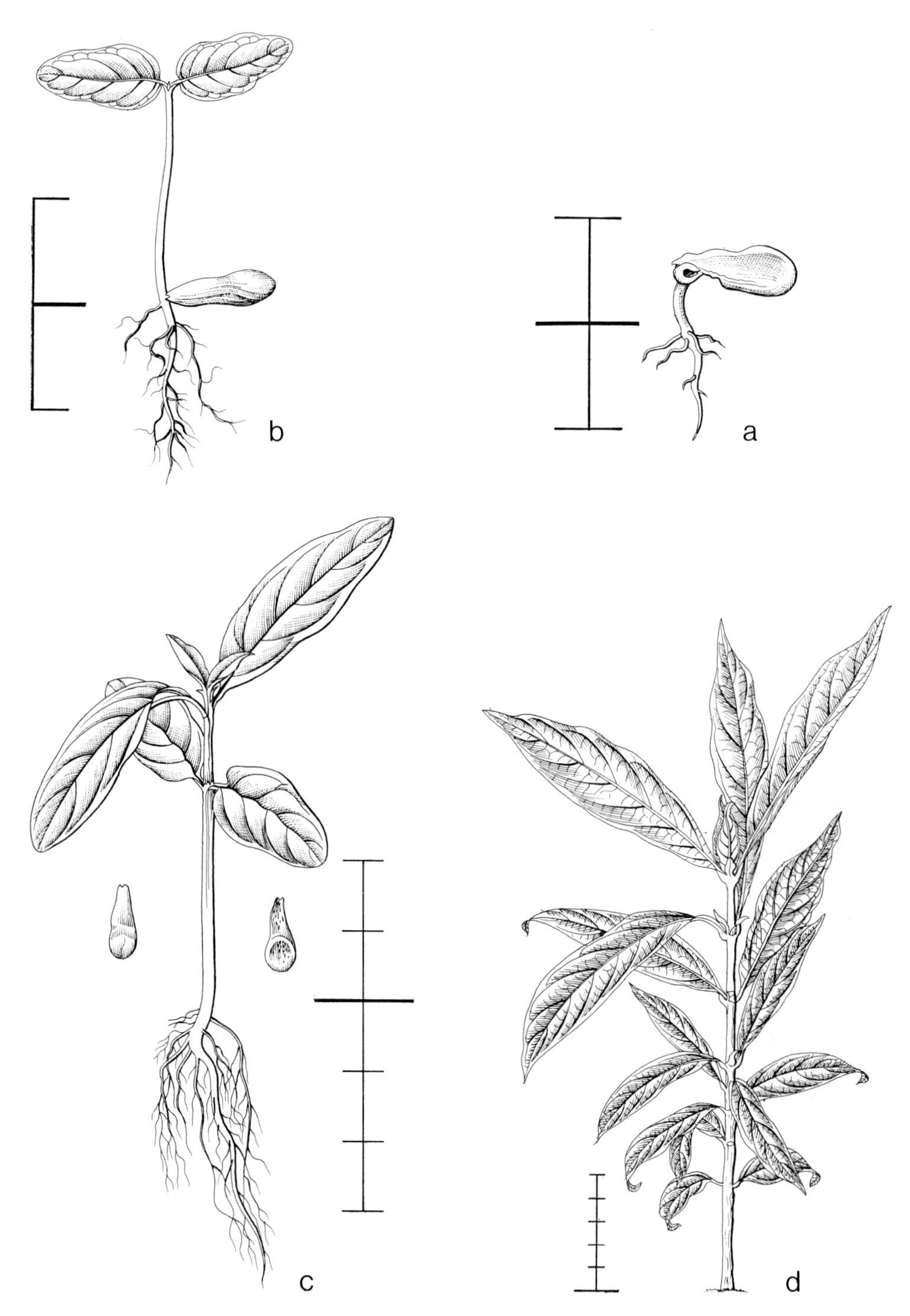

Fig. 155. *Morinda sp.* De V. 2488, collected 9-x-1973, planted 5-xi-1973. a and b. 12-xii-1973; c. 18-ii-1974; d. 3-xi-1974.

PAVETTA

Fl. Java 2: 323; Lubbock, On Seedlings 2 (1892) 70 (*P. madagascariensis*, *P. speciosa*).
Monograph: Bloembergen, Fedde Rep. 37 (1934) 1.
Genus in tropical and subtropical Africa, India and Ceylon, throughout S.E. and tropical E. Asia, all over Malesia, N. Australia, in the Pacific as far as New Caledonia and the Loyalties.

Pavetta sp. – Fig. 156.

Development: The hypocotyl and taproot emerge from the side of the seed. The paracotyledons, enclosed by endosperm and testa, are raised by the elongating hypocotyl. After a resting stage, the testa and endosperm are shed by spreading of the paracotyledons. A second resting stage occurs with paracotyledons expanded.
Seedling epigeal, phanerocotylar. Macaranga type.
Taproot long, slender, flexuous, slightly zig-zag, creamy-white, sideroots rather many, branched, as taproot.
Hypocotyl strongly elongating, terete, to 5 cm long, at the collet slightly narrowed into the root, green, cream-coloured below, glabrous.
Paracotyledons 2, opposite, foliaceous, in the 12th–16th leaf stage dropped; stipulate, petiolate, herbaceous, green, glabrous, in the first resting stage covered by the greyish hyaline, half globose, in the centre convexly hollowed endosperm and testa. Stipules interpetiolar, broadly triangular, to 1.5 by 1 mm; top cuspidate with a succulent, subulate cusp to 0.5 mm long; margin entire. Petiole in cross-section semi-orbicular, to 2 by 1.5 mm, the midrib descending on the upper side. Blade ± rhomboid, at most to 2.5 by 3.2 cm; base obtuse, narrowed abruptly in the petiole; top rounded; margin entire, undulating; nerves ± pinnate, above and below slightly raised, lateral nerves at the tip distinctly curved, forming a broadly undulating marginal nerve.
Internodes: First one terete, elongating to 10, rarely 20 mm, green, slightly hairy with minute, patent, simple, hyaline hairs; next internodes as the first one, second one to 1 cm, 7th one to 3.5 cm. Stem slender, straight.
Leaves all decussate, simple, stipulate, petiolate, herbaceous, green, glabrous. Stipules triangular, in the first leaves to 1 by 2 mm, gradually becoming larger, in 5th pair of leaves to 4 by 2 mm, interpetiolarly and intrapetiolarly connate at the base, in 10th pair to 5 by 5 mm, with few hairs along the midnerve, further as the paracotyledonary stipules. Petiole in cross-section semi-orbicular, later more terete, in first leaves to 4 by 1 mm, in 5th pair of leaves 5 by 1.5 mm. Blade in lowest leaves oblong, to 4.5 by 1.8 cm, in 5th pair more obovate-oblong, to 8.5 by 2.2 cm, in 10th pair to 17.5 by 6 cm; base narrowed into the petiole; top acuminate; margin entire; nerves pinnate, midrib above raised, below prominent, lateral nerves above sunken, below raised, at the tip curved and forming an undulating marginal nerve.

Specimens: 1840 from S.E. Borneo, alluvial flat, primary forest on deep clay, low altitude.

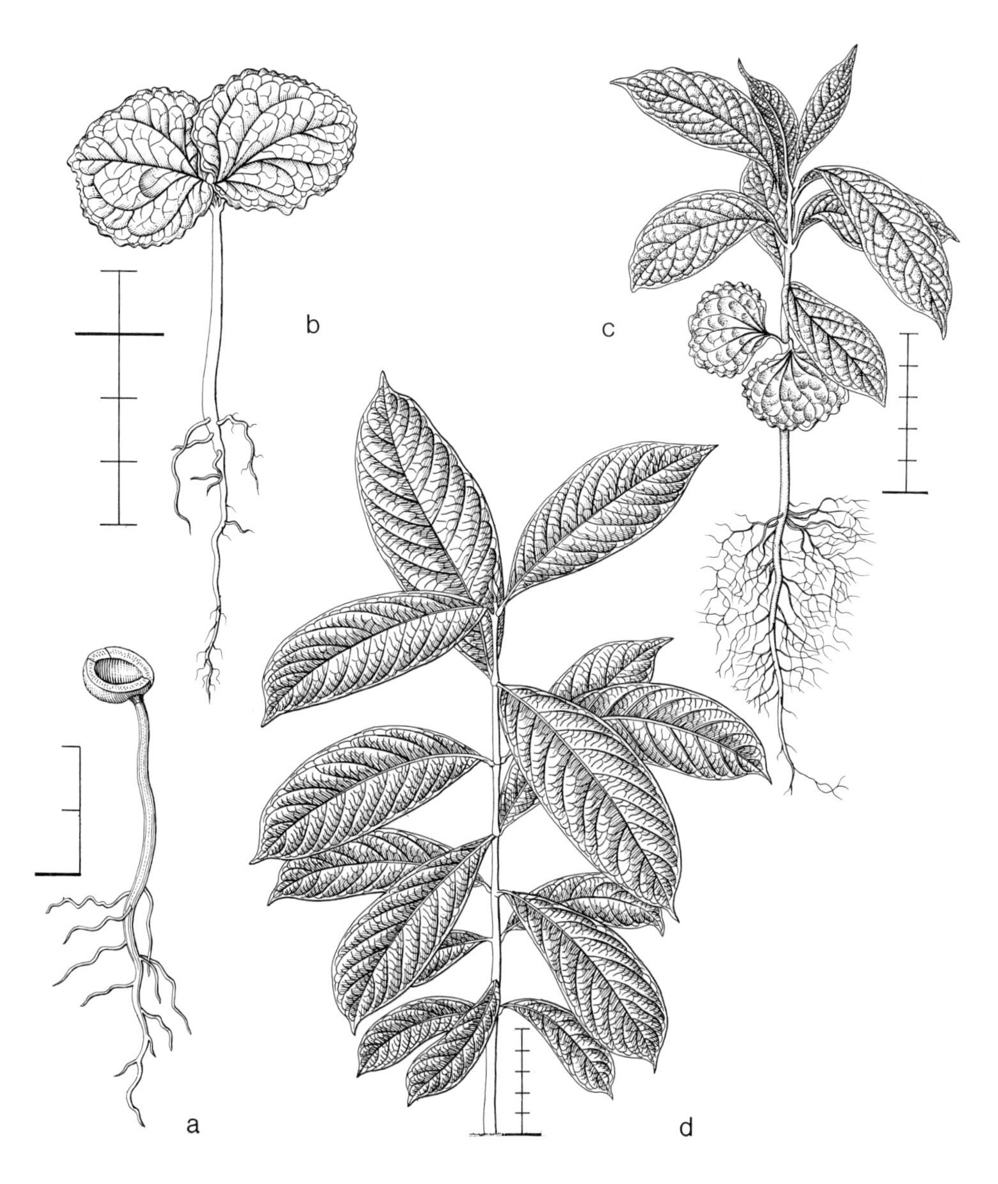

Fig. 156. *Pavetta sp.* De V. 1840, collected 24-x-1972, planted 17-xi-1972. a date?; b. 28-iii-1973; c. 15-viii-1973; d. 3-vii-1974.

Growth details: In nursery germination fair, ± simultaneous.
Remarks: All species described are similar in germination pattern and general morphology; they differ in details only.

PSYCHOTRIA

Fl. Java 2: 328; Duke, Ann. Miss. Bot. Gard. 52 (1965) 349 (*P. berteriana*); Lubbock, On Seedlings 2 (1892) 71 (*P. sp.*); Mensbruge, Germination (1966) 338 (*Grumilea venosa* = *Psychotria*).
Genus pantropic; tropical Africa, India and Ceylon, throughout S.E. and E. Asia, all over Malesia, N. and E. Australia, in the Pacific as far as the Marquesas, including the Bonins, Hawaii, Lord Howe, and New Caledonia, tropical Central and S. America, including the Galapagos.

Psychotria celebica Miq. – Fig. 157.

Development: The taproot and hypocotyl emerge from the acute pole of the seed. The hypocotyl elongates and becomes erect, by which the enclosed paracotyledons are carried above the soil. After a resting stage the testa is shed by spreading of the paracotyledons. Then follows a second temporary rest.
Seedling epigeal, phanerocotylar. Macaranga type.
Taproot rather short, slender, flexuous, cream-coloured, with rather many, in the upper part long, slender, flexuous, much-branched, cream-coloured sideroots.
Hypocotyl strongly elongating, to 4.5 cm long, slender, in the lower part terete, above somewhat angular by 4 elevated ridges from the base of the petioles, green, glabrous.
Paracotyledons 2, opposite, foliaceous, long persistent, when shed not known; simple, stipulate, shortly petiolate, herbaceous, green, glabrous. Stipules interpetiolar, sessile, narrowly triangular from a broad base, appressed to the stem, to 3 by 2 mm; top cuspidate; rather long persistent, then caducous, creamy-white to tinged red. Petiole in cross-section semi-orbicular, to 3 by 2.5 mm, slightly elevated above by the descending midrib, at the margins widened by the descending blade. Blade ± orbicular, to 2.3 by 2.6 cm; base wedge-shaped, abruptly narrowed into the petiole; top broadly rounded, irregular; margin entire, irregular; nerves pinnate, midrib below slightly raised, sidenerves hardly visible, at their tip broadly curved.
Internodes: First one somewhat elongating, to 15 mm long, but in general not exceeding 5 mm, terete, angular by a longitudinal ridge from the centre of the stipules, green, glabrous; next internodes as the first one, second one to 8 mm long. Stem straight, rather slender.
Leaves all decussate, simple, stipulate, shortly petiolate, herbaceous, green, glabrous. Stipules interpetiolate, triangular, appressed to the stem, in the first leaves to 3 by 2.5 mm, in 3rd pair to 5 by 4 mm; top acute; margin entire; one-nerved, nerve below prominent, in the higher leaves ending in a cusp below the top. Petiole in cross-section semi-orbicular, above the midnerve descending, in first leaves to 3 by 1 mm, in 3rd pair to 5 by 2 mm. Blade oblong, in the first leaves to 4.8 by 1.6 cm, 3rd

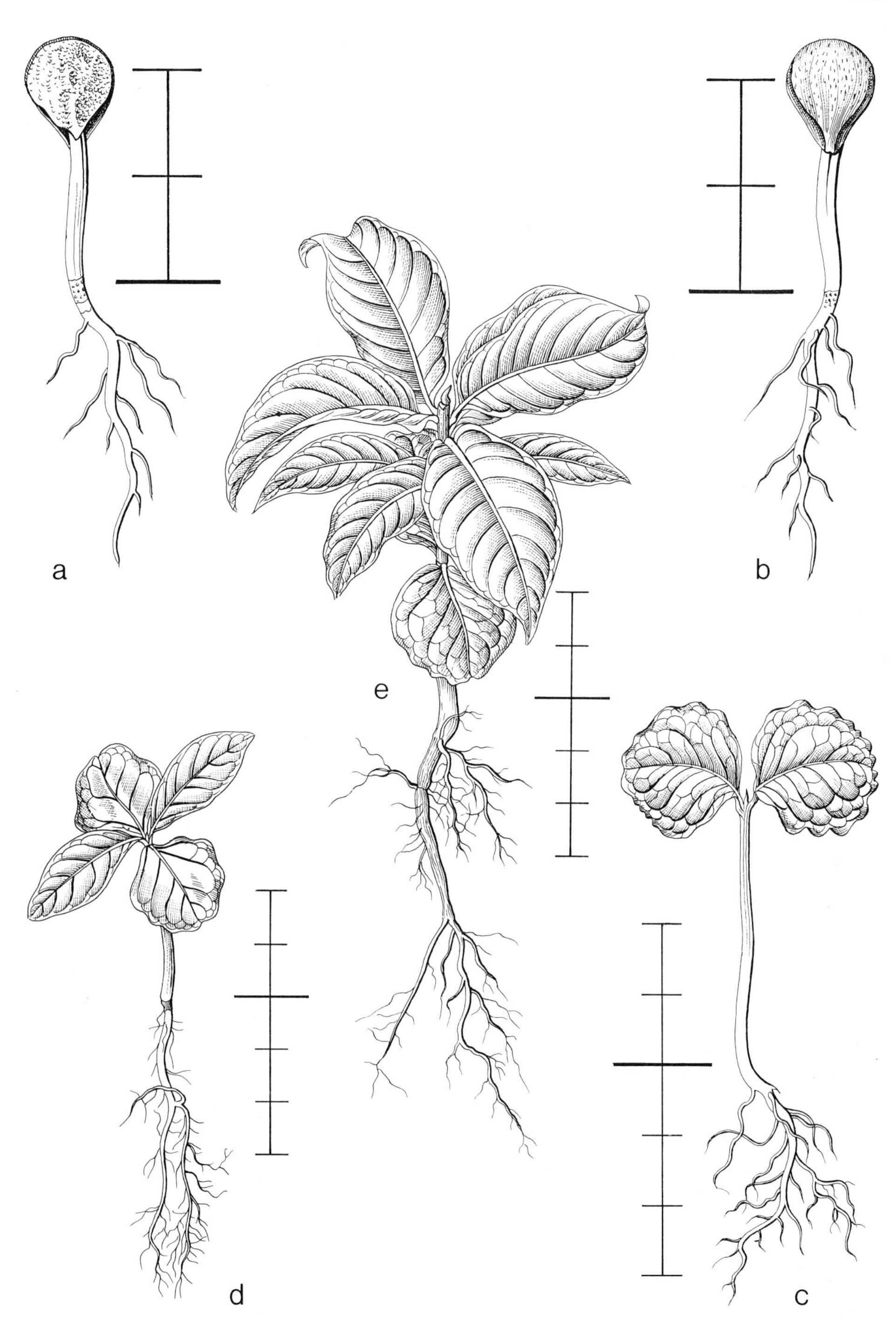

Fig. 157. *Psychotria celebica.* De V. 2437, collected 3-x-1973, planted 5-xi-1973 (a, b, and d); De V. 2510, collected 12-x-1973, planted 5-xi-1973 (c and e). a and b. 8-i-1974; c. 22-i-1974; d. 28-iii-1974; e. 5-ix-1974.

pair to 5 by 2.3 cm; base acute; top acute; margin entire; nerves pinnate, midrib above prominent, below more so, lateral nerves rather inconspicuous, near the margin with a hardly visible marginal nerve.

Specimens: 2437 and 2510 from N. Celebes, primary forest on deep clay, medium altitude.
Growth details: In nursery germination good, ± simultaneous.
Remarks: All species described show the same germination pattern and general morphology; they differ in details only.

RUTACEAE

CITRUS

Fl. Java 2: 107; Csapody, Keimlingsbestimmungsbuch der Dikotyledonen (1968) 91, 253 (*C. limon*, *C. reticulata*, *C. sinensis*); Duke, Ann. Miss. Bot. Gard. 52 (1965) 337 (*C. paradisi*); ibid. 56 (1969) 151 (*C. limonum*); Hickel, Graines et Plantules (1914) 283 (*C. triptera*); Lubbock, On Seedlings 1 (1892) 323 (*C. aurantium*, *C. decumana*).
Monograph: Engler, Abh. Wiss. Berl. Phys. Math. Cl. (1896) Taf. 3.
Genus widely cultivated in the tropics and subtropics all over the world. Native possibly in India, throughout S.E. and E. Asia to Japan, all over Malesia, N. and E. Australia, in the Pacific in the Bismarcks, the Solomons and New Caledonia.

Citrus macroptera Montr. – Fig. 158.

Development: The taproot and hypocotyl emerge from one pole of the seed. The hypocotyl becomes erect, by which the cotyledons are carried above the soil while they are still enclosed in the testa. By spreading of the cotyledons the testa is shed. No resting stage occurs.
Seedling epigeal, phanerocotylar. Sloanea type and subtype.
Taproot rather sturdy, fleshy, creamy-white, with slender, later much-branched sideroots.
Hypocotyl strongly elongating, terete, slender, to 2 cm long, green, glabrous, irregularly warty with lenticels.
Cotyledons 2, opposite, somewhat succulent, in the 3rd–4th leaf stage dropped; spreading, without stipules, almost sessile, fleshy, dark green, glabrous. Petiole in cross-section flattened, to 0.5 by 1 mm. Blade ± elliptic, to 1.4 by 0.8 cm; base slightly auriculate, auricles rounded, ± 1 by 1 mm; top rounded; margin entire; punctate with scattered hyaline dots, those along the margin crowded.
Internodes: First one strongly elongating, terete, to 2 cm long, as the hypocotyl but with scattered, minute, simple, hyaline hairs; next internodes as the first one, 2–5 mm long, beyond the 15th one gradually longer. Stem not very slender, ± straight.
First two leaves ± opposite, consisting only of the webbed petiole, simple, spreading,

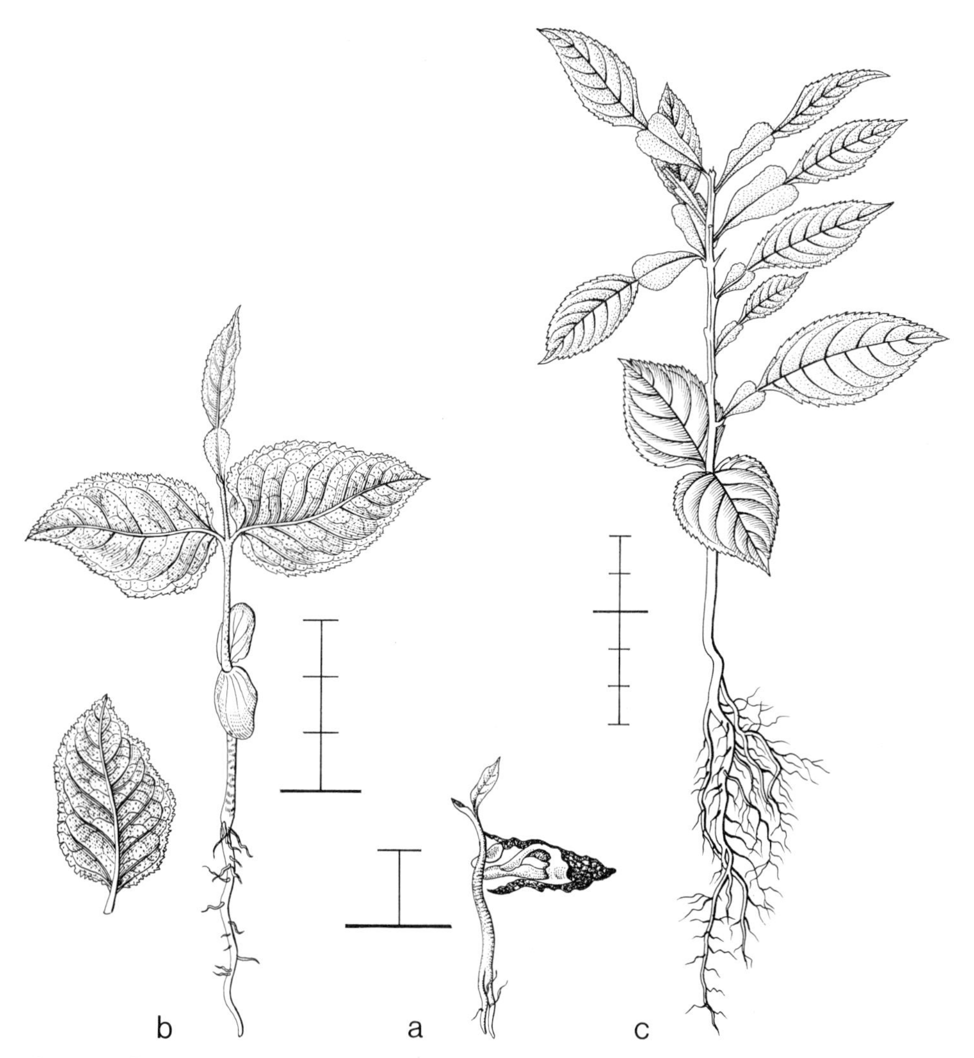

Fig. 158. *Citrus macroptera.* De V. 1417, collected 10-vii-1972, planted 27-vii-1972. a, b, and c date?.

often irregularly formed, without stipules, shortly 'petiolate', herbaceous, green, below paler. Petiolar blade ± ovate, to 3.5 by 3 cm; base obtuse to narrowed into an in cross-section semi-orbicular, to 2 by 1 mm big 'petiole'; top acute; margin irregularly double serrate; almost glabrous, punctate with scattered hyaline dots, especially along the margin.

Next leaves spirally arranged, one-foliolate, very irregular in size. Petiole distinctly winged, foliaceous, spathulate, in the lower ones 1 by 0.5 cm, in the 20th leaf to 3.5 by 1.5 cm; base narrowed; top obtuse; margin irregularly double serrate; punctate. Blade distinctly jointed, elliptic to ovate-oblong, the lower ones 2 by 1.5 cm but sometimes to 5.5 by 2.5 cm, the 20th one to 6 by 2.5 cm; base acute to obtuse; top slightly acuminate; margin irregularly double serrate; nerves pinnate, midrib above and below prominent, lateral nerves inconspicuous, ending in anastomoses; punctate as the first leaves. From 15th to 18th leaf onwards one of the buds in each axil developing into a stout, hard, green, to 15 by 1 mm big thorn, which points sidewards along the leaf.

Specimens: 1417 from N. Sumatra, somewhat depleted primary forest along river, on deep clay, low altitude.
Growth details: In nursery germination fair, ± simultaneous. Drought affects the size of newly developing leaves.
Remarks: C. macroptera is similar to *C. sinensis* in the free, fleshy cotyledons that are borne above the soil on an elongated hypocotyl. In *C. aurantium*, *C. decumana*, *C. limon*, *C. paradisi*, *C. reticulata*, and *C. triptera* the cotyledons are secund, covered by the testa, and borne at soil level.

CLAUSENA

Fl. Java 2: 103.
Genus in tropical and S. subtropical Africa, India and Ceylon, throughout S.E. and tropical E. Asia, all over Malesia, N. Australia.

Clausena excavata Burm. f. – Fig. 159.

Development: The seed splits slightly to considerably along the margin of the cotyledons. The cotyledonary petioles elongate slightly, pushing the plumule free from the envelopments, after which the shoot starts developing. No resting stage occurs.
Seedling hypogeal, cryptocotylar. Horsfieldia type and subtype.
Taproot relatively thick, hard fleshy, brownish cream, with few, thick, not branched, creamy-white sideroots.
Hypocotyl slightly enlarging, terete, to 3 mm long, green, rather densely covered with short, simple, curved, hyaline hairs.
Cotyledons 2, secund, sometimes the testa entirely splitting along the cotyledonary margins and cotyledons turning more opposite, succulent, in the 7th–9th leaf stage shed; simple, without stipules, petiolate. Whole body ellipsoid, to 9 by 6 mm, the greyish brown testa persistent. Petiole in cross-section semi-orbicular, to 3 by 1 mm, green, hairy as the hypocotyl.
Internodes: First one shortly elongating, terete, to 2 mm long, green, hairy as the hypocotyl; next internodes much differing in length, in the scaly part of the stem to 1.2 cm long, as the first one, those below in the leafy part of the stem to 8 mm long, hardly gaining in length. Stem slender, straight.
Leaves all spirally arranged, compound, lower ones reduced, without stipules, petiolate, herbaceous, green, hairy as the hypocotyl. First up to 6 lower ones much reduced, simple, herbaceous, green, hairy as the hypocotyl; blade subulate, to 2 mm long, with minute, swollen, hyaline warts, in the highest ones longer and more

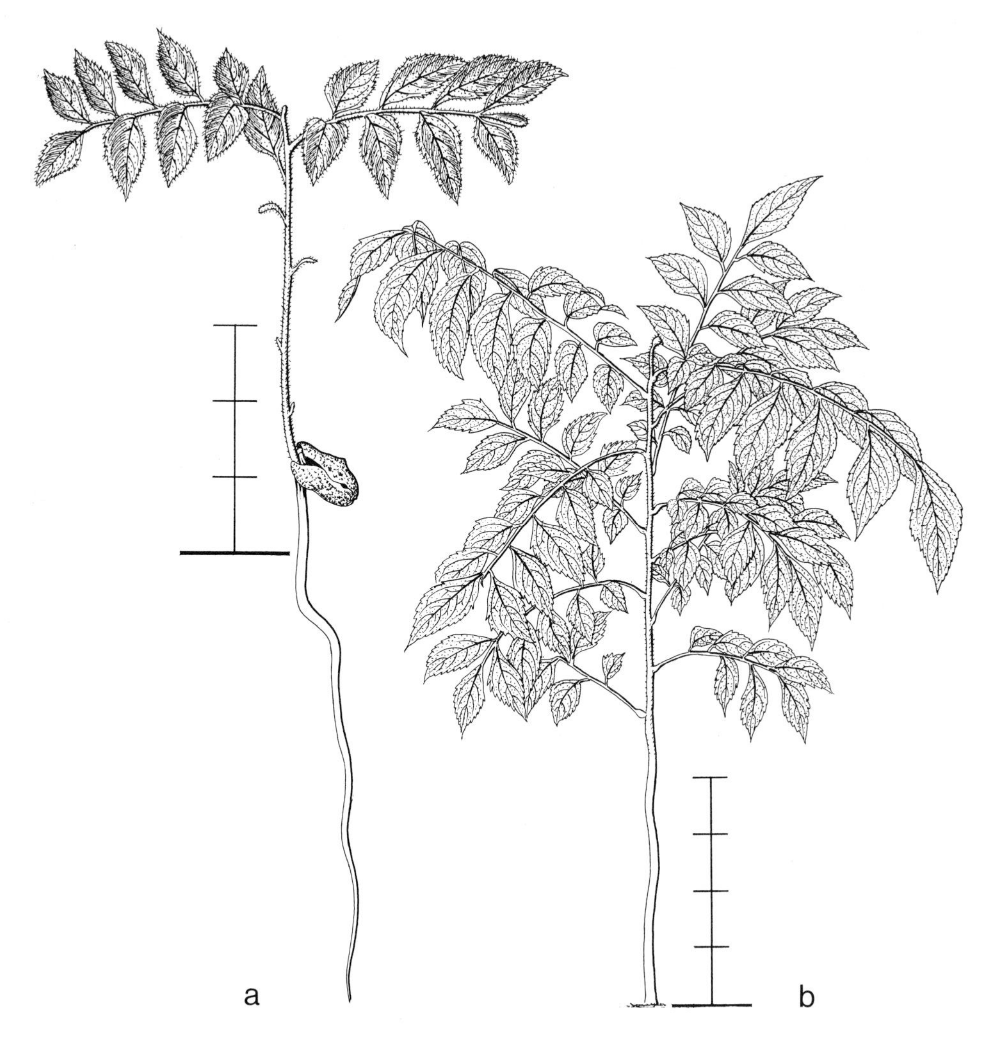

Fig. 159. *Clausena excavata*. Bogor Bot. Gard. III. F. 18a, collected and planted ix-1973. a. 26-xi-1973; b. 12-viii-1974.

leaf-like. Petiole terete, to 7 by 0.3 mm; base swollen, warty, hairy as the reduced leaves. Rhachis as the petiole, slightly webbed, especially towards the base of a leaflet, in lowest leaves to 2 cm long, in 10th one to 4 cm long. Blade imparipinnate, with terminal leaflet. Leaflets in first developed leaf 7, in second one 9, in 15th one 14, ± alternate, sometimes almost opposite, petiolulate. Petiolule in cross-section semi-orbicular, in the lowest leaves to 0.4 by 0.2 mm, in 15th one to 2 by 0.7 mm, hairy as the hypocotyl. Blade of leaflets oblique rhomboid, in the lower leaves to 13 by 5 mm, in 10th one to 2.2 by 1 cm; base distinctly asymmetric, the adaxial side acute, the abaxial side broadly obtuse, abruptly narrowed into the petiolule; top acute (to acuminate); margin entire but irregularly undulating in the plane of the blade to almost regularly serrate, above and below on the midnerve and along the margin slightly hairy, punctate with hyaline dots, along the margin bigger ones, the latter provided with a bundle of hairs; nerves pinnate, midrib above slightly prominent, below slightly more so, lateral nerves rather inconspicuous, forming an inconspicuous, much undulating marginal nerve.

Specimens: Bogor Botanic Gardens III. F. 18a, from Celebes.
Growth details: In nursery germination good, ± simultaneous.

LIMONIA

Fl. Java 2: 106; Lubbock, On Seedlings 1 (1892) 322 (*L. acidissima*).
Genus in India and Ceylon, throughout S.E. Asia, W. Malesia.

Limonia acidissima L. – Fig. 160.

Development: The hypocotyl and taproot pierce the testa at one pole of the seed, and establish. The hypocotyl becomes erect, by which the enclosed cotyledons are carried above the soil. The cotyledons spread, freeing themselves from the testa. No resting stage occurs.
Seedling epigeal, phanerocotylar. Sloanea type and subtype.
Taproot long, slender, flexuous, cream-coloured, with few, rather short, few-branched sideroots.
Hypocotyl strongly elongating, terete, slender, to 3 cm long, pale green, rather densely hairy with minute, simple, hyaline hairs.
Cotyledons 2, opposite, succulent, in the 10th–14th leaf stage dropped; simple, without stipules, petiolate, rather soft fleshy, dark green, paler below, and there with many, minute, slightly darker green dots. Petiole in cross-section semi-orbicular, to 2.5 by 1 mm, hairy as the hypocotyl. Blade elliptic, to 1.4 by 1.1 cm; base auriculate, auricles broadly triangular with obtuse tip; top broadly rounded; margin entire, minutely frayed; glabrous; nerves pinnate, inconspicuous, above slightly sunken.
Internodes: First one slightly elongating, terete, to 5 mm long, green, hairy as the hypocotyl; next internodes as the first one, higher ones becoming slightly larger, 10th one to 1 cm long, 23rd one to 1.5 cm long. Stem slender, slightly zig-zag.
Leaves all spirally arranged, compound, without stipules, petiolate, subcoriaceous, green, glabrous. First 1–4 leaves one-foliolate, next 2–4 ones 3-foliolate, later ones with 5 leaflets, after 2 years to 7-foliolate. Petiole in one-foliolate leaves semi-orbicular, to 3 by 1 mm, those in higher leaves broadly winged, in 3-foliolate leaves to 9 by 2.5 mm, in 23rd leaf to 20 by 2 mm, at the top constricted, along the margin with hyaline dots. Rhachis as the petiole of higher leaves. Blade in one-foliolate leaves broadly spathulate, to 1.5 by 0.9 cm; base narrowed; top irregularly rounded; margin irregularly serrate; punctate all over the blade with minute, along the margin with slightly larger, hyaline dots which appear darker green below. Blade in 3-foliolate leaves ± rhomboid in outline, to 2.5 by 2.5 cm, in 5–7-foliolate leaves obovate, in 23rd leaf to 6 by 4 cm. Leaflets obovate to spathulate, in 3rd leaf to 10 by 6 mm, in

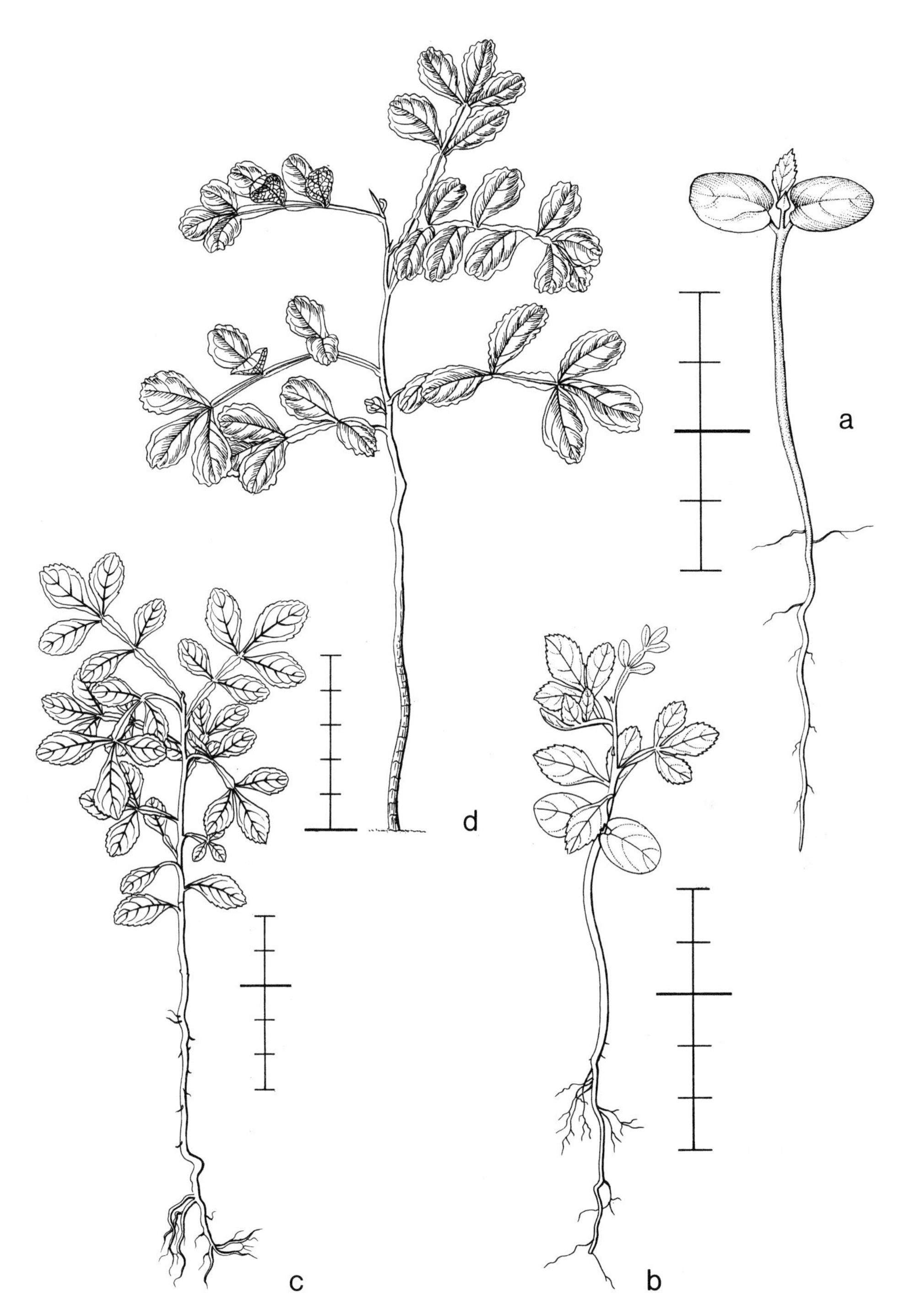

Fig. 160. *Limonia acidissima.* De V. 1358, collected and planted 17-v-1972. a, b, and c date?; d. 7-viii-1974.

23rd leaf to 24 by 15 mm; base acute to narrowed; top rounded; margin irregularly dentate; nerves pinnate, midrib at the base above slightly prominent, below prominent, lateral nerves in the lower leaves rather inconspicuous, in the higher ones above and below slightly raised, at their tip curved to the next one.

Specimens: 1358 from W. Java, cultivated fruit tree from the neighbourhood of Bandung.
Growth details: In nursery germination poor.

SABIACEAE

MELIOSMA

Fl. Java 2: 144; Duke, Ann. Miss. Bot. Gard. 52 (1965) 342 (*M. herbertii*); ibid. 56 (1969) 153 (*M. panamensis*); Lubbock, On Seedlings 1 (1892) 368 (*M. arnottiana* = *M. pinnata* ssp. *arnottiana*, *M. pungens* = *M. simplicifolia* ssp. *pungens*).
Revision: Van Beusekom, Blumea 19 (1971) 355.
Genus in India and Ceylon, throughout S.E. and E. Asia as far north as Korea and S. Japan, all over Malesia.

Meliosma simplicifolia (Roxb.) Walp. – Fig. 161.

Development: The hypocotyl and taproot emerge from the seed. The hypocotyl becomes erect by which the paracotyledons enclosed in the envelopments are carried above the soil. Then follows a first rest. By unfolding of the paracotyledons the testa is shed. A second resting stage occurs with the paracotyledons expanded.
Seedling epigeal, phanerocotylar. Macaranga type.
Taproot rather short, slender, flexuous, creamy-white, with many slender, long, branched, creamy-white sideroots.
Hypocotyl strongly elongating, terete, to 3 cm long, slender, brownish cream, rather densely covered with minute, simple, brownish hairs.
Paracotyledons 2, opposite, foliaceous, when dropped not known, but before the 11th leaf stage; without stipules, shortly petiolate, herbaceous, green, above and below on the midnerve slightly hairy as the hypocotyl, on the blade with scattered, minute, blackish, glandular hairs. Petiole in cross-section semi-orbicular, to 1.2 by 1.2 mm. Blade ovate, to 1.3 by 0.9 cm; base obtuse; top acute; margin entire; nerves pinnate, midrib above slightly prominent, below more so, lateral nerves rather inconspicuous.
Internodes: First one hardly elongating, terete, to 3 mm long, pale green, hairy as the hypocotyl; next internodes as first one, gradually higher ones larger, 10th one to 5 mm long. Stem rather slender, straight.
Leaves all spirally arranged, simple, without stipules, shortly petiolate, herbaceous, green, hairy as the paracotyledons. Petiole in cross-section semi-orbicular, to 1.2 by 1.2 mm, those of the higher leaves slightly larger, 10th one to 5 by 1.5 mm, at the base

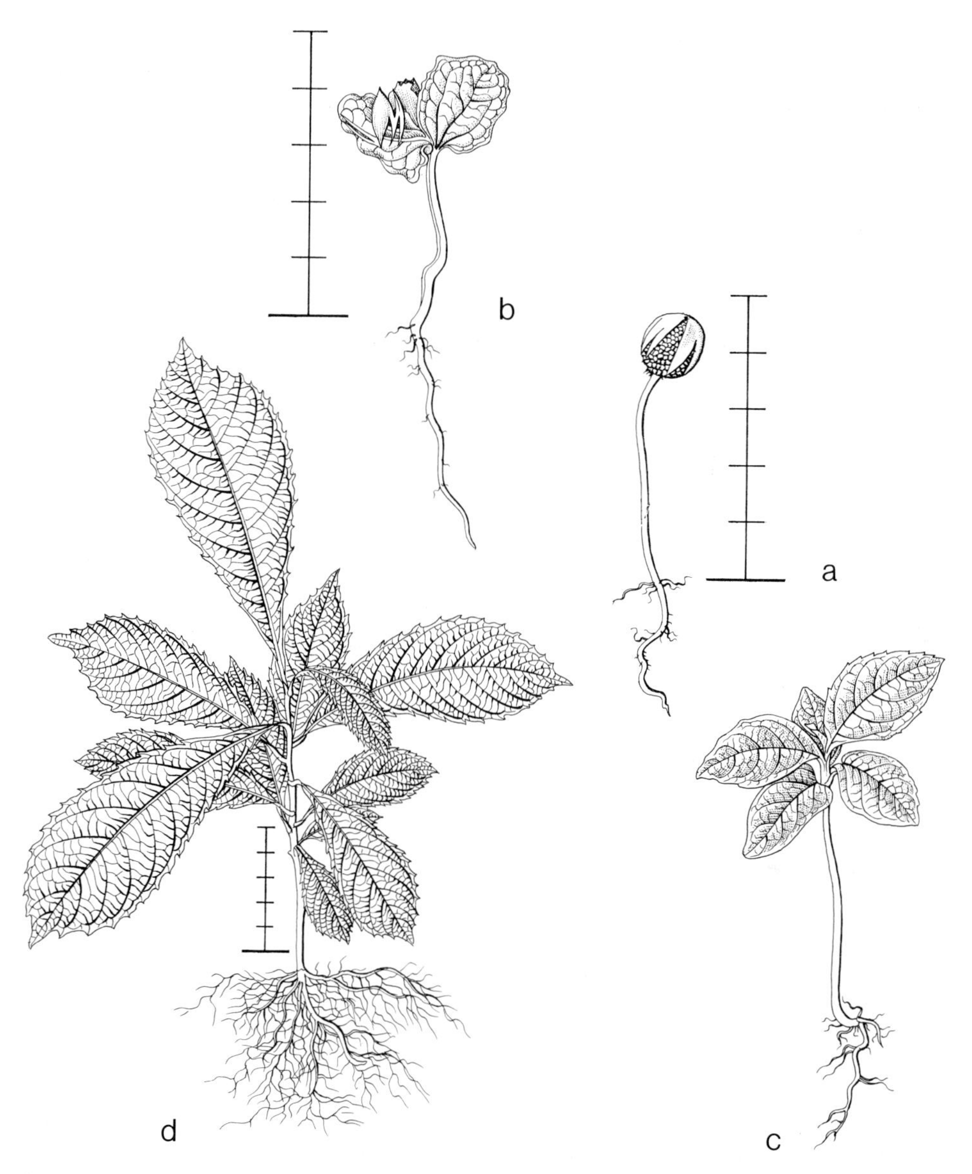

Fig. 161. *Meliosma simplicifolia.* De. V. 1437, collected 12-vii-1972, planted 27-vii-1972. a, b, and c date?; d. 7-viii-1973.

distinctly swollen. Blade elliptic, in 5th leaf to 3 by 2.2 cm, the lower ones smaller, gradually increasing in size, 10th one to 5.5 by 2.7 cm, 18th one to 10 by 4 cm; base slightly narrowed; top acute; margin widely serrate; nerves pinnate, midrib above slightly prominent, below much so, lateral nerves below prominent, ending free in the teeth.

Specimens: 1437 from N. Sumatra, alluvial flat along river, somewhat depleted primary forest on deep clay, low altitude.
Growth details: In nursery germination poor.
Remarks: Five species described, similar in germination pattern and morphology. They differ in details only.

SAPINDACEAE

CUBILIA

Fl. Java 2: 143.
Monograph: Radlkofer, Pflanzenreich IV, 165, 1 (1932) 921.
Genus in Central Malesia: Borneo, the Philippines, Celebes, and the Moluccas.

Cubilia cubili (Blanco) Adelb. – Fig. 162.

Development: The taproot emerges from the scar at the blunt pole of the seed. By elongation of the petioles the plumule is pushed out of the seed, by which the shoot is enabled to develop. No resting stage occurs.
Seedling hypogeal, cryptocotylar. Horsfieldia type and subtype.
Taproot long, slender, flexuous, pale brownish, with many, much-branched pale brownish sideroots, especially in the upper part.
Hypocotyl not enlarging.
Cotyledons 2, secund, succulent, in the 2nd–3rd leaf stage shed; simple, without stipules, petiolate. Whole body ellipsoid, to 3 by 2 by 2 cm, the dark brown, shining testa persistent, at the widest part with a faint circular ridge, around the petioles with a dull brown spot. Petioles irregular, flattened, to 3 by 3 mm, warty, pale brownish cream.
Internodes: First one strongly elongating, ± terete, to 5 cm long, green, almost glabrous; next internodes as the first one, irregular in length, 0.3–6.5 cm long, in the leafy part more regular, 2–2.5 cm long, beyond the 8th one more irregular in length, 10th one to 7 cm long. Stem slender, rather zig-zag in the leafy part.
Leaves all spirally arranged, the developed ones compound, spreading, without stipules, petiolate, herbaceous, green, almost glabrous. The first two to four leaves scale-like, dropping soon. Petiole in cross-section semi-orbicular, in higher ones more terete, at the base distinctly swollen, above with a longitudinal ridge, in second leaf to 8 by 0.2 cm, in 15th one to 13 by 0.2 cm. Rhachis in second leaf to 2.5 by 0.2 cm, in 15th one to 14 cm long, with sharp ridges along the margin. Blade imparipin-

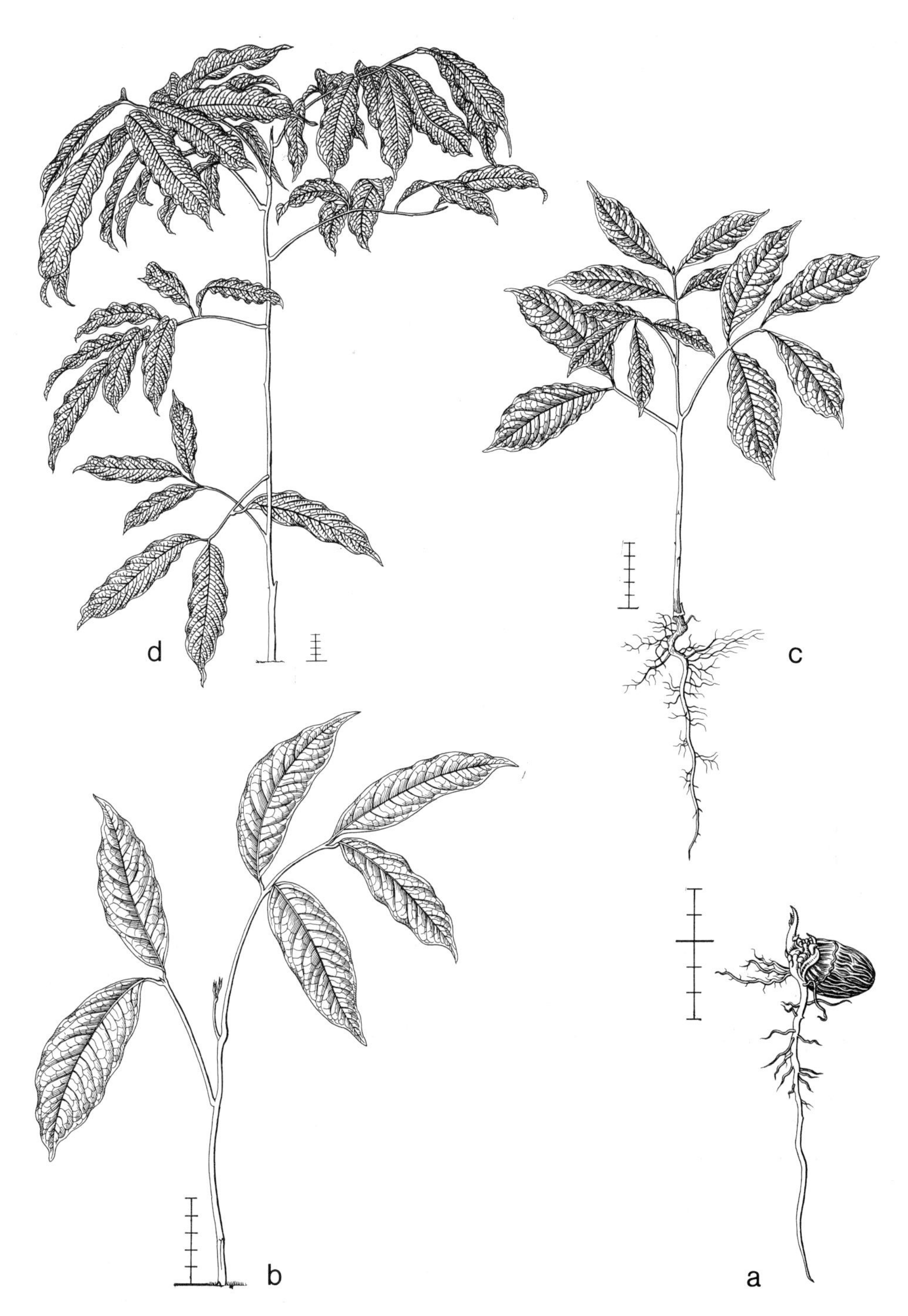

Fig. 162. *Cubilia cubili*. De V. 2386, collected 16-iii-1973, planted 22-iii-1973. a. 18-iv-1973; b. 12-v-1973; c. 7-iv-1973; d. 21-viii-1974.

nate or paripinnate, with a terminal cusp, in first leaf with two opposite petiolulate leaflets, sometimes with 3 alternating leaflets, in second leaf 3-foliolate, with terminal cusp, sometimes 2- or 4-foliolate; gradually the higher ones producing more leaflets, in 15th leaf with 4 pairs of subopposite leaflets. Petiolule to 8 by 1.5 mm. Cusp subulate, above channelled, to 5 by 1 mm, rather soon caducous. Leaflets oblong, in lowest leaves to 13 by 4.8 cm, in 15th leaf to 23 by 7.5 cm; base more or less narrowed; top acuminate to cuspidate; margin entire, irregular; nerves pinnate, above raised, below more so, lateral nerves free ending or curved back to the next one.

Specimens: 2386 from S.E. Borneo, primary forest on deep clay, low altitude.
Growth details: In nursery germination fair, ± simultaneous.

DIMOCARPUS

Fl. Java 2: 136 (*Euphoria*), 137 (*Pseudonephelium*).
Revision: Leenhouts, Blumea 19 (1971) 113; ibid. 21 (1973) 377.
Genus in India and Ceylon, throughout S.E. and tropical E. Asia including Formosa, all over Malesia except the Lesser Sunda Islands, N.E. Australia.

Dimocarpus longan Lour. var. *malesianus* Leenh. – Fig. 163.

Development: The taproot emerges from one pole of the seed. The elongating petioles push the plumule out of the seed, by which the shoot is enabled to emerge. A short resting stage with two developed leaves occurs.
Seedling hypogeal, cryptocotylar. Horsfieldia type and subtype.
Taproot long, sturdy, creamy-white, with many short, few-branched, creamy-white sideroots.
Hypocotyl not enlarging.
Cotyledons 2, secund, succulent, in the 2nd leaf stage shed; simple, without stipules, petiolate. Whole body ellipsoid, to 17 by 15 by 15 mm, the shining dark brown testa persistent. Petiole flattened, to 10 by 5 mm, pale greenish, glabrous.
Internodes: First one strongly elongating, terete, to 10 cm long, green, rather densely hairy with minute, simple, hyaline hairs; next internodes as the first one, to 6 cm long, reddish when young, turning green. Stem slender, ± straight.
First two leaves opposite, compound, without stipules, petiolate, petiolulate, subcoriaceous, red when young, turning green. Petiole in cross-section semi-orbicular, to 15 by 1.5 mm, at the base swollen, hairy as the first internode. Petiolule in cross-section semi-orbicular, to 4 by 1.5 mm, hairy as the first internode. Blade paripinnate, with terminal cusp, in outline obtriangular. Terminal cusp subulate, to 7 by 1 mm, above channelled. Leaflets 2, opposite, ovate-oblong, to 9 by 3.7 cm; base acute to obtuse; top acute to acuminate; margin entire, irregular; below and above on the nerves and along the margin hairy as the first internode; nerves pinnate, above slightly prominent, below much so, especially the midnerve, lateral nerves free ending.

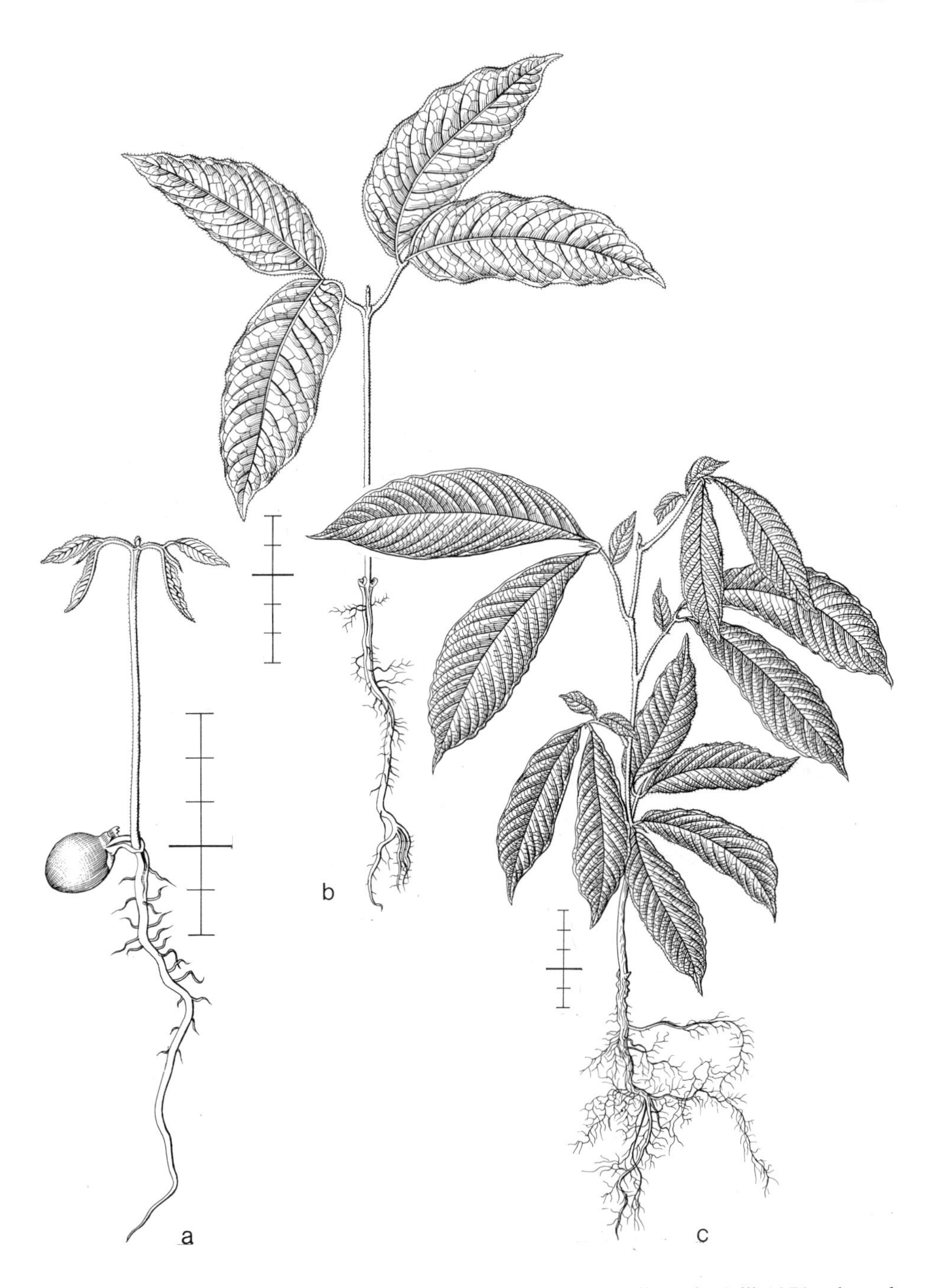

Fig. 163. *Dimocarpus longan* var. *malesianus*. De V. 2314, collected 13-iii-1973, planted 22-iii-1973. a. 9-iv-1973; b. 7-v-1973; c. 29-v-1974.

Next leaves spirally arranged. Petiole in second leaf to 25 by 1.5 mm, in 10th one to 14 by 0.4 cm. Rhachis in the second leaf to 8 by 1.5 mm, in 10th one to 10 cm long, further as the petiole of the first leaves. Terminal cusp dropping soon, in 12th leaf to 15 by 2 mm. Petiolules in second leaf to 3 by 1 mm, in 10th one to 10 by 3 mm. Blade with 2 pairs of leaflets, from ± 8th leaf onwards with 3 pairs of leaflets. Lower leaflets in each leaf rather small, in the lower leaves sometimes distinctly to entirely reduced, beyond the 4th leaf often reduced to a subulate cusp to 3 by 0.5 mm, in the second leaf to 2.8 by 1.5 cm, in 10th leaf to 16 by 8 cm; base often inequal, acute; top ± acute, apiculate; margin entire; hairy as the first internode. Upper leaflets lanceolate, in 3rd leaf to 9 by 2.8 cm, in 10th leaf to 26 by 9.5 cm; base acute; top acuminate to cuspidate; further as those of the first leaf.

Specimens: 2314 from S.E. Borneo, alluvial flat, primary forest on deep clay, low altitude.
Growth details: In nursery germination fair, ± simultaneous.

HARPULLIA

Fl. Java 2: 142.
Monograph: Radlkofer, Pflanzenreich IV, 165, 2 (1933) 1433.
Genus in India and Ceylon, throughout S.E. Asia, all over Malesia, N. and E. Australia, in the Pacific as far as Tonga Arch., including the Loyalties.

Harpullia arborea (Blanco) Radlk. – Fig. 164.

Development: The taproot emerges from one pole of the seed. The testa splits and is usually shed by the slightly to distinctly spreading cotyledons, after which the shoot starts developing from the opening. No resting stage occurs.
Seedling hypogeal, semi-cryptocotylar to phanerocotylar. Horsfieldia type and subtype to Endertia type, Chisocheton subtype.
Taproot long, slender, fibrous, creamy-white with brown splitting bark, with many long, filiform, much-branched sideroots.
Hypocotyl enlarging to 2 cm long, terete, subterranean, green, glabrous.
Cotyledons 2, ± secund, succulent, when shed not known, between 3rd and 8th leaf stage; more or less spreading, more or less unequal in size, without stipules, petiolate, hard fleshy, green, glabrous. Petiole flattened, to 5 by 4 mm. Blade semi-ellipsoid, to 1.5 by 1 by 1 mm, outside concave, inside flattened to slightly concave; base and top rounded, often with two triangular auricles to 1 by 1 mm at the base in the biggest cotyledons, margin entire. The testa remains persistent on the blade.
Internodes: First one slightly elongating, terete, to 5 mm long, green, with rather close-set, simple, hyaline hairs; next internode as the first one, irregular in size, lowest ones to 2 cm long, higher ones not much larger. Stem straight, the lower 5 internodes with scales.
Leaves all spirally arranged, compound, spreading, without stipules, petiolate, herbaceous, green. The lower 4 to 5 reduced to simple, subulate, green scales to 3 by 0.5

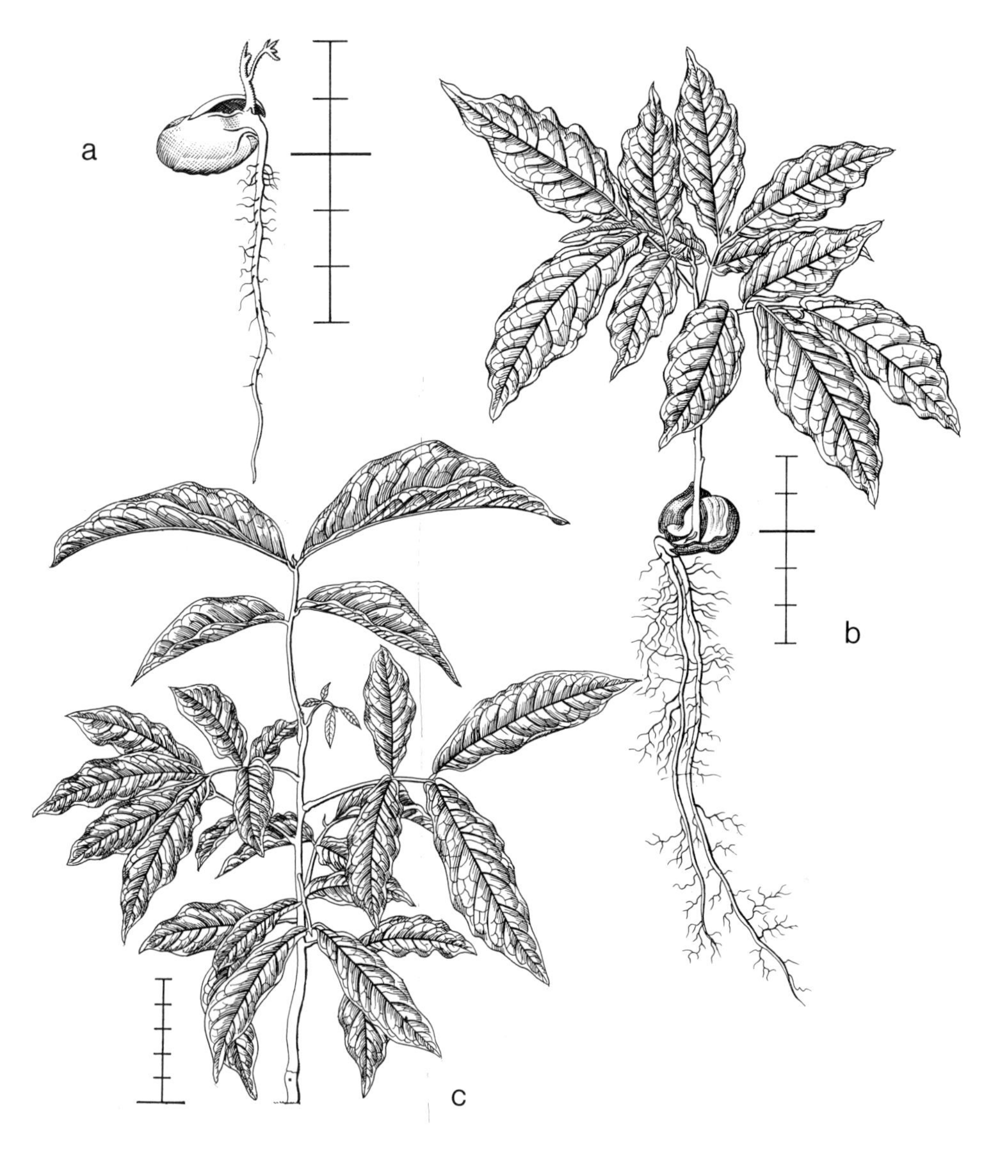

Fig. 164. *Harpullia arborea.* Bogor Bot. Gard. III. L. 62, collected and planted i-1973. a. 26-iii-1973; b. 29-iii-1973; c. 6-ix-1973.

mm. Petiole terete, in lowest leaves to 1.5 cm long, in 10th one to 3.5 cm, in 15th one to 5 cm long, at the base distinctly swollen, hairy as the first internode. Rhachis as the petiole, in 3rd leaf to 2 cm long, in 10th one to 4.5 cm, in 20th one to 9.5 cm long. Blade compound, ± obovate in outline, the lower ones 4-foliolate, higher ones gradually with more leaflets, 17th one 8-foliolate, with a herbaceous, subulate terminal cusp to 3 by 0.5 mm. Leaflets sessile to minutely petiolate, in the lower leaves opposite, in the higher ones ± alternate, the lower leaflets in each leaf ± oblong, to 4.5 by 2.2 cm, often smaller, the higher ones to 8 by 3.5 cm; base equal to distinctly unequal; top slightly acuminate, its tip apiculate; margin entire, slightly undulating; nerves pinnate, above slightly raised, below more so, lateral nerves at the tip curved to the next one.

Specimens: Bogor Botanic Gardens III.L.62 from Kei Islands.
Growth details: In nursery germination good, ± simultaneous.
Remarks: Two species cultivated. *H. thanathophora* has the lower ± 15 leaves simple, not reduced, the first two being ± subopposite.

Harpullia thanathophora Bl. – Fig. 165.

Development: The seedcoat splits irregularly. The taproot develops from one pole. The shoot emerges from between the more or less spreading cotyledons. No resting stage occurs.
Seedling hypogeal, semi-cryptocotylar to phanerocotylar. Horsfieldia type and subtype to Endertia type, Chisocheton subtype.
Taproot long, slender, fibrous, dirty creamy-white, with rather long, slender, finely branched, creamy-white sideroots.
Hypocotyl slightly elongating, ± terete, subterranean, to 5 mm long, glabrous.
Cotyledons 2, ± secund but often more or less spreading, succulent, in the 4th leaf stage shed; equal to slightly unequal, sessile, without stipules, hard fleshy, green, glabrous. Blade semi-ellipsoid, to 1.2 by 1 by 0.7 cm; top broadly rounded, sometimes distinctly apiculate or grooved; margin entire; outside concave, inside slightly convex. The testa remains persistent on the blade.
Internodes: First one strongly elongating, terete, to 4.5 cm long, green, rather densely hairy with minute, simple, hyaline hairs, glabrescent; next internodes as the first one, lower ones to 3.5 cm long, usually much shorter, 10th one to 2.5 cm, 20th one to 5.5 cm long. Stem slender, slightly zig-zag.
Leaves spirally arranged, the first two ± subopposite, without stipules, petiolate, herbaceous, green, dark bluish red when young, often the first one abortive. Lower ± 15 leaves simple. Petiole in cross-section semi-orbicular, 5th one to 4 by 1.5 mm, 15th one 15 by 1.5 mm, at the base distinctly swollen, hairy as the first internode. Blade ovate-oblong, in the lower leaves to 7.5 by 3.3 cm, 5th one to 11 by 3.8 cm, 15th one to 23 by 8 cm; base narrowed into the petiole; top acuminate, in higher ones cuspidate, its tip obtuse, margin entire; above and below on the main nerve hairy as the first internode; nerves pinnate, midrib above and below prominent, lateral nerves less so, free ending. Subsequent leaves compound, imparipinnate with terminal cusp. Petiole as those of simple leaves, in 16th leaf to 6 by 2 mm, in 20th one to 6.5

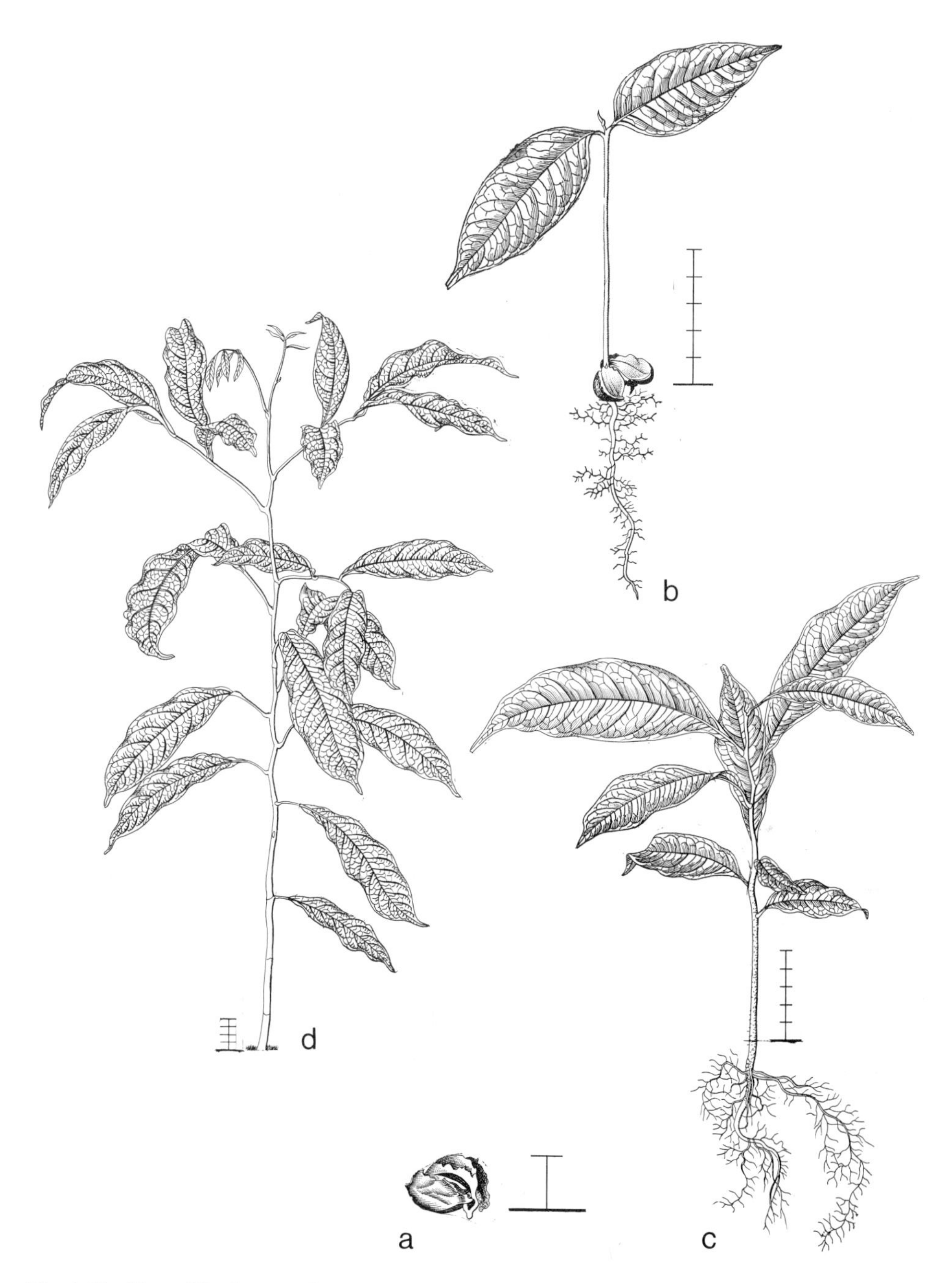

Fig. 165. *Harpullia thanatophora.* **Bogor Bot. Gard. III. I. 14, collected and planted i-1973. a** and b date?; c. 22-viii-1973; d. 9-vii-1974.

by 2 mm, at the base distinctly swollen. Rhachis as the petiole, in 20th leaf to 8 cm long. Terminal cusp subulate, in the lowest two compound leaves ± absent, in 20th leaf 3 by 0.5 mm. Leaflets in lowest compound leaves (sub)opposite, with up to 2 pairs of lateral ones, in higher ones ± alternate, in 20th leaf with up to 6 leaflets, without stipellae, petiolulate. Petiolule terete, swollen, in 20th leaf to 7 by 2 mm, further as the petiole. Blade of leaflets oblong, in 20th leaf to 17.5 by 6.5 cm; base narrowed, top cuspidate; margin entire; further as simple leaves.

Specimens: Bogor Botanic Gardens III. I. 14, from Kei Islands.
Growth details: In nursery germination good, ± simultaneous.
Remarks: See *H. arborea*.

LEPISANTHES

Fl. Java 2: 134.
Revision: Leenhouts, Blumea 17 (1961) 33.
Genus in tropical Africa and Madagascar, India and Ceylon, throughout S.E. and tropical E. Asia, all over Malesia, and N. Australia.

Lepisanthes alata (Bl.) Leenh. – Fig. 166.

Development: The root emerges from one pole of the seed. The cotyledonary petioles enlarge, bringing the plumule out of the seed, after which the shoot is able to emerge. A short resting stage occurs with the cotyledons and two developed leaves.
Seedling hypogeal, cryptocotylar. Horsfieldia type and subtype.
Taproot long, rather sturdy, fleshy, creamy-white, with numerous, rather long, slender, much-branched creamy-white sideroots.
Hypocotyl not enlarging.
Cotyledons 2, secund, succulent, in the 3rd–4th leaf stage shed; simple, without stipules, petiolate. Whole body ellipsoid, ± 2.3 by 1.8 by 1.8 cm, the leathery brown, shining testa persistent. Petiole short, slightly flattened, glabrous.
Internodes: First one strongly elongating, terete, to 10 cm long, green, with minute, scattered, simple, hyaline hairs; next internodes as the first one, slightly quadrangular with 4 longitudinal grooves, second one to 3.4 cm, subsequent ones irregular in length, to 1–3 cm long, 20th one to 5 cm long. Stem slender, straight.
First two leaves opposite, compound, without stipules, petiolate, herbaceous, purplish when young, turning green. Petiole in cross-section semi-orbicular, to 2 by 0.3 cm, distinctly winged, at the base distinctly thickened, slightly hairy as the first internode, midrib above slightly raised. Rhachis as the petiole, to 3 by 0.4 cm. Blade imparipinnate, with 2(–3) pairs of lateral leaflets and a terminal leaflet. Leaflets sessile, lanceolate-linear, to 10.5 by 1.2 cm; base acute, in the lateral ones distinctly asymmetric, with broader adaxial portion; top acute; margin entire; below at the base slightly hairy; nerves pinnate, midrib above and below raised, lateral nerves inconspicuous, forming an inconspicuous marginal nerve.
Next leaves spirally arranged, gradually higher ones larger. Petiole short, to 2 by 2.5

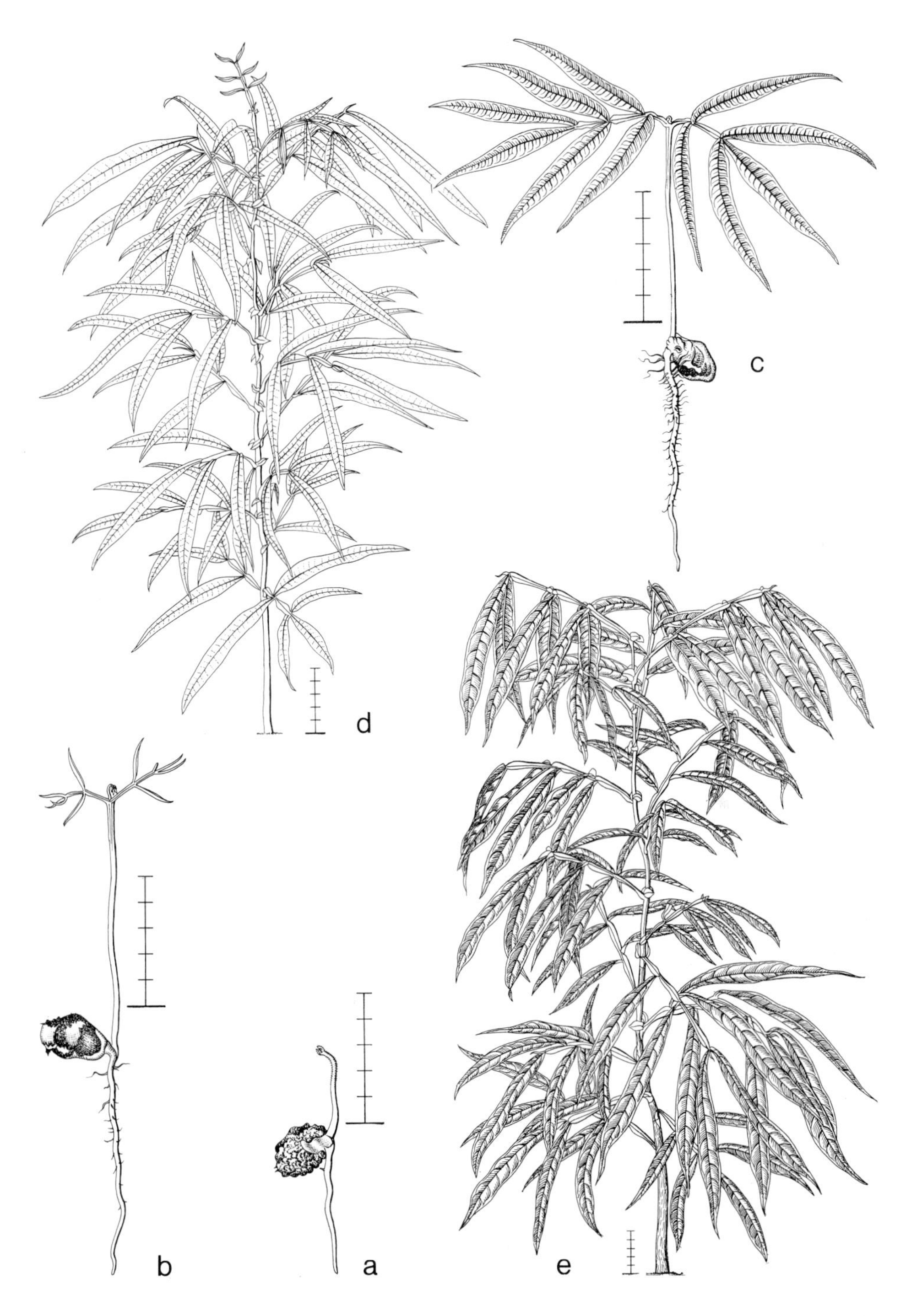

Fig. 166. *Lepisanthes alata.* De V. 2096, collected 6-xi-1972, planted 17-xi-1972. a, b, and c date?; d. 21-vii-1973; e date?.

mm, at the base thickened. Rhachis as that of the first leaves, in 10th leaf to 5 by 0.3 cm, in 20th one to 9 by 0.5 cm. Blade with 3 pairs of leaflets, after 10th–20th one with up to 4 pairs, in general paripinnate with a terminal cusp. Terminal cusp subulate, to 3–4 by 1 mm, as the rhachis. Lowest leaflets opposite, reflexed around the stem, asymmetric, ovate, to 10 by 5 mm; base acute; top rounded; margin entire; with scattered, minute glands, glabrescent. The other leaflets as those of the first leaf, subopposite, those in 10th leaf to 8 by 1 cm, in 15th one to 13.5 by 2 cm.

Specimens: 2096 from S.E. Borneo, hill ridge, primary forest on deep clay, medium altitude.
Growth details: In nursery germination good, ± simultaneous. In a sample of 100 seedlings 5 pigmentless plants were present, which died in the 2nd–3rd leaf stage.

NEPHELIUM

Fl. Java 2: 138; Meijer, Bot. Bull. Sandakan 11 (1968) 112 (*N. mutabile*).
Monograph: Radlkofer, Pflanzenreich IV, 165, 1 (1932) 950.
Genus in S.E. Asia, and all over W. Malesia, including the Philippines and Java.

Nephelium lappaceum L. – Fig. 167.

Development: The taproot emerges from one pole of the seed. The cotyledonary petioles elongate, bringing the plumule out of the seed, after which the shoot is able to emerge. A short resting stage occurs with the cotyledons and two developed leaves.
Seedling hypogeal, cryptocotylar. Horsfieldia type and subtype.
Taproot sturdy, hard fleshy, with rather many sideroots.
Hypocotyl hardly elongating, terete, slightly swollen and conical, to 5 mm long, yellowish, glabrous.
Cotyledons 2, secund, succulent, in the 2nd–3rd leaf stage shed, simple, without stipules, petiolate. Whole body ellipsoid, to 3 by 2 by 1.3 cm, the longitudinally grooved, flaking, brownish testa persistent. Petiole in cross-section flattened, to 8 by 6 mm, fleshy, reddish, hairy as the first internode.
Internodes: First one strongly elongating, terete, to 13 cm long, slender, below reddish, to the top greenish, rather sparsely hairy with minute, simple, appressed, hyaline hairs; next internodes as the first one, second one to 3.6 cm long, 5th one to 4.5 cm long. Stem slender, ± straight.
First two leaves opposite, compound, without stipules, petiolate, herbaceous, reddish turning green. Petiole in cross-section semi-orbicular, to 35 by 1 mm but often smaller, at the base distinctly swollen, above with a brownish, longitudinal ridge, densely hairy as the first internode. Blade 2-jugate with a terminal cusp. Rhachis as the petiole, to 3.5 cm long. Terminal cusp curved, in cross-section flattened, to 8 by 0.8 mm, densely hairy as the first internode, rarely developing into a leaflet. Leaflets 4, opposite to semi-opposite, almost sessile, lanceolate, to 11 by 2.2 cm; base acute; top cuspidate; margin entire; rather sparsely hairy as the first internode; nerves

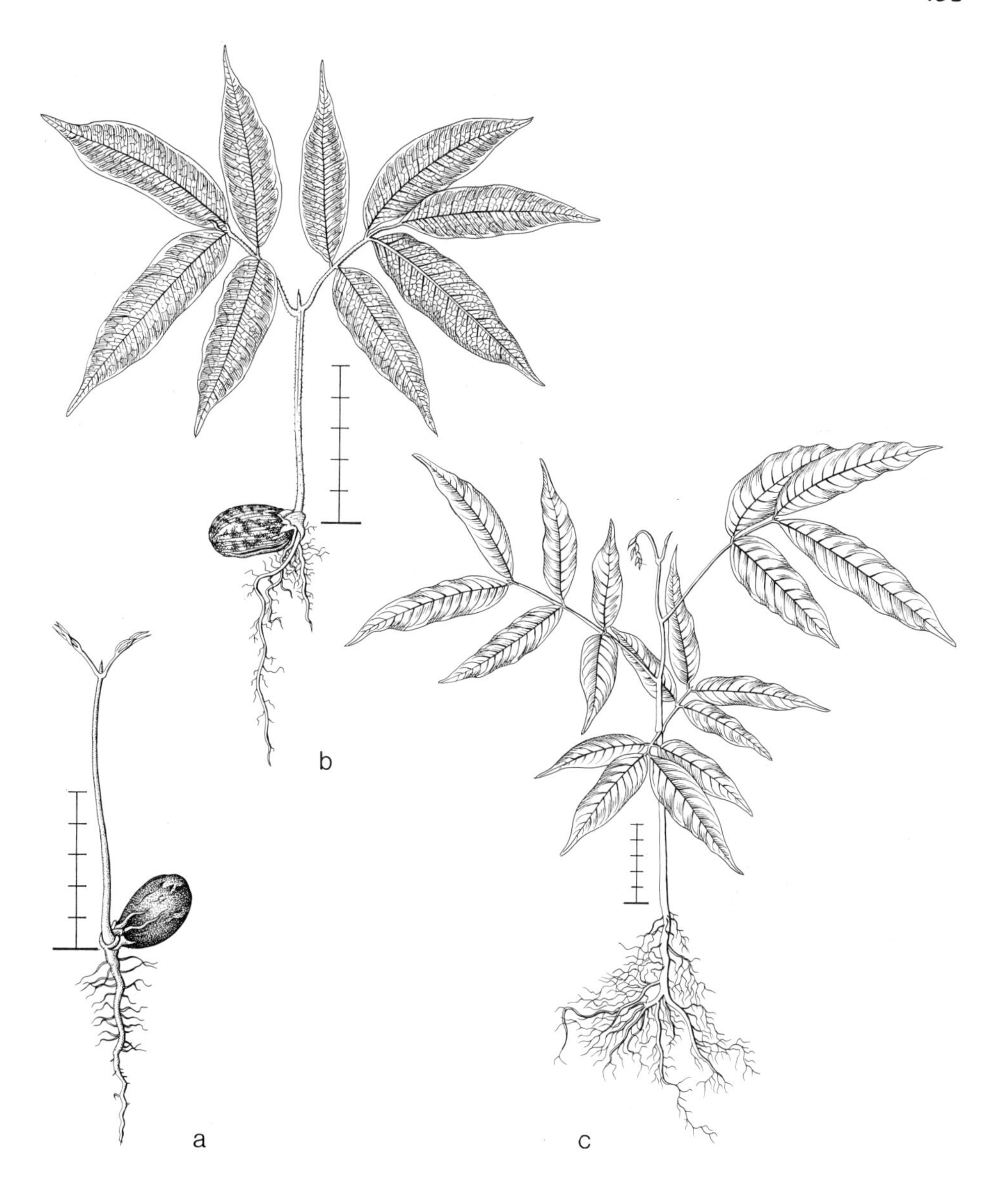

Fig. 167. ***Nephelium lappaceum.*** **De Wilde 13884, collected 29-vii-1972, planted 6-viii-1972.** a, b, and c date?.

pinnate, midrib above prominent, below much so, lateral nerves below prominent, ending in anastomoses.
Next leaves spirally arranged, compound, as the first leaves. Petiole as those of the first leaves, in 3rd leaf to 4.5 cm long, in 4th leaf to 6.5 cm long. Blade in 3rd and 4th leaf 4–6-foliolate, with terminal cusp. Rhachis in 3rd and 4th leaf to 7 cm long. Terminal cusp as that in the first leaves. Leaflets as those of the first leaves, in 4th leaf to 15 by 3.3 cm.

Specimens: De Wilde 13884 from N. Sumatra, primary forest on deep clay, low altitude.
Growth details: In nursery germination fair.

SAPOTACEAE

PALAQUIUM

Fl. Java 2: 193; Burger, Seedlings (1972) 332 (*P. amboinense*).
Revision: Van Royen, Blumea 10 (1960) 432.
Genus in India and Ceylon, throughout S.E. and E. Asia up to Formosa, all over Malesia except the Lesser Sunda Islands, N. Australia, in the Pacific as far as Samoa, including the Carolines and the New Hebrides.

Palaquium philippense (Perr.) Rob. – Fig. 168.

Development: The taproot emerges from one pole of the seed. The cotyledons curve and spread more or less, but remain secund, the testa is (partly to) entirely shed and the shoot emerges from the opening. A short resting stage occurs with the cotyledons and the two developed leaves.
Seedling hypogeal, (more or less) cryptocotylar. Sloanea type, Palaquium subtype.
Taproot long, sturdy, hard fleshy, cream-coloured, with many, rather short, slender, hardly branched, cream-coloured sideroots.
Hypocotyl ± quadrangular, subterranean, to 1 cm long, pinkish, warty with many creamy-white lenticels.
Cotyledons 2, more or less secund, succulent, in the 4th–5th leaf stage shed; simple, slightly to considerably spreading, without stipules, shortly petiolate, hard fleshy, dark greenish red, glabrous, at the lower side with sometimes partially adhering, disintegrating testa. Petiole sturdy, flattened, to 12 by 5 mm, slightly curved, warty as the hypocotyl. Blade boat-shaped, to 6 by 2.5 by 1.5 cm, distinctly curved; base and top ± acute; margin entire but corroded; below concave, very uneven by irregular pits, above convex, smooth; nerves not visible.
Internodes: First one terete, at the base slightly thicker and angular, to 15 cm long, densely covered with minute, simple, appressed, shining brown hairs; next internodes as the first one, second one to 2.5 cm, 5th one to 8 cm long, but usually smaller. Stem slender, slightly zig-zag.

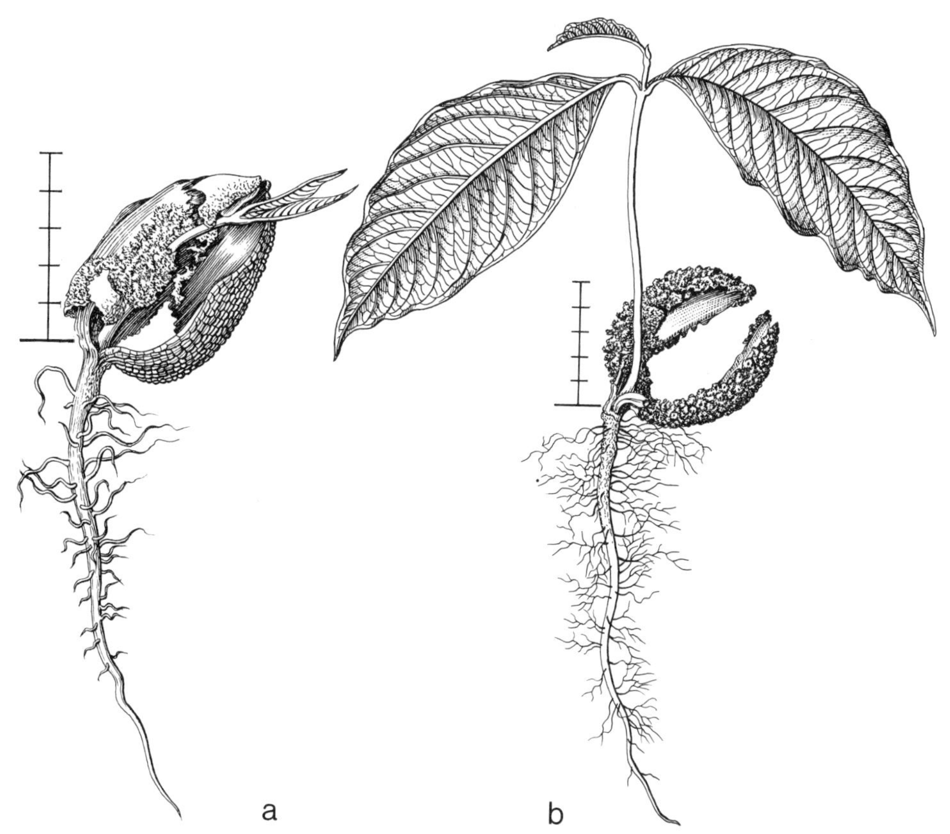

Fig. 168. ***Palaquium philippense***. **Bogor Bot. Gard., without number, collected and planted** ix-1973. a. 10-xii-1973; b. 11-xii-1973.

First two leaves opposite, simple, when young conduplicate-induplicate, spreading, without stipules, petiolate, subcoriaceous. Petiole in cross-section semi-orbicular, to 8 by 2 mm, brown and hairy as the first internode. Blade oblong to ovate-oblong, to 21 by 9 cm but usually smaller; base slightly narrowed into the petiole; top cuspidate to caudate, but often irregularly formed; margin entire; when young dark brownish red above, turning green, below brown as the hypocotyl by dense indument as on the first internode, above much less dense hairy, there glabrescent; nerves pinnate, midrib above raised, below very prominent, lateral nerves above sunken, at the tip curved, but ending free.

Next leaves spirally arranged, stipulate. Stipules 2 at the base of the petiole, sessile, appressed to the stem, slightly curved, rather soon dropping, narrowly triangular, in 10th leaf to 7 by 3 mm; top acute; margin entire; somewhat succulent, outside brown and hairy as the first internode; one-nerved, nerve prominent outside. Petioles as those of the first leaves, in higher ones more terete, but above slightly flattened, in 5th one to 15 by 3 mm, in 10th one to 15 by 5 mm. Blade as in the first leaves, obovate-oblong, in 5th one to 25.5 by 8.5 cm, in 10th one to 27 by 10 cm, further as the first leaves.

Specimens: Bogor Botanic Gardens, without number.
Growth details: In nursery germination good, ± simultaneous.
Remarks: P. amboinense and *P. philippense* show the same germination pattern and morphology. They differ in details only.

PLANCHONELLA

Fl. Java 2: 189.
Revision: Van Royen, Blumea 8 (1957) 235.
Genus pantropic; in the islands of the Indian Ocean, India and Ceylon, throughout S.E. and E. Asia, all over Malesia, N. and E. Australia, in the Pacific as far as the Tuamotus, including the Bonins, the Carolines, Hawaii, Norfolk, the Kermadec Islands, and N. New Zealand.

Planchonella sp. – Fig. 169.

Development: The taproot and hypocotyl emerge from one pole of the seed. A short resting stage occurs with the paracotyledons enclosed in the endosperm, often with the testa partially shed. The hypocotyl becomes erect, and the testa and endosperm are shed by unfolding of the paracotyledons. Then follows a second temporary rest.
Seedling epigeal, phanerocotylar. Macaranga type.
Taproot rather long, slender, fibrous, hard fleshy, creamy-white turning brownish, with few creamy-white, little branched sideroots.
Hypocotyl strongly enlarging, arched above the soil in the resting stage, terete, to 5 cm long, dull green to reddish turning brown, glabrous, warty with small, elevated lenticels.
Paracotyledons 2, opposite, foliaceous, with the development of the 8th–14th leaf dropped; sessile, simple, without stipules, herbaceous, green, glabrous. Blade elliptic, to 3.5 by 2 cm; base narrowed; top obtuse; margin entire; green, below paler; nerves pinnate, with footnerves, midrib above and below prominent, other nerves less so, ending in anastomoses.
Internodes: First one elongating to 1 cm long, terete, green turning brown, slightly hairy with simple, brownish hyaline hairs; next internodes as the first one, the lower ones in general small, the higher ones often irregular in size, the 10th one to 2 cm long. Stem ± straight, rather sturdy.
Leaves all spirally arranged, simple, spreading, with little white latex, without stipules, almost sessile, subcoriaceous, above shining dark green, below paler and dull, the higher ones gradually larger. Blade in the lower leaves ± obovate-oblong, to 5.5 by 2 cm, that in the 10th leaf obovate-lanceolate, to 20 by 5.5 cm; base acute; top acuminate; margin entire; with scattered, long, simple, appressed hyaline hairs, above soon glabrescent; nerves pinnate, midrib above raised, below much so, lateral nerves above sunken, below raised, at their tip ending in anastomoses.

Specimens: 1376 from W. Java, alluvial flat, primary forest on coral sand, low alt.
Growth details: In nursery germination poor.

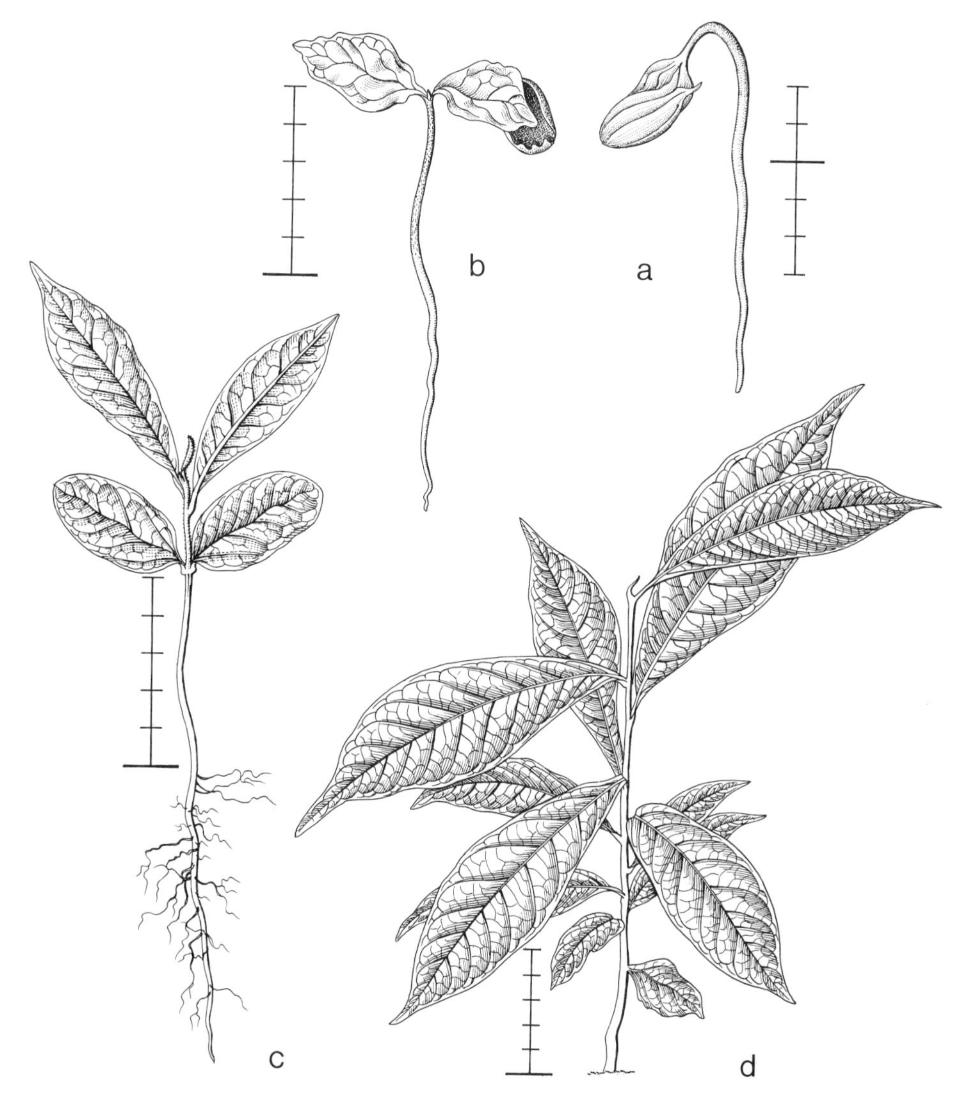

Fig. 169. ***Planchonella sp.*** **De V. 1376, collected 21-vii-1972, planted 27-vii-1972. a, b, and c** date?; d. 31-vii-1973.

SIMAROUBACEAE

IRVINGIA

Fl. Males. I, 6: 223; Mensbruge, Germination (1966) 211 (*I. gabonensis*, *I. ivorensis*); Meijer, Bot. Bull. Sandakan 11 (1968) 112 (*I. malayana*); Pierre, Flore Forestière Cochinchine 4 (1879–1899) pl. 263 (*I. oliveri*).
Genus in tropical Africa, throughout S.E. Asia and W. Malesia, not in Java and the Philippines.

Irvingia malayana Oliv. ex Benn. – Fig. 170.

Development: The fibrous endocarp splits along the margin of the valves. The taproot and hypocotyl emerge from one pole. The hypocotyl becomes erect, by which the cotyledons are withdrawn from the fruit wall, and carried above the soil. A short resting stage occurs with the cotyledons and two developed leaves.
Seedling epigeal, phanerocotylar. Sloanea type and subtype.
Taproot long, slender, fibrous, brownish cream, with many long, slender, much branched, creamy-white sideroots, of which one whorl around the collet.
Hypocotyl strongly enlarging, to 7.5 cm long, distinctly 6-angular, (bluish) green, soon turning dark reddish brown to dark brown, glabrous.
Cotyledons 2, opposite, succulent, in the 3rd–4th leaf stage dropped; sessile, without stipules, hard fleshy, green to yellowish green, glabrous. Blade ± obovate, to 3.5 by 2 by 0.3 cm; base slightly asymmetric, slightly auriculate, auricles rounded, to 2 by 3 mm; top rounded; margin entire; below slightly concave, above flat, nerves inconspicuous.
Internodes: First one strongly elongating, terete, to 6.5 cm long, (bluish) green, glabrous; next internodes as the first one, to 1.5 cm long, irregular in length, after the 5th one gradually somewhat longer, 15th one to 2.5 cm, 25th one to 3.5 cm. Stem slender, rather zig-zag, later slightly drooping.
First two leaves opposite, simple, inflexed when young, stipulate, petiolate, herbaceous, purplish red when young, turning green, glabrous. Stipules 4, or two more or less interpetiolarly connate ones, dropping soon, narrowly triangular, to 11 by 0.7 mm, the connate ones to 1.5 mm wide; top acute; margin entire; sticky, parallel-nerved. Petiole in cross-section semi-orbicular, to 6 by 2 mm, above flattened to channelled. Blade oblong, to 11 by 5 cm; base acute; top acute; margin entire; nerves pinnate, above slightly raised, below more so, lateral nerves distinctly curved back to the next ones.
Next leaves spirally arranged, beyond the 10th one turned into one plane. Stipules 2 at the base of each petiole, dropped when the leaf is fully developed, varying in length, narrowly triangular to linear, in 5th one 8 by 2 mm, in 30th one to 25 by 4 mm; base half clasping the stem; top acute; margin entire; creamy-white, parallel nerved, the one main nerve more pronounced. Petiole in 8th leaf to 10 by 2 mm, in 20th leaf to 12 by 2 mm, in the lower leaves above channelled, in the higher ones flattened. Blade obovate-oblong to obovate-lanceolate, the later ones more distinctly so, the third one to 10 by 3.7 cm, the 8th one to 15 by 4.3 cm, in 20th one to 18

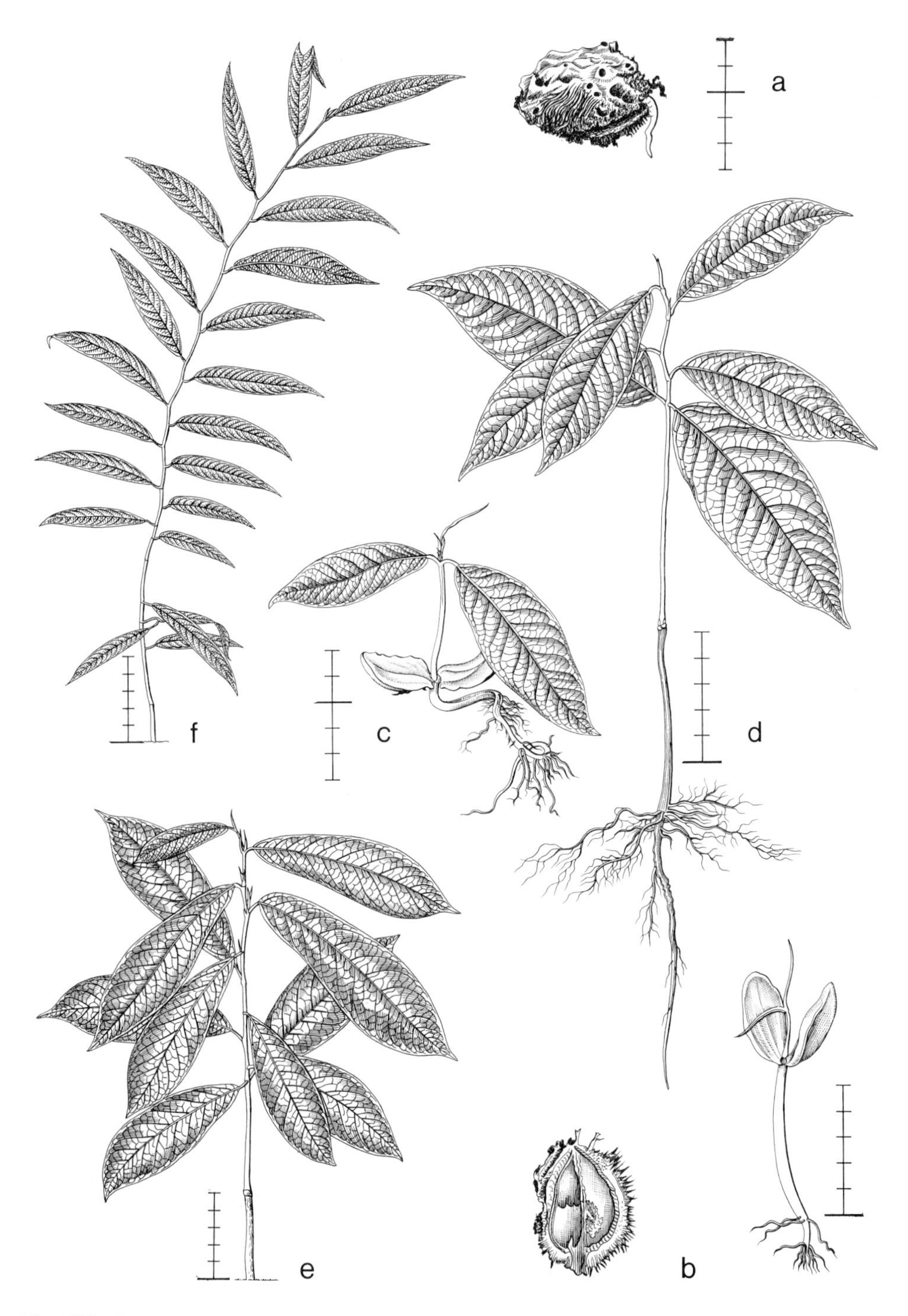

Fig. 170. ***Irvingia malayana.*** De V. 2296, collected 12-iii-1973, planted 22-iii-1973. a, b, and c. 7-iv-1973; d. 29-v-1973; e. 23-vii-1973; f. 7-iii-1974.

by 5.2 cm; base acute to rounded; top acuminate to cuspidate; further as the first leaves.

Specimens: 2296 from S.E. Borneo, primary hill forest on deep clay, low altitude.
Growth details: In nursery germination fair, over a rather long stretch of time. Sometimes polyembryony, but in general only one seedling able to establish.
Remarks: All species described are similar in germination pattern and general morphology; they differ in details only.

SONNERATIACEAE

SONNERATIA

Fl. Males. I, 4: 280; Chapman, Mangrove vegetation (1976) 316 (*S. caseolaris*).
Genus along the shores of tropical E. Africa, all over the Indian Ocean, India and Ceylon, S.E. and E. Asia up to the Ryu Kyu Islands, all over Malesia, N. Australia, in the Pacific as far as the Marshall Islands, New Hebrides and New Caledonia.

Sonneratia caseolaris (L.) Engl. – Fig. 171.

Development: The taproot and hypocotyl emerge from one pole of the seed. After establishment of the root the hypocotyl becomes erect, carrying the enclosed paracotyledons up. After a resting period the testa is shed by spreading of the paracotyledons. A second resting stage occurs with the paracotyledons unfolded.
Seedling epigeal, phanerocotylar. Macaranga type.
Taproot rather slender, faintly developed, creamy-white, with few, scarcely branched sideroots.
Hypocotyl strongly elongating, slightly quadrangular, to 2 cm long, green, glabrous.
Paracotyledons 2, opposite, foliaceous, not known when shed; equal, sessile, simple, often the tips for a long time held together by the brownish testa, herbaceous, green, glabrous. Blade ± elliptic to ovate-oblong, to 12 by 5 mm; base narrowed; top acute, margin entire; nerves inconspicuous.
Leaves not seen.

Specimens: 1367 from W. Java, beach forest along brooklet, primary forest on coral sand, low altitude.
Growth details: Germination of the seeds is good, the plantlets establish satisfactorily. However, growth beyond the stage of developed paracotyledons was not observed, the plantlets dying off in masses, possibly due to a deficiency in minerals or adverse conditions.

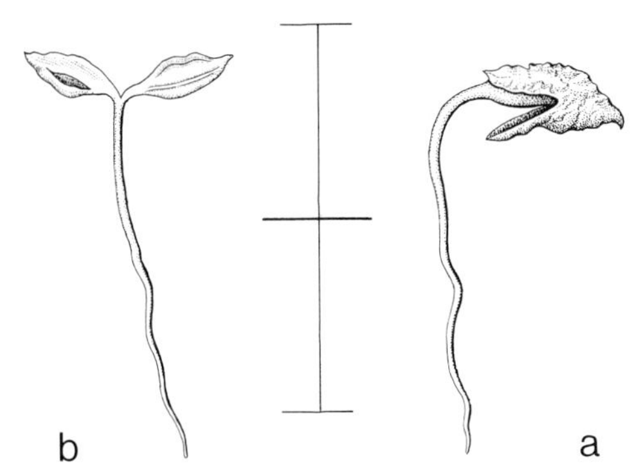

Fig. 171. *Sonneratia caseolaris.* De V. 1367, collected 21-vi-1972, planted 2-vii-1972. a and b date?.

STERCULIACEAE

SCAPHIUM

Tree Fl. Malaya 2: 373.
Monograph: Kostermans, J. Sc. Res. Indon. 2 (1953) 2.
Genus in S.E. Asia, and W. Malesia except the Philippines and Java.

Scaphium macropodum (Miq.) Beumée – Fig. 172.

Development: The taproot and hypocotyl emerge from the free pole of the winged fruit. Then follows a resting stage. The hypocotyl becomes erect and the paracotyledons spread, by which the testa and fruit wall are shed. A resting stage occurs with the paracotyledons expanded.
Seedling epigeal, phanerocotylar. Macaranga type.
Taproot rather sturdy, slender, flexuous, brownish cream, with many rather short, unbranched, brownish cream sideroots.
Hypocotyl strongly enlarging, ± terete at the base, towards the top slightly angular with rather inconspicuous ridges, to 7 cm long, green, minutely hairy with simple, brownish hairs.
Paracotyledons 2, opposite, foliaceous, not exactly known when dropped, between 8th and 12th leaf stage; simple, spreading, without stipules, shortly petiolate, herbaceous, green, paler so below. Petiole in cross-section semi-orbicular, to 4 by 1.6 mm, hairy as the hypocotyl. Blade elliptic, to 3.5 by 3 cm; base slightly retuse, with rounded auricles; top broadly rounded; margin entire, slightly thickened; glabrous; nerves palmate, main nerves 3–5, hardly prominent.
Internodes: First one terete, to 1.5 cm long, rather densely hairy with minute hairs as on the hypocotyl, and with scattered, minute, hyaline, stellate hairs; next internodes as the first one, the second one to 1 cm long, 10th one to 2.5 cm long, but often shorter. Stem rather slender, straight.

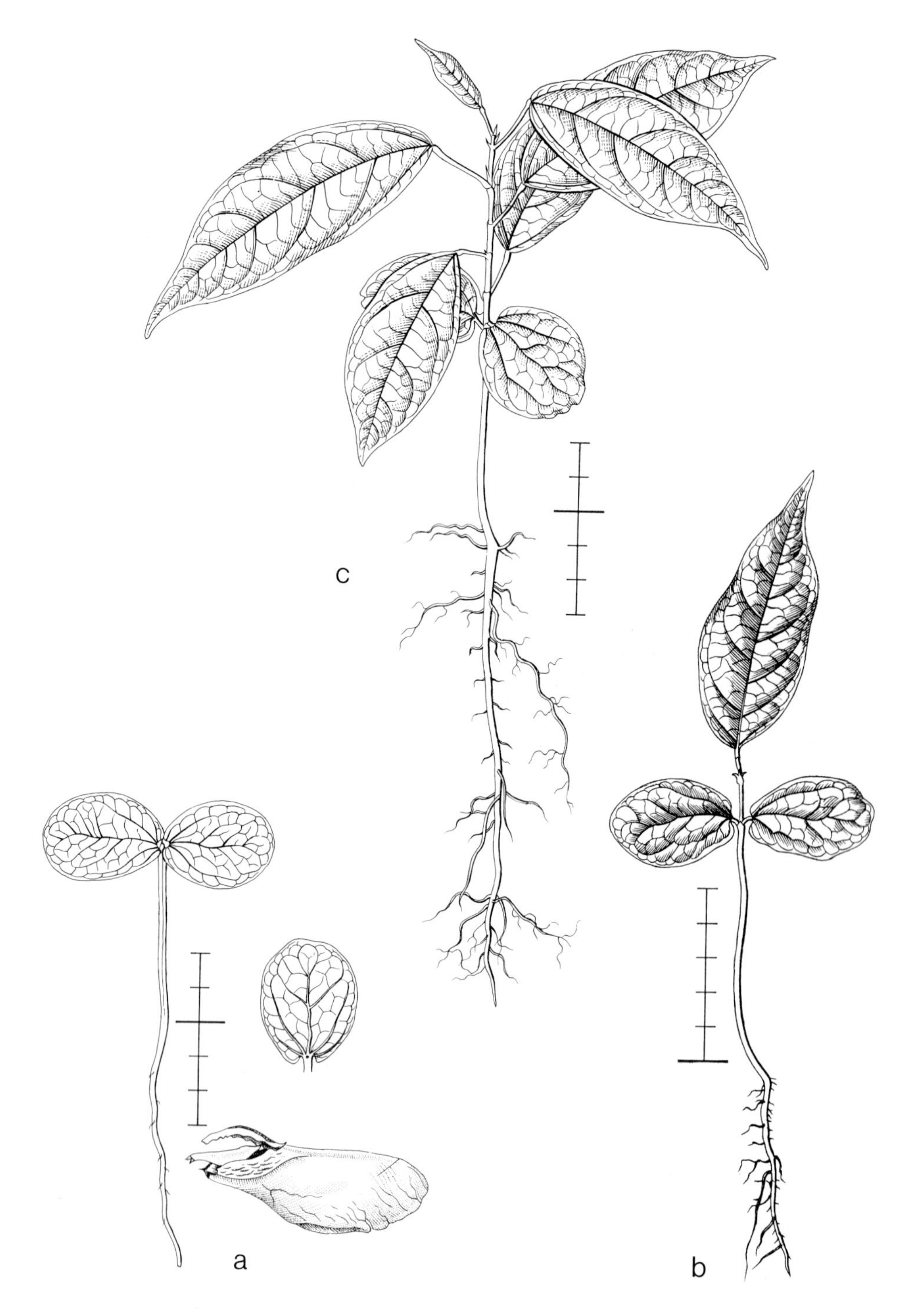

Fig. 172. *Scaphium macropodum.* De V. 1409, collected 26-vi-1972, planted 2-vii-1972. a date?; b. 26-ix-1972; c date?.

Leaves all spirally arranged, simple, stipulate, petiolate, herbaceous, green. Stipules caducous when the leaf is full-grown, narrowly triangular, in the lower leaves to 4 by 1 mm, in the 20th leaf to 5.5 by 2 mm, sessile, somewhat curved; top acute; margin entire; slightly succulent, green, hairy as the first internode; 1-nerved, nerve slightly, in higher ones distinctly prominent below. Petiole terete, in the lower leaves to 12 by 1 mm, in 10th leaf to 20 by 1 mm, in 20th leaf to 45 by 1.5 mm, at the base slightly thickened, at the top more so, hairy as the first internode. Blade elliptic, in higher ones more obovate, first one sometimes ovate, in lower leaves to 7 by 4 cm, 10th one to 11.5 by 5.3 cm; base obtuse, in lower ones sometimes more or less retuse; top acuminate, in higher leaves cuspidate; margin entire, often somewhat irregular; almost glabrous; nerves pinnate, with footnerves, midrib above raised, below prominent, lateral nerves above sunken, below prominent, at the tip curved to the next one.

Specimens: 1409 from W. Java, tertiary coral reef, primary forest on coral sand, low altitude.
Growth details: In nursery germination poor, ± simultaneous.

STERCULIA

Fl. Java 1: 411; Burger, Seedlings (1972) 334 (*S. foetida*); Duke, Ann. Miss. Bot. Gard. 52 (1965) 343 (*S. apetala, S. foetida*); ibid. 56 (1969) 155 (*S. costaricana*); Hickel, Graines et Plantules (1914) 304 (*S. acerifolia, S. platanifolia*); Lubbock, On Seedlings 1 (1892) 269 (*S. heterophylla*); Mensbruge, Germination (1966) 254 (*S. oblonga, S. rhinopetala, S. tragacantha*).
Monograph: Tantra, Comm. Lemb. Penel. Hutan 102 (1976) 1–194.
Genus pantropic; tropical Africa, throughout the Indian Ocean, India and Ceylon, throughout S.E. and E. Asia, all over Malesia, N. and E. Australia, in the Pacific as far as Tonga, including the Carolines and New Caledonia, tropical S. and Central America.

Sterculia stipulata Korth. – Fig. 173.

Development: The taproot and hypocotyl emerge from one pole of the seed. The hypocotyl becomes erect by which the cotyledons are lifted somewhat above the soil. The cotyledons with the adhering endosperm below become revolute, and the testa is shed. No resting stage occurs.
Seedling epigeal, phanerocotylar. Sterculia stipulata type.
Roots (in the few specimens present) 4, equally developed, long, little branched, with cream-coloured additional roots.
Hypocotyl little elongating, to 8 mm long, distinctly swollen, white when young turning brownish, densely covered with minute, simple, hyaline hairs.
Cotyledons 2, opposite, (sometimes having difficulties in freeing from the testa), ± succulent, in the 2nd leaf stage dropped; simple, without stipules, sessile, fleshy, creamy-white, outside of the adhering endosperm pale brownish, glabrous. Compound body elliptic, to 15 by 8 by 2.5 mm; base and top acute; margin entire; inside

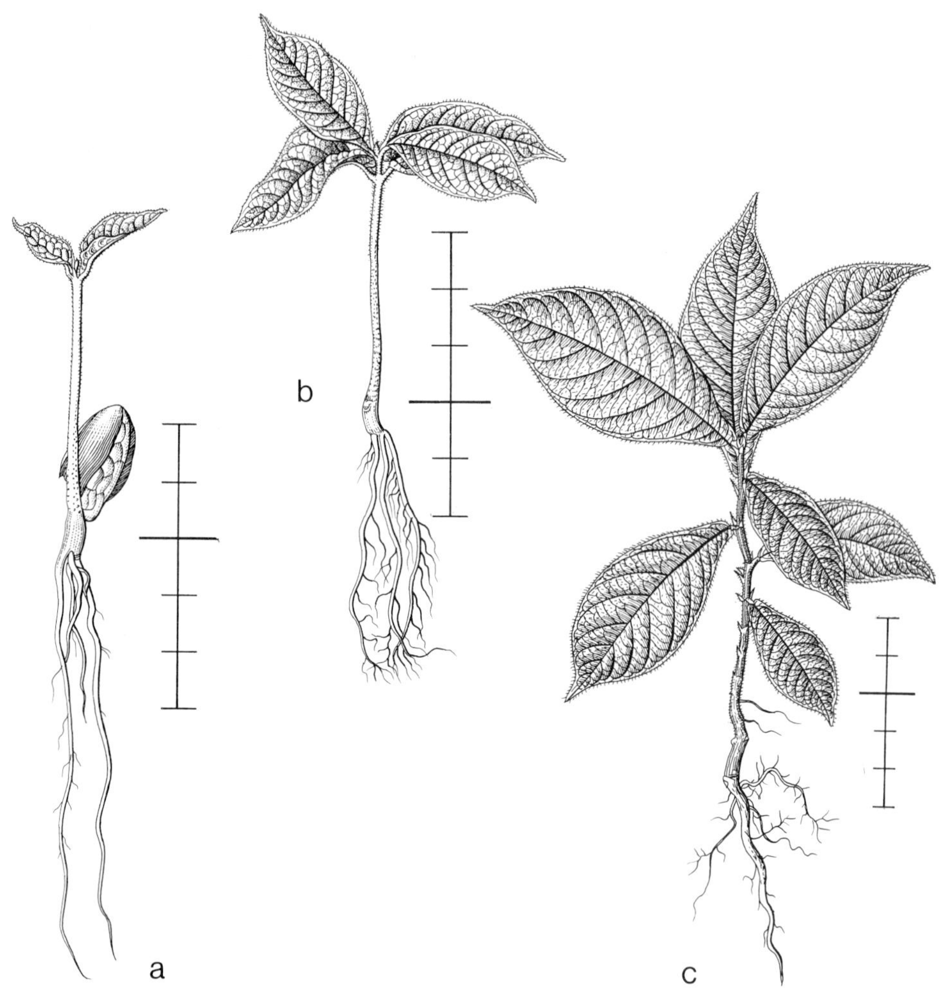

Fig. 173. *Sterculia stipulata.* De V. 2352, collected 15-iii-1973, planted 22-iii-1973. a. 14-iv-1973; b. 5-vi-1973; c date?.

concave, outside convex, nerves not visible. Endosperm adhering at the lower side of the cotyledons, when older at the margins distinctly revolute, in the centre 2.5 mm thick.

Internodes: First one strongly elongating, terete, to 4 cm long, brownish cream, densely covered with minute, simple, brownish hairs; next internodes as the first one, in the lower ones to 6 mm long, 5th one to 2 cm long. Stem slender, straight.

First two leaves opposite, simple, stipulate, shortly petiolate, subcoriaceous, green. Stipules 2 at the base of each petiole, shrivelling soon but persistent, sessile, subulate, to 1.5 by 0.2 mm, hairy as the hypocotyl. Petiole in cross-section semi-orbicular, to 3 by 1.5 mm, hairy as the first internode. Blade ovate, to 2.7 by 1.6 cm; base broadly auriculate, auricles rounded, to 1 by 4 mm; top acute to acuminate; margin entire;

with short, stiff hairs, especially on the nerves and along the margin; nerves pinnate, above sunken, below very prominent, especially the midrib, lateral nerves forming a much undulating marginal nerve.
Next leaves spirally arranged, sometimes the lower ones semi-opposite. Stipules in 5th leaf ± narrowly triangular, to 4 by 1.5 cm; top apiculate; margin entire; hairy as the hypocotyl, vaguely parallel-nerved. Blade in 5th leaf ovate-oblong, to 4.3 by 2 cm; base slightly auriculate, auricles rounded, to 1 by 0.5 mm; top acuminate; margin entire; along the margin and above with simple hairs as on the first leaves, below with stellate hairs; nerves pinnate, as in the first leaves, but less prominent below, lateral nerves ± free ending, further as the first leaves.

Specimens: 2352 from S.E. Borneo, alluvial flat, primary forest on deep clay, low altitude.
Growth details: In nursery germination poor, a high percentage of seedlings lost by damping off.
Remarks: Of the epigeal species, *S. foetida* has coriaceous cotyledons, the (sub)opposite first leaves are palmately compound (it should be checked whether the coriaceous nature of the cotyledons is not due to the persistence of the endosperm to the lower side of the cotyledonary blade, like in *S. rhinopetala* and *S. stipulata*). *S. acerifolia*, *S. heterophylla*, *S. oblonga*, and *S. platanifolia* have the cotyledons leaf-like, the leaves are all spirally arranged and simple.
In *S. stipulata* the secund, ± fleshy cotyledons are slightly raised above the soil on a short thick hypocotyl, and they have difficulties in freeing from the testa. Each cotyledon is on the outside covered by the persistent endosperm, and the first two leaves are opposite on a long first internode; in essence *S. rhinopetala* is similar but the hypocotyl is much more elongated.
The hypogeal species *S. costaricana* and *S. tragacantha* have the cotyledons secund at soil level, the first leaves are (sub)opposite and simple. *S. apetala* is similar, but the first two opposite leaves are palmately lobed.

THEACEAE

PYRENARIA

Fl. Java 1: 321.
Genus throughout S.E. and E. Asia up to the Ryu Kyu Islands, all over W. Malesia, including Java and the Philippines.

Pyrenaria serrata Bl. – Fig. 174.

Development: The taproot and hypocotyl emerge from the margin of the seed. The hypocotyl becomes erect, and the enclosed paracotyledons are carried above the soil. Then follows a resting stage. The testa is dropped by unfolding of the paracotyledons. A second resting stage occurs after the paracotyledons are expanded.

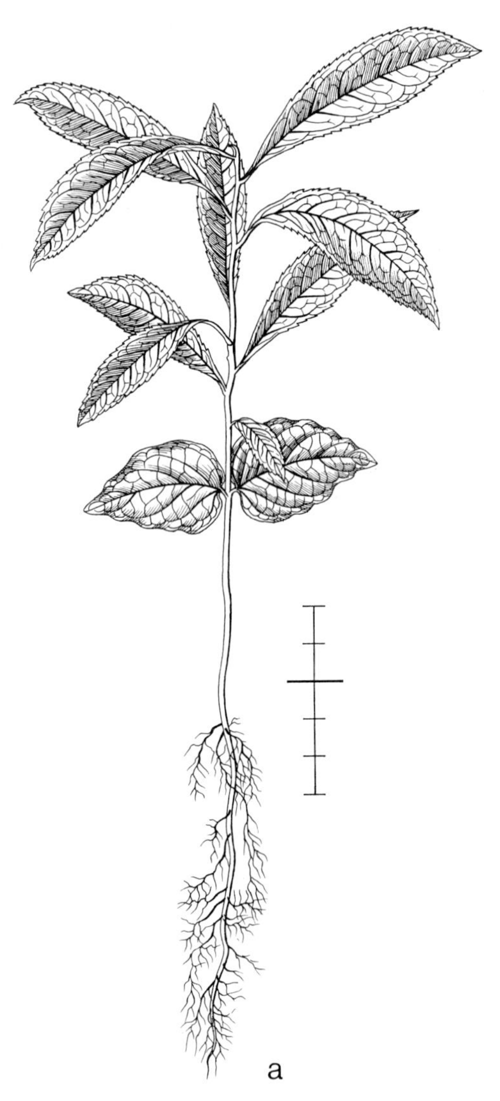

Fig. 174. ***Pyrenaria serrata.*** Bogor Bot. Gard. VI. C. 12a, collected and planted i-1973. a date?.

Seedling epigeal, phanerocotylar. Macaranga type.
Taproot long, slender, flexuous, pale brownish, with rather few, hardly branched, creamy-white sideroots.
Hypocotyl strongly elongating, in cross-section quadrangular, to 6 cm long, green, glabrous.
Paracotyledons 2, opposite, foliaceous, in the 15th–20th leaf stage dropped; without stipules, petiolate, subcoriaceous, green, glabrous. Petiole in cross-section semi-orbicular, to 1.5 by 1.5 mm. Blade ovate, to 3.5 by 2.5 cm; base truncate; top acute to obtuse, sometimes apiculate; margin irregularly undulating, almost entire; 5-nerved,

midrib above distinctly sunken, below slightly prominent, the other nerves above slightly sunken, below inconspicuous.
Internodes: First one strongly elongating, terete, to 2.5 cm long, green, rather densely covered with short, simple, slightly appressed, hyaline hairs, slowly glabrescent; next internodes as the first one, the second one to 5 mm long, gradually becoming longer, 20th one to 2.5 cm long. Stem slender, straight.
Leaves all spirally arranged, simple, involute when young, without stipules, shortly petiolate, subcoriaceous, reddish when young, turning green. Petiole in cross-section semi-orbicular, in the lower leaves to 3 by 1 mm, in 30th one to 10 by 2.5 mm, slightly hairy as the first internodes. Blade ± obovate-oblong, the second one to 4.5 by 2 cm, gradually becoming larger, the 30th one to 20.5 by 4.5 cm; base narrowed into the petiole; top acuminate; margin regularly dentate; below with scattered hairs as on the first internode, especially on the midnerve; nerves pinnate, midrib above sunken, below prominent, lateral nerves above slightly sunken, below hardly prominent, forming an undulating marginal nerve, veins ending free in the teeth.

Specimens: Bogor Botanic Gardens VI. C. 12a, from Java.
Growth details: In nursery germination good, ± simultaneous.

TERNSTROEMIA

Fl. Java 1: 321.
Genus pantropic; in tropical Africa, India, and Ceylon, throughout S.E. and E. Asia up to Japan, all over Malesia, N.E. Australia, in the Pacific in the Bismarcks, tropical S. and Central America.

Ternstroemia elongata (Korth.) Koord. – Fig. 175.

Development: The seed splits along the margin, the hypocotyl emerges. After the root is established, the hypocotyl turns erect, often for a short time carrying one of the endocarp valves up. No resting stage occurs.
Seedling epigeal, phanerocotylar. Ternstroemia type.
Taproot long, slender, rather sturdy, hard fleshy, brownish cream, unbranched in the beginning.
Hypocotyl elongating, terete, to 5.5 by 0.5 cm, fusiform, tapering towards the base and top, dark green, glabrous.
Cotyledons 2, opposite, very small and scale-like, in the 4th–5th leaf stage dropped; simple, sessile, without stipules, slightly succulent, green, glabrous. Blade ± narrowly triangular, to 5 by 1.5 mm; top rounded; margin entire; nerves not visible.
Internodes: First one hardly elongating, almost terete but slightly angular by irregular, small ridges from the bases of the petioles, to 8 mm long, green, glabrous; next internodes as the first one, the second one to 10 mm long, the following ones crowded, to 1 mm long, gradually becoming larger, 5th one to 3 mm long, 10th one to 6 mm long. Stem rather stunted, straight.
First two leaves opposite, in general much reduced, simple, ± sessile, without stipules,

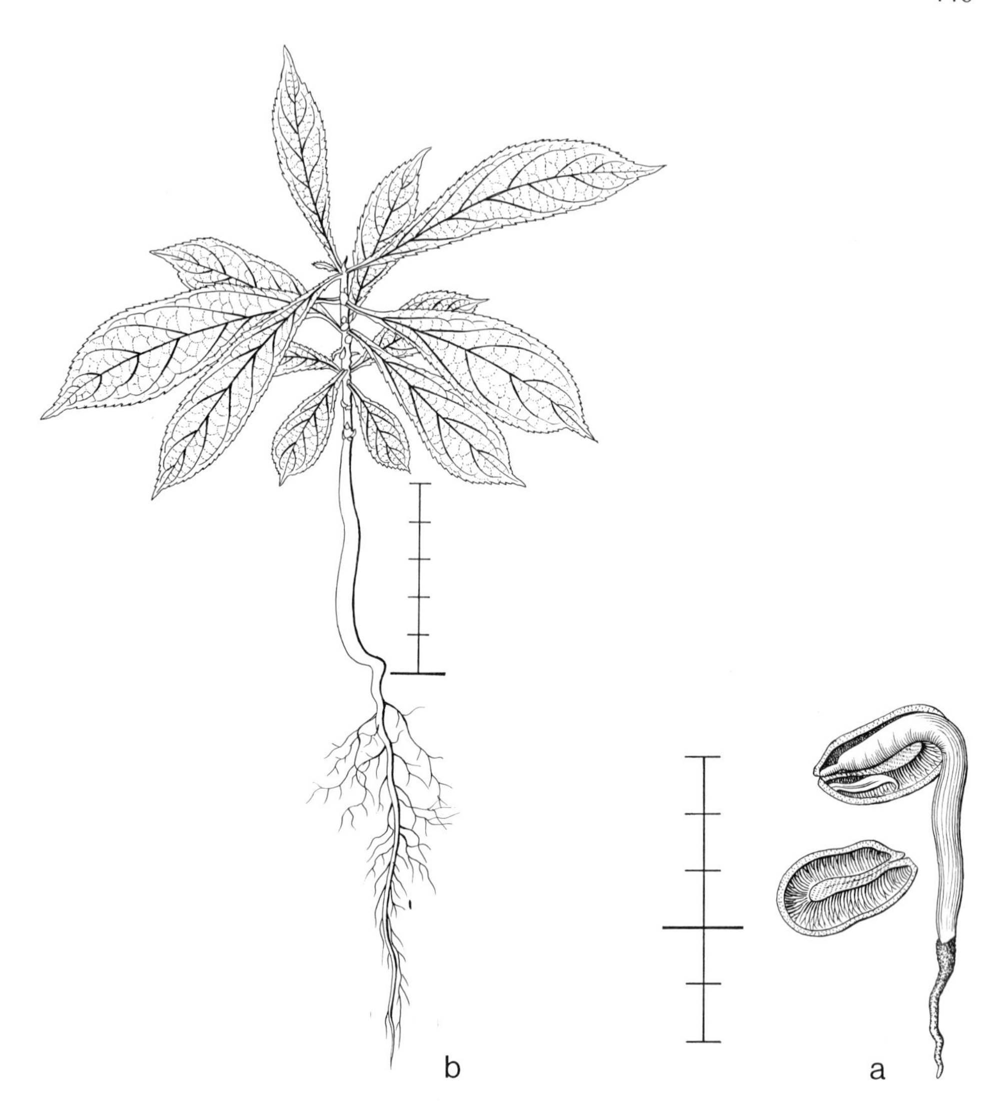

Fig. 175. *Ternstroemia elongata.* De V. 2400, collected and planted 20-iv-1973. a. 29-viii-1973; b. 18-iv-1974.

herbaceous, green, glabrous, distinctly serrate.
Next leaves spirally arranged, fully developed, convolute when young, shortly petiolate. Petiole distinctly jointed at the base, in cross-section semi-orbicular, in the lower leaves to 1 by 1 mm, in 5th one to 3 by 1 mm, in 15th one to 5 by 1.5 mm. Blade ± elliptic, later more obovate-oblong, in 7th leaf to 5 by 2.8 cm, in 20th one to 9.5 by 3 cm; base narrowed into the petiole; top acuminate to cuspidate; margin rather irregularly serrate; nerves pinnate, midrib above sunken, below raised, lateral nerves inconspicuous, hardly raised.

Specimens: 2400 from W. Java, primary forest on deep clay, medium altitude.
Growth details: In nursery germination poor, over a long stretch of time.

THYMELAEACEAE

PHALERIA

Fl. Males. I, 6: 15.
Genus in Ceylon, throughout Malesia, N. and N.E. Australia, in the Pacific as far as Samoa and Tonga Islands, including the W. Carolines.

Phaleria capitata Jack – Fig. 176.

Development: The taproot emerges from one pole of the fruit. The cotyledons spread more or less in the fruit, more or less rupturing the fibrous endocarp. The shoot emerges from the same pole as the root. Seedling growing in flushes, a short resting stage occurs with 4 developed leaves, and after each newly developed flush.
Seedling hypogeal, cryptocotylar. Heliciopsis type and subtype.
Taproot long, slender, flexuous, cream-coloured, turning pale brown, with rather few, long, slender, cream-coloured to pale brown, pale yellow tipped sideroots.
Hypocotyl not enlarging.
Cotyledons 2, more or less secund, succulent, after the 4th leaf stage shed or ineffective; simple, sessile, without stipules. Whole body heart-shaped to ovate in outline, at the base rounded, at the top acute, to 1.3 by 1.3 cm, the fibrous endocarp persistent. Blades close together, flattened where touching, outside concave.
Internodes: First one strongly enlarging, terete, to 4 cm long, green, glabrous; second internode hardly elongating, as the first one, to 3 mm long, but usually smaller; next internodes alternating, slightly elongating, terete, to 3 mm long, but sometimes the first one of these to 2.5 cm long, and strongly elongating, terete, towards the top in cross-section more elliptic, in the lower ones to 2.5 cm long, in the 8th one to 3.5 cm long.
First four leaves decussate, developed in one flush, close together, simple, without stipules, petiolate, herbaceous, green, glabrous. Petiole in cross-section semi-orbicular, to 3 by 1.5 mm. Blade oblong to obovate-oblong, to 4.5 by 2 cm; base acute; top cuspidate, margin entire; nerves pinnate, above somewhat raised, below

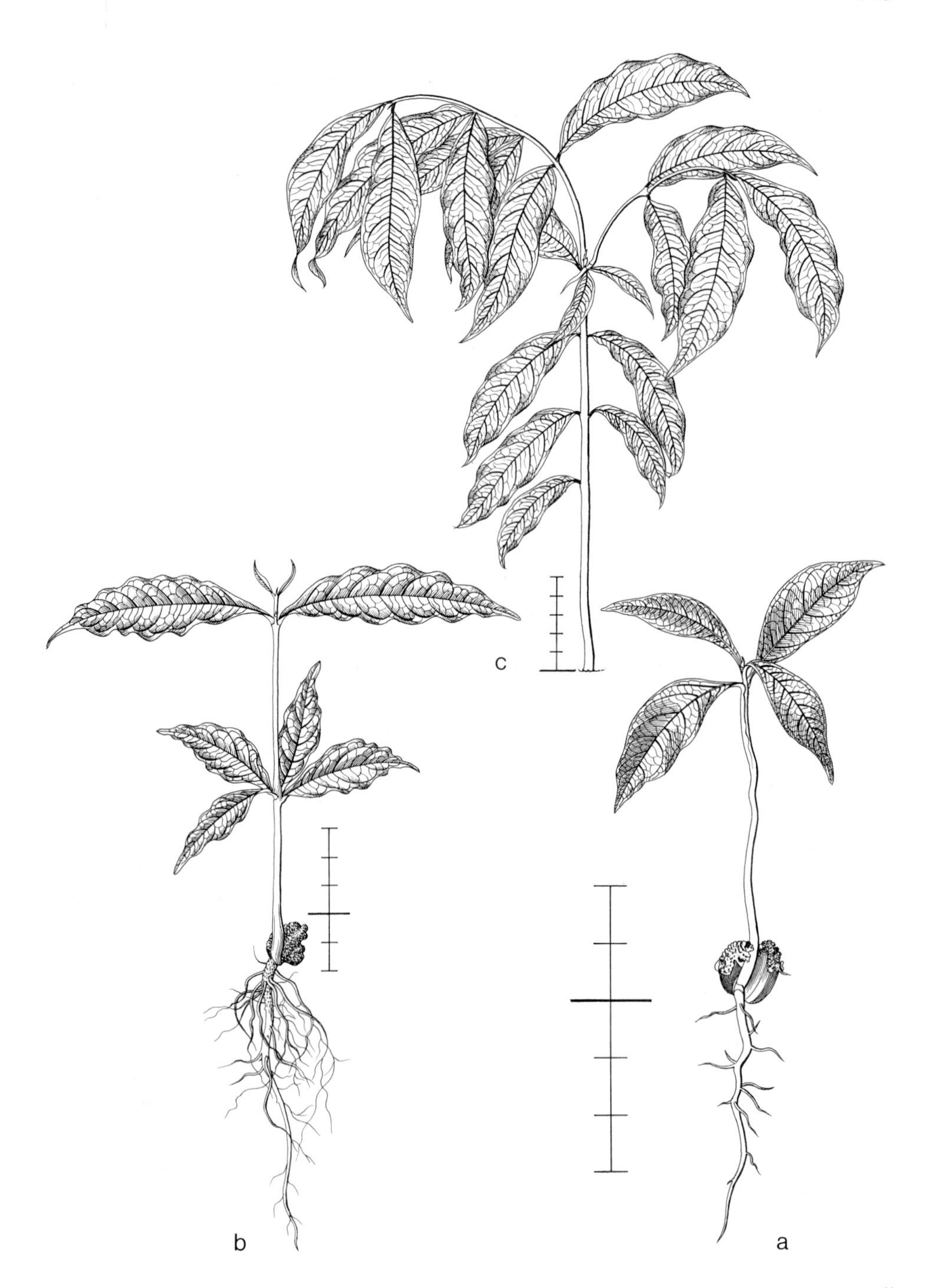

Fig. 176. *Phaleria capitata.* De V. 2654, collected 1-xi-1973, planted 5-xi-1973. a. 15-xii-1973; b. 23-iii-1974; c. 11-ix-1974.

less so, lateral nerves at the tip curved, forming an undulating marginal nerve.
Next leaves decussate, developing in flushes of 4(–6) leaves, of which 2(–4) developed ones, dark purplish green turning green, alternatingly much reduced and fully developed, giving the shoot a distichous appearance. Reduced leaves sessile, rather soon caducous, narrowly triangular, in lowest flush to 2 by 0.5 mm, in 6th flush to 8 by 1 mm; 1-nerved, nerve slightly prominent below. Petiole of developed leaves in cross-section semi-orbicular, those in second flush to 4 by 1.5 mm, in 6th flush hardly larger, usually tinged red. Blades of fully developed leaves lanceolate, the lowest ones to 7 by 2 cm, in second flush to 9.3 by 1.8 cm, in 6th flush to 12 by 4 cm; nerves pinnate, midrib above raised, below more so, lateral nerves above slightly raised, below less so, further as in the leaves of the first flush.

Specimens: 2654 from N. Celebes, primary forest on deep clay, medium altitude.
Growth details: In nursery germination good, ± simultaneous.

VERBENACEAE

TEYSMANNIODENDRON

Fl. Java 2: 602; Kostermans, Reinwardtia 1 (1951) fig. 2 (*T. coriaceum*).
Monograph: Kostermans, Reinwardtia 1 (1951) 78.
Genus in Vietnam, all over Malesia except for Java and the Lesser Sunda Islands, in the Pacific in Bismarck and the Solomons.

Teysmanniodendron coriaceum (Clarke) Kosterm. – Fig. 177.

Development: The taproot emerges from one pole of the fruit, the cotyledonary petioles elongate, bringing the plumule out of the fruit, after which this starts elongating. No resting stage occurs.
Seedling hypogeal, cryptocotylar. Heliciopsis type and subtype.
Taproot long, slender, quite sturdy, creamy-white, with rather many long, hardly branched, creamy-white sideroots.
Hypocotyl not enlarging.
Cotyledons 2, secund, succulent, in the 6th–8th leaf stage shed; simple, without stipules, petiolate. Whole body ellipsoid, to 15 by 8 by 8 mm, the dark bluish fruit wall and testa persistent. Petiole in cross-section semi-orbicular, to 4 by 2 mm.
Internodes: First one strongly elongating, terete, to 5 cm long, green turning brownish, rather densely covered with minute, simple, hyaline hairs; next internodes as the first one, with two more or less distinct longitudinal grooves, the second one to 2 cm long, the higher ones gradually larger, 5th one to 2.5 cm long, 10th one to 5 cm long. Stem slender, straight.
Leaves all decussate, but in general in the lower 5 pairs one of the two leaves of a pair entirely reduced, simple, without stipules, petiolate, subcoriaceous, green. Petiole in cross-section semi-orbicular, in the lowest one to 9 by 1 mm; base distinctly swollen;

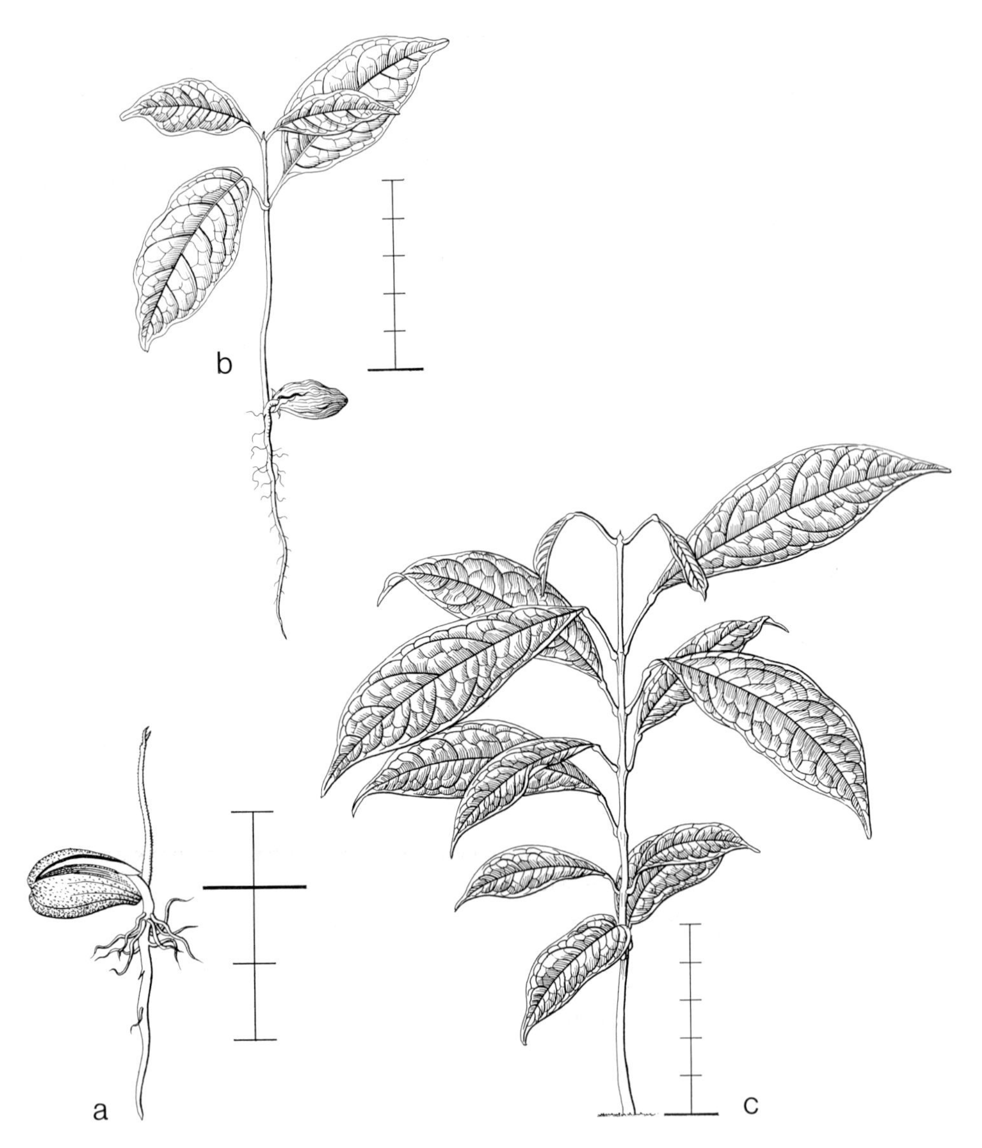

Fig. 177. *Teysmanniodendron coriaceum.* De V. 2320, collected 13-iii-1973, planted 22-iii-1973. a. 11-iv-1973; b. 7-v-1973; c. 11-iii-1974.

hairy as the first internode; higher ones as the first ones, gradually longer, in the 5th pair of leaves to 2 cm long, at the base slightly swollen, at the top abruptly so and there jointed. Blade sometimes irregular in size, in lowest leaves elliptic, to 7 by 3.8 cm, but in general much smaller, in higher ones irregular in size, but in general distinctly increasing in length, those in the 5th pair of leaves obovate-oblong to obovate-lanceolate, to 9.5 by 3 cm, in 10th pair to 10.5 by 5 cm but often smaller; base and top acute; margin entire; below slightly hairy as the first internode, soon glabrescent; nerves pinnate, midrib above slightly prominent, below much so, lateral nerves above sunken, below prominent, forming an irregular, curvy marginal nerve.

Specimens: 2320 from S.E. Borneo, primary hill forest on deep clay, low altitude.
Growth details: In nursery germination good, ± simultaneous.

VITACEAE

TETRASTIGMA

Fl. Java 2: 88.
Genus in E. India, S.E. and E. Asia as far as the Ryu Kyu Islands, all over Malesia, N.E. Australia, in the Pacific in Bismarck, the Solomons and Fiji.

Tetrastigma sp. – Fig. 178.

Development: The taproot and hypocotyl emerge from one pole of the seed, carrying the enclosed paracotyledons above the soil. After a resting period the paracotyledons unfold, rupturing the testa in two valves. A second short resting stage occurs with the paracotyledons expanded.
Seedling epigeal, phanerocotylar. Macaranga type.
Taproot not developing, accessory roots long, creamy-brown, few-branched.
Hypocotyl strongly elongating, in cross-section elliptic, to 8 cm long, pale green, glabrous.
Paracotyledons 2, opposite, foliaceous, in the 4th–5th leaf stage dropped; simple, without stipules, petiolate, herbaceous, green, glabrous. Petiole in cross-section semi-orbicular, to 7 by 2 mm, above channelled. Blade elliptic, to 5 by 2.7 cm; base slightly asymmetric; top acuminate, its tip broadly rounded; margin entire; nerves palmate, main nerves 5, above prominent, below vaguely prominent, with an inconspicuous marginal nerve.
Internodes: First one slightly elongating, terete, to 7 mm long, green, glabrous; next internodes as the first one, 5th one to 3 cm long. Stem at first congested, crooked, later irregularly zig-zag.
Leaves all spirally arranged, stipulate, petiolate, herbaceous, green, glabrous. First leaf simple. Stipules recurved, persistent, sessile, triangular, to 3 by 2 mm; top acute; margin entire, below the petiole ending in a shallow ridge, opposite the petiole descendent along the stem; vaguely 1-nerved. Petiole in cross-section semi-

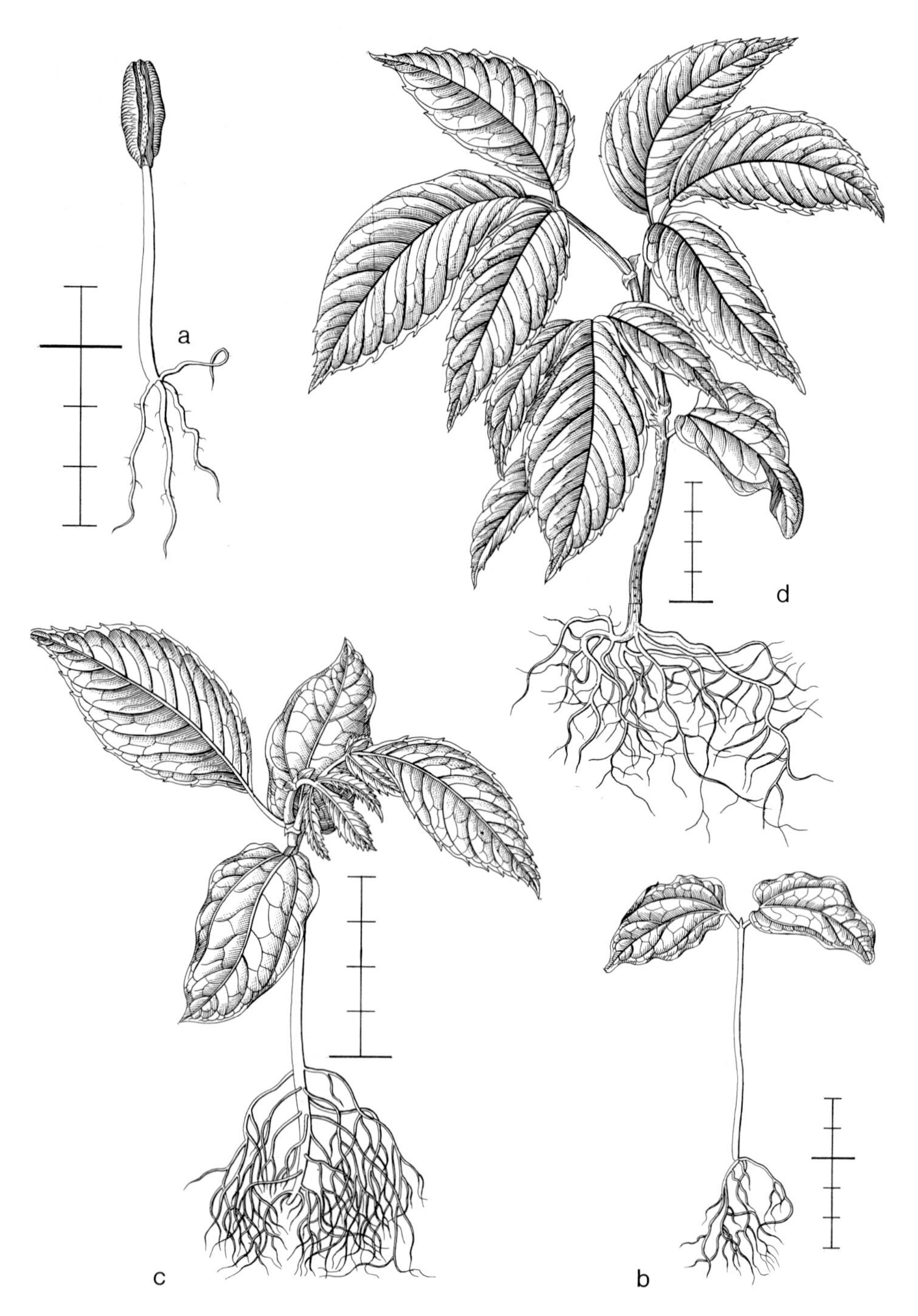

Fig. 178. *Tetrastigma sp.* De V. 2199, collected 6-iii-1973, planted 22-iii-1973. a. 1-iv-1973; b. 29-v-1973; c. 27-ix-1973; d. 21-ii-1974.

orbicular, to 3 by 0.2 cm, above with a distinct ridge formed by the descending midrib, and channelled by the slightly winged sides. Blade ovate, to 8 by 4.5 cm; base ± acute, descending into the petiole; top slightly acuminate, its tip rounded, apiculate; margin widely serrate; nerves pinnate, above prominent, midrib below much so, lateral nerves much less so, ending free in the teeth. Next leaves trifoliolate. Stipules as those of first leaf, patent at the tip, ovate, to 3 by 2.5 mm; top rounded; margin entire; slightly fleshy; 1-nerved. Petiole as that of the first leaf, in the second leaf to 3.5 by 0.2 cm, in 5th leaf to 5.5 by 0.4 cm, slightly tapering to the top. Petiolules as petioles, in the second leaf those of the lateral leaflets to 4 by 1 mm, that of the terminal leaflet to 8 by 1.5 mm, in 5th leaf the lateral ones to 8 by 2 mm, the terminal one to 15 by 2 mm. Blade of the lateral leaflets ovate-oblong, in second leaf to 6 by 2.5 cm, in 5th one to 10.5 by 5 cm; base slightly asymmetric; top acuminate; further as the first leaf. Blade of the terminal leaflet obovate-oblong, in higher ones more elliptic, in second leaf to 9.5 by 4.5 cm, in 5th one to 13 by 7.5 cm; base wedge-shaped, but rounded at its tip; top acuminate; further as first leaf.

Specimens: 2199 from S.E. Borneo, alluvial flat, primary forest on deep clay, low altitude.
Growth details: In nursery germination poor.
Remarks: Two species cultivated, similar in germination pattern and general morphology. *T.* cf. *glabratum* (De Vogel 1536) differs in details only.

Index to scientific names of plant taxa

Names of families and tribes are not included.
In the descriptive part of the book descriptions are arranged alphabetically according to families, genera, and species. Specific names in the literature references in the headings to the genera are not indexed, except where their taxonomic position has been changed.

For names mentioned in figures the page numbers are marked with *, for tables with **. The colour plates have no page number; the number of the opposite page is given in bold type.

THE LIBRARY
ST. MARY'S COLLEGE OF MARYLAND
ST. MARY'S CITY, MARYLAND 20686